# The Routledge Handbook of Mapping and Cartography

T0314533

This new Handbook unites cartographic theory and praxis with the principles of cartographic design and their application. It offers a critical appraisal of the current state of the art, science, and technology of map-making in a convenient and well-illustrated guide that will appeal to an international and multi-disciplinary audience. No single-volume work in the field is comparable in terms of its accessibility, currency, and scope.

*The Routledge Handbook of Mapping and Cartography* draws on the wealth of new scholarship and practice in this emerging field, from the latest conceptual developments in mapping and advances in map-making technology to reflections on the role of maps in society. It brings together 43 engaging chapters on a diverse range of topics, including the history of cartography, map use and user issues, cartographic design, remote sensing, volunteered geographic information (VGI), and map art.

The title's expert contributions are drawn from an international base of influential academics and leading practitioners, with a view to informing theoretical development and best practice. This new volume will provide the reader with an exceptionally wide-ranging introduction to mapping and cartography and aim to inspire further engagement within this dynamic and exciting field.

*The Routledge Handbook of Mapping and Cartography* offers a unique reference point that will be of great interest and practical use to all map-makers and students of geographic information science, geography, cultural studies, and a range of related disciplines.

**Alexander J. Kent** is Reader in Cartography and Geographic Information Science at Canterbury Christ Church University, UK. He is currently the President of the British Cartographic Society, Editor of *The Cartographic Journal* and Chair of the International Cartographic Association (ICA) Commission on Topographic Mapping.

**Peter Vujakovic** is Professor of Geography at Canterbury Christ Church University, UK. He is past Editor and current Associate Editor of *The Cartographic Journal* and expert contributor to *The Times Comprehensive Atlas of the World.*

# The Routledge Handbook of Mapping and Cartography

*Edited by Alexander J. Kent and Peter Vujakovic*

Routledge
Taylor & Francis Group

LONDON AND NEW YORK

First published 2018
by Routledge
2 Park Square, Milton Park, Abingdon, Oxon OX14 4RN

and by Routledge
52 Vanderbilt Avenue, New York, NY 10017

First issued in paperback 2020

*Routledge is an imprint of the Taylor & Francis Group, an informa business*

*British Library Cataloguing-in-Publication Data*
A catalogue record for this book is available from the British Library

*Library of Congress Cataloging-in-Publication Data*
Names: Kent, Alexander, editor. | Vujakovic, Peter, editor. | Routledge (Firm)
Title: The Routledge handbook of mapping and cartography / edited by
    Alexander Kent and Peter Vujakovic.
Description: Milton Park, Abingdon, Oxon ; New York, NY : Routledge,
    2018. | Includes bibliographical references and index. Identifiers: LCCN
                        2017013902 | ISBN 9781138831025
    (hardback : alk. paper) | ISBN 9781315736822 (ebook)
Subjects: LCSH: Cartography—Handbooks, manuals, etc.
Classification: LCC GA108.5 .R68 2018 | DDC 526—dc23
LC record available at https://lccn.loc.gov/2017013902

ISBN 13: 978-0-367-58104-6 (pbk)
ISBN 13: 978-1-138-83102-5 (hbk)

Typeset in Bembo
by Swales & Willis Ltd, Exeter, Devon, UK

Printed and bound in Great Britain by
TJ Books Limited, Padstow, Cornwall

# Contents

Contents

Contents

# Figures

# Figures

Figures

# Figures

# Tables

# Contributors

**Sabine Afflerbach-Thom** is working as a cartographer in the General and Policy Issues Unit of the Department for Geoinformation at the Federal Agency for Cartography and Geodesy, Germany. In the past, she has been involved in EU projects dealing with data harmonization of topographic data across Europe. She is now supporting the management of the department in international affairs.

**Vyron Antoniou** is an Army Officer serving in the Hellenic Military Geographical Service. He studied Surveying and Rural Engineering at the National Technical University of Athens (NTUA), Greece, and holds an MSc in Geoinformation from NTUA and a PhD from University College London (UCL) in Geomatics. His research interests include Volunteered Geographic Information (VGI), citizen science, spatial data quality, spatial databases, gamification and mapping applications.

**Peter Barber** was Head of Maps and Views at the British Library from 2001 to 2015. Specializing in medieval maps and maps and government in early modern England, he has edited popular books on the history of cartography and has been adviser to several television series on maps. He was awarded an OBE in 2012 for services to cartography and topography.

**Timothy Barney** is Assistant Professor of Rhetoric and Communication Studies at the University of Richmond in Virginia. He is the author of *Mapping the Cold War: Cartography and the Framing of America's International Power* (University of North Carolina Press, 2015), and has published a number of journal articles on the rhetoric of cartography and international affairs.

**Christopher Board** is a geographer, educated at the London School of Economics (LSE) 1952–55, followed by research in South Africa. He has lectured at Cambridge and LSE and attended summer school at Columbus, Ohio. He participated in International Cartographic Association (ICA) Commissions on Technical Terms in Cartography and Cartographic Communication and was the UK delegate to the ICA 1984–2005. He was awarded the British Cartographic Society Medal in 2004, and was awarded an OBE for services to cartography in 2005.

**Cristina Capineri** is Associate Professor of Geography at the Department of Social, Political and Cognitive Studies (DISPOC), University of Siena, Italy. Her research interests concern broadly transport and telecommunication networks, GI science, volunteered geographic information and citizen science, local development and sustainable development, environmental indicators, organic agriculture and landscape.

# Contributors

**William Cartwright** is Professor of Cartography in the School of Science at RMIT University, Australia. His major research interest is the application of integrated media to cartography and the exploration of different metaphorical approaches to the depiction of geographical information. In 2013 he was made a Member of the Order of Australia (AM) for services to cartography and geospatial science as an academic, researcher and educator.

**James A. Cheshire** is a Senior Lecturer in Quantitative Human Geography at the Department of Geography, University College London (UCL), and a Deputy Director of the Consumer Data Research Centre. Aside from his academic outputs, a wide range of his maps and visualizations have been featured in the popular print press (such as *National Geographic* and *The Guardian*) as well as online. He enjoys blogging both for *spatial.ly* and *mappinglondon.co.uk*.

**Steve Chilton** worked for 42 years at Middlesex University (as Educational Development Manager, after starting as a cartographer). His work has been published extensively, particularly in his roles as Chair of the Society of Cartographers and Chair of the ICA Commission on Neocartography. He is heavily involved in the OpenStreetMap project, co-authoring *OpenStreetMap: Using and Enhancing the Free Map of the World*. He also co-edited *Cartography: A Reader* (selected papers from *The Bulletin of the Society of Cartographers*, the Society's respected international journal, which he edited for 25 years).

**Giles Darkes** was formerly Senior Lecturer in Cartography at Oxford Brookes University, specializing in thematic mapping and cartographic project management. He now works freelance as a cartographic editor, and is the Cartographic Editor of the *British Historic Towns Atlas* series.

**Martin Davis** has worked as an Instructor in Geography at Canterbury Christ Church University since 2014, alongside his ongoing PhD research into Soviet military cartography. In 2015, Martin was awarded the British Cartographic Society's Ian Mumford Award for excellence in original cartographic research. Martin is a member of the British Cartographic Society and is Reviews Editor and Editorial Assistant of *The Cartographic Journal*.

**Catherine Delano-Smith** is a Senior Research Fellow at the Institute of Historical Research, University of London, and Editor of *Imago Mundi: The International Journal for the History of Cartography*. Since gaining her DPhil, she has taught in the universities of Durham, Nottingham, and London, and is a Fellow of the Royal Geographical Society and of the Society of Antiquaries of London. As a historical geographer, she published on rural landscape and post-Neolithic environmental change in southern Europe before turning to the history of cartography, contributing to volumes 1, 2, 3, and 4 of *The History of Cartography* (University of Chicago Press) on prehistoric maps and map signs. Her main interests are medieval and early modern maps, exegetical mapping, maps as image, and maps in travel.

**Danny Dorling** is the Halford Mackinder Professor of Geography at the University of Oxford. He grew up in Oxford and went to university in Newcastle upon Tyne. He has worked in Newcastle, Bristol, Leeds, Sheffield, and New Zealand. His work concerns issues of housing, health, employment, education, inequality, and poverty – and cartography.

**Matthew H. Edney** is Osher Professor in the History of Cartography, University of Southern Maine; he directs *The History of Cartography* (at University of Wisconsin-Madison) and edits, with Mary Pedley, Volume Four, *Cartography in the European Enlightenment*. He is broadly interested in the history and nature of maps and mapping practices.

**Corné P.J.M. van Elzakker** is an Assistant Professor at the University of Twente, Faculty ITC in The Netherlands. He is the former Chair of the ICA Commission on Use and User Issues. His current research spearhead is the implementation of methods and techniques of user research in the geodomain.

**Alison Gazzard** is a Senior Lecturer in Media Arts and Education at the UCL Institute of Education, University College London. Her books include, *Mazes in Videogames: Meaning, Metaphor and Design* published by McFarland in 2013, and *Now the Chips Are Down: The BBC Micro* published by MIT Press in 2016.

**Joe Gerlach** is a Lecturer in Human Geography at the School of Geographical Sciences, University of Bristol. His research interests span cultural and political geography, including critical cartography, micropolitics, and nature-society relations in Ecuador.

**Stuart Granshaw** is currently the Editor of *The Photogrammetric Record*, the world's only international journal devoted solely to photogrammetry. He participates in the Council meetings of the UK Remote Sensing and Photogrammetry Society (RSPSoc) and was previously a lecturer in cartography, photogrammetry, and remote sensing at Oxford Brookes University.

**Amy L. Griffin** is a Senior Lecturer in Geography at UNSW Canberra in Australia. She is the current Co-Chair of the ICA Commission on Cognitive Issues in Geographic Information Visualization.

**Muki (Mordechai) Haklay** is a Professor of Geographic Information Science at University College London (UCL). He is the founder and Co-director of the UCL Extreme Citizen Science group. He is recognized as an international expert in participatory mapping and science, usability and Human–Computer Interaction aspects of geospatial technologies, and public access to environmental information.

**Guntram H. Herb** is Professor of Geography at Middlebury College, Vermont. His major publications include *Cambridge World Atlas* (2009), *Nations and Nationalisms in Global Perspective: An Encyclopedia of Origins, Development, and Contemporary Transitions* (2008), *Nested Identities: Nationalism, Territory, and Scale* (1999), and *Under the Map of Germany: Nationalism and Propaganda, 1918–1945* (1997). He is the recipient of a Fulbright Fellowship in France and is on the editorial boards of *Geographical Review*, *Political Geography* and *National Identities*.

**Anja Hopfstock** is a GIS specialist and cartographer in the General and Policy Issues Unit of the Department for Geoinformation at the Federal Agency for Cartography and Geodesy, Germany. She has been deeply involved in several international projects dealing with cross-border harmonization of topographic and cadastral reference data. She is also active in the ICA Commission on Topographic Mapping and in the German cartographic association.

**Alexander J. Kent** is Reader in Cartography and Geographic Information Science at Canterbury Christ Church University. He is currently the President of the British Cartographic Society, Editor of *The Cartographic Journal* and the Chair of the International Cartographic Association (ICA) Commission on Topographic Mapping. His research focuses on cartographic aesthetics, intercultural map design, Soviet mapping, and the role of maps in constructing national identity.

**Miljenko Lapaine** studied Mathematics and obtained his PhD from the Faculty of Geodesy with a dissertation entitled 'Mapping in the Theory of Map Projections' at the University of Zagreb, where he has been a full professor since 2003. He has published more than 900 papers, several textbooks and monographs. Prof. Lapaine is the Chair of the ICA Commission on Map Projections, a founder and President of the Croatian Cartographic Society, and the Executive Editor of *Kartografija i geoinformacije* (the cartography and geoinformation journal).

**Radu Leca** is a Romanian art historian specializing in the perception of space in early modern Japan. As an IIAS Affiliated Fellow, Radu is surveying both Western maps of Japan and Japanese maps in Dutch collections for a monograph entitled *Myriad Countries: The Outside World on Historical Maps of Japan*.

**Paul A. Longley** is Professor of Geographic Information Science at University College London, where he also directs the Economic and Social Research Council Consumer Data Research Centre. His publications include 18 books and over 150 other refereed publications. He has supervised more than 50 PhD students and held principal or co-investigator roles on more than 50 research grants.

**Kate McLean** is a designer whose research links embodied sensory data with urban environments in the form of sensory maps, which are held by UK permanent collections and feature in numerous publications (see www.sensorymaps.com). She is Senior Lecturer in Graphic Design at Canterbury Christ Church University and PhD candidate at the Royal College of Art.

**Beata Medynska-Gulij** specializes in cartography. She has worked at the Adam Mickiewicz University in Poznan since 2001. In 2010, she became Head of the Department of Cartography and Geomatics. She has published two books (in Polish): *Cartography and Geovisualisation* (2011, reprinted in 2012) and *Cartography: Principles and Applications of Geovisualisation* (2015, reprinted in 2016).

**Mark Monmonier** is Distinguished Professor of Geography at Syracuse University. Author of nineteen books, including *How to Lie with Maps* and *Adventures in Academic Cartography*, he was editor of Volume Six (the twentieth century) of *The History of Cartography*. In 2016, he was elected to URISA's GIS Hall of Fame.

**Ian Muehlenhaus** is the Director of the University of Wisconsin-Madison's Online Master's in GIS and Web Map Programming. Prior to this position, he was an Assistant Professor of Cartography and Web Mapping at James Madison University. He earned his PhD from the University of Minnesota and his MSc from Penn State University. He is the author of the book *Web Cartography* by CRC Press.

**Kristien Ooms** is a postdoctoral researcher at the Department of Geography, University of Ghent (since 2013). Kristien focuses on cartographic user research to evaluate the usability of (static and interactive) maps using a mixed methods approach. She is specialized in eye tracking in combination with statistics and visual analytics. Kristien is currently the Chair of the ICA Commission on Use, User, and Usability Issues.

**Inge Panneels** is an artist and academic, who uses mapping in her artwork to explore notions of place and space. She is Senior Lecturer at the University of Sunderland and an AHRC-funded PhD student at Northumbria University to undertake further research into the subject of 'Why artists map?'.

**Gyula Pápay** received his doctorate from the University of Budapest 1967 and his habilitation at the University of Rostock in 1988. He was an editor at the Hermann Haack Geographic-Cartographic Institute Gotha from 1963 to 1979 before he joined the University of Rostock, where he was Professor of Theoretical History and Historical Cartography from 1996 until his retirement in 2004. He is a member of the German Society for Cartography and an honorary member of the Hungarian Geographical Society.

**Chris Perkins** is Reader in Geography at the University of Manchester. His research interests lie at the interface between mapping technologies and social and cultural practices, with ongoing research into performative aspects of contemporary mapping behaviour, an interest in sensory mapping, and an emerging interest in play. Chris was Chair of the ICA Commission on Maps and Society and is author of numerous single and co-authored books and academic articles.

**Stephen Scoffham** is a Visiting Reader in Sustainability and Education at Canterbury Christ Church University and President Elect of the Geographical Association (2018–19). He is also the chief editorial adviser for a range of primary and secondary school atlases. His current publications include the *Collins Junior Atlas* (2017) and *UK in Maps* and *World in Maps* (both Collins, 2013).

**Mary Spence** is a Cartographic Design Consultant and her career spans over 40 years in cartographic publishing. She has an MA in Geography from the University of Aberdeen and a Postgraduate Diploma in Cartography from the University of Glasgow. A Chartered Geographer and Fellow of the Royal Geographical Society, in 2004 she was awarded an MBE for services to cartographic design.

**Peter Thomas** was Principal Lecturer in Geography at Canterbury Christ Church University until his retirement in 2010. He has a particular interest in the geography of Western Europe and completed his PhD thesis on the political and economic geography of Belgium in 1996.

**Judith Tyner** is Professor Emerita of Geography from California State University, Long Beach, USA. Her MA and PhD degrees were in Geography from the University of California at Los Angeles (UCLA) with a thesis on 'Lunar Cartography' and dissertation on 'Persuasive Cartography'. She has written four books on cartographic design and map reading and a scholarly work (2015), *Stitching the World: Embroidered Maps and Women's Geographical Education.*

**E. Lynn Usery** is Senior Scientist and Director of the US Geological Survey Center of Excellence for Geospatial Information Science. He has served more than 27 years with the USGS and 17 years as a professor in the academy. He earned BS, MA, and PhD degrees in Geography and is a Fellow of CaGIS (Cartography and Geographic Information Society) and UCGIS (University Consortium for Geographic Information Science).

**Peter Vujakovic** is Professor of Geography at Canterbury Christ Church University. His research focuses on the socio-political and educational significance of cartography, covering disability access to maps and geopolitics. He is past Editor and current Associate Editor of *The Cartographic Journal* and expert contributor to *The Times Comprehensive Atlas of the World*.

**Christopher Wesson** has a passion for cartography, making sense of data, and telling compelling stories through geographic visualization. He has over 13 years of geospatial experience, as a cartographer and consultant for Ordnance Survey, producing his own maps and blog materials, and as Convener of the British Cartographic Society's Map Design Group. His industry expertise are buoyed by an experience of engineering projects at Capita and Arup, and a higher education in science, finance and management from the University of Southampton.

**Denis Wood** curated the Power of Maps exhibition for the Smithsonian and writes widely about maps. His most recent book is *Weaponizing Maps* (2015). A former Professor of Design at North Carolina State University, USA, Wood is a currently an independent scholar living in Raleigh, North Carolina.

# Acknowledgements

*The Routledge Handbook of Mapping and Cartography* would not have been possible without the knowledge, hard work and generosity of all our contributors from around the world. We owe each of them our thanks for their willingness to share their expertise in creating this substantial new volume. We would also like to thank Andrew Mould, Egle Zigaite, Sarah Gilkes and all the team at Routledge, whose diligence and professionalism have made this project an exciting, challenging and rewarding experience. Finally, in bringing this project to fruition we owe a huge debt of gratitude to our respective families, and so we offer our very special thanks to Alison, Edward, Una, Alexander and Tom for all their support and understanding.

# Introduction

*Alexander J. Kent and Peter Vujakovic*

For thousands of years, mapping has been essential for understanding the world around us and today more people are making, using and sharing maps than ever before. We rely on maps to navigate, to delineate and to decorate, and few other artefacts possess such versatility and significance within the human journey. The formalization of cartography (which we define as the art, science and technology of map-making) as an independent scientific discipline began to emerge in the late nineteenth and early twentieth centuries, when attempts were made to establish theoretical principles upon which best practice could be based. In particular, the Austrian geographer and cartographer Karl Peucker (1859–1940) had devised new ways of depicting elevation based on colour perception, while Max Eckert, in the publication of the two-volume *Die Kartenwissenschaft* (map science) in 1921 and 1925, sought to explicitly promote and establish cartography as a science.

With the intensive utilization of maps during two world wars and the introduction of new methods of survey and colour printing in photogrammetry and photolithography, cartography could address ever-increasing demands and acquire a more prominent role in industrialized societies. These demands soon galvanized a trajectory for cartography based on the values of scientific inquiry and provided purpose and direction towards the optimization of map design and production. The aspiration is clearly implied in the words on the cover of Arthur Robinson's *Elements of Cartography* in 1953, the first of six editions of what was to become the principal textbook of Western cartographic practice of the twentieth century, which 'Presents cartography as an intellectual art and science rather than as a sterile system of drafting and drawing procedures'.

By the late 1960s, various theories of cartographic communication emerged from the application of general theories of information transfer, such as those of Hartley (1928), with a view to improving map design. By identifying and examining the elements involved in the gamut of the cartographic process independently, these theories sought improvements at each stage so that information could flow between cartographer and user more efficiently. From the 1960s, geographical information systems (GIS) sought to release maps from the limitations imposed by the dominant medium of the time – paper. The transition from analogue to digital gradually brought fluidity and freedom between scales and between areas and extents, but also in the type and amount of data that could be stored and presented simultaneously. Yet, until the late 1980s, the notion of information transfer remained the dominant paradigm in cartographic theory.

The bulk of research continued in its aim to understand how to create better maps by being more effective at communicating a message to particular user groups, often drawing on theories of visual perception and cognition to offer new empirical insights. From the perspective of the practitioner, cartographic research aimed to support the creation of an optimum map.

As selective portrayals, the nature of maps means that they are never only about technology or design. The introduction of an epistemological shift through the work of Brian Harley (e.g. 1989, 1990) in particular, challenged the paradigm of cartography as an objective science by drawing on social theory, and, in particular, on the ideas of Michel Foucault and Jacques Derrida. Instead of offering new methodologies for optimizing maps, Harley introduced a new critique where the map is either a text to be deconstructed or a discourse in which knowledge as power is to be revealed (Taylor, 1992: 127). The trajectory of 'critical cartography' since Harley has sought to explore the relationship of maps in society more fully, e.g. by revealing the motivations of the cartographer and by challenging the stability of cartographic representations and their ontologies.

Following these conceptual developments, a series of technical innovations in the first decade of the new millennium brought some colossal changes in the ways that maps are made, shared and used. These include access to greater levels of accuracy with global positioning systems (GPS) (in particular, the removal of the Selective Availability error from the US constellation of GPS satellites in 2000), the emergence of OpenStreetMap in 2004 as an online global map created through the collaborative capture and rendering user-acquired data, and improved accessibility to global imagery and maps with the launch of Google Earth and Google Maps in 2005. The wider availability of mapping technologies, location-based services, governmental geographical data and volunteered geographic information (VGI) since the beginning of the new millennium has been accompanied by a resurgence of the role of art in cartography, for example, where the visualization of geographic data now extends to mapping the ephemeral and invisible.

With the diversity of approaches to mapping and cartography and the transformation in the making, use and interpretation of maps, there is a very real need for a holistic approach in reflecting upon cartographic praxis. This volume therefore takes advantage of these substantial theoretical and technological developments and sets them within a wider context. There is not scope in this Handbook to cover all facets of mapping and cartography or to deal with any of these themes comprehensively. Our aspiration in compiling this Handbook is therefore to create a starting point for further research, exploration and discovery. We aim to provide a reference point for international progress in the field that introduces the reader to a growing body of epistemological approaches to mapping and cartography and provides a critical appraisal of the theory, history, design and technology of maps and map-making. Accordingly, some chapters introduce and explain major theoretical concepts while others offer more practical guidelines that explore the principles of cartography and allow the reader to put theory into practice. We also seek to encourage dialogue across the disciplines that contribute to cartographic praxis and new theoretical perspectives, ranging from computer science to psychology, from geographic information (GI) science to fine art. The Handbook is primarily aimed at advanced-level undergraduate and graduate students, but those undertaking higher-level research will find many of these chapters useful as references.

The contributors to this volume have been selected because they are undertaking some of the most exciting work in this field. They represent leading academic institutions that excel in mapping and cartography or are leading practitioners who each have many years' experience of working in the cartographic industry. We have attempted to assemble a truly international cast, drawn from within and beyond the Anglophone sphere, enabling readers to obtain an appropriate view of how mapping and cartography is studied and practised around the world. Nevertheless, while we attempt to offer a wide-ranging single volume, the Handbook is not intended to be an encyclopaedia and is in no way exhaustive of existing cartographic scholarship. Readers are frequently referred to a wealth of relevant and significant publications in the field, such as the multi-volume work, *The History of Cartography* (Harley *et al.*, 1987–).

This volume consists of forty-three chapters and is divided into six Parts. Part I provides a critical review of the key concepts and paradigms that have shaped cartographic theory and practice, such as cartographic communication, critical cartography and performative cartography, bringing together these different approaches and providing new critical insights into their application. The starting point is Gyula Pápay's appraisal of Max Eckert's contribution to the formalization of cartography as a scientific discipline, with the publication of his two-volume treatise on 'map science', *Die Kartenwissenschaft* (1921, 1925). This is followed by a reflection on theories of cartographic communication and their legacy by Christopher Board, whose seminal chapter in Chorley's and Haggett's *Models in Geography* in 1967 epitomized this theoretical approach. The next two chapters explain how scientific methods of enquiry have brought new insights through visual perception and cognitive psychology (Amy Griffin) and user studies (Corné van Elzakker and Kristien Ooms). The second half of this first Part includes chapters which explore aspects of cartographic theory that follow the epistemological shift introduced by Harley. The first, by Matthew Edney, explains approaches to understanding the history of cartography, while Chris Perkins introduces critical cartography and its routes of enquiry. The final two chapters explore recent notions of performance (Joe Gerlach) and spectacle (Peter Vujakovic) in cartography that critically examine the notion of representation in mapping and the endurance of its relevance in society.

The second Part presents a concise critical account of major themes in the history of cartography with an emphasis on how visualization and representation can be situated both historically and culturally. It commences with a chapter by Peter Barber and Catherine Delano-Smith on maps in medieval Europe that examines the function of maps in medieval society. This is followed by Radu Leca's post-representational interpretation of the role of maps in the 'Age of discovery'. The next two chapters, by Matthew Edney, provide a critical examination of science and the state in transforming the role of maps in society during the Enlightenment. The second Part is concluded by Timothy Barney's examination of the cartographies of war and peace and the development of new global perspectives which emerged through maps in the twentieth century.

Part III is concerned with how maps model the Earth and aims to provide a concise but pragmatic approach to the scientific basis of cartography, while incorporating a survey of the key concepts, issues and technologies, such as the latest developments in data capture. The first two chapters, respectively by Miljenko Lapaine and E. Lynn Usery, introduce the basics of geodesy and provide an explanation and critical survey of map projections. The next four chapters describe, explain and critically evaluate the principles and applications of different mapping technologies. Stuart Granshaw's chapter covers photogrammetry, the creation of maps from aerial photographs, and remote sensing, while Paul Longley and James Cheshire introduce GIS. The related technologies of GPS and mobile mapping, both key technologies in the gathering and processing of data for location-based services, are explained by Martin Davis. The Part ends with Steve Chilton's chapter on neocartography and OpenStreetMap, taken from the perspective of their convergence at a conference in 2005.

The next Part acts as a guide for creating maps by focusing on the principles of cartographic design and providing an array of practical information for ensuring best practice, whether maps are made for print or for the web. It incorporates new research on aesthetics and brings together a compendium of advice on standards and conventions for practising cartographers. An introduction to map design is provided by Giles Darkes, which is followed by a chapter on cartographic aesthetics by Alexander Kent. The next three chapters then focus on specific aspects of map design, i.e. layout, balance and visual hierarchy (Christopher Wesson), colour (Mary Spence), and lettering and labelling (Christopher Wesson). The following two chapters examine the considerations of mapping for printed and online media in particular, with Judith Tyner addressing issues surrounding the former and Ian Muehlenhaus providing a survey and guide to the latter. Stephen Scoffham concludes Part IV with a chapter focusing on the design of maps and atlases for schoolchildren – a particular type of user with particular needs and abilities – and the next generation of map-makers.

Part V takes the finished map as a point of departure and provides a series of chapters which explore personal and social interaction with maps. It aims to present a critical review of map use in a range of fields. The first four chapters focus on the importance of maps for constructing a sense of place and identity, from the local, in Denis Wood's essay on 'mapping place', to the contributions from Alexander Kent and Peter Vujakovic, and Guntram Herb, who explore the relationship between maps and the construction of identities and nationalist ideologies. Kent and Vujakovic provide a strong focus on post-communist identities in Europe, while Herb applies social movement theory to understand mapping as part of a wider protest movement, not just an association with the right, during the Weimar Republic. Judith Tyner's discussion treats the wider issues surrounding how persuasive maps are created and work.

The rest of the chapters in this Part focus on more specific map formats and their role in society, particularly in education and entertainment. Peter Thomas explores the use of schematic maps and chorematic diagrams within the practice of regional geography, while Peter Vujakovic explores the wider impact on the public of maps within the news media. Turning the tables, Vyron Antoniou, Cristina Capineri and Muki Haklay explain the significance of VGI and the empowerment that maps can offer. The last four essays are closely focused on the interrelationship between maps, art and leisure. Peter Vujakovic reviews the role of maps in literature, but also how some novels can contain deep geographies and be cartographic in their own right. Kate McLean explores mapping of the invisible and the ephemeral, and provides case studies from her own innovative work in smell mapping. Inge Panneels' chapter provides an overview of art and cartography, and concludes that the critical turn in cartography, in the hands of artists, has become a 'radical cartography'. The final essay, by Alison Gazzard provides insights into virtual worlds of gaming and the maps involved. While dealing with the playful, she also reminds us that all navigation, whether for leisure or with serious intent, involves mental mapping 'as new spaces are remembered, and places are secured'.

Part VI contains a collection of insights into the recent practice and the future of cartography and GI industries from leading scholars and practitioners. It addresses future developments in terms of theory and applications in the context of an increasing accessibility of data and in the number of methods of mapping and visualization. The Part also engages with recent developments in technology and suggests how the roles of cartography and cartographers have evolved and might further evolve according to the changing needs of society. Mark Monmonier's chapter heads this Part and brings his long experience in cartography to bear on questions of the future of cartography. His title 'Hunches and hopes' suggests that 'crystal-ball gazing' is no easy thing to undertake, but that he sees the future as positive and enthusiastically advocates a national museum of cartography lest we forget where we come from. Danny Dorling's essay 'Can a map change the world?' provides a thoughtful discussion of that theme. In this he brings his own experience in mapping social and related issues to bear and regards mapping as a way of 'seeing the world anew'. The next chapter, by Beata Medynska-Gulij, explains a strategy and overview of curricula for the education of the cartographers of tomorrow and suggests how cartographers can enrich society through their work in related fields. The final essay provides an appropriate conclusion to Part VI and to the entire volume; William Cartwright uses an autobiographical approach to lead us through recent developments in cartography, even to the point of mapping leaving its physical mark on his body. In 'Drawing maps: human vs. machine', he brings us back to the core of our field: 'We draw. We draw maps that represent geography. This is what we do'.

## References

Board, C. (1967) "Maps as Models" in Chorley, R.J. and Haggett, P. (Eds) *Models in Geography* London: Methuen, pp.671–725.

Eckert, M. (1921, 1925) *Die Kartenwissenschaft: Forschungen und Grundlagen zu einer Kartographie als Wissenschaft* (2 vols) Berlin and Leipzig, Germany: Walter de Gruyter.

Harley, J.B., Woodward, D., Lewis, G.M., Monmonier, M., Edney, M.H., Pedley, M.S. and Kain, R.J.P. (Eds) (1987–) *The History of Cartography* (6 vols in 12 bks) Chicago, IL: University of Chicago Press. Reprinted online (free access) at www.press.uchicago.edu/books/HOC/.

Harley, J.B. (1989) "Deconstructing the Map" *Cartographica* 26 (2) pp.1–20.

Harley, J.B. (1990) "Cartography, Ethics, and Social Theory" *Cartographica* 27 (1) pp.1–23.

Hartley, R.V. (1928) "Transmission of Information" *Bell System Technical Journal* 7 pp.535–563.

Robinson, A.H. (1953) *Elements of Cartography* New York: John Wiley & Sons.

Taylor, P.J. (1992) "Politics in Maps, Maps in Politics: A Tribute to Brian Harley" *Political Geography* 11 (2) pp.127–129.

# Part I

# Situating cartography

## From craft to performance

# 1

# Max Eckert and the foundations of modern cartographic praxis

*Gyula Pápay*

(TRANSLATED BY SABINE AFFLERBACH-THOM AND ANJA HOPFSTOCK)

## Max Eckert (1868–1938) and the genesis of *Die Kartenwissenschaft*

Friedrich Eduard Max Eckert was born in Chemnitz, Germany on 10 April 1868 (in 1934, he changed his name to Eckert-Greifendorff to indicate the family's ancestry). During his childhood, he lived in Löbau (130 km east of Chemnitz), where his father worked as a town sergeant. From 1892 onwards, Eckert studied at the University of Leipzig and gained his doctorate under Friedrich Ratzel with a thesis on geomorphology that was entitled *Das Karrenproblem: Die Geschichte einer Lösung* (The Karren Problem: The Story of a Solution). He continued to study geomorphology for his habilitation, which he gained in 1903 under the supervision of Otto Krümmel of Kiel University (Ogrissek, 1985).

Eckert stayed in Leipzig until 1903, where he worked at the University, first as an assistant of Friedrich Ratzel and later as a senior teacher conducting exercises in cartography. Ratzel inspired Eckert by directing his attention to economic geography, paving the way for a greater focus on thematic cartography. At this time, Leipzig was one of the most important centres of cartographic publishing and learning, with cartography being taught as part of the University's course in geography. Eckert received numerous suggestions for a more detailed study of cartography, which led to the publication of his *Neuer methodischer Schulatlas* (New Methodological School Atlas) in 1898.

From 1900, Eckert worked as a private lecturer in Kiel and was Head of the Museum of Ethnology at the University until 1907. Here, the geography of trade (economic geography) became his main subject. His two-volume book *Grundriß der Handelsgeographie* (Outline of the Geography of Commerce), published in 1905, helped to develop geography as an academic discipline, which, until then, had largely focussed on geomorphology, especially by Ratzel. Eckert's work on economic geography led to publications on colonial geography, including the atlas *Wirtschaftsatlas der deutschen Kolonien* (Economic Atlas of the German Colonies) that was published in 1912. While working in the field of economic geography, Eckert studied the theory of map projections, especially equal-area projections. He made a significant contribution to the development of cartography by devising several new map projections, including his equal-area pseudocylindrical projection (Eckert IV), which is still used widely today.

Eckert's first publication on theoretical cartography was his contribution to *Petermann's Geographische Mitteilungen* in 1906. Shortly afterwards he articulated the need to formally recognize cartography as a science; first in a lecture entitled 'Scientific Cartography in University Teaching' at the 16th Geographers' Day in Nuremberg in May 1907, and, second, through his article 'Kartographie als Wissenschaft' (Cartography as a Science), which was published in *Zeitschrift der Gesellschaft für Erdkunde zu Berlin* in the same year. By this time, however, the term 'Kartenwissenschaft' (map science) had already emerged. The idea of cartography as an independent scientific discipline had been proclaimed in the late nineteenth century, especially by the Austrian cartographer Karl Peucker (1859–1940). Yet, Eckert's contribution was significant because it gained support from academic geographers as well as from map publishers. What set Eckert's work apart was his interpretation of the process and development of cartography. His aim was to promote and establish theoretical and scientific cartography as an academic discipline that possessed scientific value independently of geography (Eckert, 1907: 539).

Eckert's positive approach towards the differentiation of cartography from geography was derived partly through his geo-economic research and partly through his work with Krümmel, who had advocated that oceanography should be distinct from geography and who had also put a great emphasis on teaching cartography as a separate subject. Recognition of the growing importance of maps in geography was also reflected by the introduction of mapping to the University curriculum. Teaching in cartography at Kiel University was given during a four-semester cycle by Krümmel and Eckert together. Based on this course, they created a cartographic textbook for use in universities (Krümmel and Eckert, 1908).

In 1907, Eckert was 39 years old and the father of three children. As a profession, cartography did not offer a secure livelihood or a sufficient social status, and his efforts to obtain a professorship at a university, such as Konigsberg, were in vain. However, in that year, Eckert was appointed Associate Professor to the newly created Department of Economic Geography and Cartography at the Technische Hochschule (Technical University) in Aachen. This was interrupted by World War I, when he volunteered for military service and was transferred to the surveying corps of the 3rd Army. Shortly afterwards, he became the head of the corps of 800.

After the war, Eckert set to work on *Die Kartenwissenschaft*, and the first volume was published by Walter de Gruyter (Berlin and Leipzig) in 1921 (Figure 1.1). To gather material for the book, Eckert travelled to many cities, e.g. Hamburg, Berlin, Dresden, Munich, Nuremberg, Vienna, Paris, London, Amsterdam, Brussels, Copenhagen, Stockholm, St. Petersburg and Moscow. Eckert was not content with his job in Aachen and he had hopes of becoming a Full Professor at a university. Finally, in 1922, his dream was realized and his associate professorship was transformed into an ordinary professorship at Aachen. In 1925, Eckert published the second volume of *Die Kartenwissenschaft* and his treatise for establishing cartography as a science was complete.

Although Eckert had indirectly contributed to the foundation of a cartographic society by promoting the breadth of cartography, he was not, however, involved in the formation of the German Cartographic Society and did not participate in its founding meeting in Leipzig in 1937. After much disappointment, he was eventually invited to lead the Steering Committee for Scientific Cartography.

From Eckert's point of view, his life's work was incomplete without the institutionalization of cartography. During 1938–39, he came to the conclusion that this goal could only be reached by gaining political support. Although he was not a member of the *Nationalsozialistische*

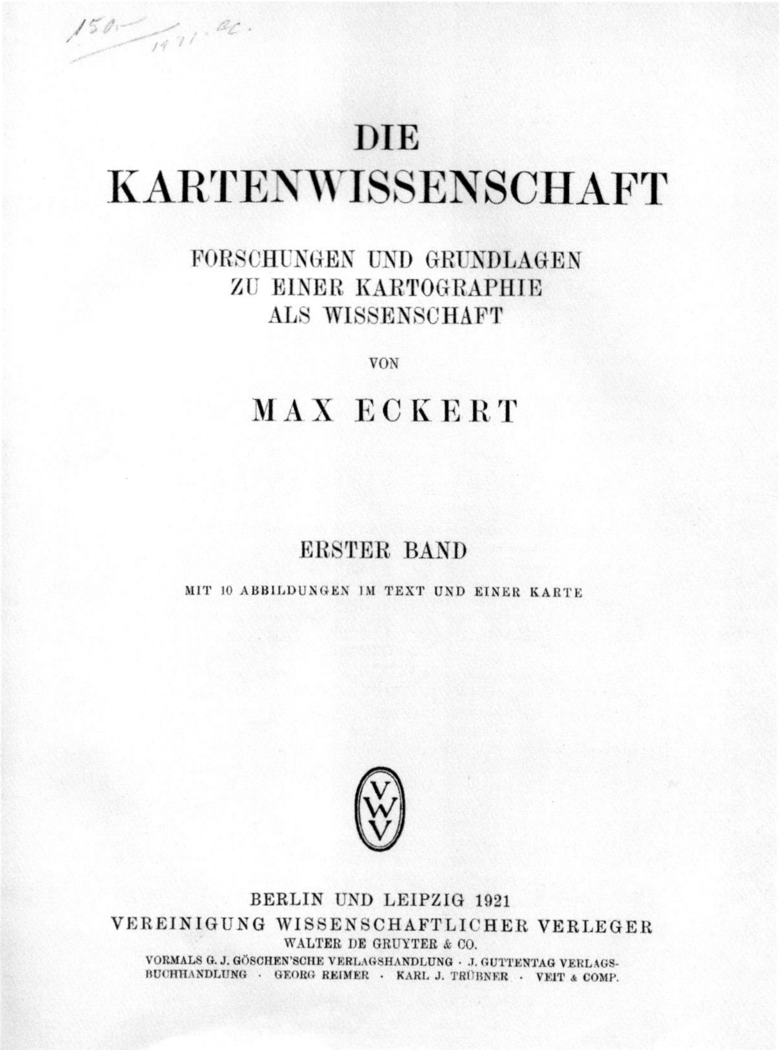

*150.—*
*1971. ac.*

# DIE
# KARTENWISSENSCHAFT

FORSCHUNGEN UND GRUNDLAGEN
ZU EINER KARTOGRAPHIE
ALS WISSENSCHAFT

VON

## MAX ECKERT

ERSTER BAND

MIT 10 ABBILDUNGEN IM TEXT UND EINER KARTE

BERLIN UND LEIPZIG 1921
VEREINIGUNG WISSENSCHAFTLICHER VERLEGER
WALTER DE GRUYTER & CO.
VORMALS G. J. GÖSCHEN'SCHE VERLAGSHANDLUNG · J. GUTTENTAG VERLAGS-
BUCHHANDLUNG · GEORG REIMER · KARL J. TRÜBNER · VEIT & COMP.

**Figure 1.1** The title page from Volume 1 of *Die Kartenwissenschaft* (1921)

*Deutsche Arbeiterpartei* (Nazi Party), Eckert wanted to demonstrate how important cartography was for National Socialist policy by publishing *Kartographie: Ihre Aufgaben und Bedeutung für die Kultur der Gegenwart* (Cartography: Its Tasks and Importance for the Culture of the Present). In the Foreword to this book, he offered two objectives: to prove the predominance of German cartography in the world and to promote the institutionalization of cartography as part of the National Socialist government. He clearly sought attention from the party's leaders, often citing Hitler and proposing the use of Nazi symbols on maps, e.g. by using swastikas to represent the German border. He even suggested renaming the 'New General Map of the Empire' the 'Adolf-Hitler-Map' (Eckert, 1939: 141).

11

*Figure 1.2* Photographic portrait of Max Eckert (as reproduced in *Kartographische Nachrichten* in 1968), reproduced courtesy of Prof. W.G. Koch

Paradoxically, his illusions that the Nazi regime would support his plan to establish a 'German Cartographic Research Institute' were strengthened instead by the institutionalization of scientific cartography in the Soviet Union. The foundation of the Scientific Research Institute of Geodesy, Cartography and Aerial Photography (CNIIGAiK) in 1929 had enormous significance for cartographic research. In 1929 and 1930, cartographic departments were established at the Geography faculties of the universities of Moscow and Leningrad (St Petersburg), and, in 1936, a Cartography faculty was founded at the Moscow Institute of Geodesy, Photogrammetry and Cartography (MIIGAiK). Eckert's *Die Kartenwissenschaft* played an important role in this process of institutionalization, as most of its volumes were ordered from within the Soviet Union. Eckert became a victim of his illusions with regard to the true political intentions of the Nazi regime and concerning the fascist policy towards the sciences. In 1938, Max Eckert died in Aachen and was buried in Löbau. His book *Kartographie: Ihre Aufgaben und Bedeutung für die Kultur der Gegenwart* was edited by his son Fritz Eckert-Greifendorff and published posthumously.

## Critically evaluating *Die Kartenwissenschaft*

It has been noted that Eckert's *Die Kartenwissenschaft* was 'almost totally unknown' in international research (Scharfe, 1986: 61) and, as yet, there is no English translation. Hence, the first objective of the following parts of this chapter is to describe the contents of Eckert's two-volume treatise on cartography before providing a critical evaluation of its relevance to the establishment of cartography as a science. The following descriptions incorporate some tables which explain the organization of material in the two volumes and translate the sections into English (see Tables 1.1 and 1.2).

Any analysis of the structure of *Die Kartenwissenschaft* is complicated by the fact that Eckert tried to introduce a modified terminology into the newly founded scientific discipline which was partially derived from traditional terms, and which Eckert himself did not always use consistently. For example, instead of his word '*Kartenwissenschaft*' (map science), Eckert used the broader term '*Die Kartographie als Wissenschaft*' (Cartography as a Science) for the heading of the first part.

In order to delineate *Kartenwissenschaft* (map science) from practical cartography, Eckert uses the same term a little later in Volume 1 (p.3) to refer to scientific or theoretical cartography. This distinction is revised again with the remark that practical cartography cannot be entirely excluded from the science of cartography (p.4). In other places, however, he is again in favour of a distinction. For example, while making a case for establishing a new scientific discipline, Eckert argues that the primary task of theoretical cartography is to instruct practical cartography (p.5). Even while highlighting the importance of historical map-making for *Die Kartenwissenschaft*, Eckert argues for delimitation: 'Research into the history of cartography expresses unequivocally the independence of theoretical cartography with respect to practical cartography' (p.25).

Historical methods of cartography are covered extensively in *Die Kartenwissenschaft*. Eckert drew from a tremendously large body of empirical material at a time when the study of the history of cartography was at an early stage. Undoubtedly, these extensive historical studies served to provide Eckert with a quasi-theoretical basis for his treatise. Obsessively, he sought for innovations in cartographic representation, but, strangely enough, he overlooked the fact that many innovations with a scientific basis had emerged from practical cartography.

## Volume 1 (1921)

Part I of the first volume is dedicated to the question of where cartography belongs amongst the sciences. Eckert's main objective was partly to delineate cartography from geography and partly to keep the close links with its parent discipline. He described cartography as the 'distinguished sister and indispensable partner of geography' (p.7). Eckert was not able to provide a clear answer to the question of the position of theoretical cartography in the organization of scientific disciplines. Due to the variety of cartographic representations, it is impossible 'to assign cartography a definite position within the structure of scientific disciplines' (p.7). Today, we might regard this as being a characteristic of cartography's interdisciplinarity.

In Part I.B, Eckert explores the nature of maps. He provides the following definition:

> A geographic map is a two-dimensional depiction of a greater or lesser part of the Earth's surface, representing the location, areal and spatial relationships as well as geophysical, cultural and natural historical facts in a clear way, enabling the features shown to be read and measured.
>
> *(p.53)*

At this point it becomes necessary to briefly explain Eckert's unconventional classification of maps in order to understand the following headings in *Die Kartenwissenschaft*. Maps are first divided into topographic and geographic maps (p.50). Geographic maps are subdivided into geographically real maps – showing reality as faithfully as possible – and geographically abstract maps – representing the substance of a phenomenon. The latter maps are further subdivided into chorographic maps (smaller scale maps) and applied maps (thematic maps). For chorographic maps, it is noted that 'in ordinary life' they are called 'maps' (p.51), and Eckert frequently uses this term in *Die Kartenwissenschaft* without providing another definition. In Part I.C the principles of present and future developments in cartography are described.

Part II (pp.115–207) is dedicated to map projections. Eckert discusses map projections for topographic and chorographic maps while the principles of their construction are only touched upon. The focus is on historical studies and on critical evaluation of their application, and there are several references to the map projections developed by Eckert. Part III 'Surveying' (pp.208–294) is structured in a similar way to Part II. Again, the focus is on the historical development of cartography and on a critical assessment of surveying methods. Specific methods of surveying are dealt with only briefly. Eckert gives some focus to aerial photography and aerial survey (pp.266–294), a new method which was developing into photogrammetry. Today, this area falls within geodesy instead of cartography.

Specific cartographic elements are analysed in Part IV (pp.295–398), which covers the representation of natural and man-made features, as well as scale, orientation, generalization, lettering and labelling. On these subjects, Eckert drew upon a large amount of material, applying his historical-critical method. It contains a few suggestions for practical cartography, i.e. the introduction of vegetation representation in atlases (p.371), but these lack any theoretical basis. In his observations on generalization, he notes the impossibility of establishing scientific rules: 'Here only an empirical approach applies. Execution depends entirely on the ability and knowledge of the cartographer' (p.332).

Eckert devoted two parts of Volume 1 of *Die Kartenwissenschaft* to the representation of relief. In Part V (pp.399–497) the history of relief representation is explored from ancient times to the then present, while in Part VI (pp.498–624), the scientific basis or theory of relief representation is explained. The hachure system of Johann Georg Lehmann (1765–1811) is discussed in some detail (pp.509–523), while its successors (pp.524–528), the sloping hachures used in France (pp.529–532), as well as shaded hachures (pp.535–539) and their implementation – as associated with the Swiss topographer Guillaume-Henri Dufour (1787–1875) – receive much less attention. The relative merits of illumination (e.g. orthogonal and from one side) are also discussed (pp.540–563).

Eckert's method of using points to depict terrain is described in detail (pp.578–591) and provides the only example of a map in Volume 1 (see Figure 1.3). He considered his point-based method to be a progression of hachures and to offer a more scientific approach than Lehmann's method because it is based on natural illumination (p. 582). Indeed, Eckert was convinced that his point method avoids the disadvantages of sloping and shaded hachures and he applied for a patent in 1898 (Pat. No.110973 Kl. 42. Instruments, 9 February 1898). Although Eckert's method was a prime example of the application of cartographic theory to practical cartography, it remained completely unnoticed by practising cartographers. By the first decades of the twentieth century, the use of hachures was coming to an end and new aspects of depicting relief were beginning to emerge (amongst them, *Farbenplastik* by Karl Peucker, which used the psychological impact of colours to achieve a greater spatial effect with the use of contours). The final part of the first volume of *Die Kartenwissenchaft* (pp.625–639) is dedicated to the work of Peucker, who, according to Eckert, was the founder of a new era in cartography.

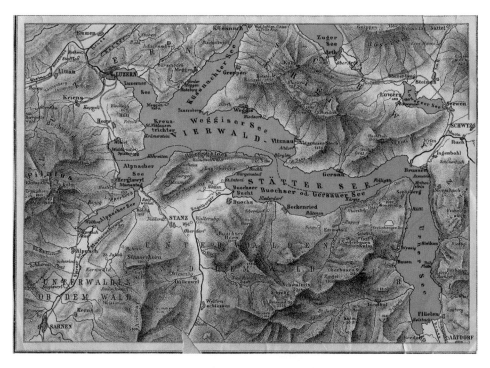

*Figure 1.3*    Map of Lake Lucerne that uses Eckert's point method of relief depiction. The map was published in black and white in *Die Kartenwissenschaft* and the two-colour version shown here was possibly made for the proposed third volume, *Genetischer Faksimileatlas* (Genetic Facsimile Atlas). This map is the only preserved copy (author's collection).

*Table 1.1* Table of contents of Volume 1 of *Die Kartenwissenschaft*

Table of contents (English translation)

*(continued)*

*Table of contents (English translation)*

## Volume 2 (1925)

The second volume of *Die Kartenwissenschaft* consists of six parts, five of which are dedicated to thematic mapping. Again, Eckert's viewpoint is almost exclusively derived from a critique of historical methods and he does not offer any discussion of cartographic theory. Part I (pp.1–125) is dedicated to hydrographic charts and oceanographic maps, which Eckert did not consider to be thematic maps unlike today. The focus of Part II (pp.126–222) entitled 'Thematic maps and their scientific method' is on population maps, but transportation and economic maps are also considered. First, the various methods of representation for thematic maps, such as diagrams and cartograms, are analysed. Part III (pp.223–384) concerns the depiction of inorganic (man-made) features on maps with a special focus on hydrographic, geological and meteorological maps. The depiction of organic (natural) features on maps is presented in Part IV (pp.385–519), which consists of biogeographic and anthropogeographic maps, maps of nations and languages as well as political and historical maps. Economic and transport maps are discussed separately in Part V (pp.520–669).

With its systematic approach to understanding these different cartographic elements, *Die Kartenwissenschaft* provided the first comprehensive and detailed exposition of thematic mapping, in which Eckert impressively revealed the enormous variety of cartography. He attempted to provide suggestions for their further development and therefore an impetus to theoretical cartography. Especially forward-looking was his advice to consider psychology. He cited Wilhelm Wundt (p.636), who had established the Institute for Experimental Psychology at the University of Leipzig in 1879.

The final part of the second volume, Part VI (pp.670–755), has the promising title 'Map aesthetics and logic'. First, the relationship between cartography and art is discussed: 'Cartography is a kind of applied art, a creative activity based on scientific rules' (p.670). Such rules are explicitly mentioned in Eckert's exposition on the aesthetics of relief representation: the 'plastic comparative-law' (p.710). These rules, however, are nothing more than two different design principles for the hypsometric tinting (layered colouring) of elevation maps: 'the higher, the brighter' and 'the higher, the darker'. This highlights again how Eckert assigned a high level of significance to the role of principles in cartographic representation and with a tendency to apply these 'rules' too broadly. Furthermore, it should be noted that the aforementioned design principles emerged from practical cartography and only later were they given a theoretical underpinning, i.e. the 'colour plastic' technique of Peucker.

The title of Section B of Part VI is 'Map logic' and is one of the most important, if not the most important, chapter on theoretical cartography in the whole work: 'Map logic is the science of the cartographic rules and principles' (p.713). This assertion carries relevance even today, but its specific interpretation has greatly changed since then, especially through the introduction of general theories such as information theory, communications theory and the theory of graphic variables. The examples cited by Eckert rarely provide any development of the design principles that emerged from practical cartography much earlier, i.e. the 'logic given by natural colours' (p.732, figure 4).

The second volume of *Die Kartenwissenschaft* contains an annex 'Military cartography' (pp.756–812), which provides a detailed overview of the state of military cartography at the beginning of the 20th century. Eckert planned to publish a third volume of *Die Kartenwissenschaft* with maps, but due to concerns over costs, the *Genetischer Faksimileatlas* (Genetic Facsimile Atlas) was not published.

With 1,520 pages, more than 1,000 references and over 2,000 map references, *Die Kartenwissenschaft* is especially useful as a valuable resource for the history of cartography.

But with his two-volume treatise Eckert also created a work that adopts an integrative approach towards all aspects of contemporary cartography. This synthesis of material, which made the wide variety of cartography explicit, was an important building block in the development of cartography as a science.

Although Eckert recognized the need for interaction between theoretical and practical cartography (p.3), his primary intention was to establish theoretical cartography as a scientific discipline that determines rules for cartographic representations from which standards for practical cartography are derived. This objective was only partially achieved by Eckert, as *Die Kartenwissenschaft* demonstrates a somewhat inverse relationship between theoretical and practical cartography. The work instead provides an empirical compilation of innovations that emerged from practical cartography. There are several reasons why Eckert was not able to elevate this rich body of empirical material to a higher level of abstraction; to derive theory from practice. On the one hand, he strongly adhered to the geographical model of thinking despite his intended secession of cartography from geography (Freitag, 1992: 309). On the other, more general theories such as semiotics, information and communication theory were not yet developed to a sufficient extent.

## Reception of *Die Kartenwissenschaft*

*Die Kartenwissenschaft* was considered to be a current textbook during Eckert's lifetime. After the Second World War, it gradually became obsolete as a guide for practising cartographers, yet its contribution towards establishing cartography as a science gained recognition. In particular, during the 1950s and 1960s, the role of *Die Kartenwissenschaft* was positively appraised by Hans-Peter Kosack and Karl-Heinz Meine (Kosack and Meine, 1955) and especially by Willy Kreisel (1960: 17):

A clarity of thought emerged … only with … Eckert's declaration! It was a solution of striking simplicity in a confusing situation. Cartography − neither a geographic nor an engineering science − but something on its own. That was something terrific at that time! You need to witness this period in order to recognize [that] … The declaration of Eckert was … a courageous act, as he encountered considerable opposition.

Others were keen to point out limits to the significance of *Die Kartenwissenschaft*, e.g. by Richard Finsterwalder, Emil Meynen and Erik Arnberger. The latter stated:

Max Eckert was in a way impressed with the unbelievable breadth of cartographic applications in the various scientific and area fields that he eventually got lost in the diversity of already published maps. Thus, the two-volume work turned from a badly needed methodology into cartology (Kartologie) as formerly − not quite wrongly − stated by Richard Finsterwalder.

*(Arnberger, 1970: 6)*

More critical opinions of *Die Kartenwissenschaft* called it a weighty and monstrous tome (Meine, 1968). It is quite conceivable that all these criticisms were a consequence of actual changes in cartography, when the evaluation of other great methodological works highlighted the difference of Eckert's approach.

In the 1980s, *Die Kartenwissenschaft* was positively reviewed several times, particularly by Konstantin A. Salishchev and Wolfgang Scharfe. Salishchev (1982: 11) noted

*Table 1.2* Table of contents of Volume 2 of *Die Kartenwissenschaft*

*Table 1.2 (continued)*

that Eckert had earned considerable attention in the Soviet Union, as substantial extracts of *Die Kartenwissenschaft* were translated into Russian and had a sizeable impact on the later development of cartography in the Soviet Union. Scharfe (1986: 66) referred to *Die Kartenwissenschaft* as a milestone:

> In spite of all critical review aspects and in spite of the relatively small consequences Eckert's Kartenwissenschaft had been a milestone and the turning-point in German cartography. There is no cartographical publication of the nineteenth and twentieth centuries which could compete with this work in richness of content, in size and in the skilful and appropriated combination of the cartographical past with the contemporary present.

Scharfe (ibid.) further pointed out that *Die Kartenwissenschaft* provided an important basis for the fundamental works of Arnberger, Imhof and Witt: 'Therefore one can say today without exaggeration that Max Eckert and his *Kartenwissenschaft* have laid the foundations of modern and autonomous scientific cartography in the German speaking area'.

Ulrich Freitag's (1992: 309) evaluation of *Die Kartenwissenschaft* is more mixed:

> In his main work Eckert mentioned, described, critically assessed and appreciated a vast amount of maps; he introduced new terms, often introduced and justified even German words for foreign terms, and thus his work remains an important reference work on the history of cartography and about the development of its terms. Beyond that it is an important compilation of all earlier works of Max Eckert, which led to the solution of individual partial problems of cartography. These are compiled here and associated here. All chapters that purport maps and their history are of interest even today. But all chapters, where Max Eckert tried to provide a theoretical basis for his mapping science, contain an incomplete surrogate of thoughts and sayings, demands and value judgements that do not meet the needs of a general theory concept.

Today, the work of Eckert is more widely appreciated, yet *Die Kartenwissenschaft* has still not been translated into English. According to Koch (2015), Eckert:

> [b]elonged to the most important representatives of [the] developing young scientific discipline [of] Cartography in the first half of the 20th century. With his two-volume work "*Kartenwissenschaft. Forschungen und Grundlagen zu einer Kartographie als Wissenschaft*" he laid a substantial foundation for the development of cartography as a science, which were expanded and consolidated significantly after 1950.

Also, Pablo Iván Azócar Fernández and Manfred Ferdinand Buchroithner give a fairly positive assessment of Eckert's contribution in their book *Paradigms in Cartography*, published in 2014. Max Eckert:

> [w]as the most important theoretician in the field of cartography. Considering these aspects, we can claim that the scientific-empirical approach in the discipline was initiated through the contributions made by Eckert. From there the literature began to include the concept of science in cartography and mapping. Also, Eckert's work initiated the writing of textbooks about cartography.
>
> *(Azócar Fernández and Buchroithner, 2014: 120)*

Also to note is that Eckert is given his own entry in *The History of Cartography* (Volume 6: The Twentieth Century); a level of recognition allotted to few other figures of theoretical cartography. Ingrid Kretschmer, author of that entry, emphasizes the special achievements of Eckert in that he 'wrote the first comprehensive German textbook on mapping science' and that he:

> [t]ogether with Karl Peucker established cartography as an academic discipline in Central Europe [...] Kartenwissenschaft served as a basis for theoretical cartography and cartographic methodology (thematic cartography) for the next generation of German-speaking authors, including Erik Amberger, Werner Witt, and Eduard Imhof.
>
> *(Kretschmer, 2015: 338–339)*

Amongst other references to Max Eckert in Volume 6, it is mentioned that *Die Kartenwissenschaft* had an impact on the development of theoretical cartography not only in the German-speaking area but also as 'an important precursor to Robinson's classic *The Look of Maps* (1952)' (Slocum and Kessler, 2015: 1505). The following aspects of Eckert's achievements are also highlighted in Volume 6: the suggestion for a logical application of colour (p.263), the provisions of psychological effects in cartography (p.1083), and the elaboration of map projections (p.1193). However, there are also historical reflections with respect to scientific cartography that do not mention *Die Kartenwissenschaft*, such as Crampton (2010) and Brotton (2012).

## Concluding remarks

The concept of Eckert's *Die Kartenwissenschaft* aims to achieve the following three objectives:

1   to make a wide variety of cartography visible, both in terms of theoretical and practical cartography;
2   to prove that theoretical cartography provided the cognitive prerequisites for the establishment of an autonomous scientific discipline of cartography; and
3   to demonstrate that by formation of an autonomous scientific discipline of cartography the theory-practice relationship can be optimized, which contributes to the further development of cartographic practice.

The first objective was fully achieved by Eckert in his *Die Kartenwissenschaft*. Its impact on cartography results first of all from this circumstance. The benefits of the work include the comprehensive review of previously neglected subjects, such as thematic cartography. Eckert did not sufficiently meet the second objective. *Die Kartenwissenschaft* primarily contains a descriptive systematization of empirical-traditional incurred elements and only a few theoretically based components, which are limited to relief representation and to map projection theory. So, he could not plausibly justify the need for secession of cartography from geography at the time, at least not for the 'guardians' of this discipline.

The conditions were not yet available for Eckert's elementary argument to achieve the third objective. His *Die Kartenwissenschaft* offers a system of knowledge with a limited basis and was methodologically ineffective. In addition, his intentions resulted from unrealistic illusions. Nevertheless, with *Die Kartenwissenschaft*, Eckert laid the cognitive foundations for the subsequent establishment of cartography as an independent scientific discipline, particularly by highlighting the variety of cartography and its distinction from other disciplines.

# References

Arnberger, E. (1970) "Die Kartographie als Wissenschaft und ihre Beziehungen zur Geographie und Geodäsie" *Grundsatzfragen der Kartographie* Vienna, Austria: Deuticke Publishers, pp.1–28.

Azócar Fernández, P.I. and Buchroithner, M. (2014) *Paradigms in Cartography: An Epistemological Review of the 20th and 21st Centuries* Heidelberg, Germany and New York: Springer.

Brotton, J. (2012) *A History of the World in Twelve Maps* London: Penguin Books.

Crampton, J.W. (2010) *Mapping: A Critical Introduction to Cartography and GIS* Chichester, UK: Wiley-Blackwell.

Eckert, M. (1907) "Die Kartographie als Wissenschaft" in *Zeitschrift der Gesellschaft für Erdkunde zu Berlin*, pp.213–227.

Eckert, M. (1921, 1925) *Die Kartenwissenschaft: Forschungen und Grundlagen zu einer Kartographie als Wissenschaft* (2 vols) Berlin: Walter de Gruyter.

Eckert, M. (1939) *Kartographie: Ihre Aufgaben und Bedeutung für die Kultur der Gegenwart* Berlin: Walter de Gruyter.

Freitag, U. (1992) *Kartographische Konzeptionen: Beiträge zur theoretischen und praktischen Kartographie 1961–1991 [Cartographic Conceptions: Contributions to Theoretical and Practical Cartography, 1961–1991]* Berlin: Berliner Geowissenschaftliche Abhandlungen.

Koch, W.G. (2015) "Eckert (seit 1934 Eckert-Greifendorff), Friedrich Eduard Max" in *Sächsische Biografie* Institut für Sächsische Geschichte und Volkskunde e. V. Available at: *www.isgv.de/saebi* (Accessed: 7 December 2015).

Kosack, H.-P. and Meine, K.-H. (1955) "Die Kartographie, 1943–1954: Eine bibliographische Übersicht" *Kartographische Schriftenreihe (Volume 4)* Lahr-Schwarzwald, Germany: Astra Verlag.

Kreisel, W. (1960) *Organisation der integralen Kartographie* Einsiedeln, Switzerland: Benziger.

Kretschmer, I. (2015) "Max Eckert" in Monmonier, M. (Ed.) *The History of Cartography (Volume 6: 1)* Chicago, IL: The University of Chicago Press, pp.338–340.

Krümmel, O. and Eckert, M. (1908) *Geographisches Praktikum für den Gebrauch in den geographischen Übungen an Hochschulen* Leipzig, Germany: Wagner & E. Debes.

Meine, K.-H. (1968) "Zum 100. Geburtstag von Max Eckert" *Kartographische Nachrichten* 18 (3), pp.77–80, 187.

Ogrissek, R. (1985) "Studium, Promotion und Lehrtätigkeit Max Eckerts an der Universität Leipzig im 19. Jahrhundert" *International Yearbook of Cartography* 25 pp.139–158.

Salishchev, K.A. (1982) Idei i teoretičeskie problemy v kartografii 80-ch godov *Itogi Nuki i Techniki: Kartografija* (10) Moscow: VINITI.

Scharfe, W. (1986) "Max Eckert's Kartenwissenschaft: The Turning Point in German Cartography" *Imago Mundi* 38 (1) pp.61–66.

Slocum, T.A. and Kessler, F.C. (2015) "Thematic Mapping" in Monmonier, M. (Ed.) *The History of Cartography (Volume 6)* Chicago, IL: The University of Chicago Press, pp.1500–1524.

# 2

# The communication models in cartography

*Christopher Board*

The cartographic communication paradigm has become one of the established frameworks for analysing how maps work. For convenience, it simplifies the processes at work in the interactions between those who design and make maps and those who have either solicited a map, or are presented with a map. It assumes that there is a real world from which a selection can be made according to some specific focus. It suggests that normally information is passed on to others in order to increase their knowledge. The paradigm ventures to make explicit all operations in the creation of maps. It also attempts to understand how the eyes and brain take in the message of those who have created the map. These attributes are thought to expand the knowledge, not merely of the map user but also that of the map creator. Such a framework encourages researchers to devise theories to explain, in most cases, part of the entire process of knowledge transfer. Experiments with maps of different characteristics are facilitated and are the better for their reference to a theory. Although the earlier representational paradigm was replaced by the communication paradigm, the latter is now being challenged by a new manifestation of the representational paradigm.

## Development and evolution of the cartographic communication paradigm

'Maps as Models', which was published in *Models in Geography* in 1967, was conceived in about 1964–65 as a contribution to a summer school in the 'New Geography' organized by geographers Chorley and Haggett. They were keen to include the idea of the map as a model of the real world as a topic in another summer school and asked me to work up a review article for a book to be published in two years' time. The essay was addressed to geographers and their use of maps in research problems, but inevitably the paper examined the nature of maps and considered literature which looked at maps from a theoretical standpoint. An opportunity to attend one of the Quantitative Institutes at Ohio State University encouraged me to seek out the views of American geographers such as Tobler and Jenks with whom I discussed the emerging system diagram of the Map-Model Cycle. An interest in landscape painting introduced me to the thoughts of E.H. Gombrich and a trawl through relevant literature in cartography including Robinson's *The Look of Maps* (1952) and introductory texts on information theory set me on a path which led beyond my original ideas. Almost by accident while scanning *Реферативный журнал* (Review Journal), an extracting bibliographic source produced in the Soviet Union, I found what was probably the earliest graphical exposition of the processing of cartographic information, by Moles (1964) (Figure 2.1). Moles was in touch with Jacques Bertin as he wrote *Sémiologie Graphique* (1967) acknowledging Moles 'to whom he expressed his most profound recognition' (translation from the French). Bertin refrained from expressing his views on Moles' diagrams. Looking at this very simple diagram by Moles, I should have seen that my Map-Model Cycle (Figure 2a) was far too complicated as a vehicle for passing on the basic information it included.

In effect, I argued that geographical analysis consists of two parts: the creation of a representation of the real world by selecting a view or area of study, reducing the phenomena down to a manageable scale, then clarifying the representation of some sets of data from which a map emerges. This gives rise to the new view, or perhaps an interesting research question. With this deliberately created model of the real world, the researcher envisages, say, a general trend, but may identify exceptions to that trend, i.e. anomalies. These trends are seen as models of an aspect of the real world whose nature may suggest an explanation, but the deviations are also real and may hint at a more complex explanation. The model of a regression cycle tabulated in Haggett (1965: 279–280) includes isopleth maps of a voting pattern and their use as data in regression analysis, throwing up a partial explanation on one variable and a further map of deviations. By judicious choice of variables, the researcher can 'improve' the explanation. Crucial in this cycle is the flow of information from the real world to the authors of maps, then to the users of maps, much as in the Moles diagram. Instead, I used the analogy of a general model of a communication system as I regarded maps as a stimulus to communication. A pattern encoded is seen as a message through a signal to a receiver, an encoded message. Noise affects the signal. In the example I quoted, the pattern is simple – the presence or absence of a phenomenon (visible traces of medieval open-field cultivation) in the English Midlands as seen in aerial photographs. The resulting map immediately invites a whole host of questions because for the first time its author could visualize a pattern, as it were, from space.

Now I can see that the Map-Model Cycle was what is now often described as a mind map, or, in broad terms, a brain-storming device, to envisage what factors are at work in the creation of maps and how they are used – expressed in the two sides of Figure 2.1. The links and directions are not as clear as they should be, as I quickly discovered when using

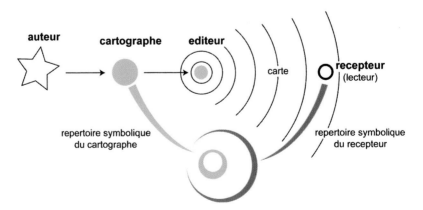

*Figure 2.1* Moles' diagram. Translation of terms into English: (map) author, data, cartographer, cartographic message, publisher, (by) diffusion to (map) reader

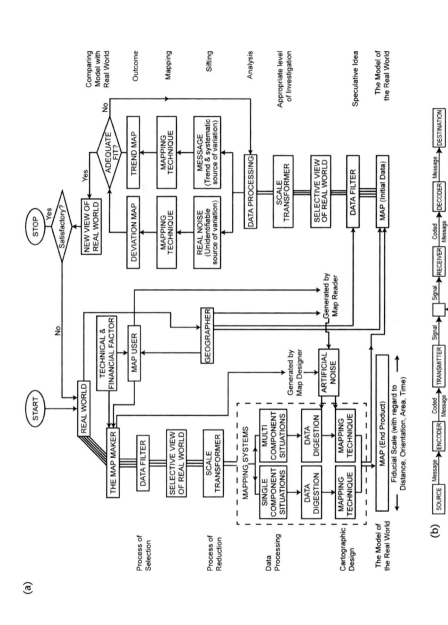

*Figure 2.2* (*a*) The Map-Model Cycle; and (*b*) Generalized communication system (redrawn from Board, 1967)

the diagram in teaching. By contrast the stark simplicity of the generalized communication system (Figure 2.2) below it encapsulated the essence of the way in which many practising and thoughtful cartographers saw their role. Most reactions have not addressed the more complex diagram. Figure 2.3 is a simplified version of this created much later for a lecture on the use of such models in cartography.

Through my contacts in the ICA (International Cartographic Association) Commission for the Definition, Classification and Standardization of Technical Terms in Cartography, I learned that a Czechoslovak cartographer Koláčný had written a theoretical paper on cartographic information which included a graphic model of much greater simplicity. This was to support a proposal to set up another group in the ICA at the forthcoming International Cartographic Conference in India in December 1968. Koláčný exchanged our papers and a group set up under his chairmanship then met in Prague in October 1969.

Despite his emphasis on cartographic information, Koláčný was keen to stress that the process described by his diagram be called 'communication of cartographic information', which led to

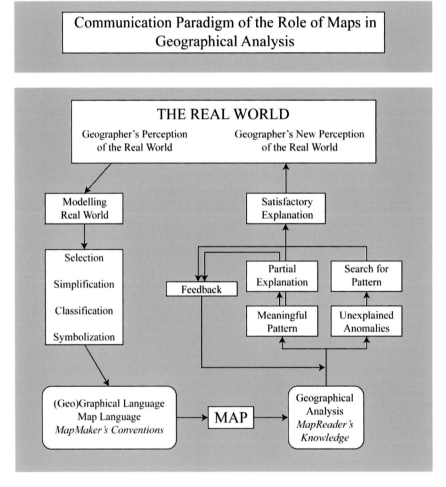

*Figure 2.3*  Communication paradigm of the role of maps in geographical analysis (a simplified version of *Figure 2(a)* used for lectures)

the shorthand term 'cartographic communication' for the process. The name of the working group and the Commission it became adopted this term as its title. Koláčný (1971) wrote a report of that meeting, but due to his 'retirement', his leadership and the continuation of theoretical work passed to the Polish academic Ratajski, who had been developing theoretical concepts on thematic map design in preparation for a book. From our first meeting in 1967, I regarded his contribution partly as a bridge between East and West, but he was also an original thinker and promoter of theoretical cartography in an era when maps and atlases were pouring off printing presses and cartographic societies were extremely active (Board, 1983).

Ratajski (1973) acknowledged his debt to Koláčný and myself in the elaboration of his research structure diagram. He was well placed to organize meetings of the new Commission and presided over several of these during the 1970s. By 1980 and after a busy programme of Commission meetings, the notion of cartographic communication was set to become a new paradigm (Ormeling, 2015). According to Ormeling, over 30 articles on cartographic theory were published in the *International Yearbook of Cartography* between 1961 and 1980. Many are illustrated with diagrams illustrating views of the way information is processed from map author to map user though the map.

Robinson's work was well known by this time and he had read a paper to the ICA technical meeting in Amsterdam in 1967. From the late 1960s, his department was engaged on a project to develop a general theory of cartography (Robinson and Petchenik, 1976: xi). The six essays authored by Robinson and Petchenik in *The Nature of Maps* were intended to redress the perceived lack of introspection among cartographers and to arrive at understanding how maps accomplished what they were supposed to do – to communicate. Notable among these essays was 'The Map as a Communication System', which formed part of a collection of important papers brought together by Guelke (1977). It appeared at a time when there was a thirst for such enquiries into what the authors called the 'map percipient'. For the first time, some comparisons could be made between the views of different authors coming from different traditions, professions or backgrounds, mostly without having to cope with non-English terminology. In an attempt to widen interest, key terms relevant to cartographic communication were listed in an Anglo-French glossary in a provisional edition of publications in that area (Board, 1976).

The Commission met in London in 1975 to discuss a wide-ranging selection of papers and there was already a great deal of common ground (Wood, 1975). Papers by Ratajski, Meine and Morrison were theoretical, while others focussed on specific parts of the entire process, such as map-reading tasks, eye movements or the experimental design of map-reading tests. Following my geographical inclinations, I chose to examine the communication process with the aim of clarifying how maps were used, but soon became involved in discussions with Morrison who was inclined to regard cartography as a communication science (see Morrison, 1974). The former employed mathematical thinking, expressing the generalization process in graphic terms as well (Morrison, 1976) (Figure 2.4). This replaces the core of Ratajski's model with a new one examining the way in which the cartographer's cognitive realm is the context for arriving at the cartographer's concept of a map, which subsequently yields information affecting the map reader's cognitive realm, in turn informing the map reader's knowledge of reality. Employing terms from the realm of natural language, he discusses the grammar of the map language using semiological concepts. By judicious use of Venn diagrams, Morrison spells out idealized steps of operations mental and practical by which the cartographer communicates information to someone else. By recognizing that information can also be created by the symbolization process itself in a manner that the whole is more than the sum of its parts, we are reminded of the power of maps over other forms of communication. The economy of the argument in favour of

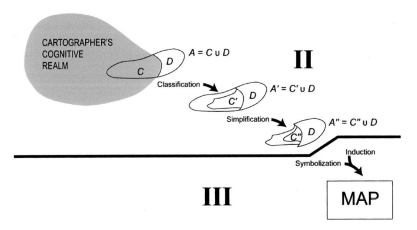

*Figure 2.4* **Processes involved in producing a map from a cartographer's cognitive realm (redrawn from Morrison, 1976)**

a scientific paradigm for cartography is forceful, leaving the student of cartography to visualize examples from a mental album of maps seen previously.

In a second paper, Morrison (1976) stresses the processes involved in map reading and interpretation, stating that cartographers must assure their readers that information has been communicated correctly: specific results of interpretation 'may be of far more concern to other scientists than to cartographers' (ibid.: 95). Morrison also elaborates the processes involved in generalization, but pointing out that map interpretation lies outside the realm of the cartographer. That individual is obliged to use the eight visual variables such that any map reader or interpreter can detect, discriminate, recognize and estimate cartographic symbols adequately (ibid.: 94). In any design process, there needs to be some iteration between the expert and the end user, depending on the purpose – until the end user is satisfied. This also formed part of the collection of important papers brought together by Guelke (1977). While Morrison was spelling out the fundamental processes of communication, by selection, classification and simplification, he allows the cartographer to communicate with himself as the initiator by a fourth process, induction. Here the cartographer can, 'by becoming a map reader', enlarge his cognitive realm. The mapping of deaths by cholera by John Snow is a classic example (see Gilbert, 1958).

Other theoretical approaches in Germany paid more attention to cartographic language. Meine, for example, reviewed cartographic communication and the concept of a cartographic alphabet. His treatment of this recalled passing references to the similarity between cartography and sign or natural language. His list of references demonstrates many publications on cartographic communication in German. At the Tokyo meeting of the ICA, Freitag (1980) proposed a new definition of cartography as 'a part of the science and technology of communication' and presented the outline of an introduction to cartography using communication theory and related studies as a framework. Combining new ideas and old experiences, he began with the history of cartography, then cartography as a system, maps as graphic signs, maps as models, maps as signals (the relationship between the map and the many pragmatics of a general sign theory), and concluded with the status and future of cartography. He acknowledges that the Map-Model Cycle was the first attempt to establish a comprehensive outline of cartography on a new conceptual basis (ibid.: 19). Similarly, he saw that I had reshaped my model considerably in 1977 and 1978 to accommodate the views of Robinson and Petchenik (1976); having published 'Maps as Models' in a non-cartographic volume, its originality, its ambiguous terminology and its complexity 'prevented the model from becoming a widely accepted concept in cartography'. This and other key papers by Koláčný and Ratajski evidently persuaded Freitag to modify his structure of cartography. Coupled with the approach adopted by Bertin and his disciples, Ormeling, reviewing the situation in 2015, suggests that semiology was rivalling the communication approach at the time.

In the USA, a new edition of *Elements in Cartography* had appeared in 1978, with Morrison joining Robinson and Sale as a third author. The impact of new thinking and new technologies was revolutionary. Communication, remote sensing and generalization were favoured, and computer-assisted cartography, graphic design and colour all received substantial treatment. Diagrams of both general and cartographic communication systems have pride of place on pages 2 and 3 (Robinson *et al.*, 1978). User feedback and projection choice also illustrate the new approach. In the UK, no such evolution was taking place in textbooks. In 1973, Keates had published his *Cartographic Design and Production* which encapsulated a decade of teaching in Glasgow. Chapter 3 on map design clearly states that 'the function of design is to communicate . . . information effectively to the user' (Keates, 1973: 29). His design principles stem from a presentation of symbolization and generalization to represent the real world, but the practicalities of compilation and production occupy by far the larger part of his book. We shall turn to him for a published critique of the communication paradigm, but this did not arrive until the publication of *Understanding Maps*, published just before the ICA conference in Warsaw in 1982, where it was eagerly awaited by Polish colleagues. Its critical stance did not affect the reception of my tribute to the late Ratajski, which was devoted to his influence on the development of cartographic theory.

In the meantime, I began to consider what geographers contribute to evaluating maps (Board, 1977a), map-reading tasks (Board, 1978) and how theories of cartographic communication can be used to make maps more effective (Board, 1977b). After 1980, the ICA Commission which had prompted so much discussion, was renewed but its:

> [c]ritics, though admitting that the wealth of literature and the accompanying diagrammatic models ... had contributed to the understanding of the process of map communication, stated that the results had very little to offer in the way of improving map design.
>
> *(Ormeling, n.d.: 56)*

He suggested that research in this field was irrelevant to the realities of map-making, but that I, as the new chairman of the Commission (following Rataksji's death in 1977), understood that there was a need 'to explore the application of the more abstract notions to the practical problems of map design, production and use' (ibid.). Another more representative group met in London in 1983, whose proceedings were published (Board, 1984) in default of accomplishing other aims. After 1984, the Commission was split into two: one on concepts and methodology under Freitag and one on map use, to be chaired by me.

Some of the most severe commentary came from Keates, who was to present an impassioned defence of art in cartography in the 1983 meeting of the Commission in London. In the much-revised second edition of *Cartographic Design and Production* (Keates, 1989), his final new part titled 'Applied Cartography' shows how cartographers approached special-subject and special-purpose maps. The very last chapter contains his view of cartographic theory and research. These carefully considered comments on the 1970s and 1980s have an authority rooted in his long, practical experience in map design, production and teaching.

Keates followed tradition by defining cartographers as those responsible for designing and constructing the graphic image of the map and objected to a wider definition which allocates more functions in the entire process: 'A good, basic model would lead naturally to the discussion of more complex details, but the multiple nature of contributions to the creation of a map is not made apparent, mainly because important terms remain highly ambiguous' (Keates, 1982: 112). Summing up, Keates argues that a comprehensive theory of cartographic communication implies a foundation of a comprehensive mapping science. He favours a narrow definition of cartography centred on maps – the making, using and studying of maps. While emphasizing the importance of cartography in map-making as a whole, such a science might help map-makers to understand the problems faced by the map user. Theorizing while avoiding the realities of map-making is an intellectual exercise. He reminds us that once published, whatever the motives of the map-maker, the map can be used by anyone as well as those for whom it was designed. The reminder is well placed because it goes to the heart of the difficulty of setting out a general model of communication by maps. Over-reliance on experimental methods which concentrate on narrowly defined questions either require very simple test maps, or have limited relevance to complex maps. Experienced cartographers or map-makers realize that they must treat their map holistically, even to the point of producing experimental maps as the most realistic presentations.

Keates (1989: 247) did, however, recognize my own interest in map use, stating that 'Maps as Models':

> [i]naugurated a period of widespread discussion about cartography and the publication of many new 'theories' about cartography. Because it seeks to examine the whole basis of cartography and the use of maps, the emergence and development of these new ideas needs to be examined in some detail.

Before Keates returned to discuss theory, he drew attention to the electronic-digital revolution affecting the 'cartographic side of map-making' eventually examining cartography and geography where he sees the danger of regarding 'cartography' and map-making as synonymous. Meeting the map user's requirements requires a quality and availability of information as well as cartographic representation: 'A comprehensive theory would have to take this into account as part of the [communication] "system" . . . and need to encompass map-making as a whole' (ibid.: 251). This would involve including the collection of primary data from surveys in communication models. Keates concludes this final chapter claiming, 'despite what is often implied by theories of cartographic communication, there is a great body of empirical knowledge about "every-day" cartographic design problems, and a considerable understanding of the principles by which they are resolved' (ibid.: 254). These principles are discussed in Part 1 of *Cartographic Design and Production*, where a clear distinction between cartography and map-making is made and put into context in Keates' own overview of map-making and cartography: factors and variables (Keates, 1989: 11) (Figure 2.5).

In the Soviet Union, further criticism was recorded by Salichtchev in *Graphic Communication and Design in Contemporary Cartography* edited by Taylor (1983: 11–35). Taylor's view was that language has been a barrier to the discussion of ideas on cartographic communication and suggested that 'Maps as Models' was omitted from the Russian translation of Chorley and Haggett's (1967) *Models in Geography*. In point of fact, Salichtchev writes: 'On the whole, this interesting though somewhat eclectic chapter is in keeping with the spirit of the monograph which examines the systemizing, generalizing of cognitive functions of models' (ibid.: 25).

By the 1990s, Keates (1996: xiii) felt the need 'to learn from and indeed to incorporate cartographic expertise into the development of artificially intelligent and expert systems applied to map production', which was promoted by rapid development of cartographic and geographic information systems. This new edition was a basis for discussion rather than a textbook or

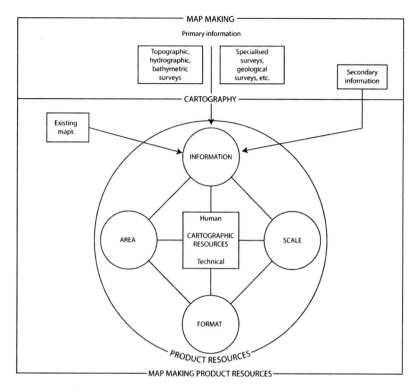

*Figure 2.5* **Map-making and cartography: factors and variables (redrawn from Keates, 1989).**

literature review. There are new elements in Part 3, called 'The Map as a Communication'. Although much of the first chapter follows its counterpart from 1982, the link with the next regards the 'desire to elevate cartography to a scientific "discipline" in its own right, complete with its own theories, is in some respects a typically academic pursuit' (p.123). Comprehensive theory needs to be based on evidence, which was lacking as far as map use is concerned. Implied here is a well-founded belief that map-making is much better understood.

Communication models are examined in the same terms as they were in 1982, but Keates newly addresses the cost of map-making in contrasting situations, leaving the reader to supply instances. In one example when creating a national atlas, the initiator (editor) may want a map for which data are not readily available for national coverage (see Clayton, 1963, who revealed the amount of research required for a single map). As for map use and cartographic research, some emphasis is given to contrasting psychological approaches: behaviourist or psycho-physical; and cognitive. Picking out Eastman's (1985) advocacy of the latter is the more effective way of understanding how the map user makes sense of a complex map, using the communication model in 'Maps as Models' as the example of that theoretical approach. Eastman (1985) reorganizes it as his first illustration (Figure 2.6) and contrasts it with the system model of visual information processing.

As a contribution to user research, the visual organization processes Eastman (1985: 100) outlines apply just as much to the cartographer as to the map user: 'The individual who reads a map brings his or her own cognitive structures, built through personal experience, with which the information thus displayed is understood. What the cartographer supplies is the spatial context of entities and the relationships between them'. What these signify to the reader

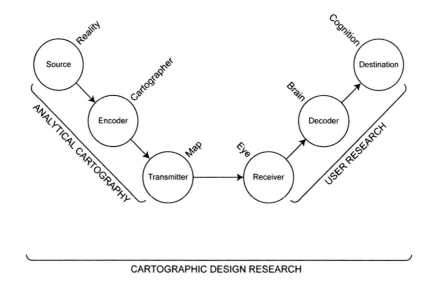

*Figure 2.6* **The cartographic communication process and cartographic design research (redrawn from Eastman, 1985)**

is unknown, 'but it is the responsibility of the Cartographer to ensure that the relationships portrayed are warranted by the original data, and are unambiguously apparent in a manner that facilitates the reader's organizational exploration (ibid.: 100). The corollary is that the cartographer has to master appropriate tools to create a map, whereas the map user need not know how to employ the tools in order to gain information. In fact, this places the cartographer in a position of great trust. Keates discusses ethics in cartography mostly in the context of Harley's espousal of a deconstructionist view of map creation. Although he devotes a short section to maps and propaganda (Keates, 1996: 107) one should remember that the codes adopted by the cartographer can readily be applied to any situation depending on the standpoint of the map author. I recall a display of results from the 1981 population of Great Britain displayed in maps in the windows of the Office of Censuses and Surveys, then in Kingsway, where population decreases since 1971 (prevalent in urban areas) were shown in red and increases were in green, to draw attention to the trend. Before leaving our mention of Keates it should be mentioned that the two final parts of *Understanding Maps* (1989) are devoted to maps as artistic work and the map as a product of skill. Keates' legacy is sharing his practical and theoretical knowledge of mapping to provide cartographers in the broadest sense with ideas for discussion, and, thus to inform further research.

Eastman had been critical of the theoretical trends and so Keates found a fellow critic in North America. Writing on theoretical work in Europe, Ormeling (2015) sees new trends in the last two decades of the 20th century. In map-making, digitization was a major challenge, particularly as it provided relatively secure and usable data models that begin to supplant the source mapping for all maps. Visualization came to have two meanings; a new one describing the output of digital mapping. Such trends raised fundamental questions about the nature of the map, including the roles of cartographer and of map user. At least the re-examination and modelling of what the map user's role was in cartography.

A year before the publication of the second edition of *Understanding Maps*, a very different study of cartography was published by the American academic geographer, Alan MacEachren, titled *How Maps Work* (which was preceded by *Some Truth with Maps* in 1994 and did use colour), which has over 500 pages and many illustrations (though not always in colour). As a result, he is able to put across his ideas more successfully, especially to a readership very familiar with many maps. It benefits from several decades of academic cartography in North America and the existence of a corpus of material on maps from which to draw.

MacEachren's (1995) preface explains that maps are spatial representations which can in turn stimulate other spatial representations and that representation is an act of knowledge construction. He aims at showing 'how meaning is derived from maps and how maps are imbued with meaning' (p.vi) and hopes the book will be a base for graduate seminars, a guide to experimental and a prompt to critical analysis. His context is an increasingly visual world, in which computers and satellites some might say are swamping us with data. Chapter 1 examines a scientific approach to improving map representation and design. It reviews approaches taken after the Second World War, starting with the functional, broadly about maps not just as art, but how they could be made to be fit for purpose. Next, maps are seen as vehicles for communicating spatial information, initiated by my 'extremely complicated flow diagram' (the Map-Model Cycle), on p.4. This and a deceptively simple and inadequate communication model was a Pandora's Box, an attractive idea from which arose many objections.

Keates is cited as one who saw the devaluation of art in cartography which with science is in the British and ICA definitions of cartography, where 'The art, science and technology of making maps together with their study as scientific documents and works of art . . . all kinds of maps . . . representing the Earth' are the key elements (for example, see British Cartographic Society, 2016). MacEachren then reviews the impact of postmodern criticism of positivist science, favouring the view of maps as reflections of culture just as much as representations of the Earth, what is now regarded as 'critical cartography' – replacing one limiting approach with another. The author then introduces his fresh approach with its emphasis on representation. Finally, in a postscript, he concludes that to:

> [m]ore fully understand how maps work, we need to investigate mechanisms by which maps both represent and prompt representation. The communication paradigm took us a step in this direction but floundered due to a fundamental assumption that matched only a small proportion of mapping situations ... maps as primarily a 'vehicle' for transfer of information.
>
> *(ibid.: 459)*

MacEachren's approach is that 'mapping and map use are processes of knowledge construction rather than transfer' (ibid.: 459). *How Maps Work* is richly illustrated by diagrams and maps, many created by MacEachren himself. However, there are two diagrams which strike me as crucial to understanding what is meant by visualization (Figure 2.7) and which modifies and extends the communication framework I helped to establish. The rise of visualization as a process is marked in a collection of essays by MacEachren and Taylor (1994, chapter 1). In another essay in MacEachren and Taylor's (1994) volume we see a diagrammatic approach to visualization reminiscent of the Map-Model Cycle: Lindholm and Sariakoski (1994: 170), show how information set, process and information flow are distinguished.

A second major book by three French cartographers published in English in 2010 sets out to provide a scientific approach to thematic cartography (Cauvin *et al.*, 2010). They cite the explosion of computer use, the proliferation of computer programs, the diversity of specialists in map production and the absence of a structured theoretical body of (relevant) knowledge as reasons. They see the map as the rest of a series of transformations. Each map should be the

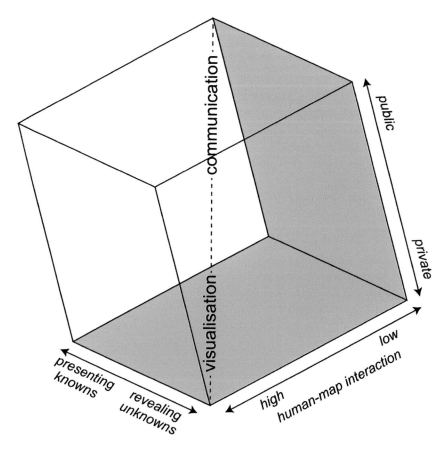

*Figure 2.7*  MacEachren's (1994) visualization cube (redrawn from MacEachren, 1995)

unique product of a potential user's question, raising the possibility of an optimal map. 'No longer can a map be designed by the two classical co-authors: the cartographer and the thematician' (Cauvin *et al.*, 2010: xxi). Expertise from others is required in a developing discipline they call geovisualization. This first volume in the set of three stresses the guiding role of cartographic design employed in creating a thematic map. Much of this could be applied to general-purpose mapping, but here the competing and conflicting requirements of users cannot all be taken into account. Readers of this book should not miss the Foreword written by the doyenne of French geographical cartography, Sylvie Rimbert. She brings us down to Earth, to current situations and problems, and commends the authors for choosing to illustrate the products of their cartographic thinking with case studies of the Duchy of Luxembourg. Part III deals with three transformations: the semiotic, the representation mode and cartographic design. Discussion of cartographic communication initiates this, using Lasswell's (1966) diagram as an organizing framework (Figure 2.8). There is some resemblance to models previously discussed, but options or choices are more obvious.

The map, a construction based on scientific reasoning, is very much the focus of this diagram and their entire approach. Room is left for alternatives to be offered to the user: for exploratory data analysis, or for effective presentation. The massive bibliography which has informed the entire work draws unusually on both English and French literature.

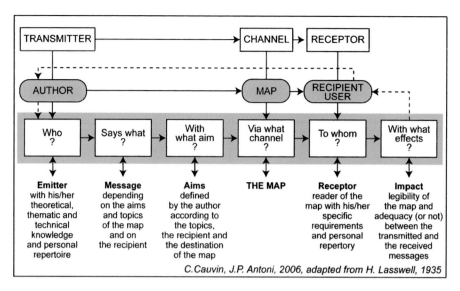

Figure 2.8   Lasswell's (1966) communication diagram applied to mapping (redrawn from Cauvin *et al.*, 2010)

Although MacEachren's (1995) book will eventually usher in a new paradigm with its re-emphasis on representation, there are still legacies of the communication paradigm to be found in texts on cartography and geographical information systems published since the mid 1990s (e.g. Kraak and Ormeling, 1996: pp.47–54; Kraak and Brown, 2003: figure 5.5). Among other text-books on mapping, that by Dorling and Fairbairn (1997) includes under 'Changing perspectives on cartographic practice', a modified cartographic communication model (Figure 2.9), which they concede has influenced cartographic practice to optimize map design in order to optimize com-munication of information. They see this being applicable more to phases of visual thinking in the private realm of the cartographer (exploration and confirmation) followed by visual communica-tion in the public realm (synthesis, presentation). It will be important not to confuse cartography with visualization, but rather to see how the latter fits into a wider scheme of ideas and practice.

It appears that the idea of a cartographic communication model may still have value, even if it has reached a dead-end and has lost its paradigmatic status. Even the telling criticism of the over-simplified communication diagram has itself encouraged more coherent theory and research on map use. Ormeling (2015) suggests that cartography has re-emerged in the new century (from the digital revolution) with its focus on theory, map use and usability.

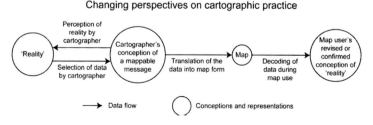

Figure 2.9   Simplified model of the cartographic communication system (redrawn from Dorling and Fairbairn, 1997: 161)

# References

Board, C. (1967) "Maps as Models" in Chorley, R.J. and Haggett, P. (Eds) *Models in Geography* London: Methuen, pp.671–725.

Board, C. (Ed.) (1976) "Bibliography of Works on Cartographic Communication" (Provisional edition) ICA Commission V.

Board, C. (1977a) "The Geographer's Contribution to Evaluating Maps as Vehicles for Communicating Information" *International Yearbook of Cartography* 17 pp.47–59.

Board, C. (1977b) "How Can Theories of Cartographic Communication Be Used to Make Maps More Effective?" *Paper presented to the Working Meeting of ICA Commission IV*. Hamburg, September.

Board, C. (1978) "Map Reading Tasks Appropriate in Experimental Studies in Cartographic Communication" *The Canadian Cartographer* 13 (1) pp.1–12.

Board, C. (1983) "The Development of Concepts of Cartographic Communication with Special Reference to the Role of Professor Ratajski" *International Yearbook of Cartography* 23 pp.19–29.

Board, C. (1984) "Higher Order Map-Using Tasks: Geographical Lessons in Danger of Being Forgotten" in Board, C. (Ed.) "New Insights in Cartographic Communication" *Cartographica* Monograph No.31, pp.85–97.

British Cartographic Society (2016) "About" Available at: *http://www.cartography.org.uk/about/* (Accessed: 1 August 2016).

Cauvin, C., Escobar, F. and Serradj, A. (2010) *Thematic Cartography and Transformations* Volume 1 London: ISTE and Wiley.

Chorley, R.J. and Haggett, P. (Eds) *Models in Geography* London: Methuen.

Clayton, K.M. (1963) "A Map of the Drift Geology of Great Britain and Northern Ireland" *Geographical Journal* 129 pp.75–81.

Dorling, D. and Fairbairn, D. (1997) *Mapping: Ways of Representing the World* Harlow, UK: Longman.

Eastman, J.R. (1985) "Cognitive Models and Cartographic Design Research" *The Cartographic Journal* 22 (2) pp.95–101.

Freitag, U. (1980) "Can Communication Theory Form the Basis of a General Theory of Cartography?" *Nachrichten aus dem Karten- und Vermessungswesen* Series II, p.38 *Paper presented at the 10th International Conference on Cartography*, Tokyo.

Gilbert, E.W. (1958) "Pioneer Maps of Health and Disease in England" *Geographical Journal* 124 pp.172–183.

Guelke, L. (1977) (Ed.) "The Nature of Cartographic Communication" *Cartographica* Monograph No.19.

Haggett, P. (1965) *Locational Analysis in Human Geography* London: Arnold.

Keates, J.S. (1973) *Cartographic Design and Production* Harlow, UK: Longman.

Keates, J.S. (1982) *Understanding Maps*. Harlow, UK: Longman.

Keates, J.S. (1989) *Cartographic Design and Production* (2nd ed.) Harlow, UK: Longman.

Keates, J.S. (1996) *Understanding Maps* (2nd ed.) Harlow, UK: Longman.

Koláčný, A. (1971) "Cartographic Information Report of the Working Group" *International Yearbook of Cartography* (IYC) 11 pp.65–68.

Kraak, M.-J. and Ormeling, F.J. (1996) *Cartography Visualization of Spatial Data* Harlow, UK: Longman pp.47–54.

Kraak, M.-J. and Brown, H.A. (2003) (Eds) *Web Cartography: Developments and Prospects* London and New York: Taylor & Francis.

Lasswell, H. (1966) "The Structure and Function of Communication" in Berelson, H. and Jonawitz, M. (Eds) *Reader in Public Opinion and Communication* New York: Free Press, pp.178–190.

Lindholm, M. and Sariakoski, T. (1994) "Designing a Visualization User Interface" in MacEachren, A. and Taylor, D.R.F. *Visualization in Modern Cartography* New York: Pergamon, pp.167–184.

MacEachren, A. (1994) *Some Truth with Maps: A Primer on Symbolization and Design* Washington, DC: Association of American Geographers.

MacEachren, A. (1995) *How Maps Work: Representation, Visualization and Design* New York: Guilford Press.

MacEachren, A. and Taylor, D.R.F. (1994) *Visualization in Modern Cartography* New York: Pergamon.

Meine, K-H. (1977) "Cartographic Communication Links and a Cartographic Alphabet" in Guelke (Ed.) *Cartographica* Mongraph No.19, pp.72–91.

Moles, A.A. (1964) "Théorie de l'information et message cartographique" *Sciences et Enseignement des Sciences* (Paris) 5 (32) pp.11–16.

Morrison, J.L. (1974) "A Theoretical Framework for Cartographic Generalization with Emphasis on the Process of Symbolization" *International Yearbook of Cartography* 14 pp.115–127.

Morrison, J.L. (1976) "The Science of Cartography and its Essential Processes" *International Yearbook of Cartography* 16 pp.84–97.

Ormeling F.J. (2015) "Academic Cartography in Europe" in Monmonier, M. (Ed.) *The History of Cartography (Volume 6)* Chicago, IL: Chicago University Press, pp. 6–12.

Ormeling F.J. (Senior) (n.d.) *International Cartographic Association 25 Years 1959–1984* Enschede, The Netherlands: International Cartographic Association.

Ratajski, L. (1973) "The Research Structure of Theoretical Cartography" *International Yearbook of Cartography* 13 pp.217–227.

Robinson, A.H. (1952) *The Look of Maps* Madison, WI: University of Wisconsin Press.

Robinson, A.H. and Petchenik, B.B. (1976) *The Nature of Maps: Essays toward Understanding Maps and Mapping* Chicago, IL: Chicago University Press.

Robinson, A.H., Sale, R. and Morrison, J. (1978) *Elements of Cartography* (4th ed.) New York: John Wiley & Sons.

Taylor, D.R.F. (Ed.) (1983) *Graphic Communication and Design in Contemporary Cartography* Chichester, UK: John Wiley & Sons.

Wood, M. (1975) "Symposium on Cartographic Communication" *The Cartographic Journal* 12 (2) p.95.

# 3

# Cartography, visual perception and cognitive psychology

*Amy L. Griffin*

Although they may be beautiful objects in their own right, maps require a map user to be made meaningful, even if in some cases the map user is the same person as the map maker. Early cartographers developed much of their understanding of which map design decisions worked and those that did not through trial and error rather than through systematic empirical study of how design decisions affected map use (Robinson, 1952). This is not to say that early cartographers did not seek any feedback on their work, but that this feedback was informal and ad hoc, rather than formal and systematic. One hallmark of twentieth-century cartography is its explicit acknowledgement of the importance of the map user (Montello, 2002).

## The beginnings of 'scientific cartography'

The nineteenth century was a period during which the organizational structures that we today call scientific disciplines began to emerge. As scientific knowledge grew, it became increasingly difficult for any one individual to master the existing state of knowledge (or even the most important knowledge) pertaining to a given topic (Weingart, 2010). This led to specialization and the differentiation of knowledge into different disciplines. As the splintering of science into different disciplines progressed, a new type of innovation emerged: the application of the theories and methods of one discipline to the object of study of another.

The first cartographer to suggest the value of a scientific approach to cartography was the German cartographer Max Eckert (see Chapter 1, this volume), who declared in 1908 that 'I should, therefore like to designate one of the most important topics that scientific cartography has to deal with: "map logic". Map logic treats of the laws which underlie the creation of maps and which govern cartographic perception' (Eckert, 1908: 348). He expanded upon these ideas in a 1500-page, two-volume exposition that argued for the application of psychological theories and research to establish the laws of map logic (Eckert, 1921, 1925). From citations of Wilhelm Wundt's work that appear in *Die Kartenwissenschaft*, it is clear that Eckert was both aware of and influenced by Wundt's establishment of the discipline of experimental psychology in Germany 1879 (p.636).

As a discipline concerned with the study of human behaviour and mental processes, psychology investigates topics such as perception and cognition, among others. Perception concerns how the brain brings meaning to sensation by organizing, identifying and interpreting sensory information. Biologically, perception is important because it provides us with information that is critical to how we interact with our environment. Cognition is also concerned with meaning making. However, its study moves beyond understanding the role of sensation in constructing meaning to also include mental processes such as attention (focus), memory, and use of existing knowledge and/or experience.

Like many other design disciplines in the twentieth century, cartographers embraced the modernist principle that 'form should follow function', implying that functional design was the highest design goal.[1] One of cartography's strongest advocates for the idea that form should follow function was Arthur Robinson, who reiterated Eckert's entreaty for a science of cartographic design, one based on the limits of human perception, in his 1952 book, *The Look of Maps*. This opened the door to research using psychological methods for defining those limits as they applied to using maps.

Implicit in the concept of functional design is that the designer can foretell and specify the designed object's intended function. Otherwise, there is no standard against which to measure the design's 'quality', that is, its efficacy and/or efficiency (Olson, 2015). In some cases, the intended 'function' of a map is to simply communicate the spatial distribution of a particular phenomenon. Indeed, in Robinson's (1952) view, scientific cartography existed to transmit scientific information from a scientist to the map user (p.17).

The process by which this information transmission occurred eventually became known as the theory of cartographic communication, a topic of intense interest in the 1960s and 1970s. A key point of focus in this research was the relationship between how the cartographer's design decisions influenced the graphical marks on the page and what information those marks allowed a map user to perceive and extract from the map. The topic of cartographic communication was considered of sufficient importance to warrant the establishment of Commission V of the International Cartographic Association in 1972. The goals of this commission included defining the principles of map language (similar to Eckert's 'map logic'), evaluating the effectiveness and efficiency of communication between the map maker and the map user afforded by this language, and establishing theory that described how information was transmitted between the two parties (Ratajski, 1974). Central to the goals of this commission were map perception experiments that sought to establish the relationship between the marks on the page and what the map user perceived.

## Visual perception: seeing map symbols

A number of authors distinguished between signal (i.e. the message being communicated) and noise (i.e. graphical marks that distract the map user from understanding the message the cartographer intends to transmit). Noise might be generated by the inclusion of unnecessary map information or failing to use the correct principles of map logic to represent map information. The concept of visual hierarchy (see Figure 3.1) and the perceptual principles, such as Gestalt theory that enable its construction within a map by a cartographer, are examples of how using appropriate (for the communication goal) map logic can improve the likelihood that a map user will receive the intended map message (Wood, 1968). Visual hierarchy refers to the apparent separation of map symbols into different depth planes within the map. Gestalt theory is a psychological theory that proposes a series of graphical characteristics that lead to the perception of grouped elements; cartographically, the grouping of map symbols into different planes of visual hierarchy.

Wood did not do any empirical testing himself, but he drew extensively on his reading of the visual perception literature from psychology. Much later, cartographers did assess the effects

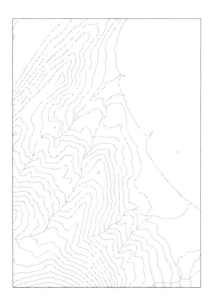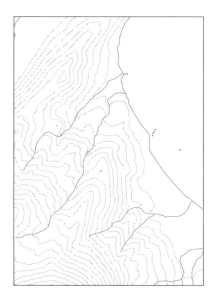

*Figure 3.1*    Left: While the Gestalt principle of good continuation makes it possible to differentiate rivers from contour lines, the visual hierarchy is weak. Right: The Gestalt principle of similarity (of colour) is added to more clearly separate the two types of line symbols onto separate visual planes, improving the map user's understanding of which line symbols belong to which real-world phenomenon

of matching map and information logics through visual hierarchy in maps used in power plant control rooms and air traffic control towers. The findings of these studies confirmed Wood's contention that appropriate visual hierarchy improves map user performance – leading to more efficient use of displays, lower perceived mental effort required to use the maps, and similar error rates to displays designed without visual hierarchy (Van Laar and Deshe, 2002, 2007).

Work within psychology has further helped cartographers to develop rules of thumb for creating visual hierarchy by identifying which visual variables are pre-attentively processed by the eye-brain system (Treisman and Gelade, 1980; Treisman, 1985; Fabrikant and Goldsberry, 2005; Fabrikant *et al.*, 2010). Pre-attentive processing leads to particular features of the display appearing to 'pop out', as was seen in Figure 3.1, without the map user needing to inspect each individual symbol in turn. While it is very easy to quickly identify all the red symbols in the map below because colour hue is pre-attentively processed (Figure 3.2), it is difficult to identify all square symbols at a glance, because shape is not pre-attentively processed. Things become more complicated when visual variables are combined. For example, it takes a long time to differentiate between blue squares and blue circles because the combination of colour hue and shape is not pre-attentively processed, and it therefore requires serial visual search in the display.

Noise could also result from incorrect messages being received by the map user as a result of human perceptual (in)abilities, such as those uncovered in psychophysical map design experiments. Psychophysical map design experiments attempt to identify the relationship between variations in some aspect of the graphical marks on the page and how map users perceive these differences. Cartographic experiment designers working in this tradition typically saw the problem of map design as a technical one that could be solved by optimizing the design to account for human perceptual capabilities – improving the efficiency and effectiveness of

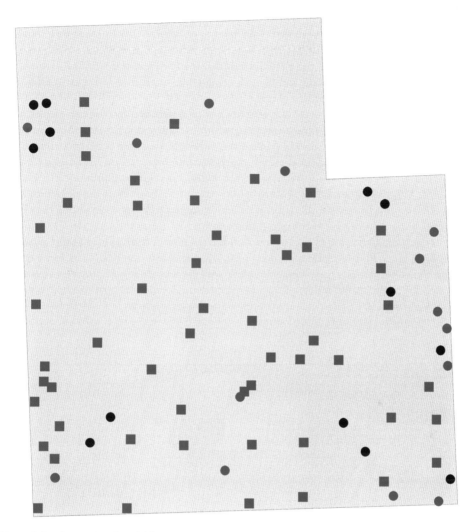

*Figure 3.2* Some visual variables are pre-attentively processed and they therefore 'pop out'. Here, the red symbols 'pop out', while the squares do not

map communication (Liben and Downs, 2015). For example, one large group of experiments, the first of which was conducted by Flannery, focused on the relationship between the area of a proportional-area symbol (often circles) and the area the map user perceived that the symbol covered (Flannery, 1956). While later research showed that these relationships were both context- and user-dependent, several cartography textbooks recommended that cartographers adjust the scaling of proportional symbols to account for map users' tendency to underestimate the area of symbols. This adjustment, whether equally effective for all individuals and contexts or not, has made its way into prominent map design software tools.

Psychophysical research on maps also provided an early focus on human discrimination of visual symbols on maps (Griffin and Montello, 2015). In these symbol discrimination experiments, the goal was to produce information that would allow cartographers to construct a set of visually discriminable map symbols, within the context of the map production technologies in use at the time. Work in this area focused on several types of symbols: gray-scale tones (e.g. Williams, 1958), colour hues (e.g. Brewer, 1997), shape (e.g. Forrest and Castner, 1985) and typeface characteristics (e.g. Shortridge and Welch, 1982). A key commonality in the findings of these studies was that map context was a critical element in determining whether two symbols would be discriminable: increasing difficulty in discrimination the 'noisier' the map context (Figure 3.3).

This difficulty in predicting the exact effect of a map design decision for a particular map and for a particular map user led some cartographers to become disenchanted with perceptual map design research (Petchenik, 1983). At the time, although computers were becoming more commonly used to design maps, the thought of personalized maps that are customized for one person and one map use context was unimaginable; today, it is a distinct possibility (e.g. Huang *et al.*, 2014; Reichenbacher, 2005). Nevertheless, computers, because of the flexibility they enabled for designing maps, did allow for more rapid progress to be made in testing the effectiveness of cartographic design conventions (Experimental Cartography Unit, 1971). Another impact of the computerization of map production was the introduction of map design 'defaults' that are built into software. Although individual differences between map users mean that a map that works well for one person may not work well for another person, implementing map design defaults based on the results of map perception studies can still lead to improvements for many map users. A practical example of this is the implementation of ColorBrewer colour schemes (Harrower and Brewer, 2003; Brewer, 2003) in many map design software packages.[2]

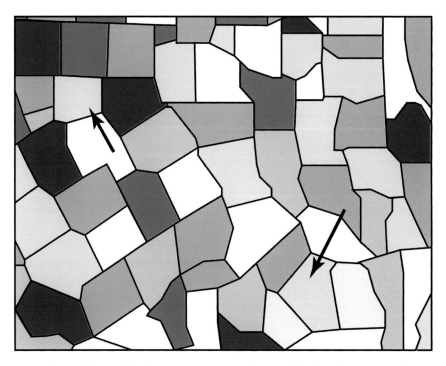

*Figure 3.3*   The problem of simultaneous contrast makes it difficult to discriminate between two colour hues, especially when there are many competing hues nearby. The two marked polygons are in fact the same colour.

## Visual cognition: making sense of map symbols

While research on visual perception can help us to understand some of the constraints within which cartographers work when designing a map, much of the early work in this area did not sufficiently acknowledge the role of cognitive processes such as attention and memory in map use. For example, what a person sees is affected not only by the character of the graphic marks on the page but also by her experience and knowledge of reading maps and the subject of the map, and her expectations of what she will see, which may influence where she looks (MacEachren, 1995).

From the 1920s until the publication of Ulric Neisser's *Cognitive Psychology* in 1967, behaviourism was a dominant paradigm within psychology. Behaviourism sought to understand behaviour by measuring observable behaviours and was assumed to be determined by conditioning; interactions with the 'environment'. Neisser's (1967) innovation was to suggest that mental processes were important for understanding behaviour and that people's mental representations were constructed, often on the fly and to suit a particular context or task. While Neisser's book was not the first writing to propose some aspects of what has become known as the cognitive paradigm, it was a cogent presentation of an alternative to behaviourism.

A significant new approach for understanding cognitive processes was the use of eye tracking to identify where visual attention was or was not focused on the map. Early eye trackers were used in fields other than cartography, to study behaviours such as reading, or viewing art or photographs (e.g. Buswell, 1935; Yarbus, 1967; Huey, 1968 [1908]). In the early 1970s, psychologists began to suggest that eye movements could be used as evidence of mental processes (Noton and Stark, 1971). This idea eventually became known as the strong eye-mind hypothesis (Just and Carpenter, 1980). Others challenged this idea by demonstrating covert attention, where a person can pay attention to something at which they are not looking (Posner, 1980). Today, researchers acknowledge that eye movements probably slightly lag behind attention, meaning that fixations (where the eye is focused for a relatively long period) provide at least some indication that the person's attention was focused on that location (Hoffman, 1998).

Eye tracking was introduced to cartographers by psychologist Leon Williams at the Symposium on the Influence of the Map User on Map Design in Canada in 1970 (Steinke, 1987). Thereafter followed a number of early experiments, in which map users' eye movements were recorded in a free viewing task – that is, participants were told to just read the map (Jenks, 1973; Dobson, 1979). These researchers were surprised to find a wide variety of viewing patterns between individuals, though most map users did seem to spend more time looking at informative parts of the map compared with less informative parts of the map.

Others identified the importance of using a map use task to prompt the map user to activate specific cognitive processes and thereby be able to come to firmer conclusions about the relationship between map design and the cognitive processes map users employed (Steinke, 1987). DeLucia (1976: 143) reiterated the importance for map use studies of link between form and function by concluding:

> [t]he most useful and meaningful standard against which all maps should be designed and subsequently evaluated is FUNCTION. From the beginning we must design our maps to enable some human user to perform some functional act or operation … As we have seen over and over again in the experimental records of this research, the nature of the task or function to be performed by the map user is the single most important factor in determining *how* [author's emphasis] he processes the information on the map.

Here, it is possible to see the beginnings of the idea that maps that are designed to support a given task will be more effective if they lessen the cognitive load of the map user – that is to say, they direct the map user's attention to the most informative parts of the map with minimum effort.

Early eye tracking experiments were technically difficult to carry out given the equipment available at the time (some were recording reflections on the surface of the eye on film), and the data they produced were very time consuming to analyse. This, in addition to the fact that they did not truly allow cartographers to see what was going on in the map user's head, led to a lull in the use of eye tracking in the 1990s. But development of cheaper, digital video-based eye trackers that measure the corneal reflection of infrared light on the eye, in combination with computer programs that can be used to preprocess eye movement data and reduce the time spent analysing data, has led to the reemergence of the technique in cartographic research. Today's eye tracking applications often combine the method with other ways of measuring cognitive processes, such as verbal protocols (e.g. Peebles *et al.*, 2007) or usability metrics (Çöltekin *et al.*, 2009; Manson *et al.*, 2012) and experiments include both those conducted in controlled laboratory settings and on mobile devices used for navigation and delivering location-based services (e.g. Kiefer *et al.*, 2014).

Cartographic researchers using eye tracking have found that there are influences of both the design of the map and the map task on visual attentive behaviour (e.g. Fabrikant *et al.*, 2010). This provides evidence that visual information processing depends both on visual perception, through what is known as bottom–up encoding mechanisms that occur in early, pre-attentive vision, and cognitive processes and strategies, through what is known as top–down encoding mechanisms that interact with a map user's existing knowledge. Neisser (1976) introduced this idea as the 'perceptual cycle'.

Cognitive processes and strategies include things like knowing where to look within a display, and interpreting information found within a map based on other knowledge stored in long-term memory. Therefore, we might expect different map users to have and use different information stored in memory when interpreting maps. For example, geologists may have detailed knowledge about characteristic 3D structures of geological formations that they draw upon when using a geological map, whereas map users without geological training may not have this knowledge. Schemata are mental structures that the map user employs to mediate between what s/he already knows and what s/he sees in a map. In other words, they are the means by which the map user constructs information from visual representations like maps. They are used to both organize information and plan behaviour and can be modified by new information.

MacEachren (1995) contends that map use probably involves using both general map schemata and specific map schemata. A general map schema might include things like understanding that legends provide explanations of the real-world phenomena for which map symbols stand, while a specific map schema could involve filling in the details of the general schema, such as understanding that blue means cool temperatures while red means warm temperatures. An individual map user may or may not have developed a particular specific map schema through training or past experience with that type of map. Following on with the geological map user example, many geological maps use a standard set of symbols for identifying specific features of geological interest, and an experienced geologist would possess the schema for interpreting these symbols, while the average layperson would likely not and would need to spend more time consulting and referring to the legend while using the map. Some cartographers have provided evidence in support of the idea that the use of specific legend designs can prompt the use of helpful map schemata and enable more effective use of the map for the task at hand (see Figure 3.4).

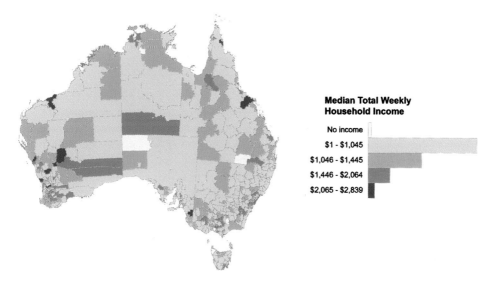

Median Total Weekly
Household Income

No income
$1 - $1,045
$1,046 - $1,445
$1,446 - $2,064
$2,065 - $2,839

*Figure 3.4*    A legend whose symbols vary in size in proportion with their prevalence on the map might prompt the use of schemata related to statistical distributions. Dykes *et al.* (2010) provide an alternative design concept for this in their 'Legend as Statistical Graphic' theme

One such example is DeLucia and Hiller (1982), who found that showing hypsometric tint symbols using a legend that modelled a hill (a 'natural' legend design) helped map users to complete map use tasks that required visualizing the shape of the terrain.

Understanding the map schemata that map users activate can help cartographers design maps that can be used with less cognitive effort, that is to say, cognitive load. Bertin (1983 [1967]) characterized such maps as 'maps to be seen' rather than 'maps to be read'. Maps to be read require effortful extraction of information from the map. The various map design guidelines such as Bertin's visual variables, and later extensions and modifications thereto can be seen as one kind of attempt to formalize specific map schemata (MacEachren, 1995). Nevertheless, many map design guidelines have not been empirically evaluated to test how well they function as map schemata (some exceptions include Garlandini and Fabrikant 2009; MacEachren *et al.*, 2012).

But not all maps have communication as their primary function. MacEachren and Monmonier (1992) coined the term geographic visualization to describe the practice of using maps for thinking and reasoning about geographic problems, most particularly when the maps could be produced at the speed of thinking via interacting with a computer.[3] In an attempt to move cartographic research beyond its (then) recent focus on maps used for communication, MacEachren and Ganter (1990) proposed a model of how thinking with maps might work. It too referenced Neisser's perceptual cycle, but also suggested that the 'seeing' produced by the combination of bottom-up and top-down encoding then interacts with knowledge schemata (both general and domain-specific ones) to produce inference and reasoning. Their model proposed iterative refinement of both the maps produced by interaction with the computer and the hypotheses produced by inference and reasoning. They further argued that to support truly original thinking and insight, it might be necessary to break existing schemata to shift the thinker's perspective on the problem. While schemata can be supportive of reasoning by enabling the map user to match their schemata to a visual display in order to interpret what s/he sees, they can also cause the map user to miss something that is unexpected because it does not fit neatly with their currently used schemata.

## Conclusions

The scientific cartography project begun by Eckert and Robinson has borne many fruit. Consideration of the map user is now firmly entrenched in the practice of contemporary cartography, at least as practised by professional cartographers.[4] Old rules of thumb and cartographic conventions have been studied and the application of psychological theories about perception and cognition have helped us to understand whether and why their design (form) enables a particular type of functional map use. Some of this research, particularly those results that can be successfully applied to maps used for communication, has made its way into everyday practice, with ColorBrewer perhaps the standout success. Much more work remains to be done, especially in understanding how maps help people to make inferences and decisions. Continued engagement with psychology, cognitive science, and neuroscience research will be critical for supporting future progress.

## Notes

1 Robinson (1952: 18) even went so far as to suggest that beauty might be undesirable in a map as it might distract from the map user's intellectual response to the map.
2 The influence of ColorBrewer extends well beyond map design software. Its colour schemes are implemented in several information graphics design and statistical analysis software packages.
3 MacEachren and Ganter (1990) originally called this idea cartographic visualization, but MacEachren later came to prefer the term geographic visualization (MacEachren, 1994), and it has since been shortened in common usage to geovisualization.
4 The rise of the amateur mapmaker presents many challenges for cartographers – chief among them the question of how they can help amateur mapmakers create more usable maps.

## References

Bertin, J. (1983 [1967]) *Semiology of Graphics: Diagrams, Networks, Maps* Madison, WI: University of Wisconsin Press.

Brewer, C.A. (1997) "Evaluation of a Model for Predicting Simultaneous Contrast on Color Maps" *The Professional Geographer* 49 (3) pp.280–294.

Brewer, C.A. (2003) "A Transition in Improving Maps: The ColorBrewer Example" *Cartography and Geographic Information Science* 30 (2) pp.159–162.

Buswell, G.T. (1935) *How People Look at Pictures* Chicago, IL: University of Chicago Press.

Çöltekin, A., Heil, B., Garlandini, S. and Fabrikant, S.I. (2009) "Evaluating the Effectiveness of Interactive Map Interface Designs: A Case Study Integrating Usability Metrics with Eye-Movement Analysis" *Cartography and Geographic Information Science* 36 (1) pp.5–17.

DeLucia, A.A. (1976) "How People Read Maps: Some Objective Evidence" *Technical Papers of the American Congress on Surveying and Mapping* pp.135–144.

DeLucia, A.A. and Hiller, D.W. (1982) "Natural Legend Design for Thematic Maps" *The Cartographic Journal* 19 (1) pp.46–52.

Dobson, M.W. (1979) "The Influence of Map Information on Fixation Localization" *The American Cartographer* 6 (1) pp.51–65.

Dykes, J., Wood, J. and Slingsby, A. (2010) "Rethinking Map Legends with Visualization" *IEEE Transactions on Visualization and Computer Graphics* 16 (6) pp.890–899.

Eckert, M. (trans. by W. Joerg) (1908) "On the Nature of Maps and Map Logic" *Bulletin of the American Geographical Society* 40 (6) pp.344–351.

Eckert, M. (1921, 1925) *Die Kartenwissenschaft: Forschungen und Grundlagen zu einer Kartographie als Wissenschaft* [The Science of Maps: Research and Foundations for a Cartography as Science] Berlin: W. De Gruyter.

Experimental Cartography Unit (1971) *Automatic Cartography and Planning* London: Architectural Press.

Fabrikant, S.I. and Goldsberry, K. (2005) "Thematic Relevance and Perceptual Salience of Dynamic Geovisualization Displays" *Proceedings of the 22th ICA/ACI International Cartographic Conference* A Coruña, Spain.

Fabrikant, S.I., Hespanha, S.R. and Hegarty, M. (2010) "Cognitively Inspired and Perceptually Salient Graphic Displays for Efficient Spatial Inference Making" *Annals of the Association of American Geographers* 100 (1) pp.13–29.

Flannery, J.J. (1956) *The Graduated Circle: A Description, Analysis, and Evaluation of a Quantitative Map Symbol* unpublished PhD thesis, Department of Geography, University of Wisconsin, Madison.

Forrest, D. and Castner, H.W. (1985) "The Design and Perception of Point Symbols for Tourist Maps" *The Cartographic Journal* 22 (1) pp.11–19.

Garlandini, S. and Fabrikant, S.I. (2009) "Evaluating the Effectiveness and Efficiency of Visual Variables for Geographic Information Visualization" in Hornsby, K.S., Claramunt, C., Denis, M. and Ligozat, G. (Eds) *Spatial Information Theory, 9th International Conference, COSIT 2009, Aber Wrac'h, France* Berlin: Springer, pp.195–211.

Griffin, A.L. and Montello, D.R. (2015) "Vision and Discrimination" in Monmonier, M. (Ed.), *The History of Cartography, Volume 6: Cartography in the Twentieth Century* Chicago, IL: University of Chicago Press, pp.1055–1059.

Harrower, M. and Brewer, C.A. (2003) "ColorBrewer.org: An Online Tool for Selecting Colour Schemes for Maps" *The Cartographic Journal* 40 (1) pp.27–37.

Hoffman, J.E. (1998) "Visual Attention and Eye Movements" in Pashler, H. (Ed.) *Attention* Hove, UK: Psychology Press, pp.119–154.

Huang, H., Klettner, S., Schmidt, M., Gartner, G., Leitinger, S., Wagner, A. and Steinmann, R. (2014) "AffectRoute: Considering People's Affective Responses to Environments for Enhancing Route-Planning Services" *International Journal of Geographical Information Science* 28 (12) pp.2456–2473.

Huey, E.B. (1968 [1908]) *The Psychology and Pedagogy of Reading* Cambridge, MA: MIT Press.

Jenks, G.F. (1973) Visual Integration in Thematic Mapping: Fact or Fiction? *International Yearbook of Cartography* 13 pp.27–35.

Just M.A. and Carpenter, P.A. (1980) "A Theory of Reading: From Eye Fixation to Comprehension" *Psychological Review* 87 (4) pp.329–354.

Kiefer, P., Giannopoulos, I. and Raubal, M. (2014) "Where Am I? Investigating Map Matching During Self-Localization with Mobile Eye Tracking in an Urban Environment" *Transactions in GIS* 18 (5) pp.660–686.

Liben, L.S. and Downs, R.M. (2015) "Map Use skills" in Monmonier, M. (Ed.) *The History of Cartography, Volume 6: Cartography in the Twentieth Century* Chicago, IL: University of Chicago Press, pp.1074–1080.

MacEachren, A.M. (1994) "Visualization in Modern Cartography: Setting the Agenda" in Taylor, D.R.F. and MacEachren, A.M. (Eds) *Visualization in Modern Cartography* New York: Pergamon Press, pp.1–12.

MacEachren, A.M. (1995) *How Maps Work: Representation, Visualization and Design* New York: The Guilford Press.

MacEachren, A.M. and Ganter, J. (1990) "A Pattern Identification Approach to Cartographic Visualization" *Cartographica*, 27 (2) pp.64–81.

MacEachren, A.M. and Monmonier, M. (1992) Introduction (to special issue on geographic visualization) *Cartography and Geographic Information Systems* 19 (4) pp.197–200.

MacEachren, A.M., Roth, R.E., O'Brien, J., Li, B., Swingley, D. and Gahegan, M. (2012) "Visual Semiotics and Uncertainty Visualization: An Empirical Study" *IEEE Transactions on Visualization and Computer Graphics* 18 (12) pp.2496–2505.

Manson, S.M., Kne, L., Dyke, K.R., Shannon, J. and Eria, S. (2012) "Using Eye-Tracking and Mouse Metrics to Test Usability of Web Mapping Navigation" *Cartography and Geographic Information Science* 39 (1) pp.48–60.

Montello, D.R. (2002) "Cognitive Map-Design Research in the Twentieth Century: Theoretical and Empirical Approaches" *Cartography and Geographic Information Science* 29 (3) pp.283–304.

Neisser, U. (1976) *Cognition and Reality: Principles and Implications of Cognitive Psychology* New York: W.H. Freeman and Company.

Neisser, U. (1967) *Cognitive Psychology* New York: Appleton-Century-Crofts.

Noton, D. and Stark, L. (1971) "Scanpaths in Eye Movements during Pattern Perception" *Science* 171 (3968) pp.308–311.

Olson, J.M. (2015) "Perception and Map Design" in Monmonier, M. (Ed.) *The History of Cartography, Volume 6: Cartography in the Twentieth Century* Chicago, IL: University of Chicago Press, pp.1063–1065.

Peebles, D., Davies, C. and Mora, R. (2007) "Effects of Geometry, Landmarks and Orientation Strategies in the 'Drop-Off' Orientation Task" in Winter, S., Duckham, M., Kulik, L. and Kuipers, B. (Eds) *Spatial Information Theory* Proceedings of the 8th International Conference, COSIT 2007, Melbourne, Australia Berlin: Springer, pp.390–405.

Petchenik, B.B. (1983) "A Map Maker's Perspective on Map Design Research 1950–1980" in Taylor, D.R.F. (Ed.) *Graphic Communication and Design in Contemporary Cartography* New York: John Wiley & Sons, pp.37–68.

Posner, M.I. (1980) "Orienting of Attention" *Quarterly Journal of Experimental Psychology* 32 (1) pp.3–25.

Ratajski, L. (1974) "Commission V of the ICA: The Tasks It Faces" *International Yearbook of Cartography* 14 pp.140–144.

Reichenbacher, T. (2005) "Adaptive Egocentric Maps for Mobile Users" in Meng, L., Zipf, A. and Reichenbacher, T. (Eds) *Map-Based Mobile Services* Berlin: Springer, pp.141–158.

Robinson, A.H. (1952) *The Look of Maps* Madison, WI: University of Wisconsin Press.

Shortridge, B.G. and Welch, R.B. (1982) "The Effect of Stimulus Redundancy on the Discrimination of Town Size on Maps" *American Cartographer* 9 (1) pp.69–80.

Steinke, T.R. (1987) "Eye Movement Studies in Cartography and Related Fields" *Cartographica* 24 (2) pp.40–73.

Treisman, A. (1985) "Preattentive processing in vision" *Computer Vision, Graphics and Image Processing* 31 (2) pp.156–177.

Treisman, A. and Gelade, G. (1980) "A Feature-Integration Theory of Attention" *Cognitive Psychology* 12 (1) pp.97–136.

Van Laar, D.L. and Deshe, O. (2002) "Evaluation of a Visual Layering Methodology for Colour Coding Control Room Displays" *Applied Ergonomics* 33 (4) pp.371–377.

Van Laar, D.L. and Deshe, O. (2007) "Color Coding of Control Room Displays: The psychocartography of Visual Layering Effects" *Human Factors* 49 (3) pp.477–490.

Weingart, P. (2010) "A Short History of Knowledge Formations" in Frodeman, R. (Ed.) *The Oxford Handbook of Interdisciplinarity* Oxford, UK: Oxford University Press, pp.3–14.

Williams, R.L. (1958) "Map Symbols: Equal-Appearing Intervals for Printed Screens" *Annals of the Association of American Geographers* 48 (2) pp.132–139.

Wood, M. (1968) "Visual Perception and Map Design" *The Cartographic Journal* 5 (1) pp.54–64.

Yarbus, A.L. (1967) *Eye Movements and Vision* New York: Plenum.

# 4

# Understanding map uses and users

*Corné P.J.M. van Elzakker and Kristien Ooms*

## Introduction

The most important starting-point for people engaged with mapping and cartography must be the awareness that the maps they produce will have to be used by human beings. If these users cannot use the maps, i.e. if they cannot obtain from the maps the required answers to their spatio-temporal questions, all design and production efforts made will be in vain. Of course, most cartographers have always been aware of this, but because of the technological difficulties involved in map creation and a lack of resources, not enough attention has been paid to doing sensible user research until quite recently.

That is not to say that in the past no map-use research took place at all. In fact, the foundation for scientific map-use research had already been laid some 65 years ago (Robinson, 1952) and since then two main types of map-use research have evolved: holistic functional map-use research and perceptual/cognitive research. The first type of research deals with the question: 'Does the map work?' and the second with the question: 'How does the map work?'. Research activities in these two domains gradually did result in better maps, as clearly demonstrated, by, for instance, maps in the news (Monmonier, 1999). However, these improvements were by far not as substantial and as fast as they potentially could be, because there were simply not enough academic cartographers and not enough resources to conduct user research. Besides, during times in which map displays were not yet commonly generated 'on the fly', it was still too expensive and too time consuming to immediately produce a new version of a map based on the outcomes of user research (van Elzakker, 2013). Finally, there was no potential or sufficient knowledge to apply advanced methods and techniques of user research outside the traditional methods of interviews and questionnaires that were most commonly implemented.

But since the turn of the last century, interest in doing map-use research has grown enormously. Before that, and concurrently, technological developments have led to a move away from the traditional simple static maps to, at first sight, more complex, interactive and dynamic digital map displays which are usually embedded into multimedia geographic information dissemination applications. The consequence is that we should no longer focus on map-use research only, but that we should broaden the scope and should also include in our research the usability of, for example, the data, the interface, the hardware and the software.

At the same time, comparable technological developments have led to an increased availability of advanced user research methods and techniques, the first application of which led to outcomes which proved to substantially increase the usability of map displays and broader cartographic applications. Probably because of these demonstrated effects, but also because of the growing interest by the academic and professional cartographic communities, which led to an increase of the available resources as well as the possibilities to quickly and cheaply produce improved application design alternatives, cartographic use, user and usability research is now booming. This development is not counteracted but rather even stimulated by the fact that more and more users are now producing their own maps.

The focus of this chapter is not on general research on cognition to help explain how map displays work, whether they are embedded in broader geo-information applications or not. To serve the needs of both scientific researchers and professional designers of cartographic applications, our focus will be on providing guidelines for the implementation of scientific methods and techniques for user-centred design and evaluation. Therefore, we will first pay attention to the successive stages of user-centred design which should lead to usable cartographic products. Thereafter, we will focus on the purposes of map use and the human map users themselves respectively. An overview of the great variety of methods and techniques of doing user research is presented next, ending with our plea to combine them into mixed-methods approaches to gain optimal insight into the uses and users of maps. We conclude by giving some practical hints for actually doing use, user and usability research in cartography.

As already indicated, technological developments have led to considerable changes in the nature of maps. In this chapter, when we are referring to 'maps', we will definitely not just refer to traditional single (static) maps. We will also include the current interactive and dynamic map displays which often embedded in more comprehensive geo-information dissemination systems or applications; such cartographic products, systems or applications will also be denoted as 'maps'.

## User-centred design

User-centred design (UCD) is not new at all in the engineering sciences (where it is also referred to as 'usability engineering'). It is an approach in which three main steps may be distinguished: requirement analysis, prototype design and usability evaluation (Figure 4.1). A map, or geographic information application in which (a) map(s) is / are embedded, may also be seen

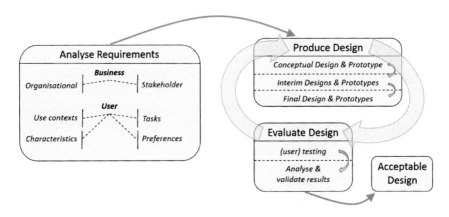

*Figure 4.1*   The user-centred design process (derived from van Elzakker and Wealands, 2007)

as a prototype. Indeed, it may be postulated that in cartography, UCD approaches have been being followed for a long time in that most map-makers were aware of the fact that the maps they produced were meant to be used by others. However, in practice, in trying to overcome technological issues in map production, map-makers sometimes lost track of the end users. This sometimes happened as well in large mapping organizations in which different employees were involved in different parts of the production process. In those situations, the production of a map itself was considered to be the goal and not the geographical needs of the users. Promoting map-reading courses was to be considered as a typical case of putting the cart before the horse. Finally, in the past, often only limited attention was paid to the analysis of the requirements and characteristics of map users.

Until the late 1990s, user research efforts were directed mainly, and often only, to testing the usability of existing maps. Of course, usability evaluation was, and still is, a very important stage in a UCD approach. (Map) usability can be defined as 'the effectiveness, efficiency and satisfaction with which specified users achieve specified goals in particular environments' (ISO 9241-11: 1998(en)) with the help of maps. This broad definition of usability (incorporating effectiveness) includes the utility concept: does the map do what it is supposed to do, i.e. does it provide the answers to the geographical questions users have? In our view, this broad definition of usability also encompasses the facets of user experience that are reflected in the so-called user experience honeycomb as shown in Figure 4.2 (Morville, 2004). However, the honeycomb clearly helps to specify particularly the comfortability and acceptability (aspects of user satisfaction) of a map for its users. For evaluating the three main aspects of usability (effectiveness, efficiency and satisfaction), different methods and techniques of user research may be applied and usually in combination (see the section on user research methods below).

The same research methods and techniques can also be used to carry out a systematic requirement analysis as the first major step in the UCD approach. Taking this step is essential for ultimately producing better maps (Haklay et al., 2010; Robbi Sluter et al., 2017). This has been proven already by various sample cases in the cartographic domain (e.g. Kveladze et al., 2013; Roth et al., 2015; Delikostidis et al., 2016). A requirement analysis is not just formulating a hypothesis about what the prospective users may need or only asking representatives of these

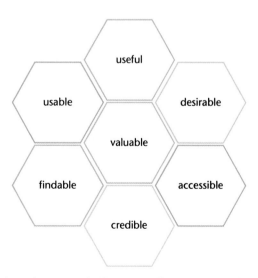

*Figure 4.2*   **User experience honeycomb (from Morville, 2004, reproduced with permission)**

users what they need through an interview or questionnaire. The requirement analysis must be a systematic investigation of the tasks to be executed with the help of maps, or, in other words, of the specific spatio-temporal questions that will have to be answered. At the same time, the characteristics and preferences of the map users need to be established carefully, as well as the context of use. Indeed, this may still be done through interviews or surveys, but the requirements may also be derived, for instance, from observing representatives of the users who interact with comparable maps or cartographic applications (see e.g. Delikostidis, 2011). Finally, institutional or organizational boundary conditions must also be established during the requirement analysis (e.g. what is the business model of a commercial production or should a house style be taken into account?).

## Map uses

During the requirement analysis stage, it is of critical importance to establish in detail the purpose or use of the map or cartographic application. A complete listing of all tasks that need to be executed with the help of the map will be the main guide for the design of the prototype. It is

*Figure 4.3*   Testing a mobile indoor navigation app in the proper context using a mixed-methods approach (mobile eye tracking, concurrent thinking aloud, video recording). Photograph by W. Kock

important to distinguish these map-use tasks from so-called map-use activities like panning and zooming (van Elzakker, 2004). Map-use tasks can be coupled directly to the spatio-temporal questions that need to be answered with the help of the map (e.g. What is there? Where is that geographic object? What is the spatial distribution of that object? How can I travel from that object to another object? Which alternative route is fastest taking into account peak travel hours?). Such questions are different for different maps and may normally be classified into different levels of complexity. The use of maps is indispensable, particularly for answering the most complex questions, usually those requiring overview.

For every prototype map to be designed we advise a detailed list of spatio-temporal questions that should be answered with the map to be drawn up, coupled to related map-use tasks and map-use activities (functionalities). In doing so, use may be made of existing taxonomies or, as Roth (2012) called it: a framework of cartographic interaction primitives. Another introduction to spatio-temporal tasks can be found in Andrienko and Andrienko (2005: chapter 3).

An important part of the map-use requirement analysis is to clearly define the context of the map use. It matters very much, for instance, whether the map is going to be used on mobile devices, in indoor or outdoor environments, or on large desktop computer screens in controlled environments (Figure 4.3). Static maps on paper are still used, but more and more they are replaced by their digital interactive and dynamic counterparts which are better suited to answer more and more specific spatio-temporal questions for individual users. For example, in this way, the large traditional paper physical planning maps for informing and engaging the stakeholders are now often replaced by interactive and dynamic participatory planning maps displayed on map (touch) tables. This is also an example of a very specific context of map use.

## Map users

As with establishing all aspects of prospective map use, in the requirement analysis stage of the UCD process, clarity must also be provided with respect to the human beings who will be affected by the cartographic product to be designed.

First of all, it is important to make a distinction between all stakeholders and the actual map users. The interests of both groups of people should be taken into account. Stakeholders may be affected by the cartographic product but they may not actually use it. For instance, people living along a road which is suddenly declared 'scenic' on a tourist map will be affected by an increase in tourist traffic. As part of a systematic requirement analysis, it is useful to establish which stakeholders are involved.

However, for the design of better maps, it is even more important to clearly define who will be the prospective map users. It may be tempting to claim that the resulting cartographic product may be used by a broad variety of users, but the quality of the map design will be improved if the characteristics and preferences of the target users are specified in detail. In the end, such specifications may be used, for instance, to not only adjust the map design itself but also to design different interfaces for target groups with different characteristics.

Knowing about the detailed characteristics and preferences of the prospective map users is also important in view of actual experiments with users in both the usability evaluation and the requirement analysis stages of the UCD process. Participants will have to be recruited who do have the anticipated characteristics and preferences. In order to check this, if only to validate the user research results, the test participants' characteristics will have to be established through an interview or questionnaire before or during the actual experiment. These characteristics may also be used to subdivide participants in different test groups, for instance for the purpose of between-subjects test design (e.g. for a comparative evaluation of design alternatives, as opposed to a within-subjects design to test one cartographic product).

When referring to test participants, the immediate question that always comes up is how many participants are required for scientific user research. The answer to this question depends, first of all, on the anticipated map uses and map users. For instance, if a broader user group is anticipated, this will require more representative test participants. However, unlike the situation in the days of the quantitative revolution in, for example, geography (Burton, 1963), there is a shared opinion now that for map-use investigations to be considered 'scientific' research, it is not always necessary to recruit tens or hundreds of test participants. Such high numbers may be required for statistical validation in quantitative research that may still be applied in, for example, the final usability evaluation of a cartographic product. However, nowadays we have witnessed a wide acceptance of more qualitative research, particularly in the earlier stages of the UCD process, for example in the requirement analysis stage or when testing the first prototype of a new cartographic product. Following the advice of Nielsen (2012), only five test participants are required to execute qualitative user research. However, this number also depends on the purpose and nature of the experiment and the research technique applied. For example, for card sorting, 15 participants are required (Nielsen, 2012). But for most qualitative research involving observation techniques (e.g. thinking aloud, interaction logging, video observation) a small number of participants allows you to discover almost the same amount of use, user and usability issues as when working with many more. To account for invalid test sessions because of, for instance, technical failures like corrupted recordings, we would advise researchers to work with six to nine test participants in qualitative user research.

One other reason for working with more than the absolute minimum number of test participants is that during the experiment it may appear that not all participants are as 'engaged' as they should be. It is extremely important that test participants are familiar with the problem the cartographic product addresses. That is, they should have had, or clearly will have, the spatio-temporal questions the map is supposed to give answers to, because they need these answers to function in their daily lives, jobs or education. Too often, it can be seen in user research studies that test participants are recruited who are just willing to do the researcher a favour but who do not 'own' the spatio-temporal task at hand. Why test, for instance, an interactive space-time cube application with subjects who will never do scientific spatio-temporal research themselves? (Kveladze *et al.*, 2013).

## User research methods

When conducting user research, data are gathered about the users' preferences, efficiency, effectiveness and other aspects related to usability or their requirements regarding a certain (cartographic) product (e.g. Slocum *et al.*, 2001; Hornbaek, 2006). Many methods and techniques are currently available to do this. Because there are so many it is beyond the scope of this chapter to explain them all. More details can be found in, for example, Roth *et al.* (2015) and on Usability Net (2006). An incomplete overview of a number of methods and a possible classification is presented in Table 4.1. Some methods (e.g. heuristic evaluation or cognitive walkthrough) are based on the input of experts, whereas others (surveys, focus groups, card sorting, observations and so on) involve representatives of the actual (prospective) users of the map. Finally, it is possible to apply research methods based on theories, i.e. without involving any human beings, except for the researchers (e.g. automated evaluation, content analysis). Most, if not all, of the methods presented originate from other research fields, e.g. software engineering

Table 4.1 Overview of user research methods (partly derived from Roth *et al.*, 2015)

| People involved | Method | Similar or related methods |
| --- | --- | --- |
| Prospective users | Surveys | Questionnaires, entry / exit surveys, blind voting, cognitive workload assessment |
| | Interviews | Structured interviews, semi-structured interviews, open interviews, contextual inquiries |
| | Focus groups | Supportive evaluations, stakeholder meeting, delphi |
| | Card sorting | Q methodology, concept mapping, affinity diagramming, brainstorming |
| | Participatory design | Co-design |
| | Observation | Ethnographies, critical incidents, milcs (multi-dimensional in-depth long-term case studies), journal / diary sessions, screen logging, interaction logging, video observation |
| | Thinking aloud | Talk aloud, introspection, retrospection, co-discovery study |
| | Eye-movement tracking | Remote/mobile eye tracking |
| | Biometric | EEG (electroencephalogram), GSR (galvanic skin conductance), facial response analysis, emotion measurement, fMRI (functional magnetic resonance imaging) |
| | Sketching | Mental map sketching, route sketching |
| | Interaction study | Performance measurement, product analysis |
| Experts | Heuristic evaluation | Rules of thumb |
| | Conformity assessment | Feature inspection, consistency inspection, standards inspection, guideline checklist |
| | Cognitive walkthroughs | Pluralistic walkthroughs, storyboarding, Wizard of Oz, task analysis |
| | CASSM | Concept-based analysis of surface and structural misfits |
| No (theory-based) | Scenario-based design | Personas, scenarios of use, use case, theatre |
| | Prototyping | Rapid prototyping, paper prototyping, functional prototyping |
| | Secondary sources | Content analysis, competitive analysis |
| | Automated evaluation | Automated interaction logs, unmoderated user-based methods |

and psychology, which have a longer and more extensive tradition of conducting user research. One of the most pressing cartographic research challenges is to implement these known methods of use, user and usability research in the unique spatio-temporal problem-solving domain in which human beings are required to link geographic reality with the representation of reality through cartographic displays and with their mental maps (van Elzakker and Griffin, 2013).

Different user research methods lead to different information about map uses, users and usability. When planning for user research, one needs to pose the critical question: 'which method(s) is (are) most appropriate and feasible for answering the specific research questions?' This is not an easy task, as one must keep in mind that no method is perfect: each method has its own advantages and disadvantages or – as stated by Carpendale (2008) – 'methods both provide and limit evidence'. The selection is a difficult puzzle to solve as it is intertwined with other aspects of a study's design (see also Figure 4.4).

## Stage of UCD

If the user research is planned for analysing the users' requirements, typically a different set of methods is then applied when an initial prototype is evaluated or when the final design iteration of the product is tested. In the first stages of the UCD lifecycle, one wants to obtain very rich data regarding the user's preferences and needs, which are often linked to a more qualitative approach. In later stages, the evaluation often shifts to a more quantitative process in order to be able to, for example, statistically evaluate the performance of the product (also see van Elzakker and Wealands, 2007; Nivala, 2007; Haklay and Nivala, 2010; Roth *et al.*, 2015).

## Qualitative or quantitative data

This distinction can be linked to the stage of the UCD lifecycle, but that does not always have to be the case. One can also distinguish between empirical testing – which is typically related to statistical tests – and (discount) usability testing. The latter requires fewer participants and is qualitative in nature as one wants to retrieve data from a rich data source (e.g. thinking aloud protocols) (also see Jick, 1979; Carpendale, 2008; Polit and Beck, 2010).

## Participants (characteristics and number)

As discussed before, map users can have a large variety of characteristics, and the selection of a method can be determined by the target group. For example, not all methods can be used when children are recruited; wearing glasses might be a problem for certain eye-tracking systems; it is also of no use to set up an online survey when (part of) the target audience is not connected to the Internet. A specific distinction can be made between users who possess a certain expertise (experts) or not (novices). Referring to Table 4.1, the selection of the method also depends on the involved of human beings (or vice versa): novices, experts or even no participants (in case of theory-based testing) (also see Montello, 2009; Ooms *et al.*, 2012a; van Elzakker and Griffin, 2013; Forsell and Cooper, 2014).

## Available time frame

Some methods require individual sessions with participants (e.g. eye tracking, thinking aloud, EEG), whereas with other methods it is possible to let a user complete the task whenever it suits him/her best (e.g. online questionnaire) or to conduct the evaluation in a group (e.g. focus groups). This can have a severe impact on the time frame needed to complete the study. Furthermore, some studies (e.g. observation, interviews, thinking aloud) require the experimenter to be present in each session, for which sufficient time needs to be foreseen. This is for example not the case for online surveys or automatic user logging. However, one should also keep in mind the availability of the target group (e.g. school period for children, exams for students, holiday period, extremely busy periods in companies) (also see Nielsen, 1993; Rubin and Chisnell, 2008; Ooms *et al.*, 2015a).

## Available budget

Research equipment can be expensive and is not available in all laboratories (for example, an eye tracker, a virtual reality module, brain imaging using, e.g. EEG), which can be a limiting factor on the options to choose from. Some of this equipment may be available for rent, which should be calculated into the total budget. Furthermore, recruiting participants can be linked to an incentive (paying an amount of money per hour, a cinema ticket and so on) that could increase the costs associated with the experiment (also see Nielsen, 1993; Rubin and Chisnell, 2008; Ooms *et al.*, 2015b).

## *Available expertise*

Each method or technique has its own characteristics, and care should be taken regarding how they are implemented in the user study. If the experimenter does not possess a sufficient level of expertise, the obtained data may be biased or not recorded properly. For certain methods, expertise can be acquired relatively fast, but this should be calculated into the available time frame to conduct the study or into the costs associated with the training (also see Nielsen, 1993; Nivala *et al.*, 2007; Rubin and Chisnell, 2008).

## *Use context*

It is of utmost importance that the user performs the research tasks in the proper context. Many experiments are conducted in a laboratory setting, which has the advantage that unwanted influencing and unpredictable factors (e.g. weather, other people walking by, noise, smells) can be controlled. However, laboratory settings typically do not reflect the actual context of use (e.g. the usual work place surrounded by other people, outside on the street). Furthermore, use context does not only refer to the location or the surroundings but also to the context of the user him- or herself: nervous, stressed, relaxed, happy and so on. Finally, cartographic products may be displayed on a different medium (desktop screen, laptop, smartphone, with or without touch input), whereas certain methods (especially in a laboratory setting) can only be used with one particular display medium (typically a desktop monitor), which may make an available method useless for, for example, evaluating applications on mobile devices (also see Van Elzakker *et al.*, 2008; Edsall and Larson, 2009; Delikostidis, 2011; van Elzakker and Griffin, 2013; Forsell and Cooper, 2014; Kiefer *et al.*, 2014).

Besides methods and techniques for data *collection*, researchers also need to consider how the data are going to be processed and *analysed*. The distinction between qualitative and quantitative user research also has consequences for selecting a method to analyse the data. However, data obtained from the application of particular methods can also be processed both in a qualitative and quantitative way. Verbal utterances recorded from a thinking aloud test can, for example, result in both qualitative (the content) and quantitative data (e.g. how often an item is mentioned) after verbal protocol analysis (also see Ericsson and Simon, 1993; van Someren *et al.*, 1994; Chi, 1997; Charters, 2003; van Elzakker, 2004). Software for both qualitative and quantitative user research data analysis may come with the data collection tools (e.g. data analysis options may be included in online survey tools or in eye-tracking solutions) or use can be made of dedicated software packages (free or proprietary), including a variety of statistical packages. Furthermore, it could be beneficial to inspect the data collected with the help of (geo-)visual analytics (also see Li *et al.*, 2010; Andrienko *et al.*, 2012; Ooms *et al.*, 2012b).

So far, we have provided some guidelines to select the most appropriate methods for a certain study. However, since no method is perfect on its own, and each method leads to different insights into uses, users and usability, it is nowadays good practice, including in our cartographic domain (van Elzakker, 2004; Bleisch, 2011; Delikostidis, 2011), to combine – or mix – multiple methods (Jick, 1979; Kaplan and Duchon, 1988; Brannen, 2005; Carpendale, 2008; Forsell and Cooper, 2014). A good combination should put the principle of 'complementary

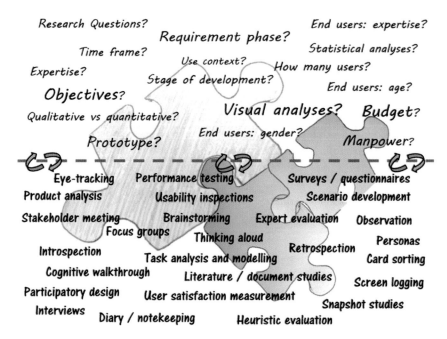

*Figure 4.4* Selecting the most optimal method of user research

strengths and no-overlapping weaknesses' into practice (Johnson and Onwuegbuzie, 2004: 18). Besides complementarity, insights can be extended or triangulated across multiple methods. This also benefits the validation of the results: will you arrive at the same conclusion when analysing data from different sources (methods)? A practical problem that may have to be solved for the analysis of the data collected is the synchronization of concurrent recordings (e.g. video with screen logging, thinking aloud and eye tracking) but the software that comes with the tools may take care of that as well.

## Doing user research

To conclude, in this section we present a number of tips and tricks for doing (cartographic) user research. They are derived from Rubin and Chisnell (2008) and Nielsen (1993) or are based on the authors' personal experience.

When *setting up* a test, it is of utmost importance that unwanted biases are avoided as much as possible. Therefore, one should try to control as many influencing factors as possible.

*Before* actually starting a user test, you should take sufficient time to work out (and document) every step of the research sequence. The decision on which method to use should be based on research questions which are linked to the research objective(s). Furthermore, tasks to be executed by test participants and use context (e.g. the test environment) should also be directly derived from these research questions. The tasks to be executed should be clear to the users to avoid misinterpretations. Finally, it is good practice to execute one or more pilot tests to check the overall set-up and determine the test duration. Test participants are a very valuable resource that should not be wasted.

To have a sufficient level of control and efficiency *during* the test, you should prepare test scenarios/protocols/checklists to create exactly similar conditions for all test participants. In the same light it is advisable to instruct all test participants in exactly the same way (based on written instructions). Furthermore, providing training blocks before the actual start of the study is helpful for letting the participant become accustomed to the tasks, stimuli, environment, context and so on.

It is important that your *participants* represent the product's target groups and that they are engaged. These participants provide very valuable information, so it is of primary importance that they are handled with care (Olson, 2009). First of all, do not force them; ask their permission (e.g. informed consent) and respect their privacy. Second, do not waste the participant's time (work efficiently) and do not expose the participant to risks (e.g. navigating through heavy traffic). In doing so, do not take for granted that the product or service you have designed or generated is usable for everyone (Forsell and Cooper, 2014).

*After* executing the study, the data should be analysed and this should provide an answer to the research questions. Moreover, validating the research method/technique is also advisable to verify whether you really measured what you wanted to measure and whether you arrived at the same conclusion using different research methods. Finally, take care to completely the report on the participants, materials, procedures and analysis (Forsell and Cooper, 2014).

## Conclusion

Since the turn of the last century, interest in better understanding map users and uses has grown enormously, both in academia and in the professional mapping domain. There are several reasons for this, but the most important one is that use, user and usability research leads to better maps, i.e. maps that are more usable in terms of effectiveness, efficiency and user satisfaction. By reporting on some results of the increased research activities, this chapter has provided several guidelines for executing user research in mapping and cartography.

Except for a number of concrete tips and tricks for doing user research, as presented in the preceding section, our main suggestion for arriving at 'maps that work' is to follow UCD approaches in which, in particular, more attention will be paid to a systematic requirement analysis to better understand the prospective map uses and map users. In doing so, we plead for the application of synchronized mixed methods of user research, including in the process of usability evaluation.

In order to be able to apply usable mixed-methods approaches (also here in terms of effectiveness, efficiency and satisfaction), the challenge for cartographic research is to learn more about the implementation of relatively new methods and techniques of research in our specific domain. In doing so, we also have to investigate successful scenarios for the integration of individual techniques in such mixed-methods approaches.

## References

Andrienko, N. and Andrienko, G. (2005) *Exploratory Analysis of Spatial and Temporal Data: A Systematic Approach* Berlin: Springer.

Andrienko, G., Andrienko, N., Burch, M. and Weiskopf, D. (2012) "Visual Analytics Methodology for Eye Movement Studies" *IEEE Transactions on Visualization and Computer Graphics* 18 (12) pp.2889–2898.

Bleisch, S. (2011) *Evaluating the Appropriateness of Visually Combining Abstract Quantitative Data Representations with 3D Desktop Virtual Environments Using Mixed Methods* London: City University.

Brannen, J. (2005) "Mixing Methods: The Entry of Qualitative and Quantitative Approaches into the Research Process" *International Journal of Social Research Methodology* 8 (3) pp.173–184.

Burton, I. (1963) "The Quantitative Revolution and Theoretical Geography" *Canadian Geographer* 7 (4) pp.151–162.

Carpendale, S. (2008) "Evaluating Information Visualizations" in Kerren, A., Stasko, J.T., Fekete, J.-D. and North, C. (Eds) *Information Visualization* Berlin, Heidelberg: Springer, pp.19–45.

Charters, E. (2003) "The Use of Think-aloud Methods in Qualitative Research: An Introduction to Think-aloud Methods" *Brock Education Journal* 12 (2) pp.68–82.

Chi, M.T.H. (1997) "Quantifying Qualitative Analyses of Verbal Data: A Practical Guide" *Journal of the Learning Sciences* 6 (3) pp.271–315.

Delikostidis, I. (2011) *Improving the Usability of Pedestrian Navigation Systems* Enschede, The Netherlands: University of Twente.

Delikostidis, I., van Elzakker, C.P.J.M. and Kraak, M.J. (2016) "Overcoming Challenges in Developing more Usable Pedestrian Navigation Systems" *Cartography and Geographic Information Science* 43 (3) pp.189–207.

Edsall, R.M. and Larson, K.L. (2009) "Effectiveness of a Semi-Immersive Virtual Environment in Understanding Human-environment Interactions" *Cartography and Geographic Information Science* 36 (4) pp.367–384.

Ericsson, K.A. and Simon, H.A. (1993) *Protocol Analysis* Cambridge, MA: MIT Press.

Forsell, C. and Cooper, M. (2014) "An Introduction and Guide to Evaluation of Visualization Techniques Through User Studies" in Huang, W. (Ed.) *Handbook of Human Centric Visualization* New York: Springer, pp.285–313.

Haklay, M. and Nivala, A.M. (2010) "User-Centred Design" in Haklay, M. (Ed.) *Interacting with Geospatial Technologies* Chichester, UK: John Wiley & Sons, pp.89–106.

Haklay, M., Skarlatidou, A. and Tobón, C. (2010) "Usability Engineering" in Haklay, M. (Ed.) *Interacting with Geospatial Technologies* Chichester, UK: John Wiley & Sons, pp.107–123.

Hornbaek, K. (2006) "Current Practice in Measuring Usability: Challenges to Usability Studies and Research" *International Journal of Human-Computer Studies* 64 (2) pp.79–102.

ISO 9241-11: 1998(en) "Ergonomic Requirements for Office Work with Visual Display Terminals (VDTs) – Part 11: Guidance on Usability" Available at: *www.iso.org/obp/ui/#iso:std:iso:9241:-11:ed-1:v1:en* (Accessed: 19 July 2016).

Jick, T.D. (1979) "Mixing Qualitative and Quantitative Methods: Triangulation in Action" *Administrative Science Quarterly* pp.602–611.

Johnson, R.B. and Onwuegbuzie, A.J. (2004) "Mixed Methods Research: A Research Paradigm Whose Time Has Come" *Educational Researcher* 33 (7) pp.14–26.

Kaplan, B. and Duchon, D. (1988) "Combining Qualitative and Quantitative Methods in Information Systems Research: A Case Study" *MIS Quarterly* pp.571–586.

Kiefer, P., Giannopoulos, I. and Raubal, M. (2014) "Where Am I? Investigating Map Matching During Self-Localization with Mobile Eye Tracking in an Urban Environment" *Transactions in GIS* 18 (5) pp.660–686.

Kveladze, I., Kraak, M.J. and van Elzakker, C.P.J.M. (2013) "A Methodological Framework for Researching the Usability of the Space-Time Cube" *The Cartographic Journal* 50 (3) pp.201–210.

Li, X., Çöltekin, A. and Kraak, M.J. (2010) "Visual Exploration of Eye Movement Data Using the Space-Time-Cube" *Geographic Information Science* 6292 pp.295–309.

Monmonier, M. (1999) *Maps with the News: The Development of American Journalistic Cartography* Chicago, IL: University of Chicago Press.

Montello, D. R. (2009) "Cognitive Research in GI Science: Recent Achievements and Future Prospects" *Geography Compass* 3 (5) pp.1824–1840.

Morville. P. (2004) "User Experience Design" Available at: *http://semanticstudios.com/user_experience_design/* (Accessed: 19 July 2016)

Nielsen, J. (1993) *Usability Engineering* San Francisco, CA: Morgan Kaufmann.

Nielsen, J. (2012) "How Many Test Users in a Usability Study?" Available at: *www.nngroup.com/articles/how-many-test-users/* (Accessed: 19 July 2016)

Nivala, A.-M. (2007) *Usability Perspectives for the Design of Interactive Maps* Helsinki, Finland: University of Technology.

Nivala, A.-M., Sarjakoski, L.T. and Sarjakoski, T. (2007) "Usability Methods' Familiarity Among Map Application Developers" *International Journal of Human-Computer Studies* 65 (9) pp.784–795.

Olson, J.M. (2009) "Issues in Human Subject Testing in Cartography and GIS" ICC 2009: Proceedings of the 24th International Cartographic Conference ICC *The World's Geo-spatial Solutions* Santiago, Chile: International Cartographic Association.

Ooms, K., De Maeyer, P., Fack, V., van Assche, E. and Witlox, F. (2012a) "Interpreting Maps through the Eyes of Expert and Novice Users" *International Journal of Geographical Information Science* 26 (10) pp.1773–1788.

Ooms, K., Andrienko, G., Andrienko, N., De Maeyer, P. and Fack, V. (2012b) "Analysing the Spatial Dimension of Eye Movement Data Using a Visual Analytic Approach" *Expert Systems with Applications* 39 (1) pp.1324–1332.

Ooms, K., Çöltekin, A., De Maeyer, P., Dupont, L., Fabrikant, S.I., Incoul, A., Kuhn, M., Slabbinck, H., Vansteenkiste, P. and van der Haegen, L. (2015a) "Combining User Logging with Eye Tracking for Interactive and Dynamic Applications" *Behavior Research Methods* 47 (4) pp.977–993.

Ooms, K., Dupont, L., Lapon, L. and Popelka, S. (2015b) "Accuracy and Precision of Fixation Locations Recorded with the Low-cost Eye Tribe Tracker in Different Experimental Set-ups" *Journal of Eye Movement Research* 8 (1) pp.17–26.

Polit, D.F. and Beck, C.T. (2010) "Generalization in Quantitative and Qualitative Research: Myths and Strategies" *International Journal of Nursing Studies* 47 (11) pp.1451–1458.

Robbi Sluter, C., van Elzakker, C.P.J.M. and Ivanova, I. (2017) "Requirements Elicitation for Geo-information Solutions" *The Cartographic Journal* 54 (1) pp.77–90.

Robinson, A.H. (1952) *The Look of Maps* Madison, WI: University of Wisconsin Press.

Roth, R.E. (2012) "Cartographic Interaction Primitives: Framework and Synthesis" *The Cartographic Journal* 49 (4) pp.376–395.

Roth, R.E., Ross, K.S. and MacEachren, A.M. (2015) "User-centered Design for Interactive Maps: A Case Study in Crime Analysis" *International Journal of Geo-Information* 4 (1) pp.262–301.

Rubin, J. and Chisnell, D. (2008) *Handbook of Usability Testing: How to Plan, Design and Conduct Effective Tests* (2nd ed.) Indianapolis, IN: Wiley Publishing.

Slocum, T. A., Blok, C., Jiang, B., Koussoulakou, A., Montello, D.R., Fuhrman, S. and Hedley, N.R. (2001) "Cognitive and Usability Issues in Geovisualisation" *Cartography and Geographic Information Science* 28 (1) pp.61–75.

Usability Net (2006) "Methods Table" Available at: *www.usabilitynet.org/tools/methods.htm* (Accessed: 20 July 2016)

Van Elzakker, C.P.J.M. (2004) *The Use of Maps in the Exploration of Geographic Data* Netherlands Geographical Studies 326. Utrecht/Enschede: Koninklijk Nederlands Aardrijkskundig Genootschap/ Utrecht University/International Institute for Geo-Information Science and Earth Observation.

Van Elzakker, C.P.J.M. (2013) "The Atlas of Experience" *Geospatial World* 4 (4) pp.34–36.

Van Elzakker, C.P.J.M. and Wealands, K. (2007) "Use and Users of Multimedia Cartography" in Cartwright, W., Peterson, M. and Gartner, G. (Eds) *Multimedia Cartography* Berlin and Heidelberg: Springer, pp.487–504.

Van Elzakker, C.P.J.M. and Griffin, A.L. (2013) "Focus on Geoinformation Users: Cognitive and Use/ User Issues in Contemporary Cartography" *GIM International* 27 (8) pp.20–23.

Van Elzakker, C.P.J.M., Delikostidis, I. and van Oosterom, P.J.M. (2008) "Field-based Usability Evaluation Methodology for Mobile Geo-applications" *The Cartographic Journal* 45 (2) pp.139–149.

Van Someren, M.W., Barnard, Y.F. and Sandberg, J.A.C. (1994) *The Think Aloud Method: A Practical Guide to Modelling Cognitive Processes* London: Academic Press Limited.

# 5

# Map history

## Discourse and process

*Matthew H. Edney*

The study of early maps has had two distinct political ends.[1] From its origins in the 1830s, the 'history of cartography' has presented cartography as a surrogate for modern civilization: modern cartography served to distinguish Westerners (we make *maps*) from non-Westerners (they make map-like objects) and similarly, within the West, rational men from intuitive women; the history of cartography was written as the intertwined and masculinist histories of exploration, discovery and science that celebrated the intellectual achievements of the West as a whole and more especially of particular nations. Progressive narratives of the ever-increasing quality and quantity of geographical knowledge delineated the triumph of Western science and reason over myth and imagination. Such ideological narratives have even been given graphic expression through sequences of historical maps, from Edward Quin's forceful atlas of 1830, delineating how the light of knowledge and reason progressively dispelled the dark clouds of ignorance and superstition (Figure 5.1) to the introductory section of 'The World in a Mirror', an exhibition installed at Museum aan de Stroom, Antwerp, 24 April to 16 August 2015. Early maps have been interpreted and evaluated accordingly.

Yet growing dissatisfaction with the inadequacy of the history of cartography's narratives and its unexamined ideology has led to the formation after 1980 of a new 'map history'. (I use this phrase to highlight the new intellectual formation, although the field is still generally called the history of cartography. Edney, 2005, explores the transition.) New studies of modern, pre-modern and non-Western mapping have demonstrated that *all* maps in *all* cultures possess significance and purpose other than as instrumental statements of spatial facts and relationships. By raising issues of social and cultural relations – for example, by drawing attention to the role of maps as intellectual resources, access to which has varied by class, gender and ethnicity – map history has stimulated critical approaches to cartography. The historical exploration of map forms and mapping strategies reveals that mapping does not constitute a singular endeavour: 'cartography' comprises only a specific suite of academic and professional discourses that by no means encompass all mapping activities. Indeed, the modern idealization of cartography has its own history dating to the 1820s, when the adoption of new technologies for systematic, state-wide mapping held out such a radically new prospect for the creation of a single, unified archive of spatial information at all scales that Parisian geographers coined a special name for it: 'cartography'. The ideal was thereafter elaborated further and entrenched within modern Western culture (Edney, 2012).

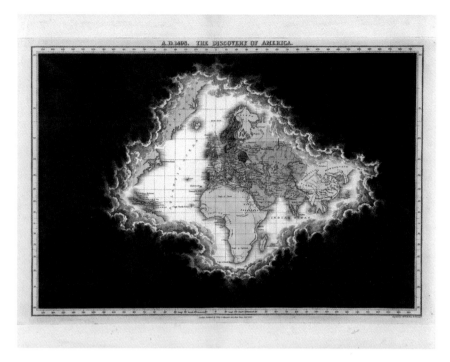

Figure 5.1   Edward Quin, 'A.D. 1498. The discovery of America', in *Historical Atlas* (London, 1830), map 16 of 21. Hand-coloured copper engraving; 33 × 47cm. Courtesy of the David Rumsey Collection (cfP4139). See www.davidrumsey.com (image 2839.016) for a high-resolution image

Ultimately concerned with the myriad spatial practices pursued by humans across cultures and throughout history, map history covers a huge intellectual territory and is disciplinarily disparate. Map historians hail from across the humanities and social sciences, each having been entranced by a particular style of map or aspect of mapping. They have brought their own disciplinary training, theories and interests to bear on this multifaceted field. Scholars from the humanities have read maps as texts in the same way that they have read novels and paintings. For example, Schmidt (2012: 1038–1043) subjected a seventeenth-century Dutch map of the Americas to an iconographical analysis (Figure 5.2). Far from being the 'newest and most accurate map of all of the Americas', as its title proclaimed, this map simply copied older geographical maps of the Americas. Its selling point was not new information – it had none – but the manner in which its upper and lower cartouches construed a 'bond between Faith and America' that paralleled the well-worn Renaissance trope of 'the inverse relation of "god" and "gold": wherever the latter flowed abundantly, it was proposed, the former was typically scarce (and the reverse)'. Humanities scholars have further analysed how maps, or references to maps, function within novels and paintings, and they have explored how novels and paintings function as maps in their own right (e.g. Burroughs, 1995; Conley, 2011).

Implicit in all such work is the need to dissolve the rigid boundaries erected by an idealized cartography around 'the map': not only do maps possess complex semiotic systems formed in large part from other systems, but maps are part of and read in conjunction with other texts and must be interpreted accordingly. Social scientists have tended to study mapping within state, imperial and nationalist contexts, with an eye to understanding how maps are deployed

*Figure 5.2* Nicolas Visscher, *Novissima et accuratissima totius Americae descriptio* (Amsterdam, c.1677). Hand-coloured copper engraving; 42 × 53cm. Courtesy of the Osher Map Library and Smith Center for Cartographic Education, University of Southern Maine (SM-1677-8). See www.oshermaps.org/map/1773 for a high-resolution image

in support of social ideologies. From geographers to political scientists, they have accordingly emphasized discourses of territorial control, resistance, ideology and propaganda. While such sociocultural studies have been revelatory, without specific consideration of spatial practices they are potentially misleading and incomplete, as both Brückner (2008) and Schmidt (2012) have argued. This chapter therefore outlines the multiple natures and histories of maps and mapping through the lens of mapping processes and spatial practices, and ends with some suggestions for how to read maps as human documents.

## Maps and mapping

There is no such thing as 'the map'. There is no single and unambiguous class of images that comprise free-standing images of the world. Setting aside the modern idealization that 'the map' must be a mimetic image of the world, Harley and Woodward (1987–2022, 1: xvi) concluded that 'maps are graphic representations that facilitate a spatial understanding of things, concepts, conditions, processes, or events in the human world' (see also Jacob, 2006). This definition has been widely adopted, yet remains inadequate. In studying the history of mapping by indigenous peoples, Woodward and Lewis argued that maps do not have to be graphic in form (Harley *et al.*, 1987–, 2.3: 1–10). They might be gestural, spoken or performed. These insights apply equally to mapping in other societies, even modern industrial ones: all cultures both incorporate meanings in speech and action and inscribe meanings in various media; differentiation occurs only in the degree to which cultures favour inscriptive over incorporative strategies (Rundstrom, 1991).

Mapping is the process of representing spatial complexity. In this respect, 'representation' is a process, the act of representing; any text, of whatever form, that represents spatial complexity is a map. People have always, of course, constructed their own 'cognitive maps' (a metaphor for internal, neurological processes). Cognitive maps are built from direct, personal experience and from maps communicated by others. People have always communicated that spatial knowledge to others. Small communities – a village of subsistence-level farmers, say, or a nucleated, urban family – hold a wealth of spatial knowledge in common. Their members can all refer to key places by verbal references, such as 'Johann's ten-acre field' or 'High Street', and they can delineate relationships between them by word and by gesture; they might perhaps make ephemeral maps, inscribed in sand or on paper. But to refer to distant places, even to know about distant places, requires a wider scope of spatial communication. Such mapping still does not necessarily produce durable, material maps (see Delano-Smith, 2006), but the likelihood of their creation increases as acts of spatial communication become less immediate and more attenuated.

Ephemerality is a function not of actual form but of intended duration. Just as durable works can easily have only a short lifespan, so it is possible, on occasion, for ephemeral works to survive for long periods. In 1896, for example, an anonymous letter writer repurposed ephemeral tourist brochures, guidebooks and railway timetables by cutting out their maps and views and pasting them into a carefully constructed letter that was clearly intended for long-term preservation (Figure 5.3).

Differences in map form and in the processes of producing and consuming maps are indicative only of differences in communicative context and not of cognitive abilities. Consider the example of the Comte de Lapérouse's 1787 encounter with the inhabitants of Sakhalin (Latour, 1987: 215–257). In an act of interpersonal communication, the locals drew for Lapérouse a map in the sand, in which they indicated that Sakhalin was an island, not a peninsula; in an act of social communication, Lapérouse then drew a durable copy of this ephemeral map in his

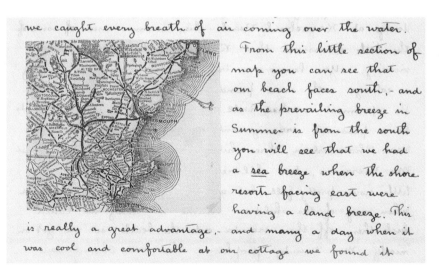

*Figure 5.3*  Detail of page 2 of an anonymous, 24-page letter to 'my dear sister', dated Springfield, Massachusetts (20 August 1896). Coloured lithograph combined with ink on paper. Courtesy of the Osher Map Library and Smith Center for Cartographic Education, University of Southern Maine (OML-1896-56). See www.oshermaps.org/map/44727.0002 for a high-resolution image

notebook, which he later incorporated together with his observations for latitude and longitude into his general chart of the northern Pacific. The mapping acts differed not because of some supposed divide between the 'primitive' and 'irrational' and the 'scientific' and 'rational', but because of the respective communicative contexts. The locals responded to leading questions from a traveller with a performance that included an ephemeral, material component; Lapérouse worked on behalf of the French government and Europe's scientific communities and sought to record knowledge durably so that it could be carried, without alteration, to France to aid in further imperial and commercial endeavours.

Like historians of cartography, map historians generally study durable, material maps. This is understandable, given how the historical records overwhelmingly comprise maps made for social purposes. Yet awareness that mapping processes generate multiple forms of maps serves to dissolve the strict boundaries that the modern ideal has traditionally erected around the material, (carto)graphic map. Material maps are all discursively interwoven within arrays of performative and material strategies for spatial representation: verbal, both spoken and written; numeric; graphic, both gestural and visual; and also physically constructed. There is no one set of characteristics that define 'the map'. It therefore makes absolutely no sense for scholars to seek the 'earliest map'; rather, the task for map historians is to define particular representational strategies within particular spatial discourses.

## Modes of mapping

A broad review both of different mapping strategies and of the communities involved in mapping reveals something of a pattern, or at least the realization that different mapping activities fall into a number of largely distinct 'modes'. These modes provide a first cut in delineating how people strategically communicate knowledge about the world and its parts. A mode is an integrated assemblage of: a distinct, scale-dependent archive of spatial knowledge; the technologies used to create and represent that knowledge; and the social institutions that require and consume that knowledge (Edney, 2011). At root, modes are delineated, in a structural-functionalist manner, by the social contexts within which maps are produced and consumed. The concept of mapping modes is a heuristic and, as a starting point for analysing mapping activities, it was fundamental in the design of the last three volumes of *The History of Cartography* (Harley *et al.*, 1987–, 6: 978–980). But modes do not exist in and of themselves.

Modes are delineated through the repeated comparison of the spatial conceptions, technologies and social function of different sets of maps, such as those reproduced in Figures 5.2 above and 5.4 below. Many groups interested in knowledge and education – civil and military officials, religious and lay educators, merchants, politicians and the general public (as it evolved after the seventeenth century) – have sought to organize their knowledge of and to construct meanings for the wider world through geographical maps (Figure 5.2). The issue of wider world is essential, for geographical maps represent more of the world than any single person can be expected to be able to observe and measure on their own. In other words, they are innately social products. These small-scale images are compiled from multiple sources: existing graphic maps, written itineraries or other descriptions of the world. Geographical maps are often prepared as part of, or in conjunction with, written geographical accounts and circulate widely within interested communities of scholars and officials; they have been of especial importance in the formation of nationalistic identities. The label 'geographical maps' stems from the interrelationship of such mapping with the practice of geography, but the mode has nonetheless encompassed, depending on the particular era, several kinds of mapping of suprapersonal space. The history of geographical mapping is thus the history of the inter-play of more precise discourses of chorography (mapping of regions), geography (mapping of the world and its parts with reference to the Earth's sphericity, as emphasized in the lines of latitude and longitude in Figure 5.2), and cosmography (mapping of the world within the cosmos, perhaps according to metaphysical conceptions).

By contrast, the creation and maintenance of real property rights provide a major arena for social and state organization and conceptualizes landscapes as fragmented into discrete parcels. Each parcel is represented by combinations of toponyms, markers placed in situ, perambulations and other performances of observation and measurement, tabular cadastres and terriers, metes-and-bounds descriptions, degrees of measurement, and graphic imagery. When produced, graphic property maps are generally grounded in plane geometry and simple surveying techniques undertaken at large scales; they rarely circulate beyond per-sonal and local archives and each is reproduced in barely a handful of copies. The history of property mapping is thus the history of how specific representational strategies have been variously implemented by people in different social institutions in support of the mainte-nance of property rights. In the case of Figure 5.4, for example, the English had transplanted to colonial North America the basic practice of describing each parcel of property by its 'metes and bounds'; the colonists eventually produced graphic property maps as indexes to those legal descriptions of the markers placed in the landscape itself. Recent scholarship has begun to reveal the ways in which early modern Europe's adoption of measured and graphic strategies for representing property was neither uniform nor coherent; instead, such strate-gies varied by, and were constitutive of, local social relations, and their development needs to be understood in terms of other strategies for representing property (Fletcher, 1995; De Keyzer et al., 2014).

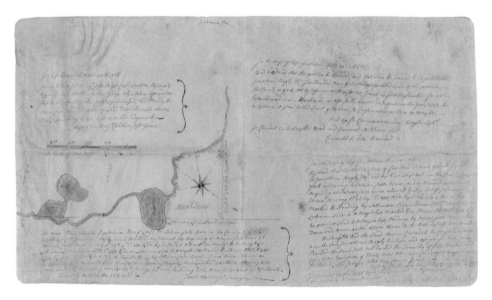

*Figure 5.4*  James Warren, Jr., 'The above Plan Discribs five Hundred Acres of Land on Salmon Falls River [. . .] Granted to Col. Jonathan Bagly' (Berwick, Maine, 13 October 1766). Manuscript; 37 × 65cm. Courtesy of the Maine Historical Society, Portland (map F 162). See www.mainememory.net/artifact/9259 for a higher-resolution image

It is obvious that – in terms of their spatial conceptions, the techniques of their production and reproduction, and the ends to which they are put by particular institutions – geographical maps and property maps represent markedly distinct modes of mapping. Repeated comparisons permit the classification of several mapping modes. They are summarized in Table 5.1.

*Table 5.1* The modes of mapping. The order is not intended to imply priority or relative importance

MAPPING PLACES – i.e. mapping discrete portions of the world whose representation can conceivably be undertaken by one individual, in land surveys, in conjunction with landscape imagery and even poetry:

| | | |
|---|---|---|
| **A** | **place mapping** | topography per se ('writing place'): the representation of the physical and cultural features of specific locales to create, perpetuate and reconfigure their distinctive meanings as places |
| **B** | **urban mapping** | the depiction of entire urban places in plans and views is distinguished from place mapping by the cultural importance granted to cities as artificial and self-regulating communities (see Kagan, 2000: 1–18) |
| **C** | **property mapping** | a subset of place mapping, but with the meaning of each place (parcel, estate, etc.) defined by the sole criterion of property rights |
| **D** | **engineering mapping** | the detailed mapping, which might be carried out over long distances (see G), in support of planning and building roads, buildings, fortifications, etc. |

MAPPING SPACES – i.e. **(E) geographical mapping** generally – which encompasses the organization of knowledge of the suprapersonal world, beyond the ability of one person to know directly by first-hand experience. While generally a single mode in the modern West, in other historical and cultural circumstances it is more appropriate to differentiate three particular modes:

| | | |
|---|---|---|
| **E$_1$** | **cosmos mapping** | cosmography per se ('writing creation') is specifically concerned with representing the interrelationships between humanity, creation and perhaps the divine, often to astrological or metaphysical ends |
| **E$_2$** | **world mapping** | geography per se ('writing the Earth') is the representation of a culture's known world (mundus or ecumene) or of the globe of the Earth |
| **E$_3$** | **regional mapping** | chorography per se ('writing region'), which although often technically distinct from world mapping, is nonetheless carried on by the same institutions of education and encompasses special-purpose mapping (such as road maps) |

MAPPING TERRITORIES – i.e. extending the large-scale mapping of places (A–D) over extensive spaces previously mapped only at smaller, geographical scales (E), generally at the hands of individual modern states, giving rise in the nineteenth century to the modern cartographic ideal:

| | | |
|---|---|---|
| **F** | **geodetic surveys** | efforts to measure the Earth's size and, after c.1670, its shape |
| **G** | **boundary surveys** | the application of the large-scale techniques for mapping a relatively narrow area for long distances along a border or frontier between states; while conceptually similar to property mapping, boundary mapping entails quite different political discourses and institutions |
| **H** | **systematic surveys** | of landscapes and cities (topography), coasts and oceans (hydrography), and perhaps also properties (cadastral), carried on by modern states in a (supposedly) consistent, coherent and uniform manner, beginning with surveys of France in the 1700s and proliferating after 1790 as one hallmark of modern cartography; eventually after 1900 these surveys extended to encompass aeronautical mapping |

MAPPING SEAS – i.e. **(I) marine charting**, mapping by mariners for mariners – which emphasizes coastlines and features in the coastal zones, even to the exclusion of interior features, and encompasses several scales of representation, from oceanic charts, to coastal charts, to hydrographic surveys (H), to harbour charts (A)

MAPPING THE HEAVENS and heavenly bodies – i.e. **(K) celestial mapping** – which like the mapping of the seas (I) encompasses several scales of representation, from star charts and cosmological diagrams (see E₁) to detailed mapping of the other planets (A, H), arrayed within a coherent set of discourses

MAPPING DISTRIBUTIONS – i.e. **(L) thematic mapping** – which is to say the modern outgrowth of geographical mapping, albeit applied at a variety of scales, intended to visualize and display variations and patterns in data about the human and physical worlds, in conjunction with modern science and governmentality. Thematic maps should not be conflated with those 'special-purpose maps' produced for narrowly specific ends within other modes

PHOTOGRAPHIC MAPPING of the Earth – i.e. **(M) overhead imaging** – with origins in the nineteenth century but formalized with fixed-wing aircraft and then satellites and space vehicles in the twentieth century, and comprising both analogue 'aerial photography' and digital 'remote sensing'

---

Understanding map history to be the intertwined and intersecting histories of mapping modes permits us to break free of the flawed and partial history of cartography (Edney, 1993; revised by Edney, 2011). We can explore the parallel development of different modes without presuming that they are necessarily interlinked, or the invention and adoption of new mapping technologies, from the chronometer to the digital computer, without presuming that they have been inevitable. Such an approach encourages us to seek complexity, for example, in the interactions within early modern Europe of chorography, geography and cosmography ($E_1$–$E_3$ in Table 5.1) that would eventually give rise to modern geographical practices. Overall, map history becomes the story of the modes and of their numerous intersections. There is much scope for new work in this arena, in addition to the more detailed re-examination of the traditional cartographic canon and in opening up new areas of mapping history. There are currently few historical overviews of particular modes (notably Harvey, 1980; Kain and Baigent, 1992) and there are as yet no new synthetic histories other than collections of essays (Akerman and Karrow, 2007; and, of course, Harley *et al.*, 1987–).

For historians turned off by the somewhat impersonal character of mapping modes themselves, map history has significant scope for more personal histories. The complex interweaving of modes within and between different human cultures seem all to have been contingent on the actions of individuals who appropriated representational strategies or spatial archives developed in one mode for use in another. Key here is the way in which some people were able to break the compartmentalization of activities around distinct modes. Christopher Columbus, for example, could not do so. He was a mariner who dabbled in cosmography: he conceived his westward voyage to Asia as an exercise in cosmography, through a detailed consideration of the size and shape of the spherical Earth and its major divisions; but once at sea, like any professional mariner, he recorded the distances and directions he sailed in a manner that ignored the Earth's curvature. The two conceptions of the world would be reconciled by Amerigo Vespucci, an Italian banker and cosmographer who dabbled in navigation and who could adjust Columbus's maritime experiences to fit a cosmographical framework of latitude and longitude. Subsequently, intellectually omnivorous geographers routinely absorbed marine information into their own maps and even presented the results as if they were 'charts', as Gerard Mercator did in his 1569 wall map of the world, on *that* projection. Indeed, subsequent geographers tended to use Mercator's projection whenever they wanted to create the *image* of a marine chart,

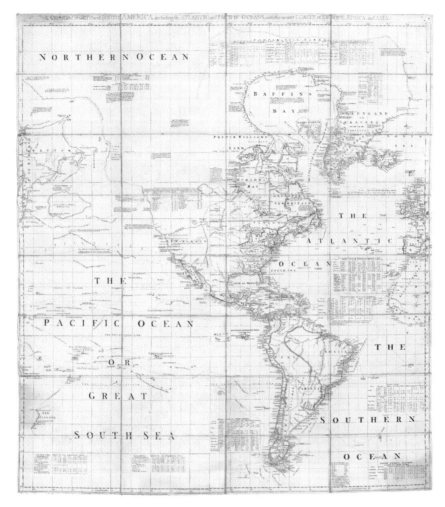

*Figure 5.5* John Green, *A Chart of North and South America* (London: Thomas Jefferys, 1753). Hand-coloured copper engraving in six sheets; dissected into 24 pieces and mounted on cloth; 126 × 110.5cm (assembled). Courtesy of the Osher Map Library and Smith Center for Cartographic Education, University of Southern Maine (OS-1753-2). See www.oshermaps.org/map/672 for a high-resolution image

as when John Green used the results of multiple nautical voyages to redefine the outline of the Americas (Figure 5.5). But even had the resultant maps circulated back amongst mariners, they were not structured as marine charts and were not used as such.

## Mapping and spatial discourses

But how do we know that the *Chart of North and South America* reproduced in Figure 5.5 was indeed intended and consumed as a geographical map and not as a marine chart? Certainty comes from understanding how maps circulated between producers and consumers. In this particular instance, the supposed chart was enmeshed in geographical discourses: its maker was a critical geographer who made no actual charts; its projection and graticule are those of geographical knowledge, not contemporary marine practice; its physical form (dissected and then pasted onto cloth) was common for large geographical maps in the mid-eighteenth century, but was never used for working charts; its publisher was located to serve London's general public, not the mariners; and subsequent reissues of the map were all made in conjunction with supplying the public with geographical information about events in North America.

By studying the circuits within which maps moved, we can move from the broad-brush, top-down, structural-functional insights offered by mapping modes to reconstructing, from the bottom up, precise and specific discourses. A discourse is a regulated network of representation and constitutes what was referred to above as 'communicative context'. Empirically rooted, the identification and analysis of discourses has explanatory potential. In particular, if one uses map evidence to argue how people in a given culture thought about the world, then one must first be clear about which members of that culture consumed which kinds of maps, and how they did so (see Figure 5.4). Spatial discourses within small, well-defined communities might be tight and compact, relying primarily on performative and verbal mapping, whereas more expansive societies supported much broader and impersonal discourses. But even in modern societies with highly specialized and articulated economies, mapping discourses are not universal and are instead shaped by socio-economic class, gender, profession, politics, educational attainment and perhaps also religious persuasion. Conversely, participation in mapping discourses gives form to social distinctions. Just as particular communities are defined in large part by the spatial knowledge held in common, so, to echo Latour's (2005) profound complaint with respect to sociology, participation in mapping is constitutive of social relations. We can thus move beyond simple explanations that 'mapping practices are social' (as Andrews, 2001: 31–32, complained), and investigate how mapping processes and spatial discourses contribute to social formation and historical change (e.g. Mapp, 2011). In this respect, the weight of historical interpretation must lie more on the consumers of maps than on the producers.

Tracing circuits of inscribed maps is a complex task. Not only are supporting materials generally dispersed, fragmentary or simply lacking, especially before the modern period, but it is difficult to track how maps moved between producers and consumers. Fortunately, we can use some surrogate information to start to parse how maps would have circulated. Of crucial importance in this regard is the form of each map. With respect to durable, material maps, we can consider several physical factors that suggest how the map was to be consulted (in a larger work, if so what is the nature of that work? Wall-mounted or dissected, and if so where mounted? Free-standing or part of a series?), its contemporary cost (degree of decoration, quality of paper and binding, etc.), specific information about its intended use and so on. Paying attention to such matters allows the researcher to see patterns and distinctions amongst an otherwise undifferentiated mass of imagery, such as the chronological bibliography of maps of New England printed before 1800 (McCorkle, 2001). Consideration of the early modern map trade (esp. Pedley, 2005) indicates that it makes sense to consider each nation's map trade as largely distinct circuits (albeit with revealing intersections). So, looking at the physical context just of English geographical maps of New England, we see that they fall into three distinct groups: small maps (no larger than A4-size) of the extent of colonial settlement within books celebratory of the potential or history of the region; within atlases, larger folio maps of New England as part of the broader array of English colonies; and a few large maps intended for wall display. Each group manifests a particular discursive context within which contemporaries would have encountered and read the maps.

## Studying early maps

The task for map historians is to define precise mapping discourses in order to establish the framework for reconstructing the contemporary meaning of particular maps. It is a recursive process. Learning about maps and their consumption – from evidence as diverse as contemporary literature to household inventories to physical marks on the maps themselves – tells us more about their discursive contexts, which in turn permits more sophisticated interpretations. Like other cultural historians, map historians must devour as wide a range of imagery as they can, to develop a repertoire of maps and to appreciate how those maps function through multiple representational strategies, and so permit valid readings of individual works.

The modern idealization of cartography and 'the map' remains so pervasive that almost everyone, not just historians of cartography, when faced with an early map, automatically compares its outlines and features with those on modern maps, reads it for its environmental information, and so evaluates its nature and significance according to its geometrical and topographical accuracy. This impulse must be resisted! The significance of any map is defined not by its content but by its context: why it was made, for whom and how did they use it? Maps are human documents, and the trick in reading them is to look past the details of their content and to consider instead their character, style and form. Particular content issues may well prove to be analytically relevant, for instance in figuring out a map's sources, but only within the map's discursive context.

The study of any newly encountered early map therefore entails the careful examination of a number of elements:

- the date and place of production, i.e. its cultural context (from imprints on printed maps, names of authorities, style or look especially contributions of its 'decorative' elements, form, perspective[s], etc.);
- the mode, i.e. its social context (from the extent of area depicted, the manner of depiction, scale, general style, etc.);
- method of reproduction and physical context (e.g. if not in a book, do old creases suggest that it was originally so, etc.);
- what the map can be understood to say about itself, in its title, text blocks, annotations, etc.;
- any other indications on the work itself of how it was used.

Together, these factors point to the work's discursive context. This is the kind of exercise that allows the identification of the *Chart of North and South America* to be a geographical map. Only by situating each map in its relevant discursive context can anything definitive be said about its particular depiction of place or space, its information content and its status with respect to other maps from the same discursive context.

The study of map history engenders a humanistic appreciation for maps and mapping that applies equally to the present as to the past. Adoption of such a historical approach allows the examination of the social and cultural significance of mapping, and provides the salutary reminder that professional and academic practices are neither universal nor technologically determined. Cartography remains a human endeavour, bound up with multiple spatial discourses and (other) mapping practices. The careful analysis of those discourses and practices offers the opportunity to develop a nuanced understanding of cartography that is socially aware and culturally open.

## Note

1 Periodization in this chapter broadly follows that of the volumes of Harley *et al.* (1987–): 'modern' is used not for the present day, but for the entire era of industrial states after c.1800 (vols 5 and 6); 'early modern', for the era of European mercantilism, c.1450–c.1800 (vols 3 and 4); 'pre-modern', for the ancient, classical, and medieval eras (vol. 1); 'non-Western', for both 'traditional' Asian societies (vols 2.1 and 2.2) and the 'indigenous' peoples of the rest of the world (vol. 2.3).

# References

Akerman, J.R. and Karrow, R.W. (Eds) (2007) *Maps: Finding Our Place in the World* Chicago, IL: University of Chicago Press.

Andrews, J.H. (2001) "Introduction: Meaning, Knowledge, and Power in the Map Philosophy of J.B. Harley" in Harley, J.B. *The New Nature of Maps: Essays in the History of Cartography* (ed. Laxton, P.) Baltimore, MD: Johns Hopkins University Press, pp.1–32.

Brückner, M. (2008) "Beautiful Symmetry: John Melish, Material Culture, and Map Interpretation" *Portolan* 73 pp.28–35.

Burroughs, C. (1995) "The 'Last Judgment' of Michelangelo: Pictorial Space, Sacred Topography, and the Social World" *Artibus et historiae* 16 (32) pp.55–89.

Conley, T. (2011) *An Errant Eye: Poetry and Topography in Early Modern France* Minneapolis, MN: University of Minnesota Press.

De Keyzer, M., Jongepier, I. and Soens, T. (2014) "Consuming Maps and Producing Space: Explaining Regional Variations in the Reception and Agency of Mapmaking in the Low Countries During the Medieval and Early Modern Periods" *Continuity and Change* 29 (2) pp.209–240.

Delano-Smith, C. (2006) "Milieus of Mobility: Itineraries, Route Maps, and Road Maps" in Akerman, J.R. (Ed.) *Cartographies of Travel and Navigation* Chicago, IL: University of Chicago Press, pp.16–68.

Edney, M.H. (1993) "Cartography without 'Progress': Reinterpreting the Nature and Historical Development of Mapmaking" *Cartographica* 30 (2,3) pp.54–68.

Edney, M.H. (2005) "The Origins and Development of J.B. Harley's Cartographic Theories" *Cartographica* 40 (1, 2) Monograph 54. Toronto, ON: University of Toronto Press.

Edney, M.H. (2011) "Progress and the Nature of 'Cartography'" in Dodge, M. (Ed.) *Classics in Cartography: Reflections on Influential Articles from Cartographica* Hoboken, NJ: Wiley-Blackwell, pp.331–342.

Edney, M.H. (2012) "Field/Map: An Historiographic Review and Reconsideration" in Nielsen, K.H., Harbsmeier, M. and Ries, C.J. (Eds) *Scientists and Scholars in the Field: Studies in the History of Fieldwork and Expeditions* Aarhus, Denmark: Aarhus University Press, pp.431–456.

Fletcher, D.H. (1995) *The Emergence of Estate Maps: Christ Church, Oxford, 1600 to 1840* Oxford, UK: Clarendon Press.

Harley, J.B., Woodward, D., Lewis, G.M., Monmonier, M., Edney, M.H., Pedley, M.S. and Kain, R.J.P. (Eds) (1987–) *The History of Cartography* (12 bks in 6 vols) Chicago, IL: University of Chicago Press. Reprinted online (free access) at *www.press.uchicago.edu/books/HOC/*.

Harvey, P.D.A. (1980) *The History of Topographical Maps: Symbols, Pictures and Surveys* London: Thames and Hudson.

Jacob, C. (2006) *The Sovereign Map: Theoretical Approaches in Cartography Throughout History* (translated by Conley, T. and edited by Dahl, E.H.) Chicago, IL: University of Chicago Press.

Kagan, R.L. (2000) *Urban Images of the Hispanic World, 1493–1793*. New Haven, CT: Yale University Press.

Kain, R.J.P. and Baigent, E. (1992) *The Cadastral Map in the Service of the State: A History of Property Mapping* Chicago, IL: University of Chicago Press.

Latour, B. (1987) *Science in Action: How to Follow Scientists and Engineers through Society* Cambridge, MA: Harvard University Press.

Latour, B. (2005) *Reassembling the Social: An Introduction to Actor-Network-Theory* Oxford, UK: Oxford University Press.

Mapp, P.W. (2011) *The Elusive West and the Contest for Empire, 1713–1763*. Chapel Hill, NC: University of North Carolina Press.

McCorkle, B.B. (2001) *New England in Early Printed Maps, 1513 to 1800: An Illustrated Carto-Bibliography* Providence, RI: John Carter Brown Library.

Pedley, M.S. (2005) *The Commerce of Cartography: Making and Marketing Maps in Eighteenth-Century France and England* Chicago, IL: University of Chicago Press.

Rundstrom, R.A. (1991) "Mapping, Postmodernism, Indigenous People and the Changing Direction of North American Cartography" *Cartographica* 28 (2) pp.1–12.

Schmidt, B. (2012) "On the Impulse of Mapping, Or How a Flat Earth Theory of Dutch Maps Distorts the Thickness and Pictorial Proclivities of Early Modern Dutch Cartography (And Misses Its Picturing Impulse)" *Art History* 35 (5) pp.1036–1049.

# 6

# Critical cartography

*Chris Perkins*

A *critical* cartography is the idea that maps – like other texts such as the written word, images or film – are not (and cannot be) value-free or neutral. Maps reflect and perpetuate relations of power, more often than not in the interests of dominant groups (Firth, 2015).

Jeremy Crampton (2011) argues that social scientific literature usually contains very little discussion of mapping, cartography or GIS. Lip service is paid to the fact that Geography as a discipline is somehow strongly associated with mapping, but the detail of mapping practice is somehow safely avoided, and left as technical domain to the GI scientists (see also Dodge and Perkins (2008) who develop this argument exploring the often ambivalent relationship between geographers and mapping). So there is a continuing gap in mainstream critical work.

Crampton also observes that the majority of published work from cartographers and GI practitioners 'has very little to say about politics, power, discourse, postcolonial resistance, and the other topics that fascinate large swathes of geography and the social sciences' (Crampton, 2011: 1). In spite of significant critique from figures such as Brian Harley, Denis Wood, Denis Cosgrove and John Pickles, most research into mapping still regards cartography as being concerned only with the scientific representation of information in mapped form. So mapping for many remains useful, neutral and apolitical.

The whole of Crampton's book is an attempt to investigate these silences and to argue for what he calls 'critical cartography', which might bring together practice and theory, and begin to fill the gaps. This chapter charts what characterizes critical cartography, exploring the theoretical and practical implications of the field, to historicize and situate its practice. It focuses upon the modes, moments and methods in the processes of writing and doing mapping in a critical manner. Like Crampton its aim is to explore the silences – in an analogous fashion to Brian Harley's (1988) influential exploration of the silences in mapping in early modern Europe, to reveal some of the forces that influence mapping and critical writing about mapping.

## What is critical cartography?

At the outset it is important to establish that a single definition for critical cartography is likely to oversimplify a complex and changing set of ways of approaching mapping. For most the term reflects ways of understanding cartography that might empower or emancipate. This might entail being radical (Bhagat and Mogel, 2008; Denil, 2011), being disruptive, tactical and playful (Hind, 2015), protesting (Wood and Krygier, 2009b), being subversive (Pinder, 1996), engaging in counter-mapping (Vujakovic and Matthews, 1994; Peluso, 1995), creating autonomous militant cartography (Counter-Cartographies Collective *et al.*, 2012), being counter-hegemonic (Perkins and Dodge, 2009), engaging in anarchist mapping pedagogy (Firth, 2014), guerrilla cartography (Anonymous and Rufat, 2015), or being experimental (Thompson *et al.*, 2008). The most frequently cited article about critical cartography highlights significant elements that are most often associated with the approach suggesting that critical cartography 'challenges academic cartography by linking geographic knowledge with power, and thus is political' (Crampton and Krygier, 2006: 11). For others, the term has been deployed to explain a more pedagogic or practical way forward, an everyday mapping practice designed to critique established orthodoxies, mapping to change the world (Firth, 2014). So critical cartography deals with the worlds of ideas and practice.

## Theory

For many the term critical cartography has come to be deployed to signify a new way of writing about maps and thinking about the academic discipline of cartography. Examining different critical ways of thinking about mapping over the last twenty-five years reveals a very diverse series of approaches, informed by different disciplinary positions and assumptions (see Kitchin *et al.*, 2009).

One of the foundational texts for critical cartography was Brian Harley's 'Deconstructing the Map' (1989). The significance of this article is reflected in a citation count that is the second highest in the whole history of *Cartographica*, and the release of a celebratory theme issue of the journal to commemorate the twenty-five-year anniversary of its publication (Rose-Redwood, 2015). Harley saw the map as a form of Foucauldian power-knowledge and critiqued the neutral view of cartography as a technical science, whilst seeking to analyse the map as a text using Derridean deconstruction. Harley's argument focused attention on the ways in which power-knowledge was at once embedded in mapping, but also impacted upon the contexts in which cartography was deployed. He and other critical cartographers saw the map as a thing that did political work. Harley delivered a strong social-constructivist attack of cartographic practice, and also on the vested interests of those involved in the cartographic project.

By way of contrast, Denis Wood (Wood and Fels, 1986; Wood, 1992) regarded mapping as a kind of Barthean sign system, and focused on its naturalizing power as a form of constructed knowledge, whose work was achieved through a shared cultural orthodoxy. Wood's emphasis continues to be on the interests embedded in mapping, for example in his recent research into the involvement of US military with indigenous mapping initiatives in South America (Bryan and Wood, 2015). John Pickles (2004) emphasized the hermeneutic power of the inscriptions in mapping, and was instrumental in calling for a socially informed critique of GI scientific claims to neutrality (Pickles, 1995). Denis Cosgrove (1999) was more interested in the wider cultural contexts of mappings, focusing in particular on the relations of mapping to visual culture (Cosgrove 2005).

The work of this first generation of critical cartographic theorists has been developed and applied by subsequent researchers. Jeremy Crampton has arguably done most to popularize the approach and sets out a detailed explanation for the field in a series of papers that were subsequently reworked into the textbook (Crampton, 2002, 2011; Crampton and Krygier 2006). Crampton's work starts with Kantian notions of critique, deploys the Frankfurt school ideas of critical theory, situates these in Foucauldian understandings of mapping as an historicized political process and concludes that critique 'examines the grounds of our decision-making knowledges; second it examines the relationship between power and knowledge from a historical perspective; and third it resists, challenges and sometimes overthrows our categories of thought' (Crampton and Krygier, 2006: 14).

More recently, post-representational approaches to cartography (described by Gerlach in Chapter 7, this volume) have questioned the ontic security of the map as a thing, and have chosen to emphasize the potential that emerges from mapping as a process. Some theorists draw on science and technology studies and the work of Latour, Serres, Callon and Law, in which mapping is seen as a material semiotic vehicle with relational force (see, for example, November *et al.*, 2010). Others have drawn insight from studies of transduction and technicity to reflect on the emergent properties of mapping. For example, Kitchin and Dodge (2007: 1) argue that 'maps are transitory and fleeting, being contingent, relational and context-dependent; they are always mappings; spatial practices enacted to solve relational problems'. Other critical post-representational approaches to mapping draw on performative approaches to identity and feminist approaches to knowledge, notably the call for hybrid approaches to mapping from Del Casino and Hanna (2005). The notion of mutability in mapping is also explored from ethnomethodologically informed research in the work of Barry Brown and Eric Laurier (2005). More phenomenological accounts of cartography are to be found in the influential work of social anthropologist Tim Ingold (2000) who argues strongly against the idea of a mental map, suggesting instead that mapping works as a narrative, weaving together series of stories of place, highlighting temporality instead of spatial structures, and invoking memory as an organizing principle. More recently non-representational theory has been deployed by Joe Gerlach (2013), whose work frequently emphasizes the minor politics of what he terms vernacular mappers.

So a critical approach depends upon different disciplinary influences. The worlds of critical theory move with the times, at once reflecting the wider epistemological and cultural context, but also the practices of cartography and mapping.

## Practice

Wood and Krygier (2009a: 344) observe that 'narrowly construed, critical cartography was . . . a professional moment in the broader history of map-making; but criticism more broadly understood has been an aspect of map-making from its earliest days'. So critical cartography also entails *making* critical maps that seek to change how things might be. People have always made maps that subvert or in other ways speak to power. Any categorization of practice simplifies often hybrid forms: here we focus on mapping protest, map art, socially networked digital mash ups, indigenous cartographies and novel everyday mappings.

Cartographies from social movements chart alternatives, revealing things often hidden in more conventional mapping, and highlighting amongst other issues eco-protest, issues relating to (dis)abilities and gender, struggles for human rights, anti-austerity and anti-military forces. Wood and Krygier (2009b) observe that maps and protest are enacted in at least three different registers: the office, street and press. Official protest maps dispute particular views, critiquing from within. More subversive mapping as a part of direct oppositional social protest is explored by Firth (2014), where street protests are planned, or recalled in map form. Real time mapping of protest as a tactic includes the now defunct mobile app Sukey, designed to facilitate demonstrators' mobility and as a means of avoiding police kettles (Hind, 2015). More abstract and artistic oppositional cartography include the psychogeographic and subversive remapping of Paris by situationists in the 1950s and 1960s (see Pinder, 1996), and Bill Bunge's radical mappings of social issues in urban ghettos in North America in the 1960s (see Johnson, 2010). Publication of alternative views in print media, and increasingly on the web, disseminates alternatives. Notable examples in the last thirty years have been published by Pluto Press, and the web facilitates the easy sharing of alternative views; see, for example, the web sites Monde Diplomatique (*www.monde-diplomatique.fr/cartes/*), Bureau d'Etudes (*http://bureaudetudes.org/*), 3Cs Counter Cartography Collective (*www.countercartographies.org/*), or Radical Cartography (*www.radicalcartography.net/*). The web and social media disseminate instantly accessible and sharable alternatives, with a capacity for wide cultural reach but equally with a sometimes hidden or marginalized ephemerality.

Other critical cartography is deployed in a rich and creative diversity of map art that continues to burgeon (see Cosgrove, 2005; Wood, 2006; Harmon and Clemans, 2009). Artists challenge taken-for-granted views of the world as well as the apparent all-seeing power of cartography and the map, across media, forms and artistic practices. Ignazio (2009) suggests three impulses characterize the artistic appropriation of cartographic strategies: a symbolic sabotage deploying the form of the map to reference novel meanings; a situated praxis making the map to explicitly change the world; and data mapping, deploying visual forms to reveal the invisible. For some artists this has involved a remapping focusing upon emotional geographies, for others an embodied mapping reflecting a problematizing of the world, the body and the territory, whilst locative art deploys digital technologies that challenge the status of the very technologies they deploy.

OpenSource collaborative cartographies also deliver novel ways of evading ownership, by hacking together different data, in mashups, layering information together and sharing critical mapping across the social network. Since the first release of mapping Application Program Interfaces (APIs) in 2005, these OpenSource data have grown apace. Crowdsourced mapping need not of itself be radical – and may indeed, like all critical cartography, be subverted by mainstream forces (see, for example, the very conventional forms and practices of OpenStreetMap documented by Perkins, 2014). But the sharing it entails offers a significant challenge to the formerly proprietary way of mapping.

Indigenous mapping seeking to reclaim lost rights or reinstate place-names in the face of colonial power hierarchies, offers another kind of critical practice. Wood (2010) has argued that many of these countermaps are strongly subverted by the funding mechanisms and often externally owned systems of control that call them into being. However, the more performative reassertion of formerly oral traditions and some explicitly political re-mappings deploying technologies to act against neoliberal power speak to a radical potential for the medium.

Other everyday and novel forms of mapping are emerging that perhaps more explicitly subvert existing norms. For example, playful mapping embedded as a central part of many genres of computer gaming calls forth new worlds of the imagination and creates new mapping communities, interested in play as against work (see Lammes, 2008; Perkins, 2009). Literary cartographies narrate and locate imaginary worlds grounding fantasy and re-imagining linguistic description, whilst placing events (Caquard, 2013). Cinematic cartographies speak across media to evoke and problematize the power of film (Caquard and Taylor, 2009; Roberts, 2012). Community mapping continues apace powered by the increased capacity of social media to bring together alterative voices from local groups (Perkins, 2007).

## Mutable critical mapping modes

The nature of the critical approach to mapping has varied, as has its frequency. So when and where critical cartography is written and practised makes a difference to what might be studied.

Matthew Edney's (1993) notion of 'modes' offers an important framing device here to help us understand changes in mapping practices and thinking. He suggests that an assemblage of people, ideas, technologies, things and meanings comes together at different times and might be deployed to understand changes in mapping and cartography. This mode brings together thought and practice and allows the map to be deployed in particular contexts. At any time different modes might coexist – for example, the printed map surviving in the era of digital mapping on mobile devices. These notions of cartographic modes have been deployed to suggest agendas for future critical cartographic research (Dodge *et al.*, 2009).

Technology is important in contemporary mapping modes as a necessary, but not sufficient means for explaining critique. In the renaissance, the gradual deployment of the printing press allowed new forms of mapping to be popularized – and the mass use of fixed maps printed on paper facilitated the use of mapping as a form of immutable mobile, carrying fixed messages to new places, and often facilitating imperial or military conquest (Latour, 1986). Nation states and other powerful forces came to deploy mapping to support their interests, to facilitate trade, help run empires, plan and administer social control and to govern citizens. So this mapping mode came to be equated with an objectifying of meaning, in which the form came to signify a naturalization of culture. Cartography in the twentieth century came to be disciplined into a series of technical processes denying any wider social or political context, and with institutions increasingly controlling and making maps as a kind of reificatory knowledge: it must be right, its *'on the map'*. It was this mode that Brian Harley (1989) so powerfully critiqued in 'Deconstructing the Map'.

The dominant mapping mode at the moment, however, is by way of contrast characterized by a shared and complex authorship, potentially traceable for perhaps the first time, with multiple opportunities for deploying information in novel cultural fashion. Map art proliferates, but is complemented by powerful but hidden infrastructural forces underpinning a frequently churning suite of interfaces, delivering more mapping than ever in human history, in novel but predominantly neoliberal contexts. A black box of code allows the whole assemblage to enrol people, maps and technologies into a mobile assemblage deployed anywhere and at anytime. This complexity demands a critical response.

The technological shifts underpinning this mapping mode arguably themselves strongly influenced the widespread use of critical cartography as an academic trope towards the end of the twentieth century and into the new millennium. Many people have argued, for example, that GIS led to a democratization of cartography opening out mapping to an ever-expanding and increasingly public series of interventions, and in so doing reducing the vested power of the cartographic and GI establishment (see, for example, Rød *et al.*, 2001). This trend it is suggested has continued apace with the collaborative sharing of mapping facilitated by the Geoweb. As data are increasingly created by users enrolled as 'prosumers', in crowdsourced and participatory systems, so the ontology and epistemology facilitated by the technological shift further changes. Warf and Sui (2010) suggest that collaborative mapping encourages a move towards a consensual and performative view of truth claims, instead of the correspondence model dominant in earlier predominant mapping modes. This performative view is much more compatible with critical cartography's recognition of mapping as a process.

So different ways of interacting with maps themselves alter the cultural significance of the mapping mode, and a shift towards critical cartography is itself influenced by technological transitions. And critical cartography allows us increasingly to reflect on the moments called into being by mapping practice, to choose when and where we might study mapping.

## Critical mapping moments

In 2009, we highlighted six instances in mapping praxis, designating these as moments that might yield useful insights, instead of focusing on more usual institutional, contextual or thematic frames that have tended to dominate in earlier research (Dodge *et al.*, 2009). In the seven years since we came up with this manifesto, critical cartographic research has indeed started to generate interesting results in these fields.

*Failure*, for example, is being theorized as a productive site, instead of the dystopian down-side of an aspirant progressive cartography. Hind and Lammes (2015) reflect on failure in the context of mapping carried out under the aegis of OpenStreetMap during the 2014 flooding of the Somerset Levels. They also deploy failure to understand the emergent and emancipa-tory potential of mobile-based protest mapping deployed in public demonstrations. *Points of change*, whether these entail political change or technological breaks, are also a fascinating field for research for critical cartography. Work by political geographers has been important here. For example, Anna Moore and Nicholas Perdue (2014) examine the status of maps in criti-cal geopolitical thought and deploy a map of Kashmir to work through their argument for a changed status of the map as object in the discourse, whilst Russell Foster (2015) explores the role of mapping in the forging of twenty-first-century European identity. Technological points of change in mapping continue to receive attention as well. The acquisition of Keyhole Inc. and the release of Google Maps and the Google Maps API continue to attract significant atten-tion (see, for example, Jason Farman, 2010), in particular because of the new tensions it called into being, between a newly hegemonic and global mapping order, and day-to-day mapping practice, with the social potential to make personal stories layered with these data. The role of the lay public in this context, creating mapping in online socio-technical controversies, such as the Fukashima nuclear disaster, is charted by Plantin (2014).

The *time space rhythms of mapping* have recently also been a very productive field for research. A forthcoming volume on temporality and digital mapping brings together twelve different papers from academics and practitioners, exploring the capacity of newly digital and conver-gent media to enact different aspects of temporality: for some this involves a stitching together of time, for others a focus on ephemerality and mobility and for yet others a stress on the (in)formalities of temporality (Gekker *et al.*, forthcoming a). *Mapping memories* are charted in narrative accounts and are indeed attracting significant attention (see, for example, Caquard, 2013). Technologies allow the encounters of mapping to be recorded and re-enacted in order to reveal formerly unseen parts of the mapping process. For example, Gekker *et al.* (forthcoming b) explore how mapping unfolds in a recent encounter in Oxford. The recollection of these moments (and others) depend upon deploying critical methods.

## Critical methods

In the past, visual studies elided mapping as a medium: the definitive guide to visual method-ologies (Rose, 2012) continued to ignore mapping across three editions published since 2001. Guides for the deployment of mapping have been commonplace in methods texts aimed at Geography students, but it is only recently that mapping has begun to appear in methods texts prepared for other discipline areas (see, for example, McKinnon, 2011 and Grasseni, 2012). But other disciplines are now increasingly adopting mapping as a method as part of their own critical encounters. Quite simply, it is technically easier to map, and the historical baggage of the map is no longer as significant as in the past. McKinnon (2011) argues mapping as a method might be useful across the social sciences. This is important for critical cartography for two reasons. First, because she focuses on different *stages* of the research process, from data collection, through analysis, to display. Methods like mapping are processual and not neutral. Second, she highlights the *diversity* of mapping methodologies that are emerging and being popularized.

In 2009, we suggested that a novel suite of methods is needed extending beyond techniques that are usually deployed by cartographic researchers (Dodge *et al.*, 2009). To draw this chapter to a close, I will briefly illustrate methodological examples since our manifesto call, which demonstrate the potential for critical methodological interventions that are processual and diverse in their focus.

Critical studies of the Geoweb have proliferated and deploy methodologies informed by political economic understandings of an unequal world. Recent examples include Haklay's (2013) critique of the democratic limits to neogeography, Glasze's and Perkins' (2015) situated analysis of OpenStreetMap and Leszczynski's (2012) placing of the emergence of the Geoweb within neoliberal political economic restructuring.

Emotional cartographies and work around effect have built upon Christian Nold's ground-breaking emotional cartography (MacDonald, 2014). Critical research into map design is also beginning to explore the affordances that might be delivered when emotions are considered in the design process. See, for example, Fayne *et al.* (2015) on transportation maps and Gartner (2012) on the design of a mobile app for recording emotional responses. A broader consideration of the emotional effect of mapping is also beginning to be enacted, for example in a recent participatory GI project in the Peruvian Amazon which argues that researchers should begin engaging in more effective and emotional thinking when constructing their research methodologies, to improve results and mitigate potential problems (Young and Gilmore, 2013).

Ethnographic and auto-ethnographic work on mapping practices and real world deployment of mapping has also increased. Rossetto (2013) for example demonstrates how photography can be used to explore how people interact with in-situ permanent map displays in the city of Istanbul, whilst Grasseni (2012) focuses upon the map-maker's role in the construction of a community mapping project in an Italian alpine valley. Techniques such as deploying video to record bodily engagement with mapping on mobile devices are being used to capture different temporalities of real world mapping (see Verhoeff, 2015).

So being critical means attending as well to the mix of methods being deployed in writing and doing mapping: research outcomes depend upon the mix and sequence of strategies deployed to apprehend the world.

## Conclusions

Instead of focusing on technologies of map production, or the aesthetics of design, or the cognitive processes deployed in map use, critical cartography potentially broadens the scope of understanding mapping, taking on board wider social concerns and re-integrating mapping with wider and more social concerns. It offers a useful framework for ongoing cartographic research and pedagogy, as well as a means of informing mapping practice, which potentially allows the formerly separate terrains of mapping theory and mapping practice to be approached from newly hybrid and unified approaches. In the past, the worlds of social theorists and practitioners were frequently seen as separate domains (see Perkins, 2003). Critical cartography makes important claims towards bringing these separate worlds together in a unified praxis (see, for example, Parker, 2006; Firth, 2014; and Krygier, 2002).

These approaches are not set in stone. In 2009, in the introduction to *Rethinking Maps* we concluded that:

> [t]he theories of mapping consist of a set of winding and contested journeys through philosophical and practical terrains. These journeys are far from over and the philosophical underpinnings of maps remain a fertile ground in which to explore issues of space, representation and praxis.
>
> *(Kitchin et al., 2009: 23)*

Critical cartography continues to inform how a changing discipline might approach its concerns, and helps to understand the burgeoning worlds of subversive and protest mapping, indigenous cartographies, everyday socially networked vernacular mapping, OpenSource geo-tagging, and artistic mapping as well as less radical multi-national and state-led mapping initiatives. It puts people back into cartography and makes the map part of life. Taken together, these are persuasive arguments for being more critical!

## Acknowledgements

I would like to thank the Charting the Digital Team for ongoing productive discussions about critical cartography. Research reported here is part funded by the European Research Council under the European Community's Seventh Framework Programme (FP7/2007-2013) / ERC Grant agreement no. 283464, and by the European Commission under the Erasmus + Key Action 2 Strategic Partnership funding framework, grant number 2014-1-UK01-KA203-001642.

## References

Anonymous and Rufat, S. (2015) "Open Data, Political Crisis and Guerrilla Cartography" *ACME: An International E-Journal for Critical Geographies* 14 (1) pp.260–282.

Bhagat, A. and Mogel, L. (Eds) (2008) *An Atlas of Radical Cartography* Los Angeles, CA: Journal of Aesthetics and Protest Press.

Brown, B. and Laurier, E. (2005) "Maps and Journeys: An Ethno-Methodological Investigation" *Cartographica* 40 (3) pp.17–33.

Bryan, J. and Wood, D. (2015) *Weaponizing Maps: Indigenous Peoples and Counterinsurgency in the Americas* New York: Guilford Publications.

Caquard, S. (2013) "Cartography I: Mapping Narrative Cartography" *Progress in Human Geography* 37 (1) pp.135–144.

Caquard, S. and Taylor, D.R.F. (2009) "What Is Cinematic Cartography?" *The Cartographic Journal* 46 (1) pp.5–8.

Cosgrove, D. (1999) *Mappings* London: Reaktion Books.

Cosgrove, D. (2005) "Maps, Mapping, Modernity: Art and Cartography in the Twentieth Century" *Imago Mundi* 57 (1) pp.35–54.

Counter-Cartographies Collective, Dalton, C. and Mason-Deese, E. (2012) "Counter (Mapping) Actions: Mapping as Militant Research" *ACME: An International E-Journal for Critical Geographies* 11 (3) pp.439–466.

Crampton, J.W. (2002) "Thinking Philosophically in Cartography: Toward a Critical Politics of Mapping" *Cartographic Perspectives* 41 pp.4–23.

Crampton, J.W. (2011) *Mapping: A Critical Introduction to Cartography and GIS* Chichester, UK: John Wiley & Sons.

Crampton, J.W. and Krygier, J. (2006) "An Introduction to Critical Cartography" *ACME: An International E-Journal for Critical Geographies* 4 (1) pp.11–33.

Del Casino, V.J. and Hanna, S.P. (2005) "Beyond the 'Binaries': A Methodological Intervention for Interrogating Maps as Representational Practices" *ACME: An International E-Journal for Critical Geographies* 4 (1) pp.34–56.

Denil, M. (2011) "A Search for a Radical Cartography" *Cartographic Perspectives* 68 pp.7–28.

Dodge, M. and Perkins, P. (2008) "Reclaiming the Map: British Geography and Ambivalent Cartographic Practice" *Environment and Planning A* 40 (6) pp.1271.

Dodge, M., Perkins, C. and Kitchin, R. (2009) "Mapping Modes, Methods and Moments" in Dodge, M., Kitchin, R. and Perkins, C. (Eds) *Rethinking Maps* London: Routledge, pp. 220–243.

Edney, M.H. (1993) "'Cartography without Progress': Reinterpreting the Nature and Historical Development of Mapmaking" *Cartographica* 30 (2,3) pp.54–68.

Farman, J. (2010) "Mapping the Digital Empire: Google Earth and the Process of Postmodern Cartography" *New Media & Society* 12 (6) pp.869–888.

Fayne, J.V., Fuhrmann, S., Rice, M.T. and Rice, R.M. (2015) "Exploring Alternative Map Products to Enhance Transportation Option Awareness" *Cartography and Geographic Information Science* 42 (4) pp.1–13.

Firth, R. (2014) "Critical Cartography as Anarchist Pedagogy? Ideas for Praxis Inspired by the 56a Infoshop Map Archive" *Interface: A Journal for and about Social Movements* 16 (1) pp.156–184.

Firth, R. (2015. "Critical Cartography" *The Occupied Times* Available at: *http://theoccupiedtimes. org/?p=13771* (Accessed: 23 April 2015).

Foster, R. (2015) *Mapping European Empire: Tabulae Imperii Europaei* London: Routledge.

Gartner, G. (2012) "Putting Emotions in Maps: The Wayfinding Example" *New Zealand: New Zealand Cartographic Society.* Available at: *www.mountaincartography.org/publications/papers/papers_taurewa_12/ papers/mcw2012_sec3_ch08_p061-065_gartner.pdf* (Accessed: 27 October 2016).

Gekker, A., Hind, S., Lammes, S., Perkins, C. and Wilmott, C. (Eds) (forthcoming a) *Time Travellers: Temporality and Digital Mapping* Manchester, UK: Manchester University Press.

Gekker, A., Hind, S., Lammes, S., Perkins, C., Verhoeff, N. and Wilmott, C. (forthcoming b) "Footage: A Collaborative Experiment About Temporality and Mapping in Oxford" *Mobilities.*

Gerlach, J. (2013) "Lines, Contours and Legends Coordinates for Vernacular Mapping" *Progress in Human Geography* 38 (1) pp.22–39.

Glasze, G. and Perkins, C. (2015) "Social and Political Dimensions of the OpenStreetMap Project: Towards a Critical Geographical Research Agenda" in Arsanjani, J.J., Zipf, A., Mooney, P. and Helbich, M. (Eds) *OpenStreetMap in GIScience* Lecture Notes in Geoinformation and Cartography Berlin: Springer International Publishing, pp.143–166.

Grasseni, C. (2012) "Community Mapping as Auto-Ethno-Cartography" in Pink, S. (Ed.) *Advances in Visual Methodology* London: Sage Publications, pp.97–112.

Haklay, M. (2013) "Neogeography and the Delusion of Democratisation" *Environment and Planning A* 45 (1) pp.55–69.

Harley, J.B. (1988) "Silences and Secrecy: The Hidden Agenda of Cartography in Early Modern Europe" *Imago Mundi* 40 (1) pp.57–76.

Harley, J.B. (1989) "Deconstructing the Map" *Cartographica* 26 (2) pp.1–20.

Harmon, K. and Clemans, G. (2009) *The Map as Art: Contemporary Artists Explore Cartography* Princeton, NJ: Princeton Architectural Press.

Hind, S. (2015) "Maps, Kettles and Inflatable Cobblestones: The Art of Playful Disruption in the City" *Media Fields Journal* 11.

Hind, S. and Lammes, S. (2015) "Digital Mapping as Double-Tap: Cartographic Modes, Calculations and Failures" *Global Discourse* 6 (1,2) pp.79–97.

Ignazio, C.D. (2009) "Art and Cartography" in Kitchin, R. and Thrift, N. (Eds) *International Encyclopedia of Human Geography* Oxford, UK: Elsevier, pp.190–206.

Ingold, T. (2000) *The Perception of the Environment: Essays on Livelihood, Dwelling and Skill* London: Routledge.

Johnson, Z.F. (2010) "Wild Bill Bunge" Available at: *http://indiemaps.com/blog/2010/03/wild-bill-bunge/* (Accessed: 12 April 2013).

Kitchin, R. and Dodge, M. (2007) "Rethinking Maps" *Progress in Human Geography* 31 (3) pp.331–344.

Kitchin, R., Perkins, C. and Dodge, M. (2009) "Thinking about Maps" in Dodge, M., Kitchin, R. and Perkins, C. (Eds) *Rethinking Maps: New Frontiers of Cartographic Theory* London: Routledge, pp.1–25.

Krygier, J.B. (2002) "A *Praxis* of Public Participation GIS and Visualization" in Craig, W. J., Harris, F. and Weiner, D. (Eds) *Community Participation and Geographic Information Systems* New York: Taylor and Francis, pp.330–345.

Lammes, S. (2008) "Spatial Regimes of the Digital Playground Cultural Functions of Spatial Practices in Computer Games" *Space and Culture* 11 (3) pp.260–272.

Latour, B. (1986) "Visualization and Cognition" *Knowledge and Society* 6 pp.1–40.

Leszczynski, A. (2012) "Situating the Geoweb in Political Economy" *Progress in Human Geography* 36 (1) pp.72–89.

MacDonald, G. (2014) "Bodies Moving and Being Moved: Mapping Affect in Christian Nold's Bio Mapping" *Somatechnics* 4 (1) pp.108–132.

McKinnon, I. (2011) "Expanding Cartographic Practices in the Social Sciences" in Margolis, E. and Pauwels, L. (Eds) *The Sage Handbook of Visual Research Methods* London: Sage Publications, pp.452–473.

Moore, A.W. and Perdue, N.A. (2014) "Imagining a Critical Geopolitical Cartography" *Geography Compass* 8 (12) pp.892–901.

November, V., Camacho-Hübner, E. and Latour, B. (2010) "Entering a Risky Territory: Space in the Age of Digital Navigation" *Environment and Planning D: Society and Space* 28 pp.581–599.

Parker, B. (2006) "Constructing Community Through Maps? Power and *Praxis* in Community Mapping" *The Professional Geographer* 58 (4) pp.470–484.

Peluso, N.L. (1995) "Whose Woods Are These? Counter-Mapping Forest Territories in Kalimantan, Indonesia" *Antipode* 27 (4) pp.383–406.

Perkins, C. (2003) "Cartography: Mapping Theory" *Progress in Human Geography* 27 (3) pp.341–351

Perkins, C. (2007) "Community Mapping" *The Cartographic Journal* 44 (2) pp.127–137.

Perkins, C. (2009) "Playing with Maps" in Dodge, M., Kitchin, R. and Perkins, C. (Eds) *Rethinking Maps: New Frontiers of Cartographic Theory* London: Routledge, pp.167–188.

Perkins, C. (2014) "Plotting Practices and Politics: (Im)Mutable Narratives in OpenStreetMap" *Transactions of the Institute of British Geographers* 39 (2) pp.304–317.

Perkins, C. and Dodge, M. (2009) "Satellite Imagery and the Spectacle of Secret Spaces" *Geoforum* 40 (4) pp.546–560.

Pickles, J. (Ed.) (1995) *Ground Truth: The Social Implications of Geographic Information Systems* New York: Guilford Press.

Pickles, J. (2004) *A History of Spaces: Cartographic Reason, Mapping, and the Geo-Coded World* London: Routledge.

Pinder, D. (1996) "Subverting Cartography: The Situationists and Maps of the City" *Environment and Planning A* 28 pp.405–427.

Plantin, J-C. (2014) *Participatory Mapping: New Data, New Cartography* Chichester, UK and New York: John Wiley & Sons.

Roberts, L. (2012) "Cinematic Cartography: Projecting Place through Film" in Roberts, L. (Ed.) *Mapping Cultures: Place, Practice, Performance* London: Palgrave Macmillan, pp.68–84.

Rød, J.K., Ormeling, F. and van Elzakker, C.P.J.M. (2001) "An Agenda for Democratising Cartographic Visualisation" *Norsk Geografisk Tidsskrift – Norwegian Journal of Geography* 55 (1) pp.38–41.

Rose, G. (2012) *Visual Methodologies: An Introduction to Researching with Visual Materials* London: Sage Publications.

Rose-Redwood, R. (Ed.) (2015) "Deconstructing the Map 25 Years On: Theme Issue" *Cartographica* 50 (1) pp.1–8.

Rossetto, T. (2013) "Mapscapes on the Urban Surface: Notes in the Form of a Photo Essay (Istanbul, 2010)" *Cartographica* 48 (4) pp.309–324.

Thompson, N., Paglen, T. and Independent Curators International (2008) *Experimental Geography: Radical Approaches to Landscape, Cartography, and Urbanism* New York: Melville House.

Verhoeff, N. (2015) "Footage: Action Cam Shorts as Cartographic Captures of Time" *Empedocles: European Journal for the Philosophy of Communication* 5 (1) pp.103–109.

Vujakovic, P. and Matthews, M.H. (1994) "Contorted, Folded, Torn: Environmental Values, Cartographic Representation and the Politics of Disability" *Disability and Society* 9 (3) pp.359–374.

Warf, B. and Sui, D. (2010) "From GIS to Neogeography: Ontological Implications and Theories of Truth" *Annals of GIS* 16 (4) pp.197–209.

Wood, D. (1992) *The Power of Maps* New York: The Guilford Press.

Wood, D. (2006) "Map Art" *Cartographic Perspectives* 53 pp.5–14.

Wood, D. (2010) *Rethinking the Power of Maps* New York: Guilford Press.

Wood, D. and Fels, J. (1986) "Designs on Signs: Myth and Meaning in Maps" *Cartographica* 23 (3) pp.54–103.

Wood, D. and Krygier, J. (2009a) "Cartography: Critical Cartography" in Kitchin, R. and Thrift, N. (Eds) *International Encyclopedia of Human Geography* Oxford, UK: Elsevier, pp.345–357.

Wood, D. and Krygier, J. (2009b) "Maps and Protest" in Kitchin, R. and Thrift, N. (Eds) *International Encyclopedia of Human Geography* Oxford, UK: Elsevier, pp.436–441.

Young, J.C. and Gilmore, M.P. (2013) "The Spatial Politics of Affect and Emotion in Participatory GIS" *Annals of the Association of American Geographers* 103 (4) pp.808–823.

# 7

# Mapping as performance

*Joe Gerlach*

## Introduction

To think 'mapping as performance' seems, at first blush, to be a tongue-in-cheek nod to cartography's occasional theatricality. If not theatricality, it is a pointed reminder of mappings' veritable dramas played out in a range of disparate registers: navigational, violent and mundane. Specifically, the uptake of 'performance' here hinges on the term's privileged status across the humanities and social sciences of late. As per the cognate term, 'performativity', performance is a conceptually modish notion that in turn has, alongside a welter of post-structuralist impulses, re-enchanted the study of cartography and mapping. Importantly, the idea of performance broadens the theoretical space in which to examine the nature of maps and mappings.

At the same time, a turn to performance also puts into sharp relief the question of 'what counts' as mapping. By way of a brief and none-too-surprising initial response, what passes today as mapping has multiplied exponentially. Indeed, it is no longer controversial to suggest that mapping goes beyond the lines, contours and legends of the paper map, leaking instead throughout many different genres of life: artistic, playful and experimental. Yet this is not to stake a claim to a particular newness or novelty in contemporary forms of mapping. On the contrary, I hope to demonstrate in what follows that mapping has always been about performance and, by extension, has always been resolutely performative, no matter the era or location with which one is concerned. Do not be under the illusion, however, that any universalizing trends or firm theoretical framework can be crafted from thinking of mapping as performance. Again, in stark opposition to such possibilities, the point of turning to performance is precisely to push against those meta-narratives of cartography which claim all too easily that maps and mappings can be deciphered by some sleight-of-hand deconstructionist or semiotic analysis. If only it were that simple! By re-thinking mapping as performance, it becomes readily apparent that mapping extends beyond the realms of Euclidean and state-led cartography. Mapping as performance is, to this end, concerned with animating the world in an infinite number of ways. At the same time, it is to also think of mapping as the ultimate expression of geography.

## Three minor cartographic performances

Of all the performances that one associates with cartography, it is perhaps the moment of 'getting lost' which infuriates the way-finder the most, despite the ostensible 'orienteering' credentials invested in a map. Subsequent face-saving claims that 'the map is out of date', or that the 'satellite-navigation isn't working properly', demonstrate both a societal obsession with Euclidean decorum and also a more prosaic point about the nature of mapping; that it generates more than mere representational outcomes of direction, route and geodesic location. Heightened blood pressure aside, it cultivates dispositions of other kinds too: orientation, disorientation, relief, mirth, bewilderment, consternation and encounter. Consider the following three vignettes.

First, the twentieth-century novelist George Orwell (1938: 38) encounters a cartographically challenged militia combatant during the Spanish Civil War in *Homage to Catalonia*:

> He was a tough-looking youth of twenty-five or six, with reddish-yellow hair and powerful shoulders. His peaked leather cap was pulled fiercely over one eye. He was standing in profile to me, his chin on his breast, gazing with a puzzled frown at a map which one of the officers had open on the table. Something in his face deeply moved me. It was the face of a man who would commit murder and throw away his life for a friend-the kind of face you would expect in an Anarchist, though as likely as not he was a Communist. There were both candour and ferocity in it; also the pathetic reverence that illiterate people have for their supposed superiors. Obviously he could not make head or tail of the map; obviously he regarded map-reading as a stupendous intellectual feat. I hardly know why.

Second, the naturalist, cartographer and proto-geographer Alexander von Humboldt (1814) reflects on the ineffable draw of maps, in his *Personal Narrative of Travels to the Equinoctial Regions of the New Continent during the Years 1799–1804* (see also Figure 7.1):

> When we begin to fix our eyes on geographical maps, and read the narratives of navigators, we feel for certain countries and climates a sort of predilection, for which we know not how to account at a more advanced period of life. These impressions, however, exercise a considerable influence over our determinations; and from a sort of instinct we endeavour to connect ourselves with objects, on which the mind has long been fixed as by a secret charm.

Finally, artist Stephen Walter (2012) explores his motivations in creating the infamous piece of map art entitled 'The Island', a critical re-working of London's traditional cartographic rendering (Figure 7.2):

> I wanted to express my fascination with the city … its contradictions, its intricacy, its mass of people all living their separate and sometimes communal lives.
>
> 'The Isles of Slough; trading estate'.
>
> I made London into an island for a number of reasons. Britain is an island; it is a collection of islands. 'It' informs our national psyche. It's a wry joke on London's self-importance.
>
> 'The Isle of Woking; dormitory island town'.
>
> It's an image of both order and chaos; and at times, an image of where the order descends into chaos. Making it I was trying to say how bewildering it can be sometimes to live in the place.

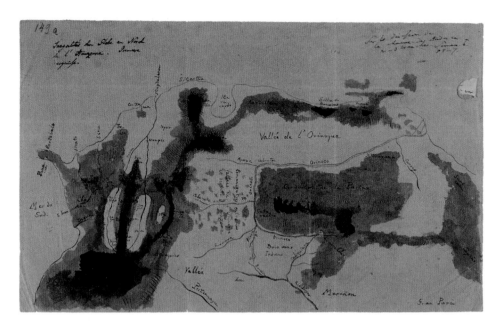

*Figure 7.1* Sketch Map of Orinoco Valley by Alexander van Humboldt. Travel Diary VIIbb/c, Berlin State Library

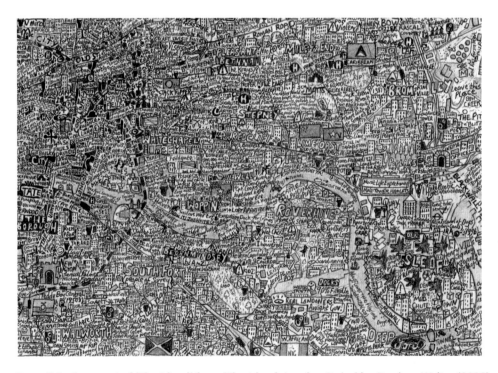

*Figure 7.2* Fragment of 'The Island' from 'The Island: London Series' by Stephen Walter (2008)

What might we draw from these minor cartographic moments in which everything from bewilderment to wonderment is expressed? Working across the three space-times, it is apparent that mapping, as a performance, cuts across a range of expressive and intellectual genres: literary, scientific and artistic, to mention nothing (yet) of the political. Indeed, such is the cross-contamination of maps and art in the burgeoning field of 'map art' that a debate rages as to whether a map actually counts as a map in instances where it has been rendered artistically, or animated in any kind of avant-garde manner. Whatever your opinion might be on this vexatious matter, what is certain is the near impossibility of attempting to police these disciplinary boundaries with any measure of success. In short ironical form, cartography, through its myriad performances and politics, is not and cannot be bound by 'abysmal' cartographic logics of border, frontier and containerization (Olsson, 2007). Mapping, then, is not the sole preserve of mappers, echoing the critical cartographer Brian Harley's (1989: 1) knowing remark that, 'it is better for us to begin from the premise that cartography is seldom what cartographers say it is'. Instead, in the broadest sense of the term, mapping lends itself as a generic device and diffuse set of performances that lend themselves to appropriation and deployment by an infinite array of actors for an equally infinite array of uses. Moreover, the quite distinct provenance of each of these vignettes also illustrates the indelible, prominent and enduring role that mapping commands past and present; a pre-eminence amongst other forms of earthly visualization that shows no sign of diminishing. On the contrary, what we are witnessing and sensing in the early decades of the twenty-first century is something of a renaissance in mapping, particularly at an everyday register of existence (Thrift, 2012).

It is on this level of the quotidian that mapping as performance and likewise the performance of mapping assumes the position of a naturalized, taken-for-granted, elemental aspect of historical and contemporary life. In the first of the aforementioned vignettes, this normalizing tendency manifests itself through Orwell's dismissive snobbery towards the militiaman's inability to read the map, as if cartographic proficiency is a mainstay of a similarly basic education. The patronizing tone struck by Orwell spotlights the saturation of maps and mappings into the capillaries of human societies, particularly in Euro-Anglo-American bodies, norms, institutions and, concomitantly, colonial ventures. Controversially, whilst it is always dangerous and spurious to speak of maps as having a determinate effect on the formation of identity, this 'carto-saturation', is according to Stephen Walter in the latter vignette, an important facet in the generation of a collective psyche. The extensive scholarship in critical cartography provides an excellent account of why this might be the case, not least owing to the political economy of cartography, namely the manner in which cartography has helped perform-into-being that self-same political economy; one which has since cemented the Westphalian accord of nation-states as the totemic unit of governance for the last 400 or so years (see key texts on critical cartography, such as Crampton and Krygier, 2006; Wood *et al.*, 2010). This of course is the same cartographically delimited unit of governance that would trigger a litany of conflicts, including the one in which Orwell and his cartographic snobbery would later find themselves embroiled.

It is not, however, simply the political economy of cartography with which the notion of performance can be invoked. The vignettes drawn from Alexander von Humboldt and Stephen Walter, respectively, demonstrate something of the romanticized performance of maps. There is even something mildly unseemly about the way in which mapping can perform; where chaos, for Walter, can be traced out of supposed order, even if it is only a speculation, or a feeling on the part of the artist themself. Nonetheless, the map performs this critical speculation in the act of its creation, and obviously, the act of mapping is here intended, explicitly, to be a performance and one with public exposure. Similarly, however saccharine we might accuse Humboldt of being in his paean to a cartographic instinctiveness – or even to a cartographic desire – he also points unstintingly to the performativity of mapping and cartography in crafting certain dispositions. In short, what is being recounted here, and to reiterate an increasingly well-rehearsed refrain in cartographic theory, the map does not simply represent reality by way of a faithful simulacra, it generates reality too. Or to be more precise, maps perform realities. A prolific cartographer and geographer, Humboldt seemed cognisant, at least in his personal diaries and correspondence, of the non-representational performance of mapping, pre-empting later post-structuralist contours of cartographic thinking at the turn of the twenty-first century in which a specific concern for performance is prominent.

To be sure, mapping for Humboldt was not solely a mechanism for contemplative, conceptual meandering. Quite the opposite. It is of course in the classic 'performance' of mapping, or put more tersely the 'power' of mapping, that Humboldt also recognized the genesis and sustenance of statecraft and the wielding of sovereign power. Take, for example, Humboldt's influential mapping of the Orinoco river basin which would in turn lay the cartographic foundations for the insurgency and subsequent political experiments of the Andean Latin American liberator, Simon Bolivar. By all accounts, Bolivar and Humboldt were well acquainted with one another, if not indeed friends. Mapping, in this instance, performed the role of networker as the Venezuelan general became increasingly seduced by the visualizations the German scientist had to offer of a future proto-Andean state. 'Mapping as performance' is not, it would seem in hindsight, quite as benign as it first sounds. This brings us to a final reflection, alongside these vignettes, upon the way in which mapping is a performance, or the manner in which mapping performs. In each of the three cases, the maps, or the implication of mapping, are performing to different ends. Orwell's tale is a reminder of mappings' uncomfortably close association with belligerence and violence; of maps as the non-human field marshals in arenas of war. Humboldt, on the other hand, strikes the romanticized chord of cartography; the soft, subtle performative snare of mapping upon human consciousness and sensibility. For Stephen Walter, the artistic intervention of a mapping performance is a form of pushing against something, a spotlighting of a particular concern, issue or anxiety, and indeed sometimes bringing that very issue into existence or into the domain of public awareness. Such mapping chimes with the long tradition of counter-mapping and subaltern cartographies across the planet; fields of action that singularly recognize the performativity of mapping and hence the notion of mapping as performance. In summary, all these invocations of mapping perform in one way or another. The idea of 'mapping as performance' does not lend itself to policing as to what counts as performance in disciplinary terms – all maps perform, whether one explicitly recognizes that contention or not. Once this performance is acknowledged, do not expect matters to become clearer. On the contrary, as the next section explores some of the theoretical matter allied with thinking about performance, only further disorientation can be guaranteed.

## On performance and performativity

The picture thus far is that maps *work*; they labour, they create, they perform. Likewise, the chapter has also pointed to the feasibility of regarding mapping as a performance. Yet how do maps and mappings perform? Here I wish to pause and consider the notion of 'performance' on its own terms, lest it be deployed lazily as a byword for action and power. Instead, how might the idea of performance be used affirmatively in the context of mapping and cartography? Performance has become a leitmotif of the social sciences and humanities of late, propelled by a growing concern for embodiment, habit and movement. In the specific field of human geography, for example, a renewed focus on performance has enabled geographers to consider and examine how the world comes into being; a quite different ontological proposition to previously tracing the world as if it were a stable, static entity. That is not to say that such approaches are universally popular amongst social scientists or social theorists and likewise; the notion of 'performance' does not serve as a stable theoretical framework of analysis. Indeed, the term performance is shared by a range of diverse conceptual and disciplinary currents, notably those preoccupied with performance-studies, non-representational theories, performativity and the dramaturgical (see Thrift, 2007; Anderson and Harrison, 2010; Gerlach, 2014). Across these subfields, a concern for performance has meant an exploration of subjectivity, expression and power in a manner that does not rely upon the application of a priori categories of existence such as 'identity'. Instead a focus on performance is necessarily a focus on practice; a focus on how things, objects, ideas, spaces come into being. To that end, it is conceptual experiments in post-structuralism and post-humanism with which the totem of performance is most associated. Of substantive salience to cartographical studies has been the invocation of ideas surrounding performativity (Crampton, 2009) and non-representational understandings of performance.

Take the pervasive concept of 'performativity', for example. This is a notion associated primarily with philosopher Judith Butler's (1997) diagramming of performativity in the context of the production and sustenance of everyday identities, especially in respect of gender. According to Butler, such identarian labels are anything but natural or pre-ordained by means of biology. No, suggests Butler, such claims are spurious and serve to uphold a dubious 'natural order'. Instead, the profoundly enmeshed propagation of gender comes about through daily performances of what counts supposedly as female and what counts as male. According to Butler, it is through mundane, quotidian reiterations and repetitions of certain practices that particular ideas, norms and politics become embedded. With every re-iterative performance of an identity, so that identity becomes ever more entrenched; hence 'performativity', the notion that identities and meanings become self-fulfilling in their existence and effect through their very performance. Butler's notion of performativity, therefore, is a useful way for thinking about the way in which maps and mappings perform in a way that does not make recourse to suspect accounts of 'power'. That is not to abrogate responsibility from mappings as having 'powerful' effects in the broadest sense of the term, but it is to think beyond the limited account that power provides in terms of cartography in which power either resides in the subject of the map-maker, or in the object of the map. Mapping as performance, then, works through performativity; not necessarily through grand narratives or belligerent action, but often through quite subtle repetitions and recapitulations. Think, perhaps, of maps as performative 'refrains', reiterations that aid the materialization of ideas, norms and societies. The quite unremarkable action of adhering Mercator global projections in school classrooms, for example, is in fact a profound geopolitical performance, one which in part disciplines the minds and bodies of those that care to glance at it. The power of the map here does not lie in the map itself, nor even its subtext, but just through its ability to be reproduced, reiterated and reperformed. Yet the performance does not end in just *looking* at a map; there is a broader set of encounters at work which does not guarantee a singular meaning, interpretation or consequence of a map's production and existence.

Mapping as performance involves more than a reliance upon visual acuity, hence why a turn to non-representational theories and approaches can be useful in re-thinking the transformational capacities of cartography. Such approaches harness the mosaic assemblage of non-representational theory (Thrift, 2007; Anderson and Harrison, 2010); an ethos that has exercised the field of cultural geography of late. In brief, non-representational theory, despite the nomenclature, is not a negation of representations but a critique of representational thought in which meaning and identity are fixed as all too stable categories of analysis. Instead, non-representational thinking (in part drawing on Butlerian notions of performativity), is interested in the performances that constitute the on-flow and becoming of everyday life. This requires exploring those performances that do not always reach the level of mental cognition; somatic performances for example. It also demands a concern for those materials that are not always visual: namely affect and the virtual. Similarly, non-representational thinking, drawing here on actor-network theory, displays a curiosity for the agency of non-humans and generative space-times of creativity. Performance, in this context, is not necessarily about the spectacle; indeed, it might even be about disappearance (McCormack, 2014). A non-representational take on performance and cartography matters for three reasons.

First amongst these reasons rests the idea of 'encounter'. Far from the map transmitting meaning, power or politics in a singular, one-way fashion, maps and mappings work instead through unfolding encounters with a range of different actors. This is best expounded by Rob Kitchin and Martin Dodge (2007) in a seminal article which contends that a map is not an onto-logically secure repository of information and meaning, but an ontogenetic (emergent) process in which meanings, ideas, behaviours and politics constantly come into being through encounter between maps and mappers. Recall for a moment, the mundane frustration of getting lost, even with the aid of a map or sat-nav device. It would be churlish to assume that the artefact of the map itself is solely responsible for one's disorientation, likewise the blame cannot be solely attributed to 'human nature'. Getting lost is an unfolding performance between map and user, one not given in advance, but charged by encounter, affect and bodily capacity.

Second amongst these reasons is a spotlight on the non-human. Mapping as performance cannot be figured without a consideration of the implication and potential agency of non-human actors. Ancient or contemporary, mapping as a performance has of course always relied upon the non-human; material and instrumental; paper, protractors, satellites and GPS devices to name but a few of such things. Whilst it seems immediately unremarkable to focus on the non-human, what matters is how the non-human *intervenes* in cartography and thereby how mapping performs. Strangely enough, we are already probably all too aware of how the non-human intervenes, given the role of the map itself; a non-human artefact or performance that has material and immaterial consequences! Importantly, what matters in this context is how the individualized components of mapping as performance act and perform themselves. For example, a decision as to what scale or resolution to render a map is a micropolitical one, or the seemingly consumer-centric choice to buy a GPS device is, unwittingly or otherwise, a geopolitical investment in the circuitry of the United States' Department of Defense. The non-human, in short, matters.

Third and finally, a non-representational take on performance also necessitates the question of who or what can participate in the performance of mapping. The previous point on non-humans adds to this conundrum, whilst also raising the ethico-political stakes of cartography as a performance. The idea and practice of 'participatory mapping' is undoubtedly popular, yet what it means to actually participate in this performance remains steadfastly perplexing. From a non-representational perspective, mere enumeration on a mapping project is not enough; participation through a weak sense of democratic numbers rather limits a political account of mapping because it says nothing of the performances involved. A non-representational anima-tion of mapping as performance would therefore seek to understand how creative practices can be simultaneously geopolitical and, moreover, how one might consider mapping as an *assemblage* of unfolding actors and desires, rather than as a rather stale accord between subject and object: the mapper and the mapped. Crucially what matters is how does a map participate itself?

How might a map intervene in the constitution of worlds: material and affective? Taking a cue from this brief dwelling on performance, I want in the following final section to consider the politics and ethics of thinking of 'mapping as performance'.

## Earthly expressions

Having now considered performance in avowedly conceptual terms, it is worth reminding ourselves of the sheer volume of mapping performances that are both currently being enacted and those that have stood before today. Boria (2013) suggests the relationship between geographers and maps is in crisis. Maybe so, but the uptake of mapping performances as a form of political articulation is in rude health. It might be tempting to regard this public enchantment with mapping as somehow recent, yet as an enduring form of earthly expression, it is anything but. Why is this the case? It is worth quoting an oft-cited passage from the philosopher Gilles Deleuze and psychoanalyst Felix Guattari (Deleuze and Guattari, 2004: 13):

> The map does not reproduce an unconscious closed in upon itself: it constructs the unconscious. It fosters connections between fields, the removal of blockage of bodies without organs onto a plane of consistency. It is itself a part of the rhizome. The map is open and connectable in all of its dimensions; it is detachable, reversible, susceptible to constant modification. It can be torn, reversed, adapted to any kind of mounting, reworked by an individual, group, or social formation. It can be drawn on a wall, conceived as a work of art, constructed as a political action or as a meditation … a map has multiple entryways.

Recall Stephen Walter's implication that maps help generate 'psyche'. Deleuze's and Guattari's contention is not too dissimilar; that maps, figuratively, produce an unconscious. They form assemblages and are themselves part of wider assemblages, or what Deleuze and Guattari term 'rhizomes'. Then arrives the second half of the quote, which could easily stand as a manifesto for activist mapping. It is precisely because of the mappings' radical mutability (Perkins, 2012) and susceptibility to modification that makes mapping as performance such a seductive political device, or 'dispositif' to borrow from Michel Foucault. 'It can be drawn on a wall, conceived as a work of art, constructed as a political action'; add to that, it can be danced, walked, performed, ignored, applauded, derided, maligned, cheered, redrawn. In short, the ontological insecurity of maps and mappings is arguably a mapping performance's most politically enabling (or dangerous) characteristic. This characteristic has been deployed in a number of ways, for example in the performances caught up in map art (Harmon, 2004; kanarinka, 2006; Bhagat and Mogel, 2007; Paglen, 2007), community mapping (Lin and Ghose, 2008), counter cartographies (Nietschmann, 1995; Peluso, 1995; Cred, 2007; Sletto, 2009; Wainwright and Bryan, 2009), narrative mapping (Pearce, 2008; Caquard, 2013) and humorous maps (Caquard and Dormann, 2008) (Figure 7.3). Elsewhere, the emergence of the crowd-sourced mapping project, OpenStreetMap is a leading example of participatory mapping that harnesses processual experience as the basis for crafting maps (Gerlach, 2015) (Figure 7.4).

Given the disparate array of mapping performances in existence, it is both impossible and undesirable to generalize with regard to the politics and ethics of such earthly expressions. Categorical labels such as counter-mapping and indigenous mapping do not always work, not least because much like a map itself, such labels cannot hope to contain or speak for the divergent and multiple interests at work in a mapping performance. Yet whilst the individualized politics of mapping performances might be diverse, there is something that can be enunciated more generally about the political and the ethical in relation to these performances. If one takes the political to be that which encourages a space for dissensus (Barry, 2002), then mapping performances have the latent potential to generate such productive spaces. Yet if one figures mapping performances as expressing a singular politics, or a singular meaning, then the vitalist and agitant potential of mappings is lost. Taking a cue from the philosopher Brian Massumi (2011), just because a map depicts political content does not automatically make it political in its expression or vocation. Sloganeering is not enough. What makes a map, and by extension, a mapping performance political is its capacity to be modified, torn and redrawn; or, to be re-imagined.

*Figure 7.3*　A signal example of a counter-cartographic mapping performance, the Disorientation Guide 2 by 3Cs: Counter Cartographies Collective (2009). Creative Commons Licence

*Figure 7.4*　Remote sensing: an OpenStreetMap gathering. Photograph by the author

The ethical injunction is in how to harness the radical mutability of mapping performances without diluting them into tired political statements, or likewise into dubious representational tropes concerning the meaning of certain signs, territories and symbols (Figure 7.3).

In ending, to recognize mapping as performance is to also know what a map is. Without doubt, it is easier to cast definitions as to what a map *is not* as opposed to anything in the affirmative. Nonetheless, the pressing deluge of mapping performances of late provides some impetus to get a handle on what a map, and what a mapping performance can do, if not what a mapping performance is. Consequently, one might draw this chapter to an ending, as ever, in the middle of things; that mapping performances are 'intensive becomings'. That is to make the claim that maps themselves are performances. Maps are mappings. A map emerges through the performance it enacts; sometimes the performance is material, and other times it is immaterial – affective and virtual. Most, if not all of the time, it is both. These performances are as intensive as they are micropolitical, placing demands on bodies that orientate as much they disorientate; that territorialize as much as they de-territorialize. A focus on mapping as performance disabuses us of the caricature of the map as a technology of capture and instead prompts a recognition of the generative, emergent and politically enactive qualities of cartography. The ethico-political injunction to mapping performances then, is to, 'not stop completing, remaking, amassing, redesigning in order to rearrange cartographic criteria in the face of the urgencies of the present' (Pelbart, 2011: 76). It does not stop at the present; one also needs to come to terms with the cartography of the past, as well as to anticipate the futures that mapping performances will at one time bring into immanent being.

## References

3Cs: Counter Cartographies Collective (2009) "Help Us Map the Economic Crisis!" Available at: www.countercartographies.org/category/dg2/ (Accessed: 30 April 2014).

Anderson, B. and Harrison, P. (Eds) (2010) *Taking-Place: Non-Representational Theories and Geography* Farnham, UK: Ashgate.

Barry, A. (2002) "The Anti-Political Economy" *Economy and Society* 31 (2) pp.268–284.

Bhagat, A. and Mogel, L. (2007) *An Atlas of Radical Cartography* Los Angeles, CA: Journal of Aesthetics and Protest Press.

Boria, E. (2013) "Geographers and Maps: A Relationship in Crisis" *L'espace politique* 21 Available at: *http://espacepolitique.revues.org/2802* (Accessed: 30 April 2014).

Butler, J. (1997) *Excitable Speech: A Politics of the Performative* London: Routledge.

Caquard, S. (2013) "Cartography I: Mapping Narrative Cartography" *Progress in Human Geography* 37 (1) pp.135–144.

Caquard, S. and Dormann, C. (2008) "Humorous Maps: Explorations of an Alternative Cartography" *The Cartography and Geographic Society Journal* 35 (1) pp.51–64.

Crampton, J.W. (2009) "Cartography: Performative, Participatory, Political" *Progress in Human Geography* 33 (6) pp.840–848.

Crampton, J.W. and Krygier, J. (2006) "An Introduction to Critical Cartography" *ACME: An International E-Journal for Critical Geographies* 4 (1) pp.11–33.

Cred, C. (2007) "Counter.Cartographies: Notes Towards a Future Atlas" *Parallax* 13 pp.119–131.

Deleuze, G. and Guattari, F. (2004) *A Thousand Plateaus: Capitalism and Schizophrenia* London: Continuum.

Gerlach, J. (2014) "Lines, Contours and Legends: Coordinates for Vernacular Mapping" *Progress in Human Geography* 38 (1) pp.22–39.

Gerlach, J. (2015) "Editing Worlds: Participatory Mapping and a Minor Geopolitics" *Transactions of the Institute of British Geographers* 40 (2) pp.273–286.

Harley, J.B. (1989) "Deconstructing the Map" *Cartographica* 26 (2) pp.1–20.

Harmon, K. (2004) *You Are Here: Personal Geographies and Other Maps of the Imagination* Princeton, NJ: Princeton Architectural Press.

Humboldt, A. (1814) *Personal Narrative of Travels to the Equinoctial Regions of the New Continent, During the Years 1799.* London.

kanarinka (D'Ignazio, C.) (2006) "Art-Machines, Body-Ovens, and Map-Recipes: Entries for a Psychogeographic Dictionary" *Cartographic Perspectives* 53 pp.24–40.

Kitchin, R. and Dodge, M. (2007) "Rethinking Maps" *Progress in Human Geography* 31 (3) pp.331–344.

Lin, W. and Ghose, R. (2008) "Complexities in Sustainable Provision of GIS for Urban Grassroots Organisations" *Cartographica* 43 (1) pp.31–44.

Massumi, B. (2011) *Semblance and Event: Activist Philosophy and the Occurrent Arts* Cambridge, MA: MIT Press.

McCormack, D. (2014) *Refrains for Moving Bodies: Experience and Experiment in Affective Spaces* Durham, NC: Duke University Press.

Nietschmann, B. (1995) "Defending the Miskito Reefs with Maps and GPS" *Cultural Survival Quarterly* 18 (4) pp.34–37.

Olsson, G. (2007) *Abysmal: A Critique of Cartographic Reason* Chicago, IL: University of Chicago Press.

Orwell, G. (1938) *Homage to Catalonia* London: Penguin.

Paglen, T. (2007) "Groom Lake and the Imperial Production of Nowhere" in Gregory, D. and Pred, A. (Eds) *Violent Geographies: Fear, Terror and Political Violence* New York: Routledge, pp.237–254.

Pearce, M. (2008) "Framing the Days: Place and Narrative in Cartography" *Cartography and Geographic Information Science* 35 (1) pp.17–32.

Pelbart, P.P. (2011) "The Deterritorialized Unconcious" in Alliez, E. and Goffey, A. (Eds) *The Guattari Effect* London: Continuum, pp.68–83.

Peluso, N.L. (1995) "Whose Woods Are These? Counter-Mapping Forest Territories in Kalimantan, Indonesia" *Antipode* 27 (4) pp.383–406.

Perkins, C. (2012) "Plotting Practices and Politics: (Im)mutable Narratives in OpenStreetMap" *Transactions of the Institute of British Geographers* 39 (2) pp.304–317.

Sletto, B. (2009) "'Indigenous People Don't Have Boundaries': Reborderings, Fire Management, and Productions of Authenticities in Indigenous Landscapes" *Cultural Geographies* 16 (2) pp.253–277.

Thrift, N. (2007) *Non-Representational Theory: Space, Politics, Affect* Abingdon, UK: Routledge.

Thrift, N. (2012) "The Insubstantial Pageant: Producing an Untoward Land" *Cultural Geographies* 19 (2) pp.141–168.

Wainwright, J. and Bryan, J. (2009) "Cartography, Territory, Property: Postcolonial Reflections on Indigenous Counter-Mapping in Nicaragua and Belize" *Cultural Geographies* 16 (2) pp.153–178.

Walter, S. (2012) *A Picture of London: The Island* BBC1 Television, 4 July.

Wood, D., Fels, J. and Krygier, J. (2010) *Rethinking the Power of Maps* New York: Guilford Press.

# 8

# The map as spectacle

*Peter Vujakovic*

## Introduction: the power to fascinate

Maps have always had the power to fascinate. If we accept that the Neolithic mural discovered at the Çatalhöyük in Anatolia, Turkey, is the world's oldest map, then maps have been invested with drama from the start. A recent study supports earlier interpretations of the image that suggest it represents the town plan of Çatalhöyük and an explosive eruption of the nearby Hasan Dağı volcano (Schmitt *et al.*, 2014). Some early modern maps and atlases are explicit in their references to theatre, from Abraham Ortelius's *Theatrum orbis terrarum* – his 'Theater of the World' (first edition 1570), to Blaeu's world map of 1635 with its flamboyant cartouche that includes allegories of the sun, the moon and the planets, and representations of the seven wonders of the ancient world. The 'map as spectacle' is a recurrent element of the history of cartography and has driven much of the allure of maps for collectors. This chapter, however, explores a deeper understanding of the concept of 'map as spectacle' – as an element of socio-political control.

The critical turn in cartography reopened debate with regard to the cultural and political roles of maps and mapping. The history of cartography was previously founded on a 'progressivist narrative' (see Chapter 5, this volume), defined by the view that cartography was largely a science, driven by technical developments, increasing accuracy and precision, and supposed objectivity. A number of authors have challenged this viewpoint, and there is now a wide acknowledgement that maps should be treated as complex cultural texts and 'practices' (see Azócar Fernandez and Buchroithner (2014) for a detailed discussion of recent paradigm changes in cartography, from Harley's deconstructionist approach to the so-called post-representational paradigm).

This chapter argues that understanding the 'map as spectacle'[1] provides a valuable insight into how maps work and what interests they serve (Wood, 1992). In this interpretation, the 'map as spectacle' is recognized to be a complex 'Baroque' artefact; rich and sometimes flamboyant on the surface but underpinned by refined and continuously developing technologies of survey, design and reproduction. In this sense, it reminds us not to ignore technical development in cartography. Here, Rundstrom's (1991) ideas are helpful. While agreeing with much that Harley had to say concerning the cultural role of maps, Rundstrom was critical of the tendency of so-called 'postmodernist' approaches that place too much emphasis on the discussion of maps as 'texts', while ignoring the wider technical *and* social processes within which they are embedded. This did not lead Rundstrom to reject or replace Harley's agenda, but to add a new dimension to the conceptual tools available to those interested in the social and political dimensions of cartography. He argued for a conceptualization which he refers to as 'process cartography', consisting of two concentric ideas. The first situates the 'map arte-fact' within the realm of technical production, and the second 'places the entire map-making process within the context of intracultural and intercultural dialogues' (Rundstrom, 1991: 6). Hence, process cartography is posited on 'the idea that maps-as-artefacts are inseparable from mapping-as-process, and that the mapping process in turn, is made necessary and meaningful only by the broader context of the cultural processes within which it is located' (Rundstrom, 1993: 21). This approach acknowledges the importance of the understanding the technologi-cal, organizational and design issues involved in map production, but stresses also the need to locate the maps and their makers within their cultural *and* historical contexts. In this chapter, it is argued that Rundstrom's two circuits are relevant to understanding mapping as a Baroque structure that generates spectacle.

It is important to point out that no case is here being made that *all* maps are implicated in 'spectacle' as a form of socio-political control. Powerful agencies have always exerted influence and control in various modes, from persuasion and propaganda to the use of overt force. Thus, what we might characterize as 'base-mapping' – sometimes detailed but unspectacular maps, for instance, cadastral maps, infrastructure maps or aviation charts – have been used to manage the state's monopoly of force and coercive regulation. It is also important to distinguish 'map as spectacle' from the mass of 'mundane maps' (road maps, simple location maps in adverts and so on). It *is* argued, however, that the concept of 'map as spectacle' does apply to a very large component of cartographies circulated by elites: the government and its agencies, 'big business', including the news media, the communications and entertainment industries, and other vested interests. The important and growing role of 'counter-mapping' (see Chapter 6, this volume) and 'tender mapping' (Caquard, 2015) is acknowledged, but remains minuscule in its impact compared to the mass of maps generated by elite groups.

## The Baroque and spectacle

This section introduces the 'Baroque' and 'spectacle' as key concepts for an understanding of socio-political control that can be applied through cartography as a process. The ideas presented here draw primarily on José Antonio Maravall's concept of the Baroque as a 'historic structure' (rather than simply an art movement), and Guy Debord's writings on 'the society of spectacle'. While the two concepts should not be conflated, they do offer related understandings of con-trol. The argument made below is that the better understanding of maps in society is supported by thinking in terms of 'Baroque cartographies' producing maps as 'spectacle'. This builds on Rundstrom's concept of process cartography, but situates it more precisely within a specific structure – the world economy – as understood within a 'world-system's' approach. This frame-work contextualizes (carto)graphic representations within long-term historical processes and a global system through an engagement with the culture of everyday political geographies (see Flint and Taylor, 2011). The world economy is not regarded as a monolithic structure, but one in which elite elements compete. Maps may, for example, be mobilized by some sectors of capital against others; for instance, the 'media' and academia critique of the military-industrial complex's role in the aftermath of the Iraq invasion of 2003, without fundamentally compro-mising the overarching system. The actions and representations of individuals, groups and states are to be understood within the larger structures of the world-system, as attempts to maintain or challenge these structures.

According to Maravall, the Baroque is a phenomenon ('historical structure') that emerged to deal with the crises and contradictions of early modern Europe. In his *Culture of the Baroque* (1986; originally published as *La Cultura del Barroco* in 1975), Maravall makes it clear that he regards the Baroque not simply as a concept of style but as an historical epoch which played a role in the repression of feelings of threat amongst the masses. The Baroque emerged to control populations facing crises of economic and social insecurity, war, a new world marked by growing populations, and the exodus from the countryside to the cities. But even for the elites, there was a tension between tradition and modernization; as Del Valle (2002: 141) notes:

> The Baroque has as one of its main axes the clash between two opposing tendencies: a traditionalist impulse that wanted to maintain the world in its existing state, and another that, in its modernizing and expansionist thrust, found itself linked to such contemporary phenomena as geographic displacement and upward social mobility.

For Maravall, the Baroque project was 'a near-seamless web of control exerted by the supreme absolutist state', in which the elites mobilized the technologies of *modernity* to maintain their *traditional* position. This has been criticized as overstating the elite's ability to impose cohesion 'by dictating culture from above' (Boyden, 1999: 762), but comparison with Debord's ideas suggests that Maravall's ideas are not overstated and have traction in contemporary society.

Responding to these crises, the monarchical-seigniorial elite maintained authority and redirected the potentially disruptive (urban) masses by two means. First, by the use of force, and second, through the captivation of minds through the arts and other displays of power (e.g. fireworks) to achieve a 'reinscription of the entire social sphere under the new authority of the state' (Godzich and Spadaccini, 1986: xviii), a state 'shaken by desires for freedom' (Maravall, 1986: xxxii). Theatre was regarded as particularly important as a means of disseminating collective ideals and creating group consciousness (Sullivan, 2010). Lighting effects and 'stage machinations' were instrumental in creating a sense of awe and wonder (Maravall, 1986).

The use and development of the 'technologies' underpinning spectacle are critical to an understanding of how the Baroque 'worked' and how it can contribute to our understanding of cartography. On the surface, Baroque theatre and other forms were characterized by 'sudden and momentary artistic effects that take spectators by surprise and sweep them away in a rush of excitement' (Harbison, 2010: 47), but the sight of kings and saints descending from heaven was entirely dependent on machinery and other technical effects. Sullivan (2010) suggests that the Baroque theatre was essentially 'the equivalent of TV, the movies, video games and legitimate theatre (all in one)', speaking directly to social tensions within the audience. It is not difficult to see how cartographic spectacles fit within such a project.

The theatricality of fireworks is perhaps the ultimate Baroque mobilization of technology and science for spectacle. As Maravall (1986: 246) notes:

> Fireworks (*fuegos de artificio*) were a very adequate sign of splendor of whoever ordered them because of their very artifice, their difficulty, the expense in human labour and in money that they implied (which the rich and powerful had at their disposal for non-productive purposes).

The genius of the Baroque over preceding periods was the recognition and application of 'the power of representation as a human artefact, and not simply a reflection of nature, [that] opens the way to its conscious political use' (Mohr, 2012: 54). The rejection of models of cartography as mimetic and objective, and the recognition of the social power of maps chimes with this aspect of representation as control. Lollini (1998) notes that Maravall, amongst others writing on the Baroque, recognized the similarities between Baroque culture and twentieth-century culture (see, for example, Barreiro López, 2014, on the use of Baroque in Franco's Spain).

A useful interpretation of the Baroque for its insights for contemporary cartography is Mary Kaldor's (1981) concept of the 'Baroque Arsenal', as applied to the twentieth-century military-industrial complex. For her, 'Baroque' as process involves the 'perpetual improvements that *fall within the established traditions* of the armed services and the armourers' (p.4, emphasis added). Her basic thesis is that 'Baroque technological change represents "improvements" to successive weapons systems which can pass through phases of invention, innovation, and integration without disturbing the social organization of the users' (Kaldor, 1986: 591). It would be easy to replace, in the last sentence, the words 'weapons systems' with 'cartographic modalities' to recognize cartography as a Baroque process.

The product of 'Baroque cartography' is the 'map as spectacle'. The recent use of Google maps and animation in news media presentations of war, for example, is an obvious example of innovation in popular representation of conservative geopolitics. Maps more generally 'enthral' their readers through their (technically controlled) exuberance and spectacle. Here the term spectacle has a special meaning and draws on Debord, who regards spectacle as a means of creating a population of compliant consumers: 'In societies dominated by modern conditions of production, life is presented as an immense accumulation of spectacles. Everything that was directly lived has receded into a representation' (Debord, 2009: 24).

People become so wrapped up in life as spectacle that it 'monopolizes the majority of the time spent outside the production process' (p.25). The 'society of spectacle' is the progeny of state and corporate power maintained through the output of the news media, official institutions, the advertising and entertainment industries. Central to this conception of society is the idea that compliance and alienation are a product of socio-economic organization that achieved its climax in capitalism and advanced technologies, as theorized by Herbert Marcuse.

As Callens (2000: 191) notes, the mass media, and more recently the 'personal computer', have turned from being a means of facilitating the exchange of information to become a 'self-enforcing dominant cultural environment' in which the mass of people have become immersed. Teurlings (2013) goes further and argues for a mutation into the 'society of the machine', based on society's growing obsession with the very machineries of representation. Technologies, it could be argued, however, now allow people to become 'active contributors' (e.g. OpenStreetMap), reconnecting society and taking control. The problem is that these platforms tend to reinforce immersion in make-believe; consider the role of geographic information systems in location-based augmented reality games that turn reality into the staging of spectacle.

Debord recognizes three types of spectacle. The first, 'concentrated spectacle', is associated with dictatorships, and is underpinned by fear and bureaucratic control of culture. The second, 'diffused spectacle' is associated with developed capitalist societies characterized by abundance and competition for the attention of the consumer, while the third, 'integrated spectacle' is associated with liberal democracies where 'spectacle is marked by incessant technological development, [and] a state of general secrecy' (Kosović, 2011: 20), a society managed and governed by 'experts'. The connection with the Baroque is obvious and Debord acknowledges its importance as a historical turning point (Roberts, 2003).

Debord also noted how spectacle can be mobilized to create compliance through fear within liberal democracies. This is particularly pertinent to the so-called 'war on terror', and the role of maps as popular geopolitics. Debord (1998: 24) states that:

[d]emocracy constructs its own inconceivable foe, terrorism. Its wish is to be judged by its enemies rather than its results. The story of terrorism is written by the state ... The spectators must certainly never know everything about terrorism, but they must always know enough to convince them, that compared with terrorism, everything else must be acceptable or in any case more rational and democratic.

Kosović (2011) argues that the 9/11 attacks on the United States changed this. He suggests that Debord's spectacle is fundamentally different from an emerging 'spectacle of fear', as the latter is totally devoid of the apparent benefits of the former. For Kosović, the spectacle of fear 'offers no betterment, its promise is that of death and destruction; terror for its own sake' (p.23). Quite clearly, however, society has not become paralysed by fear despite the increasingly graphic images of terrorist atrocities. Debord's observation on the representation of terror makes much more sense and meets the needs of authority for compliance with their approach for dealing with the phenomenon. The spectacle has to remain balanced between fear and the ability of the liberal democracies to deal with it. Maps have a major part to play in this aspect of spectacle, and increasingly so, as the war on terror is represented as a 'video game' in which control of space has been intensified by control of pace; the ability to hit where and when desired, and rapidly (Barrinha and Da Mota, 2016). Airstrikes by drones become the *fuegos de artificio* of the political and military elites, while news media maps provide the supporting metaphors of accuracy, precision (Vujakovic, 2002) and the vantage point of total power.

The case can certainly be made that the map as spectacle is deeply embedded within forms of representation as control, generated by a Baroque cartography. Maps in the mass media, including adverting, news and recreational uses, have tended to the spectacular as they vie for attention. If anything, this has reached its apogee, or perhaps we might argue its nadir, in the growth of infographics as 'infotainment', in which flamboyance masks meaning, but is the product of highly sophisticated technologies.

## The map as meme complex

Before exploring specific case studies of the map as spectacle, it is important to understand how maps *work* as spectacle and *infect* mind-sets and worldviews. It is not sufficient simply to describe the elements that make up a map (the hierarchies of symbolization, the cartographic 'silences', use of colour) and make claims about their rhetorical intent – the power exerted *on* and *through* cartography. We need to understand how these elements succeed, despite the potential for multiple 'readings' and understanding, and how they become 'selected' as units of cultural reproduction, 'memes' that can reinforce positions of power. Why, for example, did certain colours (France as blue, Britain as rose-pink or red) or map projections (e.g. the Mercator) persist as memes with positive associations with specific empires and work on behalf of the elites that reproduce them? We need to understand how these and other elements function as 'viral' entities, persisting from one representation to another. Such representations cannot be seen as artlessly generated with each new map, but should be understood as an evolutionary sequence in which survival is based on the ability of a 'meme' to serve its intended purpose. If a map or a map symbol does not effectively serve the interest it is designed to support, it will fail to be reproduced and effectively become extinct. A memetic framework takes a countervailing approach to the post-representative conceptualization of maps. While agreeing to an extent with Kitchin *et al.* (2013) that the uses and meanings of maps can be conceived as fluid, contested and 'always in the process of becoming' (p.481), a memetic approach does not accept their premise that it 'follows that maps are [fluid] as well' (ibid.). A memetic approach regards memes and meme complexes as relatively stable, *but not static.* To conceive of maps as totally fluid implies a world in which the meaning is totally renegotiated on each reading (a chaotic viewpoint if taken to the extreme), it ignores the fact that 'convention' works, and most people do read what is required of them. If they did not work, the authorities that exploit them would look for alternatives. And this does happen, as when memes no longer serve their purpose they become extinct.

The concept of the meme and the meme complex can help us to understand the power of maps as spectacle. The concept of the meme originates with the evolutionary biologist Richard Dawkins. The 'meme' – as opposed to the 'gene' – is the concept of a unit of cultural imitation and replication that thrives in brains or the products of brains (meme vehicles – maps, books, graphics, computers, web-sites and so on) (Dennett, 1995). Dawkins (1986) regards memes as a form of 'cultural DNA' and uses this concept to explain how a particular set of ideas about the world might flourish (or fail) in specific populations. He is careful, however, to state that the 'meme pool' is *not* organized in the same consistent manner as the 'gene pool'. Unlike genes, large step mutations in memes do occur because they 'arise less from random copying error than from active reinterpretation by the receiving mind; and they can be accepted, rejected, and reaccepted over the course of the lives of the organisms whose minds are their carriers' (Runciman, 2002: 10). The fact that variation in memes, unlike genes, is not random, and that they are highly targeted (Jablonska, 2002) makes them effective as viral entities. What is also fascinating is their strength of persistence, their stability.

The concept of the meme complex can clearly be applied to maps as meme vehicles and their role in influencing spatial concepts and associated worldviews. Medieval *mappa mundi* provide an obvious example of a dominant world-view (and map as spectacle) that eventually failed to thrive as a meme complex and was replaced by (temporarily) more effective representations of the world. The meme is adopted here as a convenient name for units of cultural transmission that can be applied to maps. It is acknowledged that this is a contested term, but is regarded as a helpful concept to explore the evolution, persistence and extinction of ideas about the world (see Dennett, 1995). Worldviews or 'schema' that are reproduced and spread by memes can be 'contagious' in the sense that 'infection' can mean *to influence the mood or emotions of people.* Its broader meaning *to taint; to corrupt* is also relevant to the subliminal influence of memes that support certain discourses; for instance, nationalism, empire, racism and xenophobia, or global citizenship (see, for example, Vujakovic, 2009, 2013, on memes and Eurocentrism).

Maps as meme complexes are often extremely persistent, stable structures that create or reinforce specific views. A classic map that can serve as the 'type specimen' of a cartographic meme complex is also an outstanding example of the map as spectacle, the well-known world map supplement to *The Graphic* (1886) (Figure 8.1). This map 'displays' (as a peacock parades his feathers) the 'Imperial Federation – the British Empire in 1886'. The map's 'margins' contain a 'parade' of the peoples of the empire. Britannia, as personification of the metropole, sits astride a globe, her trident an 'arrow' drawing the eye to the British Isles at *centre stage*. The native peoples and the British 'colonials' gaze on Britannia in awe; and 'art' confirms the 'natural order' – and, in effect the map *envisions* the core-periphery relationship of the capitalist world-system. This map is not fluid, its intent is clear and its memes persistent (even today), although, as a meme complex, its once politically 'valid' world-view will now be rejected by most readers.

As a world map, this image is a meme of empire, but is constructed of a multitude of memes, and warrants being described as a 'meme complex' – in the sense that any one of its various elements acts as a powerful meme in its own right; the figure of Britannia for example. Careful examination of the map reveals memes within memes; for example, a British soldier/ explorer holds chained a submissive tiger representing European dominion over both nature and Asia, while behind him, bent under bales of produce, are two 'natives' of the sub-continent.

*Figure 8.1* 'Imperial Federation, map of the world showing the extent of the British Empire in 1886' (supplement to *The Graphic*, 24 July 1886). Available at: *https://upload. wikimedia.org/wikipedia/commons/3/3c/Imperial_Federation%2C_Map_of_the_ World_Showing_the_Extent_of_the_British_Empire_in_1886.jpg* (Accessed: 6 November 2016)

The latter is a trope (rhetorical device) all too redolent of the 'march of human evolution' with the 'superior' European at the forefront (Vujakovic, 2009). And, of course, all (sea-)roads lead to the 'new Rome' at top-centre of the map, a reminder of British global hegemony. The 'all roads lead to. . .' meme, centred on the metropole, is a constant trope of empire in mapping. Even the image of Britannia herself can be unpacked as a meme complex; her trident represents Britain's rule of the seas, while she sits astride a globe – a display of authority. Her shield is a further meme complex, in which the national flags of England, Scotland and Ireland are combined as signifier of political and economic union. Of course, Britannia as meme complex can be contested, satirized, or 'hyped', but she remains a stable meme. Neither the biting satiric cartoons of Margaret Thatcher as Britannia, Britannia's recent use in Brexit cartoons, or images of 'Cool Britannia' (Britain as a soft-power) could otherwise 'work'.

Maps as meme complexes are not simply 'art', but rely on artifice, technologies, survey and the mathematics of projection – a Baroque process. Here, again, a memetic approach parts from conceptions of cartography that do not allow that cartography is 'progressive'. Progress does not imply a teleological understanding of cartography, that cartography, for example, has a goal of absolute objectivity and scientific rigour. Progress, as in biological systems, is an unfolding process in which those genes (or memes) most fit to survive do so within the ecosystems in which they find themselves through adaptation and evolution of their reproductive apparatus. The following discussion will explore how maps as meme complexes operate as spectacle and an important element in social control.

## Girdle the Earth: maps, space and pace

*I'll put a girdle round about the earth / in forty minutes.*
(Puck to Oberon, *Act II, Scene I*, A Midsummer Night's Dream)

To understand Baroque cartography and the role of map as spectacle in society, a useful case study is the representation of the technical mastering of time and space as a product of modernity. Mapping of intercontinental air routes in the early-mid twentieth century, for example, provided an opportunity (to paraphrase Maravall on fireworks) to display the magnificence of artifice, the difficulty, the expense in human labour and in money that air travel implied, but which only the rich and powerful had at their disposal. These maps were the ultimate signifiers of imperial control, connoting the core (hub) status of global cities like London within the capitalist world-system.

The ability to move people and objects with increasing velocity has created a new relationship with time and space. Advance in speed of travel and the control of pace is a key theme in the work of Paul Virilio, who coined the term 'dromology' (from *drómos*, Greek for race course, place for running):

> [t]he science of the ride, the journey, the drive, the way. To me this means that speed and riches are totally linked concepts. And that the history of the world is not only about the political economy of riches, that is, wealth, money, capital, but also about the political economy of speed. If time is money, as they say, then speed is power (cited in Barrinha and Da Mota, 2016: 4).

While some people in the global periphery remain locked into medieval structures of space-time, much of the world has seen unparalleled changes, which include not only the possibility of rapid travel, but also in the delivery of death. As Barrinha and Da Mota (2016: 4) have noted, 'compression of time and space leads to the potential creation of new environments'. Furthermore, these novel environments can and have been mapped.

During the twentieth century, the achievement of powered flight became the ultimate display of control over space and time. It provided the opportunity to celebrate human ingenuity through both the ultimate mastery of nature and control of (geopolitical) space. This was recognized by influential cartographers, geographers and graphic artists; launching the period of 'air-age globalism'. These included, for example, Samuel Whittmore Boggs, Geographer US Department of State (1924–54) who advocated a new form of mapping to depict the impact of increases in transport speed and communications (Barney, 2015), as well as the innovative air-age maps of Richard Edes Harrison and Charles H. Owens (Cosgrove and della Dora, 2005). While many pre-air-age maps tended to the 'battleship Gothic' aesthetic of the Mercator and other rectangular projections (redolent of the age of sea-power), some graphic artists were experimenting with more fluid forms. The pictorial maps created by MacDonald 'Max' Gill, for example, emphasized the sphericity of the globe, while still reinforcing the centrality of Britain as the imperial metropole by clever use of projections and constant use of the 'all roads lead to…' meme. Gill's Empire Marketing Board poster *Highways of Empire* (1927), and his *Post Office Radio – Telephone Services* (1935) and his *Mail Steamship Routes* (1937) maps for the General Post Office, use a form of sinusoidal map. These maps celebrate technology with either shipping routes or telegraphic-phonic services converging on Britain. His *Cable & Wireless Great Circle Map* (1946) uses a Great Circle or azimuthal projection centred on London to reinforce the 'all roads lead to…' meme. Here, the technology of time-space compression *is* the spectacle, with a modern cable-laying ship and mobile telegraph station featured as supports in the cartouche. Gill is generally treated, perhaps because of his cartoon-like style, with some sympathy, but his contribution to the British imperial project is also acknowledged (see Barney, 2015). In fact, many of Gill's other maps can be understood as representations of the commodification of space, from his famous *Wonderground Map of London* (1914) (Dobbin, 2011), to his war-time 'rallying cry in time of war' (Barber and Harper, 2010: 166) – *Tea Revives the World* (1940).

In the promotional air-age maps and posters that proliferated during the 1930s and 1940s, 'spectacle' is the order of the day and in many cases these representations draw heavily on the 'theatrical' mapping and cartouches from previous centuries. They also inevitably focus on the metropole (the parent state of the empire) as the *omphalos* or hub – all roads lead to the new Rome. Examples of this cartographic projection of 'global reach' were found amongst all the contenders for hegemony during the 1930s – Britain, Germany, the Netherlands, France and the United States. See, for instance, the Imperial Airways (Britain, 1937) *Map of Empire & European Air Routes,* by Laszlo Moholy-Nagy, or the range of posters produced for the Hamburg-Amerika Linie, including Anton Ottmar's poster for the Graf Zepplin's route to South America (c.1935).

Perhaps the ultimate practitioner of this 'global theatre'[2] was Lucien Boucher (1889–1971) who produced poster-maps for Air France during the 1930s to 1960s (high resolution examples of Boucher's work can be viewed at the David Rumsey Collection[3]). His maps are flamboyant meme complexes, containing, in some cases references to classical mythology or to the Baroque; see, for example, his use of *putti*, the winged cherub, a staple of Baroque art and architecture, in his first poster map for Air France (1934). Boucher's 1952 double hemisphere map bears a striking resemblance in basic format to the 1885 'Imperial Federation' discussed earlier. In this case, France is represented at top-centre by a drawing of Nôtre Dame cathedral set over a compass rose with the Air France logo. The building is flanked by other examples of high architectural culture (e.g. the Parthenon) representing European superiority. The map is surrounded by stereotype memes of European 'discovery' of the Americas, Asia and Africa, and seemingly benevolent 'first contact' with native peoples and animals. A similar theme can be seen in his 1939 poster map *Nova et Vetera*, in which the new air routes are compared with archaic modes of transport worldwide. This highly decorative style was emulated by others, for example, by James Cutter for his decorative map (1946) for Trans World Airlines (TWA).

The drama integral to these maps was often accentuated by representing the background as a blue sky with clouds, suggestive of humanity's ability to overcome its earth-bound existence (see, for example, Edward McKnight Kauffer's 'Air-mail routes', 1937,[4] for Britain's General Post Office (GPO)). The curve of the Earth is sometimes shown to enhance the drama; for example, the cover art for a 1930s Aeroposta Argentina brochure[5] uses a similar perspective view to those made famous by Richard Edes Harrison, one of the best-known cartographers of air-age globalism (Schulten, 1998; Barney, 2012). Harrison's graphics were clearly to be understood as providing humans with a new view of the world that derived from technological innovations both in air technology and map making. The maps produced by graphic artists such as Harrison, Howard Burke and Charles Owens, were, however, concerned mostly with the strategic importance of world regions and global force-projection during World War II (see discussion of threat discourse later).

As international flight has become commonplace, the spectacle of the skies has taken other forms, but is always the prerogative of the rich and powerful. So, in the twenty-first century, the spectacle of intercontinental flight has been replaced by other innovations and technologies – the ultimate being the commodification of spaceflight. Examples of map as spectacle in recent years have included news media and promotional mapping of the flights of Virgin Atlantic GlobalFlyer in 2005 and the experimental solar-powered aircraft, Solar Impulse 2, in 2015. Spectacle of 'the ride, the journey, the drive' has also been enhanced by animation, first through television and later the functionalities associated with the web-based cartographies.

## Maps and the 'spectacle of fear'

The visualization of terror to cause fear and ensure a compliant populace has long been used as a means of control exerted by elites; for example, the haunting medieval church murals of hell and damnation. As Debord (1998) suggests, contemporary narratives of fear are constructed in such a manner that 'compared with terrorism, everything else must be acceptable or in any case more rational and democratic' (p.24). We have yet to experience the paralysing 'spectacle of fear' whose 'promise is that of death and destruction – terror for its own sake', as proposed by Kosović (2011: 23). In Debord's version, 'spectacle' ensures acquiescence with the state's use of force and the erosion of civil rights as part of a culture of surveillance and control. Fear is mobilized, but tempered by the ability of the state to protect us or to intervene effectively following terror attacks. Maps play an important part in this. Maps are used in three main ways, especially in the news media: first to imply potential threats; second, to report incidents of terror, whether 'at home' or abroad; and finally, to display the state's ability to project force to punish or protect.

The representation of fear is never more potent than when it is associated with the familiar. The '7/7 terror attacks' (7 July 2005) on London's transport systems provide a compelling example. News providers mapped the location of the bomb explosions, generally shown as dramatic blast symbols. Often paired with images of the devastation, or the traumatized and scarred faces of those caught up in the attacks. These images projected a new landscape of fear into the everyday lived-world of Londoners. The tube map featured in several newspapers; originally designed by Harry Beck, this map has become an icon of London. From its origins in the 1930s (Ovenden, 2003), it has come to influence transport map design worldwide, as well as achieving cult status at home (Dobbin, 2011). Its very familiarity and utility added to the sense of anxiety as the tube is transformed into a potential death trap. *The Times* (08/07/2006: 5) and *The Daily Express* (08/07/2006: 18) both paired a road map with the tube map, and included photographs or diagrams showing the location of explosions on individual buses and tube trains. *The Daily Mail* (08/07/2006: 4) paired a map with a cut-away block diagram showing the tube system in 3D. All transform the familiar into a scene of 'carnage' and insecurity. These have, however, to be read in context and must be seen as the counter to reassurances that the state will deal with further threat.

Threat, real, perceived or fabricated, is associated with entities at various scales, from aggressive and 'rogue' states, via para-state structures, to a range of terrorist organizations. The projection of fear can take several forms. During recent years, western governments, geopolitical 'think-tanks' and the news media have used classic embodiments of the terrorist threat as a danger to the state as organism. The so-called Islamic State (IS) is portrayed as a disease; a cancerous entity insidiously spreading through the bodies of Iraq and Syria. For example, see the maps produced by the Institute for the Study of War, Washington, DC[6] which have been recycled across a range of news sites. IS control is shown in black with attack zones in red, and wider 'support' a dusky pink (see Cresswell, 1997, on geographic metaphors of disease). Also, note how this threat was made immediate (to a western audience) and exploited by the UK 'Vote Leave' campaign (UK European Union membership referendum) in 2016. Immigration was a key theme of the Vote Leave campaign that sent a leaflet containing a map to all households in which red, connoting danger, was used to group Syria and Iraq with EU candidate countries, and a dynamic arrow (also red) used to suggest threat of invasion (Kent, 2016). This was clearly intended to heighten the sense of danger to the UK and to play on the fears of the map's audience concerning not only more countries such as Turkey and Serbia potentially joining the EU, with implications for immigration, but the threat that it would also open further the floodgates for refugees from war-torn Western Asia.

The most overtly spectacular maps are those associated with the threat of death from the skies – a post-Guernica trope that lends itself to air-age globalist mappings. Maps of air bombardment and missile attacks are obvious examples of Baroque cartographies; underpinned by technical developments in map projection and presentation (e.g. colour printing), and illustrative of the ability of states to project power through increasingly sophisticated aerial delivery systems. Contemporary news maps of perceived threats from states such as North Korea or China have their origins in the dramatic 'pictorial maps' being produced during World War II. These maps by graphic artists such as Charles Owen (see Cosgrove and della Dora, 2005) and Howard Burke illustrated both threat and Allied retaliation. The influence of these 'air-age' maps influenced those published by the US Navy, for example *NavWarMap No.3* which showed the naval and air campaigns conducted in the Atlantic and northwestern Europe using dramatic dynamic symbols and the incorporation of graphics depicting military hardware (aircraft, bombs, naval vessels). This type of meme complex becomes a stock response to foreign threat. It echoes Kaldor's Baroque arsenal, in which new wars are fought with evermore sophisticated weaponry designed for old scenarios. Various recent 'police-actions' by the west have been mapped using this style (see Mijksenaar, 1996, on the Gulf War of 1990–1, and Vujakovic, 2002, on NATO intervention in Yugoslavia in 1999). Vujakovic, in particular, notes how maps, in combination with information graphics of weapons systems, are used as a guarantor of accuracy and precision in acts of 'peace-making'.

Other uses of maps in threat discourse have involved the justification of anti-ballistic missile systems as a response to developments by North Korea, China and Iran. These have been some of the most spectacular maps in the news media in recent years, with pyrotechnic displays as part of drama; for example, *The Times* map of 'North Korea's missile threat . . . and America's defence plan' (13/10/1999: 11). In this graphic, a US 'hit-to-kill' missile blasts a Korean delivery system from the sky in a fire-ball of flame, over a bizarre map of the world with a false horizon and missile ranges that make no sense at all – drama supersedes accuracy in this example of 'infotainment' (Vujakovic, 2014).

## Conclusion

While the rhetorical power of maps has been fully recognized in recent decades, the examination has often been limited to deconstructing individual maps or examining specific issues rather than considering how they might operate as part of wider structural relationships. Maravall's analysis of the Baroque, and its lessons for today, as well as Debord's examination of spectacle, offer us such a focus.

It is argued here that many forms of popular mapping are the product of a Baroque cartography producing representations that work to create compliant populations, whether through the spectacle as part of a total immersion in a commodity-orientated society or through threat discourse. It is suggested that with further immersion in new technologies of information production, dissemination and interaction, this process will become even more embedded, despite promises of freedom and individual control. This, together with 'globalization', a process of continuous (spatial) incorporation into the world economy, has not resulted in equality, but further disparities, which will continue to drive the need for Baroque cartographies of control. Threat and security issues are becoming ever more spatially 'networked', involving forms of organization and technology within and across existing territorial structures, in which state security forces engage in extra-territorial inter-ventions and non-state actors work through networks of multiple, dispersed nodes (Arguilla, 2010, 2013). As states and regions within periphery and semi-periphery fracture and threaten to further impact on the core regions (via terrorism, displaced populations and resource security issues), maps will grow in importance as a form of social control.

## Notes

1 The themes in this chapter were first presented at two conferences: Vujakovic, P. (2005) "Scary Stories: The Role of News Media Maps in Popular Geopolitics" at *Fear. Critical Geopolitics & Everyday Life* (Geographical Research Centre, University of Durham), 11–12 July; and Vujakovic, P. (2015) "'All the World's a Stage' – Maps as Spectacle in Journalistic Reporting of Theatres of War and Threat Discourse in the Twentieth Century" at *Theatre of the World in Four Dimensions* (International History of Cartography Conference, Antwerp), 12–17 July.

2 'This artist is known mainly for a series of absorbing, Air France, world-map posters, signaling a global theatre' – online biographical note from a vintage poster dealership. Available at: *www.thevintageposter.com/artist-biography/?at=LucienBoucher&InvNo=8268* (Accessed: 15 October 2016).

3 For a range of examples of Boucher's work, see the David Rumsey Collection online. For example, Boucher's first poster for Air France. Available at: *www.davidrumsey.com/luna/servlet/detail/RUMSEY~8~1~266420~90040773:Air-France-Reseau-Aerien-Mondial--L* (Accessed: 15 October 2016).

4 See the British Postal Museum and Archive, available at: *https://postalheritage.wordpress.com/tag/poster-design/* (Accessed: 15 October 2016).

5 Available at: *https://en.wikipedia.org/wiki/Aeroposta_Argentina#/media/File:Aeroposta_Argentina_Poster_aeropa37.jpg* (Accessed: 15 October 2016).

6 Available at: *http://understandingwar.org/project/isis-sanctuary-map* (Accessed: 24 October 2016).

## References

Arguilla, J. (2010) "The New Rules of War" *Foreign Policy* 178 pp.60–67.

Arguilla, J. (2013) "Twenty Years of Cyber war" *Journal of Military Ethics* 12 (1) pp.80–87.

Azócar Fernandez, P.I. and Buchroithner, M.F. (2014) *Paradigms in Cartography: An Epistemological Review of the 20th and 21st Centuries* Berlin: Springer.

Barber, P. and Harper, T. (2010) *Magnificent Maps: Power, Propaganda and Art* London: The British Library.

Barney, T. (2012) "Richard Edes Harrison and the Cartographic Perspective of Modern Internationalism" *Rhetoric and Public Affairs* 15 (3) pp.397–433.

Barney, T. (2015) *Mapping the Cold War: Cartography and the Framing of America's International Power* Chapel Hill, NC: University of North Carolina Press.

Barreiro López, P. (2014) "Reinterpreting the Past: The Baroque Phantom during Francoism" *Bulletin of Spanish Studies* 91 (5) pp.715–733.

Barrinha, A. and Da Mota, S. (2016) "Drones and the Uninsurable Security Subjects" *Third World Quarterly* 38 (2) pp.253–269.

Boyden, J.M. (1999) "José Antonio Maravall" in Boyd, K. (Ed.) (1999) *Encyclopaedia of Historians and Historical Writing* (Volume 2) London: Fitzroy Dearborn, pp.761–762.

Callens, J. (2000) "Diverting the Integrated Spectacle of War: Sam Shepard's *States of Shock*" *Text and Performance Quarterly* 20 (3) pp.290–306.

Caquard, S. (2015) "Cartography III: A Post-Representational Perspective on Cognitive Cartography" *Progress in Human Geography* 39 (2) pp.225–235.

Cosgrove, D.E. and della Dora, V. (2005) "Mapping Global War: Los Angeles, the Pacific, and Charles Owens's Pictorial Cartography" *Annals of the Association of American Geographers* 95 (2) pp.373–390.

Cresswell, T. (1997) "Weeds, Plagues, and Bodily Secretions: A Geographical Interpretation of Metaphors of Displacement" *Annals of the Association of American Geographers* 87 (2) pp.330–345.

Dawkins, R. (1986) *The Blind Watchmaker* London: Longman.

Debord, G. (1998) *Comments on the Society of the Spectacle* London: Verso.

Debord, G. (2009) *Society of the Spectacle* (trans. Knabb. K.) Eastbourne, UK: Soul Bay Press.

Del Valle, I. (2002) "Jesuit Baroque" *Journal of Spanish Cultural Studies* 3 (2) pp.141–163.

Dennett, D.C. (1995) *Darwin's Dangerous Idea: Evolution and the Meanings of Life* New York: Simon & Schuster.

Dobbin, C. (2011) *London Underground Maps: Art, Design and Cartography* Farnham, UK: Lund Humphries.

Flint, C. and Taylor, P.J. (2011) *Political Geography: World-Economy, Nation-State and Locality* Harlow, UK: Prentice Hall.

Godzich, W. and Spadaccini, N. (1986) "The Changing Face of History" in Maravall, J.A. (trans. T. Cochran) *Culture of the Baroque: Analysis of a Historical Structure* Manchester, UK: Manchester University Press (originally published in 1975 as *La Cultura del Barroco*), pp.vii–xx.

Harbison, R.C.M. (2010) "Baroque Exuberance: Frivolity or Disquiet" *Architectural Design* 80 (2) pp.44–49.

Jablonska, E. (2002) "Between Development and Evolution: How to Model Cultural Change" in Wheeler, M., Ziman, J. and Boden, A.M. (Eds) *The Evolution of Cultural Entities* Oxford, UK: Oxford University Press, pp.27–42.

Kaldor, M. (1981) *The Baroque Arsenal* New York: Hill and Wang.

Kaldor, M. (1986) "The Weapons Succession Process" *World Politics* 38 (4) pp.577–595.

Kent, A.J. (2016) "Political Cartography: From Bertin to Brexit" (Editorial) *The Cartographic Journal* 53 (3) pp.199–201.

Kitchin, R., Gleeson, J. and Dodge, M. (2013) "Unfolding Mapping Practices: A New Epistemology for Cartography" *Transactions of the Institute of British Geographers* 38 (3) pp.480–496.

Kosović, M. (2011) "Revisiting the Society of the Spectacle in the Post-9/11 World" *Contemporary Issues* 4 (1) pp.18–28.

Lollini, M. (1998) "Maravall's Culture of the Baroque: Between Wölfflin, Gramsci, and Benjamin" *Yearbook of Comparative and General Literature* 45,46 pp.187–194.

Maravall, J.A. (1986) *Culture of the Baroque: Analysis of a Historical Structure* (trans. T. Cochran) Manchester, UK: Manchester University Press (originally published in 1975 as *La Cultura del Barroco*).

Mijksenaar, P. (1996) "Infographics at the Gulf War" in Houkes, R. (Ed.) *Information Design and Infographics* Rotterdam, The Netherlands: European Institute for Research and Development of Graphic Communication, pp.55–62.

Mohr, R. (2012) "Signature and Illusion: Lessons from the Baroque from 'Truth' in Law, Arts and the Humanities" *Australian Feminist Law Journal* 36 (1) pp.45–63.

Ovenden, M. (2003) *Metro Maps of the World* Harrow Weald, London: Capital Transport.

Roberts, D. (2003) "Towards a Genealogy and Typology of Spectacle: Some Comments on Debord" *Thesis Eleven* 75 (1) pp.54–68.

Rundstrom, R.A. (1991) "Mapping, Postmodernism, Indigenous People and the Changing Direction of North American Cartography" *Cartographica* 28 (2) pp.1–12.

Rundstrom, R.A. (1993) "The Role of Ethics, Mapping, and the Meaning of Place in Relations Between Indians and Whites in the United States" *Cartographica* 30 (1) pp.21–28.

Runciman, W.G. (2002) "Heritable Variation and Competitive Selection as the Mechanisms of Sociocultural Evolution" in Wheeler, M., Ziman, J. and Boden, A.M. (Eds) *The Evolution of Cultural Entities* Oxford, UK: Oxford University Press, pp.9–26.

Schmitt, A.K., Danisik, M., Aydar, E., Sen, E., Ulusoy, I. and Lovera, O.M. (2014) "Identifying the Volcanic Eruption Depicted in a Neolithic Painting at Çatalhöyük, Central Anatolia, Turkey" *PLoS ONE* 9 (1): e84711. doi:10.1371/journal.pone.0084711.

Schulten, S. (1998) "Richard Edes Harrison and the Challenge to American Cartography" *Imago Mundi* 50 (1) 174–188.

Sullivan, H.W. (2010) "The Politics of Bohemia and the Thirty Years' War on the Spanish Baroque Stage" *Bulletin of Spanish Studies: Hispanic Studies and Researches on Spain, Portugal and Latin America* 87 (6) pp.723–778.

Teurlings, J. (2013) "From the Society of the Spectacle to the Society of the Machinery: Mutations in Popular Culture 1960s–2000s" *European Journal of Communication* 28 (5) pp. 514–526.

Vujakovic, P. (2002) "Mapping the War Zone: Cartography, Geopolitics and Security Discourse in the UK Press" *Journalism Studies* 3 (2) pp.187–202.

Vujakovic, P. (2009) "World-Views: Art and Canonical Images in the Geographical and Life Sciences" in Cartwright, W., Gartner, G. and Lehn, A. (Eds) *Cartography and Art* Berlin: Springer, pp.135–144.

Vujakovic, P. (2013) "Warning! Viral Memes Can Seriously Alter Your Worldview" *Maplines* Spring pp.4–6.

Vujakovic, P. (2014) "The State as a 'Power Container': The Role of News Media Cartography in Contemporary Geopolitical Discourse" *The Cartographic Journal* 51 (1) pp.11–24.

Wood, D. (1992) *The Power of Maps* London: Routledge.

**Part II**

# Exploring the origins of modern cartography

# 9

# Image and imagination

## Maps in Medieval Europe

*Peter Barber and Catherine Delano-Smith*

Three questions underpin any consideration of maps in medieval Europe: Why does the map in our hands look as it does? Whose map is it? And how much can we expect to know about maps in the post-Roman period and before the age of printing? While there are no easy answers to these questions, especially to the last two, they need to be addressed. Moreover, in raising fundamental issues in the long history of maps and mapping, they identify principles that apply to maps from all ages and all parts of the world. The format of the present essay does not give us space to explore all three, but we can introduce them as we consider selected aspects of maps in the medieval period in Europe.

Many medieval maps are familiar. The more striking examples, irresistibly attractive, preface every publication aimed at the general market. In these tomes, the great parchment *mappae-mundi*, originally made for display, from Hereford and Ebstorf sit on glossy pages matching in size the tiny *mappamundi* that sparkles jewel-like on the page of an English Psalter.[1] Likewise, regional maps on large sheets, showing the whole of Italy or Britain, are interleaved with reproductions of maps of the British Isles originally drawn on the page of a book.[2] All these, and many others, are also laid out in specialist books focussing on specific aspects of these maps, such as the Southern Continent, the Earthly paradise, the Holy Land and the often strange-looking animals also called the 'Marvels of the East'.[3] Yet, for all this literature and sense of familiarity, remarkably little is really known about all too few of these maps. The jury is still out on where the Hereford *mappamundi* was made (Lincoln as well as Hereford?), for instance, and the origins of the Gough Map of Britain have only recently been shown to be far more complex than hitherto suspected.[4] Apart from the obvious problem of the physical survival of books and other perishable artefacts, one of the shortcomings in traditional research on the Middle Ages stems from intellectual compartmentalization; map historians are as likely to find new evidence in medieval poetry as in chronicles and government records to help connect up what is known and to replace 'unknown unknowns' with 'known unknowns'.[5]

Medieval knowledge of the world, unlike modern knowledge for which the search is always for completely new data, consciously encompassed all that was already known. Thus, for the modern scholar, the primary stumbling block to understanding medieval maps is the problem of transmission. Only rare fragments survive of the enormous maps incised on stone or painted in a portico in the Roman period. More challengingly, almost no codex sheltering a map from

Classical or even Late Antiquity has reached us in its original format; the vast majority are copies, often incomplete, or copies of copies. For the modern scholar wanting to investigate medieval maps, the first problem is to discern in the fog of transmission just what may have been available in the Middle Ages in original as opposed to secondary format, or as knowledge passed on by word of mouth. The answer is almost certainly that medieval scholars knew, or at the very least were aware of, far more about the geography of their region than we can possibly appreciate. However paltry the medieval map corpus from Europe may seem to the student of modern cartography, the depths to which it embodies geographical and cartographical knowledge handed down from Classical times should not be under-estimated. In the absence of an autograph manuscript, with its maps, intensive research may help distinguish the work of successive copyists from that of the original author of the map.[6]

One reason for the apparent paucity of medieval maps is that the images in the literary or historical sources are not always recognized as maps or thought worth reporting.[7] Not only do the figures encountered in medieval documents not 'look' like a map (a matter to which we return later), but there was no special word for a 'map' to attract the modern historian's attention. The Latin *mappa* refers only to the medium ('mappa' = cloth) and *mundi* tells us only that the subject matter is the earthly world and/or the cosmos; nothing unambiguously implies a drawn cartographical figure. On the contrary, the heading 'mappa mundi' in the earliest history of Britain, Gildas's *De Excidio Britanniae et Conquestu* (The Ruin and Complaint of Britain; before 550) heralds a brief written description, and Gervase of Canterbury's *Mappa Mundi* (c.1200) is a list of the monasteries of England, while the great map of the world now in Hereford Cathedral (c.1300) refers to itself as a history (*estoire*), meaning a narrative of the events of human history. We also find 'list maps', on which the geography of a region is represented solely by place names positioned on the page in their spatially correct relationships; as on the map on the verso of the Psalter map (c.1265), where the names are contained within the outline of a divided circle (T-O), and three maps of the Holy Land from about 1476–8 with no lines at all.[8] Even when a drawing in a treatise is introduced by the author not as 'painted' (*depinta*) but as something 'we represent' (*repraesentamus*) or show as a 'figure' (*figura*), it remains unclear if the intended or original drawing was a plan or a picture.[9]

Another general point to bear in mind when thinking about medieval maps is their inherent connectedness. Total population in Western Europe was everywhere small compared with the modern period. There were probably fewer than five million people in England before 1348 (when the Black Death halved that total).[10] Those that could be described as literate, let alone learned, would have been a very small proportion of that total. The advantage for the latter group is that personal contacts would have been relatively easily established and maintained.[11] The acquisition of knowledge involved individual travel on an international scale and the purchase or borrowing of books to bring home.[12] Contrary to the old myth that 'nobody travelled', the roads of Europe were often packed as ecclesiastics, government officials and merchants moved about their business.[13] The habit of gesticulating towards places not in sight, or sketching an elementary plan in explanation, reaches back into pre-and proto-history, in Europe as throughout the Old World.[14] The empires of Greece, Rome and Byzantium penetrated beyond Mesopotamia far into central Asia, bringing knowledge of Middle Eastern measurement, land surveying and the representation of buildings, towns and regions together with ideas about representing the cosmos and the earth on a flat surface, to the Mediterranean lands and thence into medieval Latin Europe, in the process adding to the sum of knowledge from local experience. In due course, especially from the eleventh century, not only Arabic translations of Classical texts but also Islamic scholarship crossed the Mediterranean by way of Sicily and Andalusia through cultural exchange, especially by way of Chartres and Orleans, and trade.[15] Within Europe, all such ideas and knowledge were disseminated to the very limits of the continent – the British Isles in the northwest, Iceland and Scandinavia in the north.[16] Where European geographers could not portray their world directly from observation when on land or sailing along the coast, imagination helped transform the inherited wealth of written description, spoken accounts and older graphic representations into maps of the world, its regions and localities, some of which are introduced later in this chapter.

## Maps in books

A truly comprehensive review of all medieval maps would take into account two main, pro-digiously overlapping, arenas for the creation and deployment of maps: the scholarly and the administrative (roughly, church and government). It would allow for the possibility of personal initiatives from the merchant class. More importantly, it would include maps in books as a dis-crete genre. Here attention would be paid to the distinction between those drawn on the page and those just bound in; those provided by the original author (or descended from them) and those supplied by a later scribe or redactor as an aid to understanding that text; and those whose function is best described as illustration as opposed to elucidation of the written text. Much of the literature on medieval cartography tends to avoid such distinctions, however, with the result that different types of *mappaemundi* are often treated as a single corpus.[17] Yet the hundreds of such *mappaemundi* (1,100 is a commonly quoted total) in fact embrace four distinctive cat-egories of maps, three intimately associated with books. Of these groups, two (here referred to as 'mappae mundi') originated as textual illustration and are almost never found outside a codex or roll, while the third group contains book-sized versions (on a single or double folio) of maps made for display on walls (here referred to as *mappaemundi*).[18] The fourth group includes these great wall creations themselves, described in the next section.

Two groups of the book 'mappae mundi', many of which are of Classical origin, share their key visual characteristic irrespective of subject matter. That is, all are diagrammatic in style, in keeping with their function as explanatory drawings. Each is a simple composition, with straightened or smoothed outlines and minimal content for immediate and unambiguous com-munication from author to reader.[19] While one group concerns specific aspects of earth science, such as the climatic zones, habitable and uninhabitable zones, the winds, the structure of the cosmos, the effect of the planets on tides and in astrology, those in the second group represent the division of the world into three continents, sometimes with additional details such as their constituent countries or the names of the sons of Noah said in Judeo-Christian writings to have populated each continent after the Flood (Figure 9.1).[20] These deliberately simplified maps are usually found in works on the earth sciences and in encyclopaedias.[21] In complete contrast to these uncompromising diagrams are the images of the third group of book maps, the folio *mappaemundi*, characteristically rich in pictorial content and visually complex like their wall counterparts but drawn for the page where they might serve quite different functions. Whereas the Psalter map (c.1265), as its context implies, invited spiritual contemplation, Ranulph Higden seems to have supplied the original text of his *Polychronicon* (c.1342) with a map from a sense that some sort of map was expected. The map inserted into a copy of Henry of Mainz's ency-clopaedic chronicle is not irrelevant to the text, but does not seem to have been intended for it, whereas the map of the world found in many copies of Beatus of Lièbana's theological treatise on the Book of St John the Divine (Apocalypse) does appear to be keyed into the text.[22]

Unsurprisingly in an age when the Bible was considered relevant to all aspects of life and learning, theological treatises account for a substantial but largely unsung proportion of maps in medieval books. By no means every commentator, Jewish or Christian, who tried to arrive

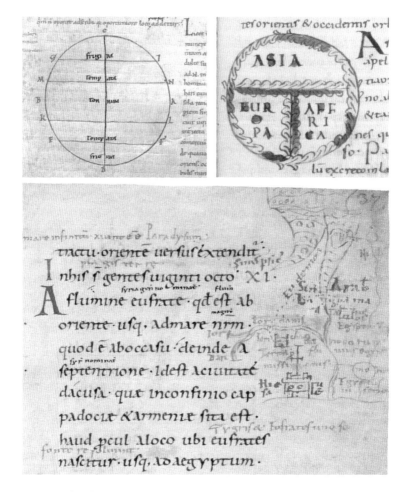

*Figure 9.1*   Maps in books

Above: two drawings in diagrammatical style to demonstrate the key points made in the adjacent text: the Earth's climatic zones (left) from an eleventh-century copy of Macrobius's *Dream of Scipio* (British Library, Harley MS 2772, folio 67 verso); and the world divided into three continents and surrounded by an outer ocean (right) in a ninth-century copy of Isidore of Seville's *Etymologies*, Book 14, Chapter 2 (St Gallen, Cod. Sang. 237, p.219). Both of Classical origin, the zonal and the T-O diagrams came to be elaborated with geographical, biblical and other historical details.

Below: a medieval reader of this ninth-century copy of Orosius's *History against the Pagans* (St Gallen, Cod. 621, p.37), possibly the monk Ekkegarts IV, has been inspired to gloss the secular description of Asia from India to the Red Sea (defined as 'Syria') in Book 1, pp.17–23, with references to paradise (not mentioned in the text) and by adding a map to show not only the Red Sea but also the route of the Exodus from Egypt to the Promised Land, the Jordan river and a prominent Jerusalem, among other places. (Reproduced with permission from the British Library and the Siftsbibliotek, St Gallen.)

at the literal meaning of the sacred texts, included an explanatory drawing in his interpretation of a particularly opaque description of a place or building, but those who did have left us with a remarkable corpus of maps and plans. The record is fragmentary and care is needed not to assume links or derivations in chronologically widely separated examples. But given a relatively stable text like the Old Testament, we find the same passages attracting the attention of Christian scholars throughout the Middle Ages and later, forming a remarkably consistent corpus in style and content. In some instances we can point to a trail of maps on a particular topic that goes back at least to the sixth century (the disposition of the Twelve Tribes around the Tabernacle in the Desert of Sinai (Numbers 2–3)) or even to the first years of the fourth century (the location of the portions allotted to the Tribes of Israel in newly conquered Canaan' for the Book of Joshua, that Bishop Eusebius said he added to his gazetteer of biblical place names).[23] The long tradition of drawing in biblical exegesis reached a medieval apex in the work of the Franciscan Nicolas of Lyra, whose influential commentary on the whole of the Bible (*Postilla literali super totam Bibliam*, 1323–32) was used in the sixteenth century by reformers such as Martin Luther and contains fifteen plans and maps.[24]

Scientific textbooks, encyclopaedic works and biblical exegesis are by no means the only kind of medieval book in which maps are found. Again, distinctions have to be made if a useful corpus is ever to be assembled. Some drawings are no more than graphic, as opposed to verbal, glosses, but it still has to be decided who was responsible, the scribe who wrote out the text, the corrector, or a student or casual reader. All three, distinguished by differences in ink and palaeography, are found on one folio of Orosius's *History against the Pagans*, now in St Gall, Switzerland, for example (see Figure 9.1).[25] The marginal maps in many manuscripts of one of the most widely read Classical texts, Lucan's history of the Civil War (*Pharsalia*), probably originated as a teacher's aids for his students that were thereafter incorporated by many copyists as part of the standard text.[26] Christoforo Buondelmonte saw no need to accompany his poetical account of the islands of the Cyclades (in the Aegean) with maps, but successive imitators inserted a map for each island described as they expanded the work.[27] Elsewhere, maps reflected a particular writer's interest in contemporary affairs; Matthew Paris of St Albans, England, is probably the outstanding example of a mid thirteenth-century map-minded chronicler for the number and range of his cartographical drawings, but many rolls recording the descent of the nation's rulers from Adam and Eve contain maps on subjects ranging from small roundels showing, for instance, the location of English bishoprics to detailed circular plans of the city of Jerusalem.[28] A French genealogy aiming to demonstrate the right of a prince of the Valois dynasty (Philip of Valois) to rule the whole of France as King – rather than the English king Edward III, the ruler of the Angevin fiefs of Aquitaine and Normandy, who had claimed the French throne after the death without heir of Charles IV in 1328 – underlined its claim with a map of France that is schematic in structure while packed with elaborate pictorial signs for towns and provinces.[29]

## Maps on walls and for walls

Maps were not to be found only in books, however. Before the Middle Ages, Roman leaders had commissioned large and highly detailed engraved marble plans of towns and rural settlements, such as Rome and Orange, which were mounted on the interior and external walls of temples, presumably for commemorative and propaganda purposes.[30] Other external walls were decorated with large bird's-eye frescoes showing the elevation of towns.[31] In the same tradition, the eighth-century Pope Zacharias is said to have commissioned a wall map fresco for his dining room in the Lateran.[32] Maps were also occasionally to be found as floor mosaics, as in the case of the sixth-century map of Palestine in the nave of a church in Madaba, Jordan.[33] More common as a subject for mural paintings after 1100 than before, maps of the world – *mappaemundi* – reached their apogee from the thirteenth century.[34]

*Mappaemundi* are perhaps the best-known European medieval maps. Whether painted directly onto a wall, or on a parchment to be used as a wall hanging, or found in books, their pictorial content extended from the spiritually profound to the mundane, with everything juxtaposed so that the weird-looking semi-mythical creatures described by Pliny in the first century AD appear close to sites associated with the Bible or with events from recent crusades. The medieval observer, however, would have perceived the mythical and semi-mythical creatures, through the filter of the moral and Christian attributes with which they had been associated by the late Roman writer Solinus and in the somewhat later descriptive text known as *Physiologus*, as part of God's creation.[35] Thus, contrary to the common presentation of these maps as geographically naive theological constructs, we should see at their core a body of solid, down-to-earth, geographical information, as emphasized by Patrick Gautier Dalché.[36] The most important geographical and urban features of the world indicated on the *mappaemundi* come from information found in the texts of the most influential writers and encyclopaedists of Late Antiquity, notably Orosius and Isidore of Seville.[37] This geographical information, however, was selected and arranged on the maps according to its chronological, historical, biblical and cultural importance in relation to the unfolding of the human history that started with Adam and Eve in the terrestrial paradise in the east (usually but not inevitably placed at the top of the map). This westward trajectory of world history, it was believed, would come to an end with the evangelization of the whole of the known world and the consequent Second Coming of Christ in Jerusalem.

Not all aspects of the *mappaemundi* would have been so easily grasped. It has been suggested that spatial relationships of some theologically significant places, such as the patriarchal seats of Constantinople, Alexandria, Rome and Carthage (whose coordinates were known from Greek astronomical observation), were intentionally distorted by creators of the original maps no later than the eighth and ninth centuries in the interests of Christian theological significance and symbolism.[38] However, the lower half of an east-oriented *mappaemundi* would have held immediate interest, for here Western Europe and, to a lesser extent, Africa, are packed with important contemporary towns and cities, popular shrines, and places on trade and pilgrimage routes. Here too are most of the references to contemporary domestic and international disputes.

Not only is the content of the *mappaemundi* varied but the imagery is multi-layered, open to different interpretations in different contexts and consequently appropriate to be used in different ways.[39] Although understandably commonly associated in the modern mind with medieval piety and learning, *mappaemundi* also played a secular role in medieval life. The largest were expensive to create and, although written evidence for their use is sparse, they seem to have been regarded as luxury objects, presented to and commissioned not only by prelates and clerics but also by princes and secular potentates wanting to impress and influence. Indeed, we can go so far as to say that in the thirteenth and early fourteenth centuries there were two distinct forms of *mappaemundi*, one for courts and the other for churches. From what can be gleaned from the few survivors and what is known of or can be deduced about their original patrons, it seems that the preferred form for palaces was that in which the world is represented as secure in Christ's embrace, as in the Ebstorf map and the world map on the verso of the Psalter map. This is in contrast to those maps with surrounds that emphasize the Last Judgement and the transience of the human world, as conveyed by the letters MORS (the Latin for 'Death') atop the thongs that appear to attach the earthly world to the divine setting on the Hereford *mappamundi*, which might be considered more suitable for an ecclesiastical setting.

We know that *mappaemundi* of varying sizes were displayed in audience chambers, noble as well as royal.[40] In this context, they would have travelled with the ruler's wardrobe as he moved about his country on the progresses that were the standard form of medieval governance.[41] Here they would have had a political function, just as the larger *mappaemundi* in cathedrals like Hereford (Figure 9.2) may have been strategically placed to make some sort of statement in ecclesiastical and – such being the nature of society at the time – also, if indirectly, local and national politics.[42]

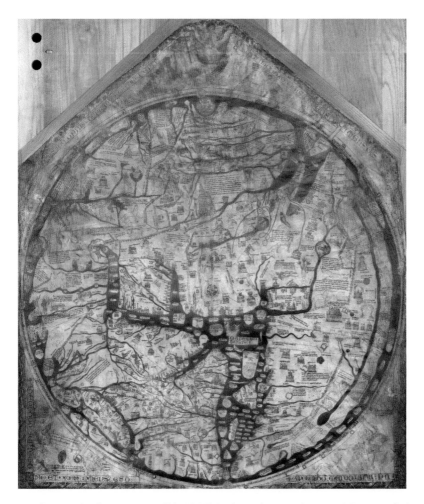

*Figure 9.2* The Hereford *Mappamundi* (c.1300) is the only complete surviving stand-alone medieval *mappamundi*. This outstanding example of the ecclesiastical stream of *mappaemundi*, the map of the world, with East at the top, is dominated by a depiction of the Last Judgement (ultimately derived from the writings of Hugh of St Victor) illustrating the contrast between the immortality of God and transience of man and his world. Amidst the Biblical, historical, ethnographic and other encyclopaedic information, there are many references to the Europe of 1300. Possibly created as part of a campaign in Hereford Cathedral to secure the canonization of Thomas de Cantilupe, a former bishop, the map's ultimate secular objective may well have been to boost Hereford's economy and bolster its bishop's political influence. (Reproduced courtesy of Hereford Mappa Mundi Trust and The Chapter of Hereford Cathedral.)

In the secular context, a *mappamundi* served a political agenda. Besides reflecting the ruler's piety and his intellectual accomplishments, the aura of God's majesty, hallmark of the religious image, was presumably intended to be transferred to the ruler, endowing him with legitimacy and power. It may also have invited comparison between Christ's suffering for humanity and those of a monarch on behalf of his subjects, while the prominence of Jerusalem on the map was a constant reminder of the Christian ruler's commitment to undertake a crusade. At the same time, there was plenty of scope for less elevated messages, especially if the map displayed or referred to a specific territorial situation involving the ruler who had commissioned, or been presented with, the map or lands ruled by his loyal relatives or allies. A case in point might be the naming of French provinces lost since 1200 by the Angevin dynasty on the Psalter map, which suggests that they had been prominent on the map from which the Psalter map, produced in Westminster in about 1265, was copied, which is likely to have been a *mappamundi* owned by Henry III.[43] Armin Wolf has argued along similar lines for the antecedent of the Ebstorf map, while Dale Kedwards has convincingly argued that two twelfth-century Icelandic mappaemundi and a geographically and administratively structured list of priest-chieftains were copied in the following century in such a way as to suggest divine sanction for continuing Icelandic independence at a time when this was threatened by Norway.[44]

After about 1300 the adoption into and adaptation of the *mappamundi* image for secular political service became more pronounced. The moral (Christian) frame derived from Hugh of St Victor[45] decreased until it disappeared as the nature of west European monarchy changed from the essentially sacerdotal to the primarily territorially based. Thus the early fourteenth-century circular world map in the audience chamber in the town hall of Siena (now lost) placed not Jerusalem but Siena at the centre of the (Sienese) world.[46] From the still-visible scoring in the wall made by the map, we may infer that it was rotated to flatter envoys from other cities as they awaited their audiences with the Sienese authorities.

Such overtly distorted and politicized *mappaemundi* were exceptional, but towards the end of the fourteenth century, in Catalonia, Italy and elsewhere in Europe, notably Germany, another new form of *mappamundi* was evolving. This was characterized by increasing geographical realism derived from the coastal outlines on portolan charts and, in the course of the fifteenth century, from maps in Claudius Ptolemy's *Geography*, newly available in Latin.[47] These trends, and an increasingly questioning attitude towards hitherto accepted sources, was typified by Fra Mauro's map, originally created in about 1450.[48] Such maps were regularly presented to and displayed by rulers and powerful individuals.

By 1520, the old world map looked very different as the American continents were added. The European monarchies, too, were not as before. Their new interests were reflected in novel world maps showing the latest discoveries on the walls of the privy galleries of their palaces, now increasingly accompanied by large regional maps of areas like the British Isles and Italy that were not at that time politically united but where there were aspirations to unity.[49] The regional maps were not necessarily all that new. In Italy, maps of different regions had been seen in the courts of the Italian states about a century earlier, while many of the maps and bird's-eye views of cities that were also appearing also dated from the mid fifteenth century.[50] The regional maps had often owed their origins to the needs of warfare and the work of military engineers, but already in the second half of the fifteenth century, if not earlier, they were reflecting the ideas of humanists who saw maps as a means not only of expressing their patriotism and establishing the geographical integrity of their cultural traditions but also, in a neo-Platonic sense, as of practical benefit to their homeland.[51]

All this is not to say that traditional *mappaemundi* were no longer created. On the contrary: it is clear that by the early sixteenth century the distinction was being made in educated circles between *mappaemundi* and the 'maps of the whole world' that showed the latest discoveries.[52] As a form of spiritual short-hand for the human and biblical world, traditional *mappaemundi* continued to adorn the walls of Henry VIII's palaces. One is known to have been placed among the religious images that decorated the gallery leading to Henry VIII's royal chapel in Hampton

Court Palace (presumably in the hope of inducing suitably pious thoughts in the minds of their beholders).[53] A little earlier, the emergence of the printed book had already made the ancient illustrated encyclopaedic texts, such as those of Pliny and Isidore, with their diagrammatic maps, available to a wider reading public.[54] It would seem that only when the Protestant tide turned against the sacrilegious display of religious images were *mappaemundi* set aside.[55]

## Manuscript maps on separate sheets

Certain types of spatial information are likely to have always been considered 'practical' in the sense they are empirical and an essential element of a professional skill that does not need to be recorded on paper. Some kinds of graphic guidelines may have been involved in the building of cathedrals, improvement of watercourses and laying out of hunting routes, for example, but formal maps were rarely produced. Where we do find a map or plan before about 1350, it seems to have been produced for a primarily commemorative or a quasi-legal purpose. This explains why the known surviving examples are not the original 'field' map but copies, which are mostly neatly drawn on the pages of chronicles, prayer books and cartularies, books that stored the copies of an institution's important documents.[56]

An exception to this generalization was the sea or 'portolan' chart (so called from written instructions (*portolani*) for navigators). Such a chart was drawn either on a single animal skin and kept rolled up, or the skin was cut into smaller pieces of similar size to be used for charts bound into an atlas.[57] The charts found today tend to be visually attractive, and it is a matter of debate whether these exemplars had ever been taken to sea or if they had been created for office use or, in the most magnificent cases, presentation.[58] Without doubt, though, the principle of a portolan chart is that it was made for practical use, primarily intended as an aid in trade and navigation in the Mediterranean and Black Seas, and their patrons and creators accordingly set a premium on the accurate depiction of coastlines. Debate also surrounds the date of their appearance and whether the structure of the charts was the fruit of centuries of accumulated practical seafaring and navigational experience or, as a few believe, was mathematical. The earliest known portolan chart, the Carte Pisane, dates from the late thirteenth century (although this has been challenged and a much later date suggested), but charts probably came into existence a century earlier in the western Mediterranean area soon after the invention of the magnetic compass.[59] The major centres of production were in Catalonia (Barcelona, Majorca) and Italy (Genoa, Venice). Portolan charts were not used by those sailing in the northern and north-western waters of Europe, but examples began to be seen in England in the course of the fourteenth century – possibly as early as the late 1320s but almost certainly by the 1370s – when elegant portolan charts and atlases were being presented to French and English monarchs by Catalan and Italian merchants seeking commercial advantages.[60]

In the course of the late fourteenth century significant changes also occurred in terrestrial mapping, although initially only in Northern Italy. Political instability – the result of economic prosperity, a relatively large population, and a fertile, generally flat terrain with a lack of natural boundaries – may explain the precocious appearance of large utilitarian sheet maps, generally from a military context in this region. By the middle of the fifteenth century there were many such maps in the Venetian and Milanese archives, several of which survive.[61]

In Europe north of the Alps, in the same period, a slow process was beginning that was to culminate in fundamental changes in map types and map use two centuries later. Change came slowly, however, and for long periods awareness of the practical potential of maps was confined to a few individuals within particular regions. We see it first in the fourteenth century in the increasing number of sketch maps of local areas that were being drawn, sometimes on loose sheets, sometimes accompanying property deeds, sometimes into notebooks, but usually in a legal context, such as disputes over land. In Britain, these sketch maps seem to have been particularly common in the regions bordering the Wash in East Anglia, significantly a region with close ties through the wool trade to Lombardy and with social and physical conditions not unlike those of that part of Northern Italy.[62]

At the same time, the character of the larger maps was also changing and a new relationship between maps and their users evolved. Possibly as a consequence of increasing literacy and the growing popularity of private study, their formality began to disappear as the script became cursive. The increasing tendency to annotate and to update a map indicates a growing intimacy between users and their maps.[63] At the same time, users were realizing that it was becoming impossible to accommodate all the information now considered relevant within the traditional confines of even the largest *mappamundi*, on which the individual countries of Western Europe were not all specifically demarcated. Britain in particular, squeezed as it was into a narrow space on the map at the extremity of the world, needed to be enlarged beyond the degree of out-of-scale enlargement that had hitherto been permissible for important regions.[64] These two factors – the increase in private study and the steady increase in mappable information – may help to explain the evolution in Northern Europe of the separate sheet map, intended primarily for consultation rather than display, propaganda and education.[65] It took more than a century, however, for such maps to become common.

The eastern orientation of the Gough Map of Britain – identical to that of Britain on a *mappaemundi* – supports the idea that the inspiration for it could have come in this way in the last third of the fourteenth century from *mappaemundi*. Not only could the island, previously depicted as in book maps (such as the Anglo-Saxton map, created between 1025 and 1050, on which the island is a small detail, and Matthew Paris's maps of the 1250s) be generously represented on a sheet measuring more than half a metre high and more than a metre long, but now improvements could be made to its outline from portolan charts, especially for eastern and south-eastern England.[66] Neither the motive for nor the location of the Gough Map's creation are known, but by the 1470s several copies had been made, one of which was owned by Merton College, Oxford.[67] In general, though, it can be said that one factor encouraging the multiplication of exemplars was the growing strength of English patriotism that was evident from about 1400.[68] The faded state of the Gough Map, and the repeated re-writings, particularly in the south-east, suggests that it was regularly consulted – rolled and unrolled – and exposed to light for extensive periods during the fifteenth century, by when it was possibly housed with the records of the royal household for reference in administration.

Whether or not the English authorities made use of the Gough Map in some way is unknown, but nothing in political events of the fifteenth century can be linked to the avail-ability of such a map. In fact, it is striking how impervious the elite in England seems to have been to the cartographic innovations outside Britain throughout the fifteenth and during the opening years of the sixteenth centuries, even when English leaders were presented with the 'new' maps. Between 1437 and 1464, John Hardyng (who, significantly, had spent time in Italy early in his career) repeatedly tried to encourage first Henry VI and then Edward IV to capture Scotland by creating for them simple but effective maps illustrating possible invasion routes, but there is no evidence that *either* king, or their counsellors, understood what they were looking at, still less that they acted on what Hardyng's map showed them.[69] Similarly, while the same kings of England are known to have owned *mappaemundi* with updated, por-tolan chart-inspired, coastal outlines, and that from at least the 1450s influential members of the royal family also owned copies of Ptolemy's *Geographia*, commissioned from Italy, it seems unlikely that any members of the ruling group realized the administrative and military potential of separate sheet maps of their own or other countries such as Scotland and France.[70]

The situation in France was quite different. Here, it was Charles VIII himself who commissioned a military map of Italy from Jacques Signot in the early 1490s in anticipation of his invasion of the peninsula which took place in 1494.[71] It was only from the late 1520s, and possibly under the influence of, above all, Thomas Cromwell, who had served as a mercenary in the Italian wars and was thus likely to have been familiar with the practical value of sheet maps as planning tools, that the English government's perceptions of the utility of maps began to change, with radical results for the mapping of England later in the century.

## Notes

1 The Hereford map measures 1.65 × 1.35 metres and the Ebstorf 3.58 × 3.56 metres. In contrast, the Psalter map is less than 10 centimetres across.

2 Unless otherwise indicated, reproductions of all maps referred to in this essay are widely available. Essential starting points include: P.D.A. Harvey, *Medieval Maps* (London, The British Library, 1991); P.D.A. Harvey, 'Local and Regional Cartography in Medieval Europe', and David Woodward, 'Medieval *mappaemundi*', in J.B. Harley and Woodward, Eds, *The History of Cartography*, Volume 1 (Chicago, IL and London, University of Chicago Press, 1987), pp.464–501; Peter Barber, Ed., *The Map Book* (London, Weidenfeld & Nicolson, 2005).

3 Alfred Hiatt, *Terra Incognita. Mapping the Antipodes before 1600* (London, The British Library, 2008); Alessandro Scafi, *Mapping Paradise. A History of Heaven on Earth* (London, The British Library, 2006); ibid., *Maps of Paradise* (London, The British Library, 2013); P.D.A. Harvey, *Medieval Maps of the Holy Land* (London, The British Library, 2012); Wilma George, *Animals and Maps* (London, Secker and Warburg, 1969); Chet Van Duzer, *Sea Monsters on Medieval and Renaissance Maps* (London, The British Library, 2013).

4 For the revisionary study of the Gough Map, see Catherine Delano-Smith and Nick Millea, Eds, et al., 'New light on the medieval Gough Map of Britain' *Imago Mundi* 69:1 (2017), pp.1–36.

5 The need for cross-disciplinary scholarly collaboration, in this instance to learn about references to maps from the specialist field of medieval poetry, is exemplified by the case of Baudri de Bourgueil, abbot of Bourgueil and later Archbishop of Dol, whose poem honouring William the Conqueror's daughter, Adela of Blois, contains a detailed description of the tripartite map of the world displayed in her audience chamber: Patrick Gautier Dalché and Jean-Yves Tilliette, 'Un nouveau document sur la tradition du poéme de Baudri de Boureuil à la contesse Adèle', *Bibliothèque de l'Ecole des Chartes x144/2* (1986), pp.241–257; David Woodward, 'Medieval *mappaemundi* (see note 2), p.339; and most recently, Natalia Lozovsky, 'Maps and Panegyrics: Roman Geo-Ethnographical Rhetoric in Antiquity and the Middle Ages' in Richard Talbert and Richard W. Unger, Eds, *Cartography in Antiquity and the Middle Ages: Fresh Perspectives, New Methods* (Leiden, The Netherlands and Boston, MA: Brill, 2008), pp.169–188 and particularly pp.182–186. Whether this particular map ever existed is a matter of debate, but the fact that it is mentioned suggests that readers of the poem were expected to be familiar with both the concept of a map in such a location and also, probably, with a map of this nature.

6 See, for example, Alfred Hiatt, 'The Map of Macrobius Before 1100', *Imago Mundi*, 59:2 (2007), pp.149–176. The same can be said of texts describing maps, as Patrick Gautier Dalché has shown in, for just one example, 'Décrire le Monde et Situer les Lieux au XIIe Siècle: L'Expositio Mappe Mundi et la Généaologie de la Mappemonde de Hereford', *Mélanges de l'Ecole Française de Rom*, 113/1 (2001), pp.343–409. See also Catherine Delano-Smith, 'Some Contemporary Manuscripts of Nicholas of Lyra's, *Postilla litteralis* (1323–1332): Maps, Plans and Other Illustrations' in Nathalie Bouloux, Aca Dan and Georges Tolias, Eds, *Orbis Disciplinae: Hommages en l'Honneur de Patrick Gautier Dalché* (Turnhout, Belgium, Brepols, 2017), pp.199–232.

7 P.D.A. Harvey, 'Medieval Local Maps from the German-Speaking Lands and Central Europe', in Gerhard Holzer, Thomas Horst and Petra Svatek, Eds, *Die Leidenschaft des Sammelns, Edition Woldan 3* (Vienna, Verlag der Österreichischen Akademie der Wisseschaften, 2010), pp.113–123. See also, Pilar Chiás Navarro, 'Two Thirteenth-Century Spanish Local Maps', *Imago Mundi*, 65:2 (2013), pp.268–279.

8  For the second Psalter map, see Evelyn Edson, *Mapping Space and Time. How Medieval Mapmakers Viewed their World* (London, The British Library, 1997), pp.136–137. For the three anonymous Holy Land list maps (Holy Land, Jerusalem, Temple), possibly by Hermonn von Sina, see the facsimile *Prologus Arminensis* (Geneva, 1885) with an introduction by W.A. Neuman, and P.D.A. Harvey, *Medieval Maps of the Holy Land* (London, The British Library, 2012), p.147, plate 75.

9  Nicholas of Lyra was possibly the most consistent medieval exegetical writer in his use of 'figure' in introducing the 54 individual explanatory drawings in his commentary on the Bible, *Postilla literalis super totam Bibliam* (1323–1332), yet the same word is used for a map of Canaan, plans of buildings, elevations of buildings and portrayals of temple furnishings and the high priest. Other words common in the later Middle Ages include *pictura*, *descriptio* and *tabula*. For maps and plans in medieval exegesis, see Thomas O'Loughlin, 'Map and Text: A Mid Ninth-Century Map For the Book of Joshua', *Imago Mundi*, 57:1 (2005) pp.7–22; Catherine Delano-Smith, 'Maps and Plans in Richard of St Victor, *In visionem Ezechielis*', in Ann Matter and Lesley Smith, Eds, *From Knowledge to Beatitude: St Victor, Twelfth-Century Scholars, and Beyond* (Notre Dame, IN, University of Notre Dame Press, 2013), pp.1–45; ibid., 'Some Contemporary Manuscripts of Nicholas of Lyra's *Postilla litteralis* (1323–1332): Maps, Plans and Other Illustrations', in Nathalie Bouloux, Anca Dan and Georges Tolias, Eds, *Orbis disciplinae: Hommages en l'honneur de Patrick Gautier Dalché* (Turnhout, Belgium, Brépols, 2016), pp.199–232.

10  Figures from S.N. Broadberry, B.M Campbell and B. van Leewen, 'English Medieval Populations: Reconciling Time Series and Cross-Sectional Evidence', available at: *www2.warwick.ac.uk/fac/soc/economics/staff/sbroadberry/wp/medievalpopulation7.pdf* (Accessed: 27 October 2016). Elsewhere, only Italy (10 million) and France with the Low Countries (19 million) stood out as densely populated: Fordham University, 'Medieval Sourcebook: Tables on Population in Medieval Europe', available at: *http://legacy.fordham.edu/halsall/source/pop-in-eur.asp*. Note the latter are based on J.C. Russell's, on which see previous reference. As today, only a small proportion of the literate would have been inclined to use drawing as an aid to explanation.

11  On medieval literacy in general, see Janet Coleman, *English Literacy in History: 1350–1400, Medieval Readers and Writers* (London, Hutchinson, 1981), esp. pp.25–26; as applied to maps, see Catherine Delano-Smith, 'Maps and Literacy I: Different Users, Different Maps' and 'Maps and Literacy II: Learning, Education and Training', in David Woodward, Catherine Delano-Smith and Cordell Yee, Eds, *Plantejaments I objectuis d'una història universal de la cartografia/ Approaches and Challenges in a Worldwide History of Cartography* (Barcelona, Institut Cartogràfic de Catalunya, 2001), pp.223–239, 241–262. Available at: *www.icc.cat/eng/Home-ICC/Publications/Books/Books-about-old-cartography* (Accessed: 27 October 2016).

12  Catherine Delano-Smith, 'Map Collections and Libraries in England and Their Place in the History of Cartography', in C. Delano-Smith and Roger J. P. Kain, Eds, *La Cartografia Anglesa* (Barcelona, Institut Cartogràfic de Catalunya, 1997), pp.253–269. Available at: *www.icc.cat/eng/Home-ICC/Publications/Books/Books-about-old-cartography* (Accessed: 27 October 2016).

13  Catherine Delano-Smith, 'Milieus of Mobility: Itineraries, Route Maps and Road Maps', in J.R. Akerman, Ed., *Cartographies of Travel and Navigation. Twelfth Nebenzahl Lectures in the History of Cartography, Newberry Library, 1996* (Chicago, IL, Chicago University Press, 2006), pp.16–68, 294–309. See also Delano-Smith and Kain, *English Maps* (see note 12), p.142 et seq.

14  See Harley and Woodward, *The History of Cartography* (see note 2), Volume 1. For prehistoric maps as fossilized prayers, see Barber, *The Map Book* (see note 2), pp.10–11.

15  Charles Burnett, *The Introduction of Arabic Learning into England* (London, The British Library, 1997).

16  Dale Kedwards has made a study of maps in medieval Iceland: 'Cartography and Culture in Medieval Iceland'. PhD thesis, Department of English and Related Literature, University of York, 2014.

17  Woodward, 'Medieval *mappaemundi*' (see note 2), pp.286–370.

18  The distinction between book and wall maps was clearly implied by Marcel Destombes, from whom Woodward (see note 2) took much of his corpus. See Marcel Destombes, *Mappemondes. A.D. 1200–1500* (Amsterdam, N. Israel, 1964), where Chapters II and III list the maps under the name of the author of the text in which they are found and Chapter IV contains a 'Catalogues des mappemondes manuscrites isolées'. Confusingly, *mappaemundi* also serves as the universal generic term.

19  The principles of drawing in diagrammatic style are given in Catherine Delano-Smith, 'For Whom the Map Speaks: Recognising the Reader', in Gestelt, Paula van and Peter den der Krogt, Eds, *Liber Amicorum Guenter Schilder* ('t Goy-Houten, HES & de Graaf, for the Faculty of Geosciences, University of Utrecht, 2006), pp.627–636. Failure to understand these principles led many early and mid twentieth-century writers to dismiss *mappaemundi* as 'crude' or, derogatively, schematic. On

medieval writers' appreciation of the problem of representing three-dimensional space on a plane surface (*in plano*), see John E. Murdoch, *Album of Science: Antiquity and the Middle Ages* (New York, Charles Scribner's Sons, 1984), pp.125–126 ff.

20 See Destombes, *Mappemondes* (see note 18) Chapter 1; Woodward, 'Medieval *mappaemundi*' (see note 2) with illustrations; and Edson, *Mapping Time and Space* (see note 8), especially pp.36–51. For a different slant, see Murdoch, *Album of Science* (see note 19) in chapters 12 and 21 to 26.

21 Diagrammatic 'mappa mundi' are sometimes found out of context. The example from San Gallen MS 237, illustrated in Woodward 'Medieval *mappaemundi*' (see note 2), p.303, figure 18.14, and there ascribed, as in Konrad Miller, *Mappaemundi: Die ältesten Weltkarten* (Stuttgart, J. Roth, 1895–98, 6 vols), p.58, to Isidore of Seville's *Etymologia*, is in fact on the cover of the codex. Moreover, the drawing and various labelling are in different hands, all thought to be much later than Miller's 'late seventh or early eighth century' (possibly as late as eleventh century). The second manuscript in the codex, which could have been written in the late ninth or tenth century, contains Isidore's text. The description of the world, starting with Asia, is appropriately illustrated by a simple T-O diagram naming the three continents placed adjacent to the text to which it applies (folio 219).

22 Peter Barber, 'The Evesham World Map: A Late Mediaeval English view of God and the World', *Imago Mundi* 47 (1995), pp.15–17. Henry of Mainz's map is known from a single exemplar, drawn on a separate folio, in the copy of his work once in the library at Sawley Abbey, Yorkshire: Harvey, *Medieval Maps* (see note 2), p.25. As regards Beatus of Liébana's map, however, Sandra Sáenz-López Pérez, *The Beatus Maps. The Revelations of the World in the Middle Ages* (Burgos, Siloé, 2014), p.272, points out that the plural form of *picturarum* could be taken to imply that the reference is to the four heads of the apostles shown on the map rather than to the map as a whole.

23 The map for Eusebius' *Onomasticon* (c.325) has not survived, but for an idea how it may have looked, see O'Loughlin (see note 9), Map and Text, figure 1, plate 1.

24 Discussed by Delano-Smith, 'Some Contemporary Manuscripts' (see note 9).

25 Folio 37 contains the text in a large clear hand in black ink; brief interlinear glosses in a small hand in dense black ink; and brief interlinear glosses in pale brown ink, all three in different eighth/ninth-century Carolingian minuscule hands. The map filling the margin, upper right, is in two inks (red and pale brown). The sketch depicts the area between the Persian Gulf and the Mediterranean coasts of Palestine and Egypt. The Exodus route, with Red Sea Crossing, the Jordan with its two sources and the Nile are marked, and places such as Jerusalem and Sinai prominently indicated.

26 Eva M Sandford, 'The Manuscripts of Lucan. Acessus and Marginalia', *Speculum* IX (1934), pp.293–295. For a list of the maps, see Destombes, *Mappemondes* (see note 18), p.39 and pp.74–78.

27 See George Tolias, '*Isolarii*, Fifteenth to Seventeenth Centuries' in David Woodward, Ed., *The History of Cartography*, Volume 3 (Chicago, IL and London, University of Chicago Press, 2007, 2 vols), 1, pp.263–284.

28 For an introduction to the role of maps and diagrams in such genealogies, see Andrea Worm, 'Ista est Jerusalem'. Intertextuality and visual exegesis in Peter of Poitiers' *Compendium historiae in genealogia Christi* and Werner Rolevinck's *Fasciculis temporum* in Lucy Donkin and Hanna Vorholt, Eds, *Imagining Jerusalem in the Medieval West* (Oxford, UK, Oxford University Press for The British Academy, 2012), pp.123–161. The literature on medieval plans of Jerusalem is extensive but see, in the same volume (pp.163–199), Hanna Vorholt, 'Studying with Maps: Jerusalem and the Holy Land in Two Thirteenth-Century Manuscripts' and her 'Touching the Tomb of Christ: Notes on a Twelfth-Century Map of Jerusalem from Winchcombe, Gloucestershire', *Imago Mundi*, 61:2 (2009) pp.244–255.

29 Camille Serchuck, 'Picturing France in the Fifteenth Century: The map in BNF MS Fr. 4991' *Imago Mundi* 58:2 (2006), pp.133–149.

30 It has been suggested (notably by Kai Broderson, *Terra Cognita: Studien zur römischen Raumerfassung* (Hildesheim, Germany, Olms, 1995) that there were no maps at all in the Roman period. For evidence to the contrary, see the essays by Michael Lewis, Richard Talbert and Benet Salway in Richard J.A. Talbert, Ed., *Ancient Perspectives. Maps and Their Place in Mesopotamia, Egypt, Greece & Rome* (Chicago, IL and London, University of Chicago Press, 2012), pp.129–234, For further fragmentary maps, apparently of Spain, on a papyrus roll dating from the first century AD, see Claudio Gallazzi and Bärbel Kramer, 'Artemidor im Zeichensaal. Eine Papyrusrolle mit Text, Landkarte und Skizzenbüchern aus späthellenistischer Zeit', *Archiv für Papyrusfuorschung* 44 (1998), pp.189–208 and more recently for a set of essays on the whole manuscript, but particularly the maps, Kai Broderson and Jas Elsner, Eds, *Images and Texts on the "Artemidorus Papyrus": Working Papers on P. Artemid* (Stuttgart, Germany, Steiner, 2009). On the surviving fragments of the wall maps, see the Stanford Digital Forma Urbis Romae Project on *http://formaurbis.stanford.edu* and for the Orange Cadaster *http://orange.archeo-rome.com*.

31 Eugenio La Rocca, 'The Newly-Discovered Fresco from Trajan's Bath, Rome', *Imago Mundi* 53 (2001), pp.121–124. Although this fresco is as yet unique among known Roman wall frescoes (which tend to portray rural scenes), its position on an external wall suggests a functional affinity with the Roman marble plans.

32 Mark Rosen *The Mapping of Power in Renaissance Italy: Painted Cartographic Cycles in Social and Intellectual Context* (Cambridge, UK, Cambridge University Press, 2015) p.44. There is a reference in the catalogue of the books at the Abbey of Reichenau dating from 821–2 to a free-standing map. Its size and content is unclear, but it might be the earliest reference to a *mappamundi* that was not created on a wall or a floor: Patrick Gautier-Dalché, 'Maps in Words: The Descriptive Logic of Medieval Geography from the Eighth to the Twelfth Century', in P.D.A. Harvey, Ed., *The Hereford World Map. Medieval World Maps and their Context* (London, British Library, 2006), p.224.

33 See the website *www.christusrex.org/www1/ofm/mad/* (Accessed: 27 December 2015); also M. Picirillo and E. Alliata, Eds, *The Madaba Map Centenary 1897–1997. Travelling through the Byzantine Umayyad Period. Proceedings of the International Conference Held in Amman, 7–9 April 1997* (SBF Collectio Maior 40) (Jerusalem, 1999). See also Zev Vilnay, *The Holy Land in Old Prints and Maps* (Jerusalem, 1993, transl. Ester Vilnay and Max Nurock), pp.2–3, 85, 93, 95, 120; 200 and P.D.A. Harvey, *Medieval Maps of the Holy Land* (London, The British Library, 2012), pp.19–20. For a possible fourth-century mosaic map of the islands of the Mediterranean, see Fathi Bejaqui, 'Iles et villes de la Méditerranée sur un mosaïque d'*Ammaedara (Haidra, Tunisie)*', *Académie des inscriptions et belles-lettres. Comptes rendus de la séance de 1997 juillet-octobre,* year 141, volume 3 (1997), pp.825–858. For a general survey of classical and early medieval display maps, see Peter Barber and Tom Harper, *Magnificent Maps: Power, Propaganda and Art* (London, The British Library, 2010), pp.12–13.

34 Patrick Gautier Dalché, 'Maps in Words' in Harvey, *Hereford World Map* (see note 32), pp.223–252 and particularly p.234.

35 Margriet Hoogvliet, *Pictura et scriptura: textes, images et hérmeneutique des mappae mundi (XIIIe-XVIe s.)* (Turnhout, Belgium, Brepols, 2006), particularly chapter 5; Uwe Roberg, 'Die Tierwelt auf der Ebstorfer Weltkarte im Kontext mittelalterlicher Enzyklopedik' in Helmut Kugler and Michael Eckhard, Eds, *Ein Welbild vor Columbus. Die Ebstorfer Weltkarte. Interdisziplinäres Colloquium 1988* (Weinheim, Germany, Acta Humaniora, 1991), pp.319–346. See also Margriet Hoogvliet, 'Animals in Context: Beasts on the Hereford Map', in Harvey, *Hereford Map* (see note 32), pp.155–165.

36 Patrick Gautier Dalché, '"Réalité" et "Symbole" dans la géographie de Hugues de Saint-Victor' in *Ugo di San Vittore: Atti del xlvii Convegno storico internazionale, Todi. 10–12 ottobre 2010*, (Spoleto, Fondazione Centro italiano di studi sull'alto Medioevo, 2011), pp.359–381.

37 Paulinus Orosius, *Seven Books of History Against the Pagans* composed 416–7; Isidore of Seville *Of the Nature of Things De natura rerum); Etymolgia (Origins)*, both composed c.600. See also Edson, *Mapping Time and Space* (see note 8), pp.31–50.

38 Brigitte Englisch, *Ordo Orbis Terrae. Die Weltsicht in den Mappe Mundi des frühen und hohen Mittelalters* (Berlin, Akademie Verlag, 2002).

39 Thus the scene in the bottom right corner of the Hereford world map, showing a knight looking back at the world while his page urges him to move forward, can be read as a symbol for the transience of the human world, although Valerie Flint has argued that it is also a reference to a judgement on the Ledbury Chase in the Malvern Hills in Herefordshire, made in (St) Thomas Cantilupe's favour and against the earls of Gloucester in April 1278: Valerie Flint, 'The Hereford Map: Its Author(S), Two Scenes and a Border' *Transactions of the Royal Historical Society* (6th series, VIII) (1998), pp.38–39. Flint's argument has been challenged by P.D.A. Harvey and others: see P.D.A. Harvey, Ed., *The Hereford World Map* (London, Folio Society, 2010), p.29.

40 P.D.A. Harvey, 'Maps of the World in the Medieval English Royal Wardrobe', in Paul Brand and Sean Cunningham, Eds, *Foundations of Medieval Scholarship: Records Edited in Honour of David Crook* (London, National Archives & Borthwick Institute, 2008), pp.51–55, citing references to world maps on cloth and parchment accompanying Kings Edward I, II and III on their progresses between 1296 and 1341. To these royal *mappaemundi* should be added Henry III's world maps in Winchester and Westminster. For *mappaemundi* owned by the nobility, see or the map commissioned by Richard or (more likely) his son Edmund, earls of Cornwall c.1290, Graham Haslam, 'The Duchy of Cornwall Map Fragment', in Monique Pelletier, Ed., *Géographie du Monde au Moyen Age et à la Renaissance* (Paris, Editions du Comité des Travaux Historiques et Scientifiques, 1989), pp.33–44; and, for the map once owned by Thomas, 2nd Earl of Lancaster (d. 1322) see Harvey, 'Maps of the World', pp.53–54. It is rare to glean any idea of dimensions although Harvey (ibid.) notes that the *mappamundi* being used by Edward II and his son was probably on a single animal skin (like the Hereford *mappamundi*) and Haslam (p.36) estimates that the

Duchy of Cornwall map, of which only a fragment survives, is likely to have been slightly larger than the Hereford map. The location of *mappamundi* in audience chambers is not mentioned in the surviving documentation; conceivably the largest could have been displayed behind the throne. See Daniel Birkholz, *The King's Two Maps: Cartography and Culture in Thirteenth-Century England* (New York and London, Routledge, 2004), pp.9–10. The tradition of displaying world maps in audience chambers may have gone back to late Imperial Rome: see R.J.A. Talbert, *Rome's World. The Peutinger Map Reconsidered* (Cambridge, UK, Cambridge University Press, 2010), pp.144–157.

41 Catherine Delano-Smith, 'Milieus of Mobility' (see note 13), pp.18–19 has remarked that for medieval European kings and emperors 'travelling was virtually synonymous with ruling ... the royal progress was primarily a propaganda exercise', a comment particularly relevant in the context of the *mappaemundi*. On Henry III's travels, see Graham D. Keevill. *Medieval Palaces. An Archaeology* (Stroud, Tempus, 2000), pp.82–85. For the progresses of Charlemagne, who according to the chronicler Einhard, owned silver tables containing maps of the world, Rome and Constantinople, presumably for display during the progresses (see Einhard and Notker the Stammerer, *Two Lives of Charlemagne*, translated with an introduction by Lewis Thorpe) (Harmondsworth, UK, Penguin, 1969), p.89. See Norbert Ohler, *The Medieval Traveller* (Woodbridge, UK, Boydell, 1989, transl. Caroline Hillier), pp.153–160, 162.

42 In this vein, it has been suggested, although challenged by Harvey, *The Hereford World Map* (see note 39, p.33), that the Hereford *mappamundi* was intended to form part of the pious and scholarly adornment of the area in the immediate vicinity of Thomas Cantilupe's tomb in Hereford Cathedral. This adornment played a significant promotional role in the successful campaign of his successor as bishop of Hereford, Richard Swinfield, between 1282 and 1316 to secure his canonization: see Daniel Terkla 'The Original Placement of the Hereford *Mappa Mundi*' *Imago Mundi* 56:2 (2004), pp.131–151. Part of the overall objectives of the canonization was to raise the profile and wealth of Hereford as a pilgrimage centre: in connection with the role of the map in pilgrimage, see Thomas de Wesselow, 'Locating the Hereford *Mappamundi*', *Imago Mundi* 65:2 (2013), pp.180–206, who suggests a different location for the map within the cathedral, on the way to Cantilupe's tomb rather than close by it.

43 Peter Barber and Michelle Brown, 'The Aslake World Map', *Imago Mundi* 44 (1992), pp.31–32; Paul Binski *The Painted Chamber at Westminster* (London, Society of Antiquaries, 1986), pp.16–17, 43–44; Birkholz (see note 40).

44 Armin Wolf, 'Ikonologie der Ebstorfer Weltkarte und politische Situation des Jahres 1239. Zum weltbild des Gervaisius von Tilbury am welfischen Hofe', in Kugler and Eckhard, Eds, *Ein Welbild* (see note 34), pp.41–116 (a shorter English version of this article was published as 'News on the Ebstorf World Map: Date, Origin, Authorship' in Pelletier, *Géographie du Monde* (see note 40), pp.51–68). In a similar vein, Ingrid Baumgärtner has commented on the political significance of the number of lions – the symbol of the Welf dynasty – shown on the Ebstorf map in her 'Erzählungen, kartieren: Jerusalem in mittelalterlichen Kartenräumen' in Sonja Glauch, Susanne Köbele, Uta Störner-Caysa, Eds, *Projektion, Reflexion, Ferne: räumliche Vorstellungen und Denkfiguren im Mittelalter* (Berlin, de Gruyter, 2011), p.219. The copies of the Icelandic maps were inserted into a book; Kedwards (see note 16), pp.256–283 and particularly pp.270–281.

45 For Hugh of St Victor, see particularly, Patrick Gautier Dalché, *La 'Descriptio Mappae Mundi' de Hugues de St Victor* (Paris, Etudes Augustiniennes, 1988), and Danielle Lecoq, 'La "Mappemonde" du *De Arca Noe Mystica* de Hugues de Saint-Victor (1128–1129)' in Pelletier *Géographie du Monde* (see note 40), pp.9–31.

46 Marcia Kupfer, 'The Lost Wheel Map of Ambrogio Lorenzetti', *Art Bulletin* 78 (1996), pp.286–310; Thomas de Wesselow, 'The "Wall of the Mappamondo" in the Palazzo Pubblico of Siena" (Unpublished doctoral thesis, University of London, 2000) and ibid., 'Ambrogio Lorenzetti's "Mappamondo": A Fourteenth-Century Picture of the World Painted on Cloth' in Caroline Villiers, Ed., *European Paintings on Fabric Supports in the Fourteenth and Fifteenth Centuries* (London, Archetype Publications, 2000), pp.55–65.

47 Evelyn Edson, *The World Map 1300–1492. The Persistence of Tradition and Transformation* (Baltimore, MD, John Hopkins University Press, 2007); Woodward, 'Medieval *Mappaemundi*' (see note 5), pp.314–318.

48 Angelo Cattaneo, *Fra Mauro's Mappa Mundi and Fifteenth-Century Venice* (Turnhout, Belgium, Brepols, 2011); Piero Falchetta, *Fra Mauro's World Map: With a Commentary and Translation of Inscriptions* (Turnhout, Belgium, Brepols, 2006) and most recently Pieor Falchetta, *Fra Mauro's World Map: A History* (Rimini-Bologna, Italy, Imago, 2013).

49 Barber and Harper, *Magnificent Maps* (see note 33), pp.20–47. Molly Bourne, 'Francesco II Gonzaga and Maps as Palace Decorations in Renaissance Mantua' *Imago Mundi* 51 (1999), pp.51–81.

50  Jurgen Schulz, 'Jacopo de Barbari's View of Venice. Map Making, City Views and Moralized Cartography before the year 1500' in *Art Bulletin* LX (1978), pp.425–474; P.D.A. Harvey, 'Local and Regional Cartography in Medieval Europe', in Harley and Woodward, Eds, *The History of Cartography* (see note 2) 1987, pp.464–501 and particularly pp.476–482, 498.

51  See for instance Marica Milanesi's analysis of early maps of Italy including the 'Cotton' map of Italy of c.1425–50 in the British Library (Cotton Roll xiii.44), the surrounding text of which is simultaneously a eulogy of Italy's past greatness and an appeal for its unification under Venetian rule: Marica Milanesi, 'Antico e Moderno nella Cartografia Umanistica: le Grandi Carte d'Italia nel Quattrocento' in *Geographia Antiqua* xvi/xvii (2007–8), pp.153–176.

52  David Starkey, Ed., *The Inventory of King Henry VIII: The Transcript* (London, Harvey Miller for the Society of Antiquaries, 2008), nos.10749, 10752, 10761, 10762, 10775, 12298, 12334, 13804, 14554. See also Peter Barber, 'Maps, Plans and History Paintings' in Maria Hayward, David Starkey *et al.*, Eds, *The Inventory of King Henry VIII* (London, Harvey Miller for the Society of Antiquaries) 3, *Commentary on the Inventory* (forthcoming).

53  Starkey, *The Inventory of King Henry VIII* (see note 52), no.12334 (Long Gallery leading to Chapel, Hampton Court). The inventory of prints collected by Ferdinand Columbus in the same period also included several printed *mappaemundi*: see P.M. Barber, 'The Maps. Town-Views and Historical Prints in the Columbus Inventory' in M.R. McDonald, Ed., *The Print Collection of Ferdinand Columbus 1488–1539* (London, British Museum, 2004) 1, pp.251–254.

54  Tony Campbell, *The Earliest Printed Maps 1472–1500* (London, The British Library, 1987), lists 30 symbolic T-O mappae mundi of various degrees of elaboration accompanying texts by Isidore, Macrobius, Ailly, Holywood and Caxton's translation of the *Miroir du Monde*, together with those of more recent writers such as Dati and Rollewimck. An example of the use of a T-O mappa mundi in a religious context in the mid fifteenth century can be seen in Rogier van der Weyden's oil painting of St Luke painting the Virgin in the Holy Land of about 1435, in which the inverted T-O serves to locate the scene (Boston Fine Arts Museum).

55  The surviving fragment of the 'Duchy of Cornwall' *mappamundi* was discovered as a binding for the records of the Court of Augmentations, the government body responsible for the dissolution of the monasteries. The implication is that the map remained intact until the dissolution in 1539 of the Augustinian college of Bonhommes in Ashridge, Hertfordshire, to which it had been presented in about 1290: Barber and Brown, 'The Aslake World Map' (see note 43); Barber, 'The Hereford Map: Context and History' in Harvey, *The Hereford Map* (see note 32), p.22. As noted earlier, *mappaemundi* were to be seen at Henry VIII's court, but there is no evidence of them at the courts of his successors.

56  The four earliest known English local plans are in prayer books and a chronicle, and chronicles and cartularies contain several of the other surviving early local plans: see R.A. Skelton and P.D.A. Harvey, Eds, *Local Maps and Plans from Medieval England* (Oxford, UK, Clarendon Press, 1986).

57  There is a voluminous and growing literature on portolan charts, but the essay by Tony Campbell, 'Portolan Charts from the Late Thirteenth Century to 1500' in Harley and Woodward, *The History of Cartography* (see note 2), istroy of Cartography (see note 2) pp.371–463, and his 'Census of Pre-Sixteenth Century Portolan Charts' *Imago Mundi* 38 (1986), pp.67–94, remain fundamental. To these should now be added Tony Campbell's website pages on portolan charts, containing additions and amendments as well as further research: *www.maphistory.info/portolan.html* (Accessed: 28 December 2015) and Ramon Poujades, *Les cartes portolanes: la representació medieval d'una mar solcada* (Barcelona, Institut Cartogràfic de Catalunya, Institut d'Estudis Catalans, and Institut Europeu de la Mediterrània, Lunwerg, 2007). Poujades' book is in Catalan and Spanish, with an English translation, 'Portolan Charts: The Medieval Representation of a Ploughed Sea', pp.401–526.

58  Poujades is convinced that portolan charts were used at sea (see his article 'Les cartes de navigation: premieres cartes a large diffusion sociales' in Catherine Hoffman, Hélène Richard and Emmanuelle Vagnon, Eds, *L'age d'or des cartes marines. Quand l'Europe découvrait le monde* (Paris, Seuil and Bibliothèque nationale de France, 2012), pp.60–65. However, Patrick Gautier Dalché has questioned this premise: Patrick Gautier Dalché, *Carte marine et portulan au XIIe siècle. Le Liber de existencia rivieriarum et forma maris nostri Mediterranei, Pise, circa 1200* (Rome, École Française de Rome, 1995).

59  For the twelfth-century origins, Gautier Dalché, *Carte marine et portulan* (see note 58). For the debate over the dating of the Carte Pisane, with Campbell supporting the traditional late thirteenth-century dating and Poujades proposing that it is a much later and inferior copy of an earlier original. Available at: *www.maphistory.info/CartePisaneTEXT.html#intro* (Accessed: 28 December 2015).

60  Marino Sanudo Torcello claimed in a letter to John, Duke of Lorraine and Limburg of 1326 that he had sent a copy of his tract *Liber Secretorum*, containing the blueprint for the invasion of the Holy Land

by way of Egypt, to Edward II: Sherman Roddy, 'The Correspondence of Marino Sanudo Torcelli' (Unpublished Ph.D. thesis, University of Pennsylvania, 1976), p.154, quoted by Peter Lock in the introduction to his translation *Marino Sanudo Torcello: The Book of Secrets of the Faithful of the Cross* (Farnham, UK, Ashgate, 2011), pp.12–13. However, Sanudo did not always forward the accompanying portolan maps, drawn in the workshop of Petrus Vesconte of Venice, and there is no evidence, even had he done so in this instance, that the novel style of the map was noticed by the King or his counsellors. For further on Sanudo, see Edson, *Mapping Time* (see note 20), pp.60–74; for a discussion of Sanudo's lobbying techniques, see Christopher J. Tyerman, 'Marino Sanudo Torcello and the Lost Crusade: Lobbying in the 14th Century', *Transactions of the Royal Historical Society*, 5th ser. 52 (1982), pp.57–73. The Aslake map of about 1365–85 contains portolan chart outlines that, Barber has argued, were probably copied from a Catalan world map with portolan chart outlines dating from c.1350–1367, which the author of the Aslake map had most likely seen at Westminster: see Barber and Brown, 'Aslake Map' (see note 43), pp.33–34. The Catalan Atlas (BNF Manuscrits espagnol 30), created in Majorca in 1375 and probably a gift from the king of Aragon to the king of France, was in the French royal collections by 1378: see Edson, *The World Map 1300–1492* (see note 47), pp.74–86. The Atlas is illustrated in Hofmann, Richard and Vagnon, *Age d'or* (see note 58), pp.42–55. For a proposed presentation of richly decorated portolan chart (probably a portolan chart-influenced mappamundi) to Richard II while in Ireland in July 1399 by an Italian merchant, see Jenny Stratford, *Richard II and the English Royal Treasure* (Aldeburgh, UK, Boydell, 2013), p.35.

61  Harvey, 'Local and Regional Cartography' (see note 2), pp.478–482.

62  Skelton and Harvey, *Local Maps and Plans* (see note 56), p.xvi, and Harvey, 'Local and Regional Cartography' (see note 2), p.484 and figure 20.19.

63  Anna-Dorothea von den Brincken, *Kartographische Quellen Welt-, See- und Regionalkarten* (Typologie des sources du moyen age occidental Faasc. 51; A-V3*) (Turnhout, Belgium, Brepols and the Université catholique de Louvain, 1988), p.68; Barber and Brown, 'The Aslake World Map' (see note 43), p.29.

64  See, for example, the grossly enlarged England depicted on the Evesham world map of about 1390: Barber, 'The Evesham World Map (see note 22), pp.13–33.

65  Despite its size and likely use for display and propaganda, the text of the Hereford *mappamundi* explicitly states it was also intended for consultation, while Barber has argued that the Evesham World Map was also probably intended for educational purposes in much the same way that wall maps were in schools in the nineteenth and twentieth centuries (Barber, see note 64 above, pp.27–28).

66  The Gough map measurements are approximately 55 × 116 cms. For a useful if now slightly dated summary, see Nick Millea, *The Gough Map. The Earliest Road Map of Great Britain?* (Oxford, UK, Bodleian Library 2007). For the essential updating, see Wikipedia (see note 4).

67  Rodney Thomson, 'Medieval Maps at Merton College Oxford', *Imago Mundi*, 61:1 (2009), pp.84–90, particularly pp.86–89.

68  Barber, see note 64 above.

69  Alfred Hiatt, 'Beyond a Border: The Maps of Scotland in John Hardyng's "Chronicle" in Jenny Stratford and Shaun Tyas, Eds, *The Lancastrian Court: Proceedings of the 18th Harlaxton Symposium* (Spalding, UK, Paul Watkins Publishing, 2003), pp.78–94.

70  Henry V's brother Humphrey, Duke of Gloucester and founder of the original Bodleian Library, commissioned a manuscript copy of the *Geographia* from Milan: Patrick Gautier Dalché, 'The Reception of Ptolemy's *Geography*', in Woodward, *History of Cartography 3* (see note 2), p.319. Gautier Dalché points out (pp.285–364) that it took several decades for the mathematical and geographical qualities of Ptolemy's co-ordinates and maps to be appreciated even in mainland Europe.

71  David Buisseret, 'Monarchs, Ministers, and Maps in France before the Accession of Louis XIV' in David Buisseret, Ed., *Monarchs, Ministers and Maps. The Emergence of Cartography as a Tool of Government in Early Modern Europe* (Chicago, IL, University of Chicago Press, 1992), p.101. Just a few years later, Lodovico 'il Moro', Duke of Milan, for reasons of commerce as well as defence, ordered a copy of Konrad Türst's manuscript map of the Swiss Confederation of 1495–7 from its author: Andrea Gamberini, 'Il ducato di Milano e gli Svyceri: uno sguardo d'insieme' in *Bolletino della Società Storica Locarnese* 16 (2013), p.14.

# 10

# Cartography and the 'Age of Discovery'

*Radu Leca*

## Deconstructing the paradigm

A comprehensive study of mapping in this period globally would be beyond the scope of this volume. In many ways, *The History of Cartography* series has already achieved that and remains an excellent reference work available online. The admittedly vast chronological and geographical scope of this theme makes it necessary to focus on some issues at the expense of others. The present chapter is an orienting presentation of recent critical and historiographical developments in the study of mapping in this period. It undertakes three tasks: first, addressing some of the inconsistencies of the 'Age of Discovery' paradigm resulting from attempts to impose modern constructs of bounded geo-political space on the rhetorics and 'realities' of the early modern world; second, rethinking maps as integral parts of an intellectual conversation about geographical knowledge; and third, analysing some of the ways in which maps acted as centring devices within a given cultural context.

Until the mid 1980s, both the history of cartography and historical studies in general were shaped by a Eurocentric and empiric paradigm. The study of maps was seen to accompany the recollection of European 'voyages of discovery' and confirm the rise of European powers on the world stage. The narrative usually began in the fifteenth century with the Portuguese exploration of the west coast of Africa, culminating in Vasco da Gama's inauguration of a direct sea route to India, and continuing with the gradual expansion of trade routes in Asia. It then turned to Christopher Columbus's discovery of islands in the Caribbean and the efforts of the newly unified Spanish crown to compete with the Portuguese. The next major moment was the first circumnavigation by the expedition initially led by Ferdinand Magellan. Meanwhile, the Spanish were conquering the Inca and Aztec empires in South America, while North America and East Asia were becoming the object of various colonial interests. The newly surveyed territories were visualized by publishers across Europe on decorative maps of evolving formats, culminating in the achievements of Dutch cartographers such as Abraham Ortelius; their prominence signalling a general shift towards Dutch achievements after the beginning of the seventeenth century.

However, recent decades have brought a series of innovations to the history of cartography, expanding the scope of map study by enlarging its definition, by highlighting their use and materiality, and by steering away from aesthetically pleasing maps and from the emphasis on progress and the elites (Edney, 2007). The resulting relativization of historiographical narratives has been summed up by the term 'cartography without progress' (Edney, 1993). Overall, the field has come to recognize that social, political and cultural phenomena must all be taken into account when studying the production and use of maps (Wood, 1984; Harley, 1989). For instance, Renaissance cosmography has been characterized as a 'mode', or a historically specific set of social and technical relations that determine representational practices (Cosgrove, 2007: 55).

In the case of European mapping, the tendency has been to emphasize the increasingly precise world view resulting from the confluence of geographical discoveries, empirical investigative methods proposed by Francis Bacon and Galileo Galilei, and innovative cartographic techniques such as Gerardus Mercator's loxodrome projection (Reitinger, 2007: 441). However, the term 'discovery' itself has been critiqued as Eurocentric, and it is worth considering that even within European writings of the time, the term did not have a unitary meaning (Edwards, 1985). Conversely, European cartography was by no means accurate, as shown by the variability of names for the Indian Ocean on European maps (Mukherjee, 2013: 234) or by the Zeno map, widely believed at the time to be a trustworthy rendering of Frisland, an island now known to be non-existent (Johnson, 1994; Fiorani, 2012).

A more notorious example is the Line of Demarcation defined by the 1494 Treaty of Tordesillas, 370 leagues west of the Cape Verde Islands. Spain was allocated the territories west of this line, and Portugal the ones east (Alegria *et al.*, 2007: 996–997). However, this was an arbitrary line, since until the late eighteenth century it was impossible to measure longitude precisely. It has therefore been called a 'line of anticipation' (Monmonier, 2010: 52) and 'a treaty in search of a map' (Earle, 2013). One of the reasons why this did not concern the treaty's signatories was that dry land was not known to exist below the Caribbean at the time. After Magellan's circumnavigation and the clash of Spanish and Portuguese interests in the 'spice islands' of Southeast Asia, the ratification of an antimeridian line by the Treaty of Saragossa of 1525 was inconclusive. Spanish interests in Manila maintained the tension even after the Spanish and Portuguese crowns were united in 1580 (Sandman, 2007: 1108–1115; Subrahmanyam 2007: 1360). Related to these controversies over longitude was Spanish Jesuit José de Acosta's support of the idea, illustrated on many sixteenth-century European maps, that America and Asia were connected by land, which would mean that territories in Northeast Asia belonged to the Spanish crown (Gemegah, 2005).

Besides being limited in its selection of canonical maps, the 'age of discovery narrative' generally occluded the cartographic traditions of Islamic and Asian cultures, as well as the spatial knowledge of the cultures of Africa and the Americas. A crucial improvement in this direction has been the field's opening up both to interdisciplinary approaches and to non-European mapping practices, paralleling the call in historical studies for paradigm revision in the direction of a 'global turn' (Hunt, 2014). An especially fruitful concept in early modern historiography has been that of 'connected histories' (Subrahmanyam, 1997: 737). Applied to the history of mapping, this means recognizing first of all that geographic knowledge from Islamic and South Asian cultures informed almost all the geographical activities of European explorers. For instance, the travels of Marco Polo, which compiled Islamic and Chinese sources, inspired the initiatives of Henry the Navigator and Christopher Columbus (Park, 2012: 4). Likewise, events in the Islamic world precipitated the European age of exploration. The 1453 Ottoman conquest of Constantinople as well as political fragmentation among Islamic states led to the disruption of trade routes from India and China to Europe (Park, 2012: 7). This created an economic emergency which determined the Portuguese to begin their exploration of an alternative route around Africa.

Another form of connection was through the sharing of intellectual and technical knowledge. Islamic and European tradition often drew from the same Greek treatises, such as *Ptolemy's Almagest*, which 'circulated widely and was for many centuries the fundamental astronomical treatise in Islam, as it was in the West' (Kunitzsch, 1974; Savage-Smith, 1992: 43). Also shared were the tools of the cosmographer and astronomer such as the astrolabe, described by Ptolemy but first produced in Syria and then transmitted to Europe (Savage-Smith, 1992: 27–29). In the Western consciousness, the astrolabe's origins were characteristically occluded, to the point where a late 1580s print by Stradano attributed its invention to the Florentine 'discoverer' Amerigo Vespucci (Markey, 2012: 421). On the other hand, the Arabic origins of the quadrant – another important tool for European maritime exploration – was illustrated on such objects as Christoph Schissler's *Quadraticum geometricum* of 1569 (Mercier, 1992: 185). Celestial cartography was also shared between European and Arabic sources. Among the latter, the most influential were the maps of al-Sufi (903–986), still circulating in Arabic and Persian manuscript copies either on their own or integrated into al-Qazwīnī's *Ajā'ib al-makhlūqāt wa-gharā'ib al-mawjūdāt* ('Marvels of Things Created and Miraculous Aspects of Things Existing'). At the same time, European copies of al-Sufi's star maps informed Albrecht Dürer's 1515 woodcut *Imagines coeli Septentrionales*. Dürer recognized the legacy of Arabic cosmography by placing alongside Aratus of Cilician Soli, Claudius Ptolemy and Marcus Manilius a turbaned figure labelled Azophi Arabus as one of the patrons of celestial mapping (Savage-Smith, 1992: 61).

European cartographic thought was also being adapted to neighboring traditions: in one of the legends to his 1513 world map, the Ottoman sailor and cartographer Pîrî Re'îs describes the Atlantic Ocean:

> [the infidels] have given it the name Ovo Sano, that is to say, sound egg. Before this, it was thought that the sea had no end or limit and that at its other end was darkness. Now they have seen that this sea is girded by a coast; because it is like a lake, they have called it Ovo Sano.
>
> *(Reis, 1992: 12, cited in Emiralioğlu, 2014: 134)*

Although the tendency is to dismiss such statements as naïve, they reflect European thinking at the same time. Although there were significant advances in geographic knowledge, in a large sense European mapping of non-European territories proceeded along classical models and formats, with Ptolemy's rediscovered writings playing an important prescriptive role (Shalev and Burnett, 2011). Most European cartographers were working with the Mediterranean as their base template, and the 'closed sea' proved to be a persistent cartographic format, which migrated to depictions of all newly discovered oceans. The first was the Indian Ocean, which emerged as a cartographic space equivalent and competing with the Mediterranean on fifteenth-century world maps such as the one by Fra Mauro (O'Doherty, 2011: 46). Likewise, the Atlantic Ocean gained a new centrality and made possible the cartographical invention of the 'New World' – in a sense a transposition of the lands beyond Europe in maps of the Mediterranean (Lestringant, 1994: 27, 106; Eklund, 2015: 5–6). The Pacific Ocean was the last to be defined cartographically: in 1589 Abraham Ortelius included the map of Maris Pacifici closed off by the Terra Australis in his *Theatrum orbis terrarum* (Padrón, 2009: 20). Ortelius includes a depiction of the only surviving ship from the Magellan expedition, with presumably Magellan himself shown navigating with the astrolabe. The depiction of the island of Japan is very similar to the depiction of Columbus's Hispaniola in, for example, Pîrî Re'îs's maps (McIntosh, 2000: 91). This shows that while integrating the latest geographical knowledge, even Dutch cartographers reproduced echoes of Columbus's belief that he had indeed discovered the islands of Asia described by Marco Polo.

## Maps as conversations

After having understood that European mapping in this period was just a part of a larger interconnected history of mapping, we can now go further, starting with the integration of alternative theoretical vehicles. Kitchin and Dodge (2007) have recently made a useful distinction between maps as discrete items and mapping as a practice or process: they argue that maps 'are always *mappings*' (p. 335), and support a view of maps not as fixed entities but as always in a state of becoming. Similarly, Del Casino and Hanna (2006) argue for a methodological move beyond 'an implicit duality between production and consumption, author and reader, object and subject, design and use, representation and practice' (p.35). In other words, highlighting the processual and performative nature of maps enables the recovery of their role as intermediaries, as forms of knowledge negotiation. This 'post-representational cartography' (Azócar Fernández and Buchroithner, 2013) goes beyond Harley's understanding of maps as a 'graphic discourse, a frozen talk . . . between a patron and a drudge' (Wood, 2002: 141). Wood (2002: 141) proposes instead the notion of 'the map as a *discourse function*, that is, as one of the ways available for people to affect the behavior of others in a communication situation'.

To illustrate this dense theoretical intervention, we can examine a notorious example: the map of the world made by the Venetian Fra Mauro around 1450. Usually discussed as a late medieval *mappamundi*, it is nevertheless influenced by portolan charts in its corroboration of written and oral information and its engagement with the reader. Fra Mauro states that 'in my own day I have been careful to compare the texts with practical experience investigating for many years and frequenting persons worthy of faith, who have seen with their own eyes what I faithfully report above' (Cattaneo, 2011: 260). He also advises the readers to 'opt for that which seems to them most reasonable and probable, both to the eye and to the intellect' (ibid.: 250). In this way, the author 'constructs a dialogue between the cosmographer, his sources and his readers in the first person.' (ibid.: 259, 376). Thus, in addition to the erudite compilation of classical sources which shaped medieval *mappaemundi*, Fra Mauro appropriated the mapping project as expressing his personal understanding of the world, aware of its selectivity and its limits (ibid.: 249). This 'combination of received authority and empirical observation' came to characterize Renaissance natural philosophy (Del Soldato, 2015). Until the late sixteenth century, cosmographers such as André Thevet customarily employed 'creative patchwork' in their work (Lestringant, 1994; Brentjes, 2008: 182; 2011).

A parallel example to Fra Mauro's world map is that produced by Pîrî Re'îs in 1513. Just like Fra Mauro, the Ottoman cartographer made use of portolans and oral information: in the 1526 verse preface to his sailing guide *Kitâb-I baḥrie* ('The Book of Things of the Sea'), he claims to have obtained some of the information on the Americas and the Indian Ocean from a Spanish sailor (Soucek, 1992: 270). Indeed, it has been shown that for his 1513 map, Pîrî Re'îs used a map directly related to Christopher Columbus's second voyage to the Caribbean and probably captured by his uncle Kemâl Re'îs from a Spanish ship near Valencia in 1501 (McIntosh, 2000: 73, 131–140). As gifts to the sultan, Pîrî Re'îs's maps were part of the sixteenth- and seventeenth-century Ottoman project of becoming a world empire (Emiralioğlu, 2014; Brummett, 2015). However, their political context did not translate into the imposition of a singular ideological vision. Rather, they testify to 'a thriving dialectical tradition in framing the changing world' (Zoss, 2010: 213). Both Mercator's grand atlases and the various versions of Kâtip Çelebi's *Kitab Cihannüma* ('Book of the View of the World') encouraged 'their readers to engage dialectically with these conflicting views, rather than passively accept[ing] a "codified" presentation of the world's form' (Zoss, 2010: 207–208). Within this context, the territories of New World maps emerged as 'graphic zones of contention in which cartographic sources and traditions were by necessity confronted and reconciled' (Zoss, 2010: 203).

This process of negotiation involved a shared knowledge space, and this is most apparent in navigation mapping (Turnbull, 1996; Brentjes, 2012). Sea routes, which continued to function when overland routes were blocked, dominated trade in the period from the fifteenth to the seventeenth centuries. Centuries of sea trade between the Arabian Peninsula, the west African and Indian coast, Malacca, Java and southern China had gradually increased mutual geographic knowledge. European merchants and navigators were late guests to this scene. The Portuguese, for example, relied extensively on 'forms of cooperation' with local experts who provided specialized geographical knowledge (Barros, 2013: 150–151). Thus, in 1498, Vasco Da Gama was guided from Malindi to Calicut by the Arab pilot Šihāb ad-Din, known also as Ibn Mājid, a prolific author of nautical books, some of which informed Pîrî Re'îs's maps (Kreiner, 1996: 18). This was at odds with the official ideology of Portuguese expansionism, for which Muslims were the ultimate enemy. In practice, however, cross-denominational collaboration was common – the same Vasco da Gama, for example, made extensive use of the services of Gaspar da Gama, an interpreter (*lingua*) of Jewish origin (Couto, 2003). Cartographically, this is visible on portolan charts of the Asian trade routes, which exemplify 'the processes of cross-cultural cartographic conversations that produced sea-centered visions of the world' (Shapinsky, 2016: 17).

The above examples show that both critical and practical knowledge took part in the conversation. However, oftentimes European voices drowned out their interlocutors. This is all the more obvious in the Americas, where local populations accumulated very detailed oral knowledge of their geographic area but did not record it until their contact with European explorers. Initial interactions were innocent and reciprocal (Lewis, 1998b: 14–19). For example, when the Spanish explorer Hernando de Alarcón met with a Yuma Indian while ascending the Colorado River in 1540, the Indian made a map of the local area for him, asking for Alarcón to make in turn a map of the place he came from (Lewis, 1998a: 5, note 3). However, although de Certeau (2000) praises Jean de Léry's 1578 account of Brazilian orality, indigenous mapping practices were usually ignored, and in many cases traces remain only in European maps of the Americas (Harley, 1992).

This obscuring of origins was possible because European maps of the New World aimed to distance themselves from the violent and hybrid processes through which they came into being, and claim an empirical objectivity which motivated political conquest. One way for present researchers to revert this anesthesia is to consider maps as 'both concrete material objects and as exchange objects whose meaning and value depended on the mobility of maps' (Brückner, 2011: 147). We can therefore think of 'connected material histories' in which objects such as maps materialize 'the memory of direct connections' (Bleichmar and Martin, 2015; Clunas, 2016: 73).

One such map is the recently discovered 'Selden Map', probably made around 1608 in Bantam (Brook, 2013). It is a rare record of merchant knowledge of trade routes centred on Quanzhou in the South China Sea. Although direct trade had ceased by then, the inclusion of place names such as Calicut and of sailing instructions to the ports of Aden, Dhofar and Hormuz points to a memory of naval voyages led by Zheng He in the first half of the fifteenth century. Along with portolan charts which were being adapted to the Asian maritime space (Shapinsky, 2006), the Selden Map advances our understanding of a dynamic period in East Asia in which the demise of the Ming dynasty and the establishment of a unified polity in Japan was accompanied by an intense trade network. This led to diverse reconfigurations of regional imaginaries of how the Asian region was interconnected and how it should be ruled.

One of these reconfigurations is recorded on a fan associated with the late sixteenth-century ruler of the Japanese archipelago, Toyotomi Hideyoshi (Nanba *et al.*, 1973: cat. 4).

On one side is a map of East Asia combining Chinese conventions with a local map of the Japanese archipelago. On the other side are listed several expressions in Japanese and their Chinese translation – effectively a conversation guide, possibly used by Hideyoshi upon receiving a Ming envoy at his Osaka castle (Nanba *et al.*, 1973: 148).

## Centring maps

Although Hideyoshi's fan commended itself to a conversational setting, its Chinese phrases were brushed in vernacular Japanese script. The fan thus concomitantly advanced a vision of East Asia united under Japanese rule. As seen in other examples, although maps are conversational, they tend to emphasize one side of the conversation – they have a monologue effect. This is most apparent in maps which organize a given geographical area around a central element, thus exhibiting what has been called the 'omphalos syndrome governing territorial organization' (Mignolo, 1994: 23). For example, the long Arabic tradition of *qibla* maps and charts used complex mathematical measurement to locate the precise direction towards Mecca for prayer. They often featured a simplified worldview centred on the Ka'ba and including city names from the Iberian Peninsula to East Asia (King, 1999).

While the *qibla* maps were centripetal, there were also centrifugal maps nevertheless structured around a centre, such as the seventeenth-century Mao Kun map which depicts the naval routes taken by the Zheng He expeditions (Ma, 1970). Although some of these coincide with the trade routes in the Selden Map, the Mao Kun map differs in being the result of a centralized political decision involving more military activity than trade (Brook, 2013: 101).

Its description as a 'strip map' (Church, 2008: 2355) is an example of modern categories being applied retroactively to historical maps. The description would benefit from acknowledging the fact that East Asian cartographers used map spaces which were tailored to the formats in which they were materialized, mostly scrolls and fan-folded albums (Yee, 2015: 101). Thus, although originally the Mao Kun map would have existed in a handscroll format, it would have been first painted in separate sections, which would then be easily translatable to separate printed pages. Conversely, the Mao Kun map is better described as displaying a 'route-enhancing cartographic space' (Yee, 2015: 101).

Shortly before Zheng He's first expedition, cartography also played an important political role on the Korean peninsula: the newly established Chosŏn dynasty ordered the compilation of two maps, one of the heavens and one of the earth (Ledyard, 1994: 225; Robinson, 2007). The latter, entitled *Honil Gangni Yeokdae Gukdo Ji Do* ('Map of Integrated Lands and Regions of Historical Countries and Capitals' – Kangnido for brevity) of 1402, integrates a variety of cartographical sources: Chinese, Arabic and Japanese – the latter specifically sourced through diplomatic trips (Ledyard, 1994: 247). A similar phenomenon occurred in 1584 when, at the request of the Ming emperor Wanli, the Jesuit Matteo Ricci adapted Western cartographic sources to the local epistemology: unlike the Selden Map, Ricci's *Kunyu Wanguo Quantu* world map was centred on continental China (Mignolo, 1994: 17–23; Reichle, 2016).

Meanwhile, in Japan, Hideyoshi's interest in conquering East Asia conjoined with Jesuit proselytism and led to both Ricci's map and the Kangnido (Robinson, 2010: 92) being combined with Western maps and military iconography on *nanban* folding screens. Interestingly, these complex cartographic objects probably meant different things to the local military elite, to whom they configured a visual rhetoric for a utopian space of authority (Unno, 1994: 376–380), and to the Jesuits to whom they visualized a 'complex radial system of vectors that departed from and arrived to [sic] a major fulcrum: the port city of Macao' (Cattaneo, 2014). The patchwork of sources thus enrolled in visualizing the Japanese archipelago within a larger world could therefore serve both to start a conversation and to end it. Nevertheless, this was the beginning of the spread of images of 'Japan' from strictly Buddhist circles to the ruling military elite, eventually reaching the medium of the woodblock print. In this more affordable medium, the meaning of space was negotiated by each agent that appropriated it, a process which has been called 'the spatial vernacular' (Yonemoto, 2000, 2003).

Another intriguing case is that of the map of the Aztec capital Tenochtitlan, sent by Fernando Cortés to Charles I of Spain and then printed in Nuremberg in 1524. The sacred centre of Tenochtitlan was used to reinforce the centrality of the Spanish empire and justify its colonial activities. For the European audience 'the spectacle of civic and social order set up by the picture implodes at the centre, where the words make present the human sacrifice that corrodes this utopia' (Mundy, 1998: 27). An alternative understanding of space was made visible to Europeans precisely through its disappearance, a process that characterizes the colonial enterprise. However, the Hapsburg banner marking Cortés's headquarters is placed on the margin, outside the sacred centre – perhaps an indication that although conquered, the indigenous epistemology remained impregnable to its conqueror.

Not all worldviews were affected by this age of unprecedented connectivity (Clunas, 2016: 63). An often-dismissed case is that of the Korean *Ch'onhado* ('Map of All Under Heaven'). Also called wheel maps because of their circular structure, they presented a concentric world view featuring an array of imaginary spaces mostly borrowed from the *Shanhaijing* ('Classic of Mountains and Seas'). Although widely circulated from the seventeenth century onwards, until recently these maps have been ignored by academic scholarship because they did not fit the theoretical frame which employed scientific accuracy as the main criterion for assessment. Recent scholarship has suggested the Kangnido format as a possible source (Ledyard, 1994: 266), that the outer circle 'represents not the earthly world but the sky' (Oh, 2008: 19) as well as homologies with Jesuit maps (Lim, 2011: 293). Research on other types of Korean maps is also yielding new concepts, such as the division of the composition of pictorial maps into the 'panoramic aerial perspective' shared with traditional landscape painting, the 'blooming flower' where the spatial elements 'appear to spread out in different directions', and 'closed-flower' where the 'objects, including mountains, appear to point inward toward the centre of the picture' (Han *et al.*, 2009: 145). The latter two compositions correspond to what Yee (2015: 101) calls a 'centre-enhancing space'.

## Conclusion

The two themes which have structured this chapter are similar to the concept of itinerant and radial space proposed by Leroi-Gourhan or to the related one of smooth and striated space proposed by Deleuze and Guattari. This does not imply a hierarchy between these binaries, although a previous bias towards 'striated' elite maps has created the necessity for a corrective insistence on vernacular mapping. This is where 'connected histories' and the study of sea-centred maps come into play. However, these concepts must be recognized as heuristic devices intended to map out a cross-pollinating historical landscape. For example, the Treaty of Tordesillas is an imposition of authority but can also be seen as a form of conversation, and the map of Tenochtitlan is testimony to an admittedly flawed conversation between Spanish and Aztec polities.

The 'New World' was, in a sense, already plotted on the world map and the geographical imagination of Europeans. The period is dominated by 'the reduction of the unknown to the known' (Lestringant, 1994: 63). In this sense, mapping involved speculation, but not in the modern sense of the word, but with the original Middle English meaning of exploration and observation. Cartographers and their audience were aware of the limits of their knowledge as well as those of cartographic representation. Therefore, the makers and users of maps employed them as tools for thinking, for negotiating their ideas about the world.

A theme present throughout has been that of 'cartographic encounters' (Short, 2009). The phrasing corresponds to the 'Age of Encounters' which has been proposed instead of that of the 'Age of Discoveries'. I do not think that 'encounters' is sufficiently free of Eurocentrism to help us move forward – it is no more than a euphemistic reformulation of European conquest. Latour's illustration of the otherwise useful discussion of maps as 'immutable mobiles' with a description of the exchange between a Chinese and a French cartographer is representative in this regard (Latour, 1986: 5–6). Framing this example in terms of 'differences between a savage geography and a civilized one' ignores the richness of the cartographical tradition in China, and therefore reinforces the bias towards a history of Eurocentric progress. With research increasingly more diverse and accessible, the task at hand is to decentre the Eurocentric narrative by acknowledging the synchronicity and intermeshing of multiple 'movable centers' (Mignolo, 1994: 30). This can only be achieved by moving beyond binaries and across disciplines to achieve a fuller picture of 'the at times fragile threads that connected the globe, even as the globe came to be defined as such' (Subrahmanyam, 1997: 762).

# References

Alegria, M., Daveau, S., Garcia, J. and Relaño, F. (2007) "Portuguese Cartography in the Renaissance" in Woodward, D. (Ed.) *The History of Cartography (Volume 3)* Chicago, IL: University of Chicago Press, pp.975–1068.

Azócar Fernández, P.I. and Buchroithner, M.F. (2013) *Paradigms in Cartography: An Epistemological Review of the 20th and 21st Centuries* Berlin: Springer.

Barros, A. (2013) "The Portuguese in the Indian Ocean in the First Global Age: Transoceanic Exchanges, Naval Power, Port Organization and Trade" in Mukherjee, R. (Ed.) *Oceans Connect: Reflections on Water Worlds across Time and Space* Delhi: Primus Books, pp.143–202.

Bleichmar, D. and Martin, M. (Eds) (2015) "Objects in Motion in the Early Modern World" (Special Issue) *Art History* 38 (4) pp. 604–609.

Brentjes, S. (2008) "Revisiting Catalan Portolan Charts: Do They Contain Elements of Asian Provenance?" in Forêt, P. and Kaplony, A. (Eds) *The Journey of Maps and Images on the Silk Road* (Brill's Inner Asian Library 21) Leiden, The Netherlands: Brill, pp.181–201.

Brentjes, S. (2011) "Patchwork: The Norm of Mapmaking Practices for Western Asia in Catholic and Protestant Europe As Well As in Istanbul Between 1550 and 1750?" in *Science between Europe and Asia: Historical Studies on the Transmission, Adoption and Adaptation of Knowledge* (Boston Studies in the Philosophy of Science, No.275) pp.77–101.

Brentjes, S. (2012) "Medieval Portolan Charts as Documents of Shared Cultural Spaces" in Abdellatif, R., Benhima, Y., König, D. and Ruchaud, E. (Eds) *Acteurs de Transferts Culturels en Méditerranée Médiévale* (Alteliers des Deutschen Historischen Instituts, Paris, Band 9) Munich, Germany: Oldenbourg Verlag, pp.135–146.

Brook, T. (2013) *Mr. Selden's Map of China: Decoding the Secrets of a Vanished Cartographer* New York: Bloomsbury.

Brückner, M. (2011) "The Ambulatory Map: Commodity, Mobility, and Visualcy in Eighteenth-Century Colonial America" *Winterthur Portfolio* 45 (2/3) pp.141–160.

Brummett, P. (2015) *Mapping the Ottomans: Sovereignty, Territory, and Identity in the Early Modern Mediterranean* Cambridge, UK: Cambridge University Press.

Cattaneo, A. (2011) *Fra Mauro's Mappa Mundi and Fifteenth-Century Venice* Turnhout, Belgium: Brepols.

Cattaneo, A. (2014) "Geographical Curiosities and Transformative Exchange in the *Nanban* Century (c.1549–c.1647) *Études Épistémè* 26 Available at: *https://episteme.revues.org/329* (Accessed: 11 April 2016).

Church, S. (2008) "Zheng He" in Selin, H. (Ed.) *Encyclopaedia of the History of Science, Technology, and Medicine in Non-Western Cultures* Dordrecht, The Netherlands: Springer, pp.2354–2357.

Clunas, C. (2016) "Connected Material Histories: A Response" *Modern Asian Studies* 50 (1) pp.61–74.

Cosgrove, D. (2007) "Images of Renaissance Cosmography, 1450–1650" in Woodward, D. (Ed.) *The History of Cartography (Volume 3)* Chicago, IL: University of Chicago Press, pp.55–98.

Couto, D. (2003) "The Role of Interpreters, or Linguas, in the Portuguese Empire during the 16th Century" *E-journal of Portuguese History* 1,2 Available at: *www.brown.edu/Departments/Portuguese_Brazilian_Studies/ejph/html/Winter03.html* (Accessed: 10 April 2016).

De Certeau, M. (2000) "Speech, or the Space of the Other: Jean de Léry" in Ward, G. (Ed.) *The Certeau Reader* (trans. by T. Conley) Oxford, UK: Blackwell Publishing, pp.129–149.

Del Casino, V. and Hanna, S. (2006) "Beyond the 'Binaries': A Methodological Intervention for Interrogating Maps as Representational Practices" *ACME: An International E-Journal for Critical Geographies* 4 (1) pp.34–56 Available at: *http://ojs.unbc.ca/index.php/acme/article/view/727* (Accessed: 14 April 2016).

Del Soldato, E. (2015) "Renaissance Natural Philosophy in the Renaissance" in *Stanford Encyclopedia of Philosophy* Stanford, CA: Stanford University Available at: *http://plato.stanford.edu/entries/natphil-ren/* (Accessed: 11 April 2016).

Earle, R. (2013) "Mapping the New World" (Seminar) *Facing the Other: European Encounters with the New World, 1500–1650* (Audio recording) Available at: *www2.warwick.ac.uk/fac/arts/history/students/modules/facing_other/seminars/t01w08/* (Accessed: 15 April 2016).

Edney, M.H. (1993) "Cartography without 'Progress': Reinterpreting the Nature and Historical Development of Mapmaking" *Cartographica* 30 (2,3) pp.54–68.

Edney, M.H. (2007) "Mapping Parts of the World" in Akerman, J. and Karrow, R. (Eds) *Maps: Finding Our Place in the World* Chicago, IL: The Field Museum and the University of Chicago Press, pp.117–158.

Edwards, C. (1985) "Discoveries of Mexico and the Meaning of Discovery" *Terrae Incognitae* 17 pp.61–67.

Eklund, H.C. (2015) *Literature and Moral Economy in the Early Modern Atlantic: Elegant Sufficiencies* Burlington, VT: Ashgate.

Emiralioğlu, M.P. (2014) *Geographical Knowledge and Imperial Culture in the Early Modern Ottoman Empire* Burlington, VT: Ashgate.

Fiorani, F. (2012) "The Enduring Power of Forgery and Imagination" in Dym, J. (Ed.) *Cartographic Conversation* Providence, RI: The John Carter Brown Library Available at: *www.brown.edu/Facilities/John_Carter_Brown_Library/exhibitions/cartographic/pages/fiorani.html* (Accessed: 9 April 2016).

Gemegah, H. (2005) "Did the Idea about the Asian Origin of the American 'Indians' Develop from 16th century Spanish Political Geography?" *Studies in Historical Anthropology* 2 (2002) pp.3–16 Available at: *www.antropologia.uw.edu.pl/SHA/sha-02-01.pdf* (Accessed: 9 April 2016).

Han, Y., Ahn, H. and Sung, B (2009) *The Artistry of Early Korean Cartography* (trans. by C. Byonghyon and edited by A. Akin)** Larkspur, CA: Tamal Vista Publications.

Harley, J.B. (1989) "Deconstructing the Map" *Cartographica* 26 (2) pp.1–20.

Harley, J.B. (1992) "Rereading the Maps of Columbian Encounters" *Annals of the Association of American Geographers* 82 pp.522–536.

Hunt, L. (2014) *Writing History in the Global Era* New York: W.W. Norton & Company.

Johnson, D. (1994) *Phantom Islands of the Atlantic* Fredericton, New Brunswick, Canada: Goose Lane.

King, D.A. (1999) *World Maps for Finding the Direction and Distance to Mecca: Innovation and Tradition in Islamic Science* London: Al-Furqān Islamic Heritage Foundation/Leiden, The Netherlands: Brill.

Kitchin, R. and Dodge, M. (2007) "Rethinking Maps" *Progress in Human Geography* 31 (3) pp.331–344.

Kreiner, J. (1996) "Notes on the History of European-Ryukyuan Contacts" in Kreiner, J. (Ed.) *Sources of Ryūkyūan History and Culture in European Collections* Munich: Iudicium Verlag, pp.15–47.

Kunitzsch, P. (1974) *Der Almagest. Die Syntaxis Mathematica de Claudius Ptolemäus in Arabisch-lateinischer Überlieferung* Wiesbaden, Germany: Otto Harrassowitz.

Latour, B. (1986) "Visualization and Cognition: Thinking with Eyes and Hands" in Kuklick, H. (Ed.) *Knowledge and Society: Studies in the Sociology of Culture Past and Present* 6 Greenwich, CT: Jai Press, pp.1–40. (Partial republication in Dodge, M., Kitchin, R. and Perkins, C. (Eds) (2011) *The Map Reader: Theories of Mapping Practice and Cartographic Representation* Chichester, UK: John Wiley & Sons, pp.65–73.)

Ledyard, G. (1994) "Cartography of Korea" in Harley, B. and Woodward, D. (Eds) *The History of Cartography (Volume 2)* Chicago, IL: University of Chicago Press, pp.235–345.

Lestringant, F. (1994) *Mapping the Renaissance World: The Geographical Imagination in the Age of Discovery* (trans. by D. Fausett) Cambridge, UK: Polity Press.

Lewis, G.M. (1998a) "Introduction" in Lewis, G.M. (Ed.) *Cartographic Encounters: Perspectives on Native American Mapmaking and Map Use* Chicago, IL: The University of Chicago Press, pp.1–6.

Lewis, G.M. (1998b) "Frontier Encounters in the Field: 1511–1624" in *Cartographic Encounters: Perspectives on Native American Mapmaking and Map Use* Chicago, IL: The University of Chicago Press, pp.9–25.

Lim, J. (2011) "Matteo Ricci's World Maps in Late Joseon Dynasty" *The Korean Journal for the History of Science* 33 (2) pp.277–296.

Ma, H. (1970) *Ma Huan: Ying-yai Sheng-lan: The Overall Survey of the Ocean's Shores (1433), Translated from the Chinese text Edited by Feng Ch'eng-chün* (trans. and edited by J.V.G. Mills) The Hakluyt Society, Extra Series No.42 Cambridge, UK: Cambridge University Press.

Markey, L. (2012) "Stradano's Allegorical Invention of the Americas in Late Sixteenth-Century Florence" *Renaissance Quarterly* 65 (2) pp.385–442.

McIntosh, G.C. (2000) *The Piri Reis Map of 1513* Athens, GA: The University of Georgia Press.

Mercier, R. (1992) "Geodesy" in Harley, B. and Woodward, D. (Eds) *The History of Cartography (Volume 2)* Chicago, IL: University of Chicago Press, pp.175–188.

Mignolo, W. (1994) "The Movable Center: Geographical Discourses and Territoriality During the Expansion of the Spanish Empire" in Cevallos-Candau, F.J., Cole, J., Scott, N. and Suárez-Araúz, N. (Eds) *Coded Encounters: Writing, Gender and Ethnicity in Colonial Latin America* Amherst, MA: University of Massachusetts Press, pp.15–45.

Monmonier, M. (2010) *No Dig, No Fly, No Go: How Maps Restrict and Control* Chicago, IL: University of Chicago Press.

Mukherjee, R. (2013) "Oceans Connect/Fragment: A Global View of the Eastern Ocean" in Mukherjee, R. (Ed.) *Oceans Connect: Reflections on Water Worlds across Time and Space*. Delhi: Primus Books, pp.215–238.

Mundy, B. (1998) "Mapping the Aztec Capital: The 1524 Nuremberg Map of Tenochtitlan, Its Sources and Meanings" *Imago Mundi* 50 (1) pp.11–33.

Nanba, M., Muroga, N. and Unno, K. (1973) *Old Maps in Japan* Osaka, Japan: Sōgensha.

O'Doherty, M. (2011) "A Peripheral Matter? Oceans in the East in Late-Medieval Thought, Report, and Cartography" *Bulletin of International Medieval Research* 16 pp.14–59.

Oh, S. (2008) "Circular World Maps of the Joseon Dynasty: Their Characteristics and Worldview" *Korea Journal* 48 (1) pp.8–45.

Padrón, R. (2009) "A Sea of Denial: The Early Modern Spanish Invention of the Pacific Rim" *Hispanic Review* 77 (1) pp.1–27.

Park, H. (2012) *Mapping the Chinese and Islamic worlds: Cross-cultural Exchange in Pre-Modern Asia* Cambridge, UK: Cambridge University Press.

Reichle, N. (Ed.) (2016) *China at the Center: Ricci and Verbiest World Maps* San Francisco, CA: Asian Art Museum.

Reis, P. (1992) *Piri Reis Haritası Hakkında İzahname* (Ed. by Y. Akçura) Ankara, Turkey: TTK.

Reitinger, F. (2007) "Literary Mapping in German-Speaking Europe" in Woodward, D. (Ed.) *The History of Cartography (Volume 3)* Chicago, IL: University of Chicago Press, pp.438–449.

Robinson, K. (2007) "Chosŏn Korea in the Ryūkoku Kangnido: Dating the Oldest Extant Korean Map of the World (15th Century)" *Imago Mundi* 59 (2) pp.177–192.

Robinson, K. (2010) "Daoist Geographies in Three Korean World Maps" *Journal of Daoist Studies* 3 pp.91–116.

Sandman, A. (2007) "Spanish Nautical Cartography in the Renaissance" in Woodward, D. (Ed.) *The History of Cartography (Volume 3)* Chicago, IL: University of Chicago Press, pp.1095–1142.

Savage-Smith, E. (1992) "Celestial Mapping" in Harley, B. and Woodward, D. (Eds) *The History of Cartography (Volume 2)* Chicago, IL: University of Chicago Press, pp.12–70.

Shalev, Z. and Burnett, C. (Eds) (2011) *Ptolemy's Geography in the Renaissance* London: The Warburg Institute/Turin, Italy: Nino Aragno Editore.

Shapinsky, P. (2006) "Polyvocal Portolans: Nautical Charts and Hybrid Maritime Cultures in Early Modern East Asia" *Early Modern Japan* 14 pp.4–26.

Shapinsky, P. (2016) "The World from the Waterline" in Wigen, K., Sugimoto, F. and Karacas, C. (Eds) *Cartographic Japan: A History in Maps* Chicago, IL: University of Chicago Press, pp.16–19.

Short, J.R. (2009) *Cartographic Encounters: Indigenous Peoples and the Exploration of the New World* London: Reaktion Books.

Soucek, S. (1992) "Islamic Charting in the Mediterranean" in Harley, B. and Woodward, D. (Eds) *The History of Cartography (Volume 2)* Chicago, IL: University of Chicago Press, pp.263–292.

Subrahmanyam, S. (1997) "Connected Histories: Notes towards a Reconfiguration of Early Modern Eurasia" *Modern Asian Studies* 31 (3) pp.735–762.

Subrahmanyam, S. (2007) "Holding the World in Balance: The Connected Histories of the Iberian Overseas Empires, 1500–1640" *The American Historical Review* 112 (5) pp.1359–1385.

Turnbull, D. (1996) "Cartography and Science in Early Modern Europe: Mapping the Construction of Knowledge Spaces" *Imago Mundi* 48 pp.5–24.

Unno, K. (1994) "Cartography in Japan" in Woodward, D. and Harley, J.B. (Eds) *The History of Cartography (Volume 2)* Chicago, IL: University of Chicago Press, pp.346–477.

Wood, D. (1984) "Cultured Symbols: Thoughts on the Cultural Context of Cartographic Symbols" *Cartographica* 21 (4) pp.9–37.

Wood, D. (2002) "The Map as a Kind of Talk: Brian Harley and the Confabulation of the Inner and Outer Voice" *Visual Communication* 1 (2) pp.139–161.

Yee, C. (2015) "Maps and Map-Making in China and Korea" in Parmentier, J. (Ed.), *The World in a Mirror: World Maps from the Middle Ages to the Present Day* Leuven, Belgium: BAI, pp.100–105.

Yonemoto, M. (2000) "The 'Spatial Vernacular' in Tokugawa Maps" *The Journal of Asian Studies* 59 (3) pp.647–666.

Yonemoto, M. (2003) *Mapping Early Modern Japan: Space, Place, and Culture in the Tokugawa Period (1603–1868)* Berkeley, CA: University of California Press.

Zoss, E. (2010) "An Ottoman View of the World: The Kitab Cihannuma and its Cartographic Contexts" in Gruber, C. (Ed.) *The Islamic Manuscript Tradition: Ten Centuries of Book Arts in Indiana University Collections* Bloomington, IN: Indiana University Press, pp.194–219.

# 11

# Mapping, survey, and science

*Matthew H. Edney*

After 1500, European mapping practices acquired a new, geometrical aspect (Harley *et al.*, 1987–, 3: 477–508; Gehring and Weibel, 2014). Those practices increasingly configured the world's features as sets of points, lines, and shapes that are then graphically and intellectually manipulated independently of the Earth itself. This geometricality is a primary support of the conviction that modern cartography is properly a single, logical, and 'scientific' endeavour. Yet the geometricality of European mapping was not uniformly implemented. Different scales of mapping and map functions entailed three markedly different forms of geometry: cosmographical, Euclidean, and Cartesian. While the development and elaboration of mathematical concepts and techniques constitute a major element in the history of early modern and modern mapping in the western tradition, that history needs to be told in terms of the several modes for measuring and mapping space (geography and charting; i.e. modes E and I in Table 5.1 in Chapter 5), place and region (topography and chorography; A–D and $E_3$), and territory (F–H).

   The increasing geometricality of mapping during the early modern era was a major part of the attempts by 'mathematical practitioners' to apply mathematics to all walks of life. Despite their own claims to the contrary, mathematical practitioners were never a uniform community. Not only was there significant variation in mathematical practice by profession, there was also a persistent divide between practitioners per se, who often followed craft practices, and the patrons and academics who advocated the use of higher forms of mathematics. That is, the degree of cartographic mathematicality was shaped as much by social standing as by map function, and implementation of higher mathematics served as an efficient means for men of skill but humble origins to climb the social ladder (Edney, 1994).

## Measuring space

Mathematical geography developed within Renaissance cosmography as the analysis of the mathematical foundation of creation within astrological and metaphysical studies of humanity's place within the cosmos. It focused on the geometrical structure of the spherical Earth at the centre of the apparently spherical cosmos by detailing the great and small circles on each – the meridians and parallels, the equator and ecliptic, the tropics and polar circles – that related the one to the other. These relationships were demonstrated in the classroom and library by pairs of globes

145

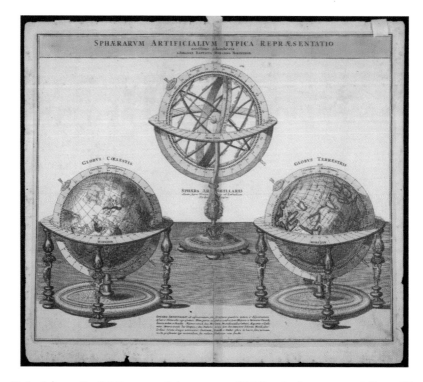

*Figure 11.1*    Johann Baptist Homann, *Sphaerarum artificialium typica repraesentatio noviβime adumbrata* (Nuremburg, 1716). A common image of the 'artificial spheres' that together represented the 'natural spheres' of the cosmos: the terrestrial and celestial globes, and the armillary sphere that defined their mutual geometrical structure. Armillary spheres were, however, expensive and so not as common in early modern libraries or studies as pairs of globes. Courtesy of the Osher Map Library and Smith Center for Cartographic Education, University of Southern Maine (SM-1716-5); see *www.oshermaps.org/map/1824* for a high-resolution image

and perhaps also armillary spheres, whether physical or figurative (Figure 11.1). Students of any age could use globes to undertake a wide variety of mathematical tasks, for example converting between longitudinal and temporal differences or between latitudes and the lengths of the longest day. Because these geometrical relationships were defined with respect to the Earth, and to the Earth's rotation about its axis, the practices of mathematical geography were largely unaffected by the Copernican revolution.

One of the major innovations of Renaissance geography was the integration of cosmography with geographical and even chorographical mapping. Jacopo Angeli's c.1411 translation into Latin of a late, Byzantine copy of Claudius Ptolemy's *Geography* gave western geographers a means to frame and structure new geographical information in the form of graphically constructed networks of meridians and parallels (Harley *et al.*, 1987–, 1: 316–317, 3: 285–364). New designs for map projections soon proliferated as scholars sought to capture a world newly exposed to European mariners, but by 1600 geographers had pragmatically settled on a few projections whose straight lines and arcs of circles made them easy to draw (Figure 11.2). Some attention was given to the mathematics and properties of these projections. For example, Edward Wright in 1599 determined the trigonometry behind Gerard Mercator's famous 1569 projection, and in 1696 Edmond Halley proved that the stereographic projection was indeed conformal. But new projections were

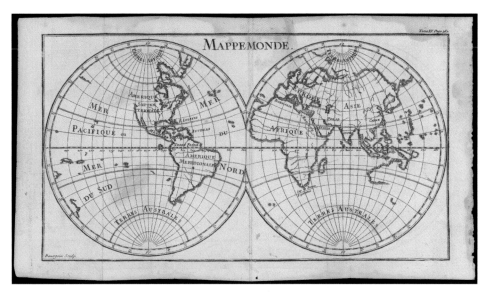

*Figure 11.2   Mappemonde,* engr. Pierre Bourgoin (Paris, c.1740). One of innumerable double-
hemisphere world maps made from the late sixteenth century until well into
the nineteenth; this simple map was originally part of an unidentified book
or journal. Before 1800, most such maps were constructed on the transverse
(equatorial) aspect of the conformal stereographic projection; after 1800,
geographers generally used an equidistant 'globular' projection, as introduced
by Aaron Arrowsmith in 1794. The curves in both projections were all arcs of
circles (Snyder, 1993: 27–28, 40–42). A few eighteenth-century geographers
implemented the oblique-aspect stereographic projection, whose construction
was first explained by Bernhard Varen (Varenius) in his influential *Geographia
generalis* (1650). To be clear, rectangular world projections such as Mercator's
became common only about 1850. Courtesy of the Osher Map Library and
Smith Center for Cartographic Education, University of Southern Maine
(OS-1740-16); see *www.oshermaps.org/map/11934* for a high-resolution image

not developed analytically until the publication in 1772 of Johann Heinrich Lambert's account of
five new projections, including the transverse aspect of the Mercator. Thereafter, descriptions of
new projections once again proliferated as mathematicians and mathematically inclined geographers sought to model, control, and minimize distortions (Snyder, 1993).

It was no easy task to hang geographic information onto the scaffold of meridians and parallels drawn across a plane surface for the map, at least for those few geographers who constructed
new maps and who did not simply copy those made by others. A few key places with known
latitude and perhaps known longitude were first plotted within the scaffold; other information,
from itineraries and other maps, could then be graphically fitted between and around the control
points, steadily building up the map image. The latitudes and longitudes of the control points
came from either calculation or direct determination. Calculation required the conversion of
itinerary distances and directions to differences of latitude and longitude, a process fraught with
uncertainty, not least over the actual size of the Earth. The determination of terrestrial coordinates was implicitly cosmographical as, for the most part, it relied on celestial observations.

Latitude had since Antiquity been determined directly by measuring the altitude of Polaris at
night. In the fifteenth century, as Portuguese mariners drew close to the equator and so progressively lost sight of Polaris in the atmospheric haze close to the horizon, cosmography supplied a

new method, using simple arithmetic to combine the noon-time altitude of the Sun with solar declination, read from a table. Several refinements were made to these observational practices, especially after 1650, in order to permit the increasingly accurate calculation of precise values.

The basic method of determining longitude was obvious to cosmographers – simply determine the difference in time between two locations by timing a mutually observable celestial event – but implementation proved wildly difficult. Solar eclipses, that were used by some ancients, were simply too infrequent. The first actual solution was to compare the observer's local time against the time predicted for the eclipses of Jupiter's primary moons, or satellites, as they passed behind the planet's body. Galileo had proposed this technique soon after he had first observed those moons in 1610; astronomers in France had successfully implemented it by 1690 (Figure 11.3). Land-based agents of Europe's overseas empires soon routinely applied this technique so that, by the early 1800s, continental outlines on the world map had acquired a

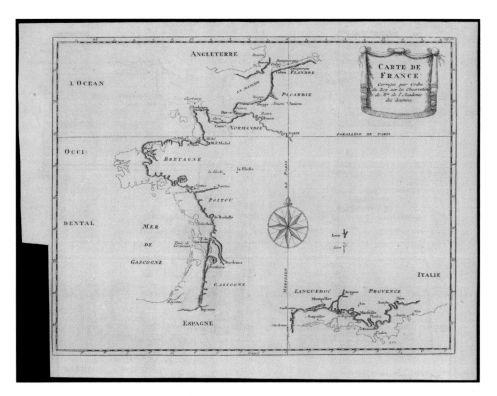

*Figure 11.3  Carte de France corrigée par ordre du Roy sur les observations de Mss. de l'Academie des Sciences* (Paris, 1729); this map, first published in 1693, compares two coastlines of France: that from Nicolas Sanson's 1679 map (drawn lightly) and the 'new' outline (drawn thickly) as 'corrected' by simultaneous observations of the eclipses of Jupiter's satellites made in 1679–81 by Jean-Dominique Cassini, at the Paris Observatory, and in the field by Jean Picard and Philippe de La Hire. Only in 1690 was Cassini's work sufficiently refined that he could publish annual tables of predicted times of eclipses (the *Connoissance des temps*); armed with these tables, observers in the field could determine their longitude with respect to Paris. Courtesy of the Osher Map Library and Smith Center for Cartographic Education, University of Southern Maine (OS-1729-1); see *www.oshermaps.org/map/622* for a high-resolution image

thoroughly modern look. But, even if eclipses of Jupiter's moons could be seen from the heaving deck of a ship, their eclipses occurred too infrequently to determine longitudes while at sea. For that purpose, two techniques were developed over the course of the eighteenth century. One technique – the method of lunar distances – was strictly cosmographical: the mariner measured the position of the Moon against the field of fixed stars and then undertook extensive and complex calculations to determine longitude. Tobias Mayer's precise tables of the Moon's motion (1755) and Nicolas de Lacaille's computational techniques (1759) together made the technique workable; in widespread use by 1800, it was still in use by aeronautical navigators during World War II. John Harrison perfected the other technique – using clocks to define a standard time – when in 1759 he completed a chronometer that could run accurately and consistently at sea. Mariners could henceforth compare local to clock time and so define their longitude almost directly, although such chronometers were not made cheaply enough for widespread adoption until the 1790s (Andrewes, 1996).

Before 1790, therefore, few mariners could determine their longitude at sea. But this was not a problem because they generally followed well-known routes and could use latitude observations and calculations of their position derived from records of their speed, time, and direction (Cook, 2006). Mariners instead used 'plane charts' that had only a tenuous relationship to the Earth's curvature; longitude scales were occasionally added to the charts, but these were only ever approximate. Nonetheless, even as early as the 1500s (Harley *et al.*, 1987–, esp. 3: 1095–1142), the educated merchants and gentry who managed marine enterprises, such as the officers of the Casa de la Contratación in Seville or the Muscovy Company in London, pushed for navigational practices to be reconciled with the cosmographical framework of latitude and longitude. Mercator designed his now famous projection in order to reconcile the two sets of geometry. Historians have traditionally lauded this achievement, but mariners could not actually use charts on the projection until they could easily determine longitude at sea and had detailed and correct coastal outlines against which they could refer their locations, which is to say not until after 1790.

Subsequently, in the mid-nineteenth century, the spread of the telegraph permitted the precise determination of longitudes for terrestrial locations through the transmission of timed signals. But by then, the prosecution of geodetic triangulations had already begun to provide a far denser network of locations with precise and accurate coordinates than astronomical or telegraphic determinations could ever sustain. Geodetic triangulations were the outcome of an alternate form of geometricization of the world, through the surveyed measurement of place and, more importantly, of territory.

## Measuring place

In contrast to the scholarly and highly geometricized character of early modern cosmography and geography, but just like marine navigation, early practices of place mapping were originally craft-based. They were grounded in apprenticeships and experience, rather than formal schooling, and they entailed as much the visualization of the landscape as its measured reconstruction. The several modes for mapping regions and places have often been collapsed into an apparently single endeavour of 'land surveying', and certainly many early modern surveyors used the same basic techniques to create geometric frameworks at a medium or small scale for states and provinces as well as for estates and properties at a large scale (Gehring and Weibel, 2014). Yet there has been considerable variation in the respective complexity and prominence of the geometricization of landscapes versus their sketching and drawing, and the history of place mapping does not comprise a narrative of the increasing dominance of the one over the other.

Of all the modes of mapping, property mapping has been most responsive to local conditions and craft practices. The identification and demarcation of property boundaries has always had to recognize the traditions of each community, including customary units of length (rather than standardized units defined by statute and promoted by central state authorities), local techniques for monumenting boundaries in the landscape, the acceptable level of precision as defined by local land values (from rough-and-ready approximation to instrumental exactness), and expectations for representing the results (with or without graphic plans). Surveying manuals proliferated after 1600, but they were not intended to educate apprentices; rather they explained the work of the surveyor to the landowner in educated terms, taught geometry through its application to real-world problems and promoted the author's own business. Most early modern property mapping entailed simple instruments that were easily wielded (Bennett, 1991) and the simplest forms of geometry and area calculation (Figure 11.4).

Regional, chorographical surveys in the early modern era generally entailed a mix of low-precision techniques. A basic geometrical frame would be constructed from a series of traverses along routes, whether roads, rivers or coastlines. The lengths of each route could be

*Figure 11.4*    This image, from a student copy of a teacher's exemplar manual, demonstrates the simplicity of most early property mapping. As the accompanying text explained, 'just one single case . . . is sufficient to get a clear perception of the use of the [surveyor's] compass and to be able to use this instrument for all possible cases'. The case – the time-honoured problem of determining the distance to a far or inaccessible place – entails a single triangle imagined in the world, recorded as magnetic azimuths in the surveyor's field book (left-hand detail) and one side BC, and eventually drawn to scale on the plan (right-hand detail); the surveyor did not even have to do any trigonometrical calculations to determine AB, just measure it on the scale plan. The resolution of the geometry of a single triangle is often misleadingly called 'triangulation', but that term is properly applied only to networks such as that in Figure 11.5. Fridrick Wilhelm von Beust (?) after Johann Christoph Schlönbach, 'Practische Anleitung zum Feld-Messen' (1759); translation by Imre Demhardt. Courtesy of the Osher Map Library and Smith Center for Cartographic Education, University of Southern Maine (OS-1759-1); see *www.oshermaps.org/map/44151.0165* for a high-resolution image

measured – if not already recorded in itineraries and local lore – by time of travel, by counting paces or by fitting a mechanical counter to a wheel; the bearing of each section of route
could be estimated, taken by a magnetic compass, estimated by the Sun's position, or simply
approximated. Great precision was not necessary for bearings because the linear measurement was known to be uncertain: river measures had to account for variable currents; road
surveys measured surface distance, up and down hills, not horizontal distance, while the need
to swerve around mud patches and potholes on roads introduced still greater errors. Even so,
once the routes were sketched out on paper they provided a graphic framework within which
the surveyor could fill in the details of the landscape, as seen from the tops of hills and towers
and from traversing the district, or taken from other maps. This was the system that seems to
have been used, for example, by Christopher Saxton in mapping the counties of England and
Wales in 1573–78 (Harley *et al.*, 1987–, 3: 506, 1628–1629). The same process was used to
construct bird's-eye views of districts or cities, except that the framework was often drawn
out to mimic linear perspective rather than plane geometry (Harley *et al.*, 1987–, 3: 680–704).

In 1533, Gemma Frisius proposed a coherent method of determining the locations of places
across a region: triangulation. A network of imaginary triangles is formed by sight lines between
high places (Figure 11.5); the surveyor measures the angles of each triangle and the length of
one side of one triangle. From this information, the triangles can be plotted out on paper, or the

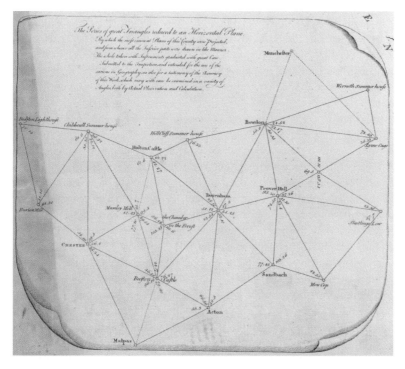

*Figure 11.5* 'The series of great triangles reduced to a horizontal plane', inset on P.P. Burdett,
*This Survey of the County Palatine of Chester* (London: William Faden, 1794, 2nd
ed.; 1st ed., 1777), sheet 3. The claim to have reduced the triangles from the
Earth's surface to the plane was egregious: not even geodetic observations were
sufficiently precise at the time to discriminate the difference. Courtesy of the
Biblioteca Virtual del Patrimonio Bibliográfico (Colección Mendoza, Biblioteca
Nacional de España, Madrid). See *www.europeana.eu* for a high-resolution image

lengths of the other triangle sides can be calculated via trigonometry and then plotted. The result was a relatively dense array of locations from which the details of a landscape could be located by further observation, through intersections of sight lines from more than one location, or sketched in directly (Harley *et al.*, 1987–, 3: 483–485). Philipp Apian used triangulation, at least in part, for his detailed survey of Bavaria in 1554–1561 (Harley *et al.*, 1987–, 3: 1223–1224), but the technique was too complex to be commonly used for local regional surveys.

It was the requirement of civil and military engineers to map projects for land reclamation or urban development, and to map proposed and existing fortifications, buildings, canals, and roads, that drove greater precision in surveying instrumentation and practice (e.g. Gerbino and Johnston, 2009). Increasingly precise angle measurement was achieved in several varieties of instrument – variously theodolites, Holland circles, surveyor's compasses, and graphometers – in which sights (alidades) were fitted to precisely graduated circles. The precision of the gradations increased significantly in the eighteenth century, when telescopic sights were also added to these instruments (Bennett, 1987). More precise length measurements could be achieved when traditional rods and ropes were replaced, starting in the mid-sixteenth century, with chains formed from consistent-length links. Engineers used these instruments to measure traverses and triangles, sometimes even actual triangulations, to map each particular site; if they had levels or instruments equipped with vertical circles, they could do so in three dimensions.

The geometry and art of mapping places came together in the plane table, often called the *mensula Praetoriana* after the German mathematician Johannes Prätorius, who had supposedly invented it in about 1590 (Harley *et al.*, 1987–, 3: 498–500). With a plane table, the surveyor could construct the geometrical framework graphically on the sheet of paper set atop the table, and then sketch in features directly (Figure 11.6). The speed of the plane table would make it a favoured instrument for landscape mapping. Most notably, in 1763–1799, military topographers used the instrument to map 570,000 km² of Austrian Habsburg territory with little or no geometrical control.

Several factors led to the transfer of the engineers' explicitly geometrical and instrumental techniques to other mappings of property, place, and region. Expectations among the land-owners and officials who commissioned surveys rose, as more and more complex mapping was undertaken. Eighteenth-century instrument makers worked to refine their high-end instruments for astronomers, refinements that they then incorporated into the more quotidian instruments that formed the bulk of their trade and whose cost dropped steadily (Bennett, 1987). Better instruments in turn required more refined practices. The proliferation of official and corporate mapping endeavours (see Chapter 12, this volume) led to the increasing professionalization of surveyors and topographical engineers, so that by 1800, the formal institutional divide between architects and civil engineers was complete, concretizing the latter's geometrical and measured foundations for place mapping.

Furthermore, the professionalization of place mapping was also accompanied in the eighteenth century by the appropriation of elements of cosmographical practice. This sporadic trend stemmed from both the surveyors' need to meet clients' expectations and their own increasing exposure to cosmographical principles through formal education. For a few socially important places, such as the new US federal capital in the District of Columbia, laid out by Andrew Ellicott in 1791–92, surveyors undertook the extra work of defining local magnetic variation so that they could delineate the places with respect to true rather than magnetic north. A few chorographers indicated longitudes with respect to a local origin, defined by calculations rather than astronomical observations. Several county surveyors in England and Wales added meridians and parallels to their county maps, but their relation to the chorography remained only approximate and unconnected to the actual survey work.

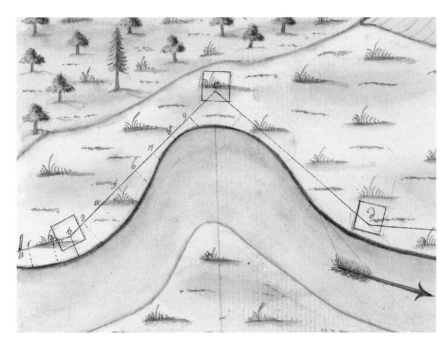

*Figure 11.6* Detail of a graphic explanation of using a plane table and chain in a traverse to map the bends in a river bank. The surveyor places the plane table at several locations along the edge of the river (the squares at 'b', 'c', and 'd') and measures the distances between them, together with offsets to the actual river edge. The instructions concluded: 'Finally draw a line between the points on the plane table, and the bends will appear on the paper'. Fridrick Wilhelm von Beust (?) after Johann Christoph Schlönbach, 'Practische Anleitung zum Feld-Messen' (1759); translation by Imre Demhardt. Courtesy of the Osher Map Library and Smith Center for Cartographic Education, University of Southern Maine (OS-1759-1); see *www.oshermaps.org/map/44151.0125* for a high-resolution image

## Measuring territory (and the Earth)

Otherwise distinct – even today – the geographical geometry of cosmographical space and the Euclidean geometry of local and regional place have been merged only within modern extensive, systematic territorial surveys. The first such survey was that of France directed by several generations of the Cassini family through the eighteenth century. A foundational triangulation was completed by 1744 (Figure 11.7). The angles of the triangles were measured to just one-to-three minutes of arc; the lengths of carefully measured baselines were then carried trigonometrically through the network. Equally careful astronomical observations fixed the whole to the Earth's surface and permitted the calculation of the latitudes and longitudes of numerous locations across France. These, in turn, were projected to create a Cartesian coordinate system with its origin at the Paris observatory, which then permitted the whole of France to be divided into 182 sheets, each 25,000 by 40,000 toises (48.725 x 77.96 km). In other words, the detailed, Euclidean observations at each triangulation station were scaled up to geographical space, which was then reconfigured *in toto* as a projected, Cartesian space. After this remarkable intellectual achievement, the actual detailed topographical survey of most of those 173 sheets by 1789 seems anti–climactic (Pelletier, 2013).

*Figure 11.7*   Detail of Giovanni Domenico Maraldi and César François Cassini de Thury, *Nouvelle carte qui comprend les principaux triangles qui servent de fondement à la description géométrique de la France* (Paris, c. 1783; originally published, 1744). Having tied the triangulation to the graticule, the academicians could calculate the latitude and longitude, and distance from Paris, of 442 major towns and other places (as listed in the table on either side of the map). The grid covering the map is that of Cartesian coordinate space; latitude and longitude appear only as marginal ticks. Courtesy of the Osher Map Library and Smith Center for Cartographic Education, University of Southern Maine (OS-1783-7); see *www. oshermaps.org/map/854* for a high-resolution image

The first stage in the survey of France had been Jean Picard's 1669–71 chain of triangles along a meridian from Paris north to Amiens. From the measured distance (153.68 km) and the difference in latitude (1°22'55") subtended by the arc of meridian, Picard calculated the size of the spherical Earth, expressed as either 57,060 toises (111 km) for one degree of the meridian, or a terrestrial circumference of 20,541,600 toises (40,036 km). Yet even as Picard worked, pendulum observations made at several places in Europe and the Caribbean to gauge the variability of gravity were already suggesting that the world was not actually spherical. Isaac Newton argued in his *Philosophiae naturalis principia mathematica* (1687) that the Earth was slightly flattened at the poles, such that the length of degrees of latitude would increase towards the poles. However, in 1720, Jacques Cassini concluded, from the southward continuation of the triangulation along the Paris meridian, that degrees of latitude actually shortened towards the pole, suggesting that the world was squeezed at the equator. The stage was set for a long and complex debate in the Académie des Sciences, between those who disdained and those who embraced Newton's strictly mathematical and metaphysics-free approach to understanding gravity. (Despite a persistent myth, the debate had nothing to do with proving or disproving Descartes' metaphysics.) In the end, two expeditions triangulated arcs of meridians at the equator (near Quito, in the viceroyalty of Peru, 1735–45) and under the Arctic Circle (at the head of the Gulf of Bothnia, in Swedish Lapland, 1736–38), producing empirical proof that the Earth is indeed an 'oblate spheroid' (Greenberg, 1994; Terrall, 2002).

Although there was still no agreement as to the precise parameters of the Earth's shape, there were no major efforts to make further systematic, geodetic-quality, triangulation-based surveys until after the French Revolution in 1789, for the simple reason that extensive triangulations were expensive to equip and conduct. The revolutionary government commissioned a new geodetic survey: seeking to overturn the eternal verities of *ancien régime* Europe, it created a universal system of weights and measures which defined the new 'metre' as one ten-millionth of the distance from the equator to the pole. This required a new measurement of the Earth's size, executed in 1792–98 through the re-triangulation of the Paris meridian. Furthermore, the age of mass military mobilization ushered in by the revolution reconfigured the relationships of European states to their territories and populations, producing a wave of systematic, triangulation-based territorial surveys. French military engineers began many; others, notably the Ordnance Survey in England and Wales, were begun in response to the French military threat; all were continued after 1815, establishing the distinctly modern era of 'national' surveys (see Chapter 12, this volume).

The need to control observational errors, especially in the more extensive of the new systematic surveys, required them to have a geodetic core. Before 1815, geodesists had used a variety of ad hoc redundancies in their observations to control the quality of their triangulations, in a manner that relied more on intuition than logic. Carl Friedrich Gauss, however, sought to statistically model the entire body of error and uncertainty that unavoidably plagues any triangulation. He implemented his methodology of least squares – so-called because the observations in a system were adjusted such that the sum of the squares of all potential errors is minimized – when computing the locations of stations in his 1818 triangulation of Hanover. Gauss's methodology permitted dense areal, secondary triangulations, needed to control detailed topographical mapping, to be themselves controlled by the more precisely measured primary triangulation of geodetic arcs of meridians and parallels (Figure 11.8). Moreover, Gauss's work on the geometry of non-Euclidean surfaces (i.e. neither planes nor spheres) enabled geodesists to abandon spherical geometry and calculate distances and bearings directly on the spheroidal surface of the regular ellipsoids that approximate the shape of the Earth's equipotential surface (the 'geoid'). Primary triangulations were the subject of great care and highly complex calculations, but once completed they permitted the detailed work of secondary triangulations and topography to go ahead with less accurate instruments, less careful measurements and less well-trained staff. Ultimately, Gauss's geometry and statistics underpinned not only the determination of datums that best fit the surface of the Earth local to each national survey but also, in the present age of artificial satellites, the definition of global datums. His mathematics solidified geodesy as a science entailing the highest mathematics, albeit the mathematics of uncertainty; this status has helped enshrine the conviction that cartography as a whole is necessarily 'scientific' and, at least within the scope of daily life, certain.

Several of the post-1815 surveys featured major geodetic arcs. In Britain, the Ordnance Survey extended the Paris meridian north to the Shetland Isles, even as the French extended it via Spain and the Balearics to Algeria. A regional survey of southern India by the East India Company evolved into a geodetic survey of the central meridian of South Asia in 1799–1847; in Russia, in 1816, Otto von Struve began an arc along the meridian of Tartu, in Estonia that by 1855 reached from the Arctic to the Black Sea.

Over the course of the early nineteenth century, such geodetic surveys became enmeshed in a new, 'Humboldtian' approach to the field sciences. In his *Essai sur la géographie des plantes* (1805), Alexander von Humboldt had demonstrated that the bands of vegetation up the slopes of the Andes mimicked the bands of vegetation from the equator towards either pole, thereby correlating ecosystems to climate. The sheer brilliance of this insight led to the widespread attempt to determine how other natural phenomena varied across the Earth's surface.

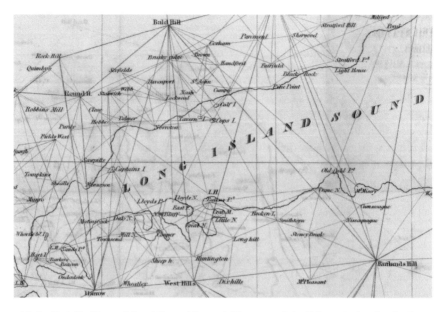

*Figure 11.8* Detail of Long Island Sound from *A Diagram of the Triangulation for the Survey of the Coast of the United States, Made in 1817 and 1833, and the Secondary Triangles Made in 1833 & 1834 in Connecticut & upon Long Island* (Washington, DC: U.S. Coast Survey, n.d.). The sides of the primary triangulation ran between hills distant from the coast, such as Bald Hill in Connecticut and West Hill on Long Island; these stations were then connected to specific points along the coastline by the more numerous secondary triangles, to serve as the basis for the detailed hydrographic survey of the coastal waters. The graticule is drawn at 20′ intervals. This impression was owned by Millard Fillmore, a US representative from New York in the 1830s and later US President (1850–1853). Courtesy of the Geography and Map Division, Library of Congress (G3781.B3 1834. U5); see *www.loc.gov/item/2012593342* for a high-resolution image

In particular, terrestrial observatories were established around the world to measure subtle variations in gravity and magnetism, while gauges installed at major ports sought to define the complex cycles of the tides. Geodetic measures were crucial in this work, as scientists sought to model the constantly fluctuating, and perhaps even stochastic, character of the Earth as a geophysical system. Inevitably, this work required international cooperation. In particular, in 1862, the Prussian government convened the first meeting of the 'central European degree measurement' with the goal of combining existing and future triangulations into a coherent geodetic system for measuring both arcs of meridians and parallels. The Prussians supported this institution until 1886, when it was transformed into the International Geodetic Association, funded by its member countries (Brown, 1949: 280–309).

A further element of nineteenth-century geodesy was the increasingly precise measurement of the third dimension. Triangulations permit the calculation of differences of height between stations; tied to an approximation of mean sea level as defined by a tidal gauge and perhaps filled in by careful levelling surveys, a dense triangulation constitutes a terrain model that in turn supports the precise graphic modelling of relief, whether constructed directly as systematic hachures or contours, or indirectly through stereoscopic photographic models. Either way, older artistic forms of sketching relief on plane tables gave way in the nineteenth century to the exact and measured depiction of relief (Imhof, 1982).

That the promise offered by the Cassinis' *Carte de France* – that the projected geometry of cosmographical space could become interchangeable with the Euclidean geometry of place, via Cartesian coordinate systems – began to be actually realized by the middle of the nineteenth century is indicated by the growing use by map makers of 'graticule' to distinguish the network of meridians and parallels from the grid of Cartesian coordinates. Thereafter, the promise was fulfilled during the Franco-Prussian War (1870–71) and then World War I (1914–18), when the deployment of large artillery pieces that could hurl shells far beyond the gunners' horizons required a new technology for long-distance targeting. The solution was to overlay topographical maps of the front with a rectilinear coordinate grid of 'artillery squares' that would permit the easy calculation of distance and, if the topographical maps were now projected on a conformal projection, direction of fire. The initially localized implementation of artillery squares was progressively extended to national, civil surveys – in the form, after 1936, for example, of the British National Grid – in turn promoting the practice of coordinate surveying.

## Conclusion

The history of the increasing geometricization of mapping in the European tradition since 1500 – of how mapping became increasingly 'scientific' – therefore features a complex interplay between no less than four different types of geometry. The early modern era can be characterized as one of competition between, on the one hand, the cosmographical geometry of global geography advocated by academics (and by those who sought social advancement) and, on the other, the craft-based Euclidean geometries used by property mappers, topographers, and engineers who together comprised the community of 'surveyors'. It was the prosecution of new, triangulation-based, systematic surveys that led to the combination of those geometries within new coordinate geometries, in which Cartesian coordinates define locations not only within the survey but implicitly on the globe. Finally, high-level geodetic works have been carried on with respect to the non-Euclidean surfaces of ellipsoids and the unique geoid, permitting more quotidian engineering and property surveys to be undertaken on a Cartesian plane. Yet, even as territorial surveying was grounded on a global framework, strictly geographical mapping largely continued to function in accordance with the cosmographical, global geometry, at least until the later twentieth-century proliferation of digital computers facilitated conversions between the different geometries.

## References

Andrewes, W.J.H. (Ed.) (1996) *The Quest for Longitude: The Proceedings of the Longitude Symposium, Harvard University, Cambridge, Massachusetts, November 4–6, 1993*. Cambridge, MA: Harvard University Press for the Collection of Historical Scientific Instruments, Harvard University.

Bennett, J.A. (1987) *The Divided Circle: A History of Instruments for Astronomy, Navigation, and Surveying* Oxford, UK: Phaidon.

Bennett, J.A. (1991) "Geometry and Surveying in Early-Seventeenth-Century England" *Annals of Science* 48 pp.345–354.

Brown, L.A. (1949) *The Story of Maps* Boston, MA: Little, Brown & Co.

Cook, A.S. (2006) "Surveying the Seas: Establishing the Sea Route to the East Indies" in Akerman, J.R. (Ed.) *Cartographies of Travel and Navigation* Chicago, IL: University of Chicago Press, pp.69–96.

Edney, M.H. (1994) "Mathematical Cosmography and the Social Ideology of British Cartography, 1780–1820" *Imago Mundi* 46 pp.101–116.

Gehring, U. and Weibel, P. (Eds) (2014) *Mapping Spaces: Networks of Knowledge in 17th Century Landscape Painting* Munich: Hirmer for ZKM Karlsruhe.

Gerbino, A. and Johnston, S. (2009) *Compass and Rule: Architecture as Mathematical Practice in England, 1500–1750* New Haven, CT: Yale University Press.

Greenberg, J. L. (1994) *The Problem of the Earth's Shape from Newton to Clairaut: The Rise of Mathematical Science in Eighteenth-Century Paris and the Fall of 'Normal' Science* Cambridge, UK: Cambridge University Press.

Harley, J.B., Woodward, D., Lewis, G.M., Monmonier, M., Edney, M.H., Pedley, M.S. and Kain, R.J.P. (Eds) (1987–) *The History of Cartography* (6 vols in 12 bks) Chicago, IL: University of Chicago Press. Reprinted online (free access) at *www.press.uchicago.edu/books/HOC/*.

Humboldt, A. von (1805) *Essai sur la géographie des plantes* Paris: Levrault, Schoell et Cie.

Imhof, E. (1982) *Cartographic Relief Presentation* (Trans. Steward, H.J.) New York: Walter de Gruyter.

Pelletier, M. (2013) *Les cartes de Cassini: La science au service de l'état et des provinces* Paris: Éditions du CTHS.

Snyder, J.P. (1993) *Flattening the Earth: Two Thousand Years of Map Projections* Chicago, IL: University of Chicago Press.

Terrall, M. (2002) *The Man who Flattened the Earth: Maupertuis and the Sciences in the Enlightenment* Chicago, IL: University of Chicago Press.

# 12

# The rise of systematic, territorial surveys

*Matthew H. Edney*

---

The character of states has varied widely according to the degree and nature of their control over their territories and other institutions. The feudal system of medieval Europe permitted the small apparatus of monarchical administration to govern via hierarchically arranged coteries of nobles and churchmen, who all challenged royal authority. Early modern monarchs sought, with varying degrees of success, to curtail the autonomy of the aristocracy and the church and to extend authority more directly over their territories. The burgeoning bureaucracies of the modern state further sought to regulate even the private, economic lives of individual subjects and citizens. Maps have played a crucial role in this steady escalation of centralized authority. Indeed, the post-1500 conception that states are necessarily territorial entities, rather than complex networks of feudal relationships, depended on the ability to *see* those territories in small-scale geographical maps. The modern formation of nations, of the imagined communities of people who share a common language and heritage, was similarly assisted by outline or logo maps, deployed within popular discourses to stand in for the nation's territory and character (Anderson, 1991: 163–185).

States have undertaken chorographical and topographical surveys for a variety of military and administrative purposes. Systematic territorial surveys – often called national surveys, although they are not limited to modern nation-states (Edney, 2009) – represent attempts to extend such spatially limited surveys to encompass entire states. The development of systematic surveys has always depended on the capacity of governments to exert their authority with sufficient finesse to map their entire territories in fine detail. The variety of surveys that resulted defies easy generalization. The only general account remains Brown's (1949: 241–309) outdated and overly generalized history; the many regional entries in Kretschmer *et al.* (1986) provide useful but brief summaries.

## Surveys of European states before c.1800

Mapping first acquired administrative functions within European governments during the sixteenth century (Buisseret, 1992). The initial appreciation of the bureaucratic and instrumental value of maps developed in parallel with the post-1550 marketplace for maps in Rome, Venice, and Antwerp: as the burgeoning communities of lawyers and increasingly professional bureaucrats consumed printed regional and world maps, they also began to commission graphic maps and plans for their official work. In the Italian city-states, map use flourished not within long-established bureaucracies but within the recently created offices that managed each city's territories (Harley *et al.*, 1987–, 3: 854–974). In England, each new generation of bureaucrats was more cartographically aware than its predecessors and more consistent in its use of maps in administering taxes and fortifications (Harley *et al.*, 1987–, 3: 1589–1669). Most of this early official cartography entailed either highly localized surveys, especially of fortresses and certain boundaries, or very general and broadly strategic mapping of entire states, oceans, and colonial regions (Harley *et al.*, 1987–, 3: 661–679).

Early modern territorial mapping was strictly chorographic in nature, carried on province by province. In the Holy Roman Empire and France, which were still largely organized in a feudal manner, it proved impossible to organize statewide surveys. Without financial support from the emperor, Tilemann Stella's attempt after 1560 to map all of Germany proved futile. Similarly, Catherine de' Medici's 1570 commission to Nicolas de Nicolay to map all of France was ineffective; his successor, François de La Guillotière, had by 1584 collected material sufficient only for a nine-sheet map of France at only a small scale (c.1: 1,000,000; published 1613). Surveys were instead organized by provincial authorities. The first officially commissioned survey – and one of very few regional surveys to be based on triangulation before 1800 – was Philipp Apian's survey of Bavaria (1554–1561; see Figure 12.1). In France, the chorography of provinces and

*Figure 12.1*   Philipp Apian, *Chorographia Bavariæ* (Ingolstadt, 1568), sheet 18, with Munich. This twenty-four-sheet printed map of the duchy of Bavaria, c.1:135,000, was reduced from Apian's original manuscript completed in 1563, c.1:45,000, which measured five metres on a side. An anonymous owner of this particular copy annotated the continuation of each river with the number of the relevant adjoining sheet. Courtesy of the Harvard Map Collection, Harvard University (MA 1620.568 pf*); see nrs.harvard.edu/urn-3:FHCL:5348421 for high-resolution images of entire map

dioceses was undertaken by a wide variety of officials and individuals. All but three of the maps that Maurice Bouguereau included in his atlas, *Le theatre francoys*, which he presented to Henri IV in 1594, had been previously published; the atlas's apparent coherence thus 'anticipated the creation of a unified France that did not yet exist' (Harley *et al.*, 1987–, 3: 673 [quotation], 1223–1224, 1213–1214, 1485, 1490–1495).

More centralized states had some success with comprehensive mapping projects. In Spain, in 1538–1545, Alonso de Santa Cruz compiled multiple sources into the (unfinished) 'Escorial atlas', a twenty-sheet map of the country at a consistent scale of c.1:400,000; Pedro de Esquivel's later survey of Spain for Philip II was left unfinished on his death in 1565 (Crespo Sanz and Maroto, 2014). Nicola Antonio Stigiola and Maria Cartaro drew up a manuscript atlas of the kingdom of Naples in 1583–1595, six copies of which survive, each containing twelve provincial maps drawn at the common scale of c.1:500,000 (Harley *et al.*, 1987–, 3:962–65). Perhaps the most famous of such early modern surveys was that by Christopher Saxton, who surveyed the English and Welsh counties in 1573–1578 for Thomas Seckford, a client of Elizabeth I's chief minister, Lord Burghley. Saxton compiled existing maps into thirty-five printed maps – of varying size, scale (1:250,000–1:350,000), and combinations of counties (some single, several grouped together) – that were eventually reissued as an atlas in 1579. The smallness and poverty of the English government meant that Saxton's work was not the product of 'a calculated strategy' but 'was marked at almost every stage by improvization and inconsistency' (Harley *et al.*, 1987–, 3: 1623–1631, esp. 1623 [quotation]).

The tradition of individual, provincial chorographies persisted well into the nineteenth century. The sixteenth-century surveys were reprinted, and perhaps updated, and some new provincial surveys produced. In the eighteenth century, the rapidly developing engineering and military needs of the burgeoning European economies sustained a new round of intensive chorographical mapping at much larger scales than previous work.

The pursuit of extensive engineering works led to substantial regional surveys, especially in Britain and France. In Britain, private capital and local authorities sponsored a series of new, large-scale surveys of individual counties, many with triangulation frameworks (see Chapter 11, Figure 11.5), starting with Joel Gascoyne's surveyed map of Cornwall at 1:63,360 (1700). That these county surveys were intended to promote improvements, from drainage projects to canal construction, is clear from the prize offered after 1759 by the Society of Arts to encourage their production (Delano-Smith and Kain, 1999: 75–111). In France, state reforms and centralization during the seventeenth century led infrastructural improvements to become a function of central government. In particular, the engineers of the Ponts et Chaussées undertook extensive surveys for roads and canals, many of which were collected between 1730 and 1780, in manuscript, in a grand atlas of France (Blond, 2014).

The primary motive for new regional surveys after 1690, however, was the dramatic shift in the practice of war, as the strategic importance of fortresses and the tactical emphasis on sieges gave way to the direct control of territory and open battles. Ever larger and more mobile armies required complex support and knowledge of the terrain. By the 1740s, as increasingly permanent army staffs began to plan and prepare for future wars, and as newly permanent corps of military engineers needed productive peacetime employment, the former set the latter to undertake large-scale territorial surveys. The engineers were themselves something of an international and even mercenary community, trading their skills for employment across Europe and its overseas colonies (e.g. Veres, 2014). The Austrian *Josephinische Landesaufnahme* (1763–1799) eventually covered, province by province, some 570,000 km$^2$ (Mapire, n.d.). The British undertook similar territorial surveys in Scotland (1746–1755: Fleet, n.d.), coastal North America (1764–1775:

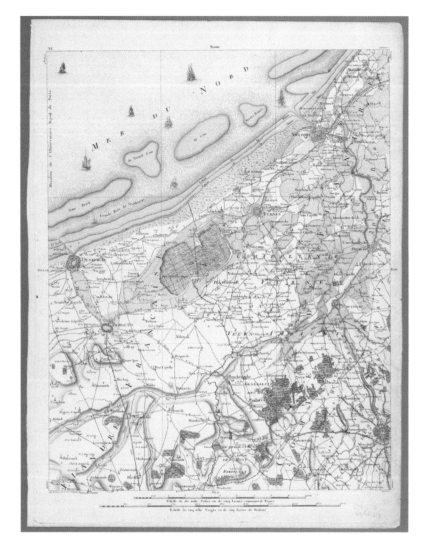

*Figure 12.2* Joseph Jean François, Comte de Ferraris, *Carte chorographique des Pays-Bas autrichiens*, 1:86,400 ([Mechelen], 1777), sheet 6. One of twenty-five irregularly sized sheets (56 × 89 cm or smaller), reduced from the plane-table surveys of Ferraris' 275-sheet 'Carte de Cabinet des Pays-Bas autrichiens' at 1:11,520 (1771–1777) that was based on the French triangulation of the southern Netherlands. Courtesy of the Universiteitsbibliotheek Vrije Universiteit, Amsterdam (80.4.220, 80.4.230, 80.4.210.400, 81.273.3); see http://imagebase.ubvu.vu.nl/cdm/ref/collection/krt/id/1908 for high-resolution images of the entire map

Hornsby, 2011), parts of India after 1757, and so on. There were also many military surveys within the various German states, such as the Prussian surveys of Silesia (1743–1756) and Saxony (1756–1763). Of all this military work, only the Comte de Ferraris's survey of the Austrian Netherlands (1771–1777), one component of the *Josephinische Landesaufnahme*, was either based on a triangulation or published at a reduced scale (Figure 12.2).

## The systematic mapping of France, 1669–1793

The uniqueness of Ferraris' survey as a military chorography reflects his emulation of the first modern systematic territorial survey: the *Carte de France* (Konvitz, 1987; Pelletier, 2013). Ultimately, the *Carte de France* was a civil exercise. While its depiction of relief was too parsimonious and vague to be useful to engineers or the military (Figure 12.3), it gave those who participated in governing the French state, whether in Paris or the provinces, an image of state territory that was suitable for understanding the nature of the country and for general planning. It complemented rather than supplanted the many surveys undertaken for particular projects, such as those of the Ponts et Chaussées. It was not born fully fledged, and its complex developmental history belies the ease with which Ferraris could first extend the triangulation network that French engineers had created during sporadic military occupations of the southern Netherlands and then publish his own survey at what then seemed the standard scale.

Jean-Baptiste Colbert, Louis XIV's finance minister, founded the Académie des Sciences in 1666 in large part to correct the existing provincial maps of France, in support of his ambitious economic reforms and the Paris observatory in 1667 to implement Jean Dominique Cassini's (I) new method for determining longitude. The Academicians advocated using the new longitude methods to refine and combine the existing provincial surveys (see Figure 12.3) together with a

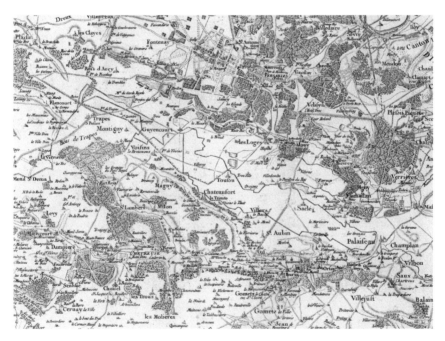

*Figure 12.3*    Detail from *nombre* 1, *feuille* 1, showing areas southwest of Paris, including Louis XIV's grand palace at Versailles, of César-François Cassini de Thury, *Carte de France*, 1:86,400 (Paris, 1762, originally published 1756). The *nombre* indicated each sheet's location in the statewide grid of sheets, with *nombre* 1 centred on Paris, then counting outwards; the *feuille* indicated the sequence of publication, starting with the Paris sheet, which defined the sequence of sheets when bound in large atlases. Note the simplicity of relief depiction: just slight escarpments to mark the edges of river valleys. Courtesy of the David Rumsey Collection (5694.019); see www.davidrumsey.com for a high-resolution image of the entire map

short triangulation (1669–1671) to measure the Earth's size; data needed to convert itineraries to differences in latitude and longitude. Realizing that a triangulation provides much more precise and rigorous control than astronomical observations (see Edney, 1997: 106–108), and despite intermittent funding, the Academicians steadily extended the triangulation: Jacques Cassini (II) finally completed the triangulation of the meridian through the Paris observatory in 1720; the triangulation of the perpendicular to the meridian at the observatory, begun in 1733, was followed by a comprehensive network of triangles covering the rest of the country. The entire work was completed and all locations computed by 1744 by Çésar François Cassini (III) de Thury and Giovanni Domenico Maraldi (see Chapter 11, Figure 11.7).

It had long been apparent that the existing provincial maps were too inaccurate or idiosyncratic to be fitted to the triangulation. The value of a consistent, triangulation-based survey was demonstrated when, in 1746, Cassini III extended the triangulation into Flanders and produced several large-scale topographical maps for the military. In 1750, Cassini III acquired a royal warrant and funding to undertake a systematic survey to create a single map of the entire country: the *Carte de France*. The map would require 180 sheets at 1:86,400 (one *ligne* [2.25 mm] to 100 *toises* [194.9 m]). Ten teams, each of two engineers, would run traverses between triangulated points; at a rate of one sheet per team per year; the survey would take just eighteen years. But a lack of trained staff delayed publication of the first sheets until 1756 and the loss of state funds on the outbreak of the Seven Years' War (1756–1763) threatened the entire survey. Cassini III accordingly formed a private company with subventions from the king and the aristocracy. The provinces, which would eventually benefit from the published maps, were supposed to reimburse the king, but those distant from Paris refused to do so and instead paid for their own surveys, which were then incorporated into the final work. Fieldwork was almost complete when Cassini III died in 1784 but, before the *Carte de France* could be fully published, the revolutionary government confiscated the entire project from Jean Dominique Cassini (IV) in 1793.

## Modern systematic surveys

In emulation of the *Carte de France*, French authorities undertook triangulations and detailed military surveys of conquered territories throughout the Napoleonic wars (1803–1815). Most of the states covered, especially in Germany and Italy, subsequently took advantage of this work and perpetuated the systematic surveys. The *Carte de France* thus spawned a profusion of military surveys across much of Western Europe, but its principal legacy was as a proof of philosophical, political, and technological concepts.

Philosophically, the concept of a triangulation-based survey held out the prospect that all mapping endeavours might be unified within a single technical process. With the Earth measured in precise and consistent detail, any map could be produced at any scale, whether of land or sea. By the 1820s, members of the Société de Géographie in Paris began to refer to this new conception of unified mapping practice as 'cartography' (Van der Krogt, 2015).

Politically, the *Carte de France* proved that it was indeed beneficial to map an entire country as one pillar of a well-organized state, assuming that politicians and bureaucrats had the desire and authority to pursue such a goal. The prosecution of such surveys was coincident with the very philosophy of modern civil government: the one did not produce the other; the proliferation of systematic surveys after 1790 was coeval with the formation of the recognizably modern state (Winichakul, 1994). Moreover, the modernizing state's self-examination was not limited to each country's surface features: the state provision and regulation of marine navigational aids, from harbour buoys to lighthouses, was intertwined with systematic, inshore hydrographical surveys (Chapuis, 1999); the regulation of populations and economies took the form of organized demographic censuses and statistical surveys (Nadal and Urteaga, 1990); and, by the mid-1800s, modern states sought to map geological strata and surficial deposits for economic ends.

Technologically, the *Carte de France* proved that it was indeed possible to map an entire country as a whole, as long as there was sufficient funding and organizational capacity. That capacity came generally from the military; the one element of early nineteenth-century states that had the size, discipline, and hierarchical structure needed to carry out any organized endeavour over an extensive territory. Military engineers were especially important, as they provided within each army a kernel of experienced mapmakers with the mathematical skills necessary to undertake extensive, triangulation-based surveys. Naval officers and their ships' crews similarly provided the skills and labour to map detailed inshore hydrography. Yet the military's mapping needs were markedly different from those of the civilian authorities: whereas civil authorities focused on property (ownership and taxation) and economic activities, the military sought information to aid the waging of conflict on sea and on land. These multiple factors mean that each modern state had its own, unique pattern of topographical, cadastral, geologic, and hydrographical surveys that manifests a particular combination of territorial structure, political will, institutional capacity, and tolerance of the costs involved.

Inadequacies in one or more areas led many systematic surveys to proceed slowly or to fail outright. Thus, Napoleon's grand military and intellectual conquest of Egypt, in 1798–1801, had insufficient time and staff to undertake the necessary triangulation framework. In British India, politicians had accepted the principle of triangulation-based, systematic topography by the

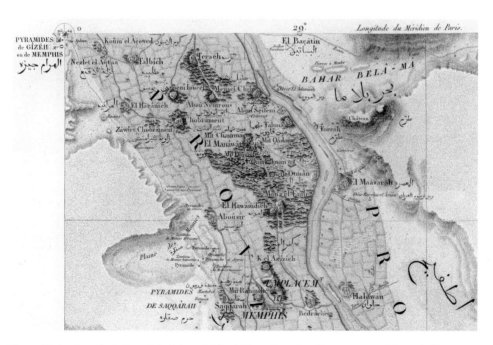

*Figure 12.4*  Detail of sheet 21, 'Memphis', of Pierre Jacotin, *Carte topographique de l'Egypte*, 1:100,000, 2nd ed. (Paris: C. L. F. Panckoucke, [1826]; 1st ed., 1818). Jacotin constructed the forty-seven-sheet topographical map of Egypt and the Holy Land on an ordered set of sheet lines with an origin at the pinnacle of the main pyramid at Giza (top-left corner). Size of original (entire): 54 × 82 cm. Courtesy of the David Rumsey Collection (3964.027); see www.davidrumsey.com for a high-resolution image of the entire map

1830s, especially as a means to modernize India and impose European rationality on South Asia, but the project was foiled by the sheer extent of the territories to be mapped. The results in both contexts, whether Pierre Jacotin's *Carte topographique de l'Egypte* (Figure 12.4; see Godlewska, 1988) or the huge array of topographical maps produced by the Survey of India (Edney, 1997; Harley *et al.*, 1987–, 6: 1451, 1473–77), were accordingly compilations of wildly different surveys with only a loose association with a triangulated foundation.

Insufficient resources inevitably led to systematic surveys dragging on for a long time. The Spanish government began a triangulation in 1854 to serve as the foundation of a topographical map at 1:50,000: topographical work began in 1870 and the first sheet appeared in 1875, the last only in 1965; in the meantime, the autonomous government of Catalunya undertook its own topographical map at 1:100,000, in 1914–1941 (Montaner, 1998). In the USA, constitutional distinctions between federal and state authorities, together with the institutional weakness of the federal government before 1860, significantly confused mapping efforts. Some states of necessity undertook their own surveys, such as Simeon Borden's 1830–1838 triangulation of Massachusetts, which was intended as a framework to combine existing maps into a single work.

*Figure 12.5*  US Coast and Geodetic Survey, map 328, *York River Harbor, Maine*, 1:20,000 (late reprint of a map surveyed in 1851–1853 and first published in 1854). The mapping of near-shore topography as well as inshore hydrography gave USCS coastal and harbour charts a distinctive look. The USCS was renamed as the US Coast and Geodetic Survey in 1878. Courtesy of the Osher Map Library and Smith Center for Cartographic Education, University of Southern Maine (OML-1854-25); see www.oshermaps.org/map/12143 for a high-resolution image of the entire map

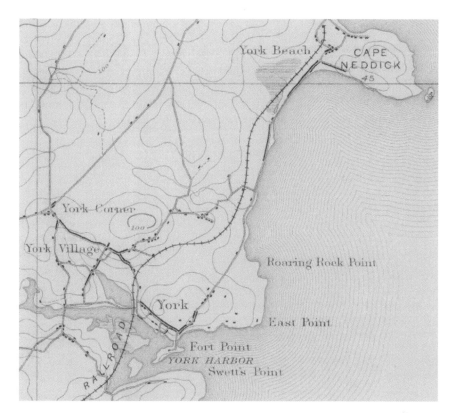

*Figure 12.6*  Detail of US Geological Survey, *Maine–New Hampshire, York sheet*, 1:62,500 (surveyed in 1888 and published in 1893), showing the same area as Figure 12.5. The USGS's primary function was to determine the geological character and mineralogical resources of the nation; Henry Gannett, the first head of what would become the National Mapping Division, used contours rather than hachures so as not to obscure overprinted thematic information, as in the portfolios of the 'National Geologic Atlas'. Note that the blue lines covering the water are not isobaths but simply colouring (copper engraving does not permit the printing of solid shades). Courtesy of the USGS, '*The national map*: Historical topographic map collection'; see nationalmap.gov/historical/index.html for a high-resolution image of this map

Work on the US Coast [and Geodetic] Survey's triangulations of the eastern and western coasts was fitful until after 1865 (Figure 12.5, and see also Chapter 11, Figure 11.8); an effort by the USC[G]S in 1871–1895 to provide detailed triangulation coverage for certain individual states failed almost completely. Only in 1884 did the US Geological Survey begin to work in collaboration with individual states to create a systematic territorial survey of the country at 1:62,500 (Figure 12.6; see Edney, 1986).

Proponents of systematic surveys initially saw them as a grand, universal solution. The strong mapping impulse of revolutionary France – which included both the country's territorial reorganization and the definition of a new, standard linear measure by measuring the Earth – extended to a new, comprehensive, joint military and cadastral survey of the country, begun in 1804. The idea was for a detailed cadastral survey at 1:2,500, which would underpin the standardization of property taxes; the military would then add terrain information to reduced-scale maps to make a topographical series suitable for military use. Yet the logistical hurdles proved too great: in the end, the cadastre was implemented piecemeal, commune by commune, while the military would eventually resurvey the country after 1818. The 273 sheets of the *Carte de l'état-major*, surveyed at 1:40,000 and published at 1:80,000, were completed in 1875. In early nineteenth-century Portugal, the prospect of a combined survey was just too ambitious and became a casualty of the political struggles between reformers and reactionaries (Branco, 2005). Such failures meant that systematic surveys in Europe, North America, and Europe's colonies remained on two tracks until well into the twentieth century (e.g. Mapire, n.d.).

The only nineteenth-century cadastral survey to be grounded on a triangulation was that of the United Kingdom. The United Kingdom's achievement in this regard was a function of the slow development, both institutional and cartographic, of the Ordnance Survey (OS), often seen as the exemplary systematic survey institution (Oliver, 2014). The Board of Ordnance, the organization comprising the Royal Artillery and Royal Engineers, in 1791 began a military survey of southern and eastern England, in case of invasion by the French. Work initially progressed along two fronts: first, William Roy's geodetic connection between the Greenwich and Paris observatories in the 1780s was expanded into a triangulation of Great Britain; second, a series of detailed topographical surveys were undertaken, county by county, mostly at two miles to the inch (1:31,680) and published at 1:63,360, beginning with the survey of Kent

*Figure 12.7* Detail of William Mudge, *General survey of England and Wales. An entirely new & accurate survey of the county of Kent, with part of the county of Essex*, 1:63,360 (London: William Faden, 1801). This map represents the first portion of the systematic survey of England and Wales. The Board of Ordnance originally undertook the survey, and sold the maps, by county. In this instance, the four sheets of the map of Kent were dissected and mounted on cloth to be sold as portable wall map. The fieldwork was subsequently re-engraved on four new sheets for incorporation into the OS's 'first series' territorial survey maps. Courtesy of the David Rumsey Collection (8534.002); see www.davidrumsey.com for a high-resolution image of the entire map

(Figure 12.7). The entire operation was moved to Ireland in 1824, to provide both a military topographical map to assist in keeping state control and a cadastre to assist in reforming property tenures and taxes. The director, Thomas Colby, initiated a strictly hierarchical process of triangulation control to permit a small army of simply trained surveyors to conduct the cadastral survey at six inches to the mile (1:10,560) while the officers mapped topography for publication at the standard 1:63,360; in the process, the survey acquired the name of 'Ordnance Survey'. The fieldwork was successfully completed because of a strongly hierarchical division of labour and because of the relatively small scale used for the cadastral work.

With the completion of fieldwork for the Irish survey in 1840, the OS initiated similar, six-inch surveys in northern England. But this work was inadequate both for civil government, which sought parish maps at c.1:2,500 for tithe commutation and the assessment of poor law rates, and for the construction of new urban infrastructure necessitated by rapid industrialization. OS leadership argued that its experience, capacity, and military hierarchy made it the appropriate organ of government to undertake detailed surveys for civil administration; to demonstrate its capacity, the OS undertook a detailed survey of the city of York, at 1:1,056, published in 1852 (Figure 12.8). Many government officials nonetheless disagreed, on both philosophical and financial grounds. The OS eventually emerged victorious from the parliamentary 'Battle of the Scales' and by 1860 was installed as the United Kingdom's sole national mapping agency. It used its extensive triangulation to make systematic surveys at 1:63,360 ('one-inch') for general and military purposes; the county series of c.13,400 sheets

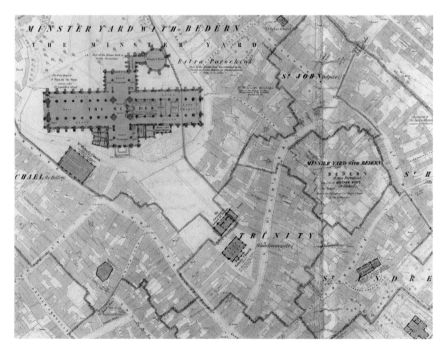

*Figure 12.8*  Detail of Henry Tucker, *Plan of York, 1852,* 1:1,056 (60 inches to the mile) (Southampton: Ordnance Map Office, 1852), sheet 9, hand-coloured, from surveys undertaken in 1849–1851. Historic sites, buildings, and monuments are indicated in black letter ('gothic'). The entire city took twenty-one sheets. Courtesy of the David Rumsey Collection (6740.009); see www.davidrumsey. com for a high-resolution image of the entire map

at 1:10,560 ('six-inch'); and the parish series of c.80,000 sheets covering Britain (i.e. not Ireland), at 1:2,500. To these were soon added urban mapping of any town with a population of at least 4,000, at 1:500. In 1870, the OS was transferred to civil authority, but it remained a fundamentally military organization until 1974.

The OS broadly exemplifies the general trend, at least in industrialized nations, for systematic surveys over time to grow more detailed and more integrated with the needs of civil government. A key marker of this transition was the shift in the depiction of relief from hachures to contours (compare Figures 12.5 and 12.6). They progressively supported geological and statistical surveys; in the twentieth century, these came to include detailed analyses of land use (Whitehead *et al.*, 2007: 86–116). After the Second World War, with the technological capacity offered by aerial photogrammetry, most systematic surveys adopted a process of continual updating and correction. Initially justified in the nineteenth century as one-off endeavours, state systematic surveys became persistent and pervasive elements of government.

The steady intensification of territorial mapping was sustained by the manner in which maps came to permeate not only the apparatus of each state but also the daily lives of the citizenry. Consumption of territorial maps contributed profoundly, during the nineteenth century, to the formation of the modern ideological systems of nationalism and imperialism. In the case of the new 'nation-states', territorial maps gave the citizenry a means to understand and appreciate *their* land in fine detail; by contrast, distribution of territorial maps was restricted in imperial settings, so that their differential consumption was a marker of difference between the imperial community, able to see the land they ruled, and the actual inhabitants, who were denied such a comprehensive and empowering perspective (Edney, 2009). This national/imperial distinction has not been an either-or proposition. In some modernizing states, the differential consumption of territorial information has sustained sharp economic and ethnic divides, which in turn serve to expose the imaginary nature of the 'nation' (e.g. Branco, 2005; Lois, 2014: 173–189).

## Conclusion

Territorial surveys provide a mechanism for government officials and private individuals to imagine that they can see distant landscapes, understand them, assess their worth, and determine means to improve them. This functionality was made explicit in a proposal to encourage the economic development of the interior of the US state of Maine by harnessing waterpower for industrial mills. An 1868 report proclaimed:

> The ordnance surveys of Great Britain, so many years in progress and nearly completed, give at a glance an accurate view of every physical feature of the country; every road, every house, every hedge row, almost every tree in the kingdom, is delineated in a style that enables the traveller or proprietor of real estate to see at a glance, not only its mountains, rivers, plains and valleys, but the exact position of each, and the character of the natural productions of the soil. By this means the remotest district of the kingdom from the capital is made familiar to every one, and the value of landed property comprehended at a glance throughout the entire realm. In such a work as this, our State would form a most striking picture from the number and beauty of the lakes of the interior.
>
> *(quoted by Edney, 2009: 27)*

This passage exemplifies the political and social function of territorial surveys. Territorial maps would let bankers in Boston and New York comprehend the distant Maine landscape and assess its economic value as collateral for loans to build mills to be driven by the state's many natural reservoirs. In the end, the capacity to fulfil Maine's cartographic potential was lacking; without a territorial survey, Maine's economic potential remained barely exploited. This explicit commentary nicely reveals that the underlying vision fostered and encouraged by territorial surveys was not only one of pragmatic need but also one of political and economic desire.

# References

Anderson, B. (1991) *Imagined Communities: Reflections on the Origin and Spread of Nationalism* (2nd (revised) ed.) London: Verso.

Blond, S. (2014) *L'Atlas de Trudaine: Pouvoirs, cartes et savoirs techniques au siècle des lumières* Paris: Éditions du Comité des travaux historiques et scientifiques.

Branco, R.M.C. (2005) "The Cornerstones of Modern Government: Maps, Weights and Measures and Census in Liberal Portugal (19th Century)" unpublished Ph.D. thesis, European University Institute.

Brown, L.A. (1949) *The Story of Maps* Boston, MA: Little, Brown & Co.

Buisseret, D. (Ed.) (1992) *Monarchs, Ministers, and Maps: The Emergence of Cartography as a Tool of Government in Early Modern Europe* Chicago, IL: University of Chicago Press.

Chapuis, O. (1999) *A la mer comme au ciel. Beautemps-Beaupré et la naissance de l'hydrographie moderne, 1700–1850: L'émergence de la précision en navigation et dans la cartographie marine* Paris: Presses de l'Université de Paris-Sorbonne.

Crespo Sanz, A. and Maroto, M.I.V. (2014) "Mapping Spain in the Sixteenth Century: The Escorial Atlas and Pedro de Esquivel's Notebook" *Imago Mundi* 66 (2) pp.159–179.

Delano-Smith, C. and Kain, R.J.P. (1999) *English Maps: A History* Toronto, ON: University of Toronto Press for the British Library.

Edney, M.H. (1986) "Politics, Science, and Government Mapping Policy in the United States, 1800–1925" *American Cartographer* 13 (4) pp.295–306.

Edney, M.H. (1997) *Mapping an Empire: The Geographical Construction of British India, 1765–1843* Chicago, IL: University of Chicago Press.

Edney, M.H. (2009) "The Irony of Imperial Mapping" In Akerman, J.R. (Ed.) *The Imperial Map: Cartography and the Mastery of Empire* Chicago, IL: University of Chicago Press, pp.11–45.

Fleet, C. (n.d.) "Roy Military Survey of Scotland, 1747–1755" *National Library of Scotland* Available at: *http://maps.nls.uk/roy/index.html* (Accessed: 30 October 2016).

Godlewska, A. (1988) "The Napoleonic Survey of Egypt: A Masterpiece of Cartographic Compilation and Early Nineteenth-Century Fieldwork" *Cartographica* 25 (1, 2) Monographs 38 and 39, Toronto, ON: University of Toronto Press.

Harley, J.B., Woodward, D., Lewis, G.M., Monmonier, M., Edney, M.H., Pedley, M.S. and Kain, R.J.P. (Eds) (1987–) *The History of Cartography* (6 vols in 12 bks) Chicago, IL: University of Chicago Press. Reprinted online (free access) at *www.press.uchicago.edu/books/HOC/*.

Hornsby, S.J. (2011) *Surveyors of Empire: Samuel Holland, J.F.W. Des Barres, and the Making of 'The Atlantic Neptune'* Montreal, Canada: McGill-Queen's University Press.

Konvitz, J.W. (1987) *Cartography in France, 1660–1848: Science, Engineering, and Statecraft* Chicago, IL: University of Chicago Press.

Kretschmer, I., Dörflinger, J. and Wawrik, F. (Eds) (1986) *Lexikon zur Geschichte der Kartographie von den Anfängen bis zum ersten Weltkrieg* Part C of *Die Kartographie und ihre Randgebiete: Enzyklopädie*. 2 vols. Vienna, Austria: Franz Deuticke.

Krogt, P. van der (2015) "The Origin of the Word 'Cartography'" *e-perimetron (www.e-perimetron.org)* 10 (3) pp.124–142.

Lois, C. (2014) *Mapas para la nación: Episodes en la historia de la cartografía Argentina* Buenos Aires, Argentina: Biblos.

Mapire. (n.d.) "Historical maps of the Habsburg empire" Budapest, Hungary: Arcanum Adatbázis Kft. Available at: *http://mapire.eu/en/* (Accessed: 30 October 2016).

Montaner, M.C. (1998) "'Topographic Cartography from a Regional Government: The Geographic Map of Catalonia, 1914–1941" *Cartographica* 35 (3,4) pp.81–88.

Nadal, F. and Urteaga, L. (1990) "Cartography and State: National Topographic Maps and Territorial Statistics in the Nineteenth Century" *Geo critica: Cuadernos críticos de geografía humana* 88 (English parallel series, 2).

Oliver, R. (2014) *The Ordnance Survey in the Nineteenth Century: Maps, Money and the Growth of Government* London: Charles Close Society.

Pelletier, M. (2013) *Les cartes de Cassini: La science au service de l'état et des provinces* Paris: Éditions du CTHS.

Veres, M.V. (2014) "Unravelling a Trans-Imperial Career: Michel Angelo de Blasco's Mapmaking Abilities in the Service of Vienna and Lisbon" *Itinerario* 38 (2) pp.75–100.

Whitehead, M., Jones, R. and Jones, M. (2007) *The Nature of the State: Excavating the Political Ecologies of the Modern State* Oxford, UK: Oxford University Press.

Winichakul, T. (1994) *Siam Mapped: A History of the Geo-Body of a Nation* Honolulu, HI: University of Hawaii Press.

# 13

# Cartographies of war and peace

*Timothy Barney*

In June 1947, mere months after President Truman outlined his canonical statement of Cold War intervention through the Truman Doctrine, the Official Geographer of the Department of State, S.W. Boggs, wrote a curious manifesto about maps for the popular *Scientific Monthly*. Boggs was an academic and a bureaucrat, known for his pioneering work in political geography (particularly borders and sea claims) – but here he traversed popular culture by worrying aloud about the public's engagement with cartography. Boggs warned forcefully against the potential of what he termed 'cartohypnosis', or a condition where the 'map user or the audience exhibits a high degree of suggestibility in respect to stimuli aroused by the map and its explanatory text' (Boggs, 1947: 469). As the nation's 'official geographer', Boggs also represented a powerful vantage point of political space in a globalizing world. In his position since the late 1920s, Boggs saw the transformation of world space through the phenomenon of 'air-age globalism', a discourse in which the airplane's ability to shrink world distances into minutes not miles captured the imagination of academic, government, and popular culture (Ristow, 1944). Like the bi-polar world of the burgeoning Cold War around him, where one ideological system was pitted against another, Boggs split maps into two categories: those that could 'delude the public' and those that 'may be used to awaken people to an intelligent understanding of the world and the problems of our times' (Boggs, 1947: 469). 'Cartohypnosis', therefore was part of a complex context of blurred war and peace, where maps were being marshaled more and more frequently to document and represent the idealism of a flexible and rapidly changing world, but also a world where that flexibility and dynamism heightened the fear of proximate enemies more than ever.

This chapter is not about Boggs per se, but about the kinds of cartographic conundrums of war and peace that he interfaced with as world space became 'closed' while the amount of interdependent global relationships grew exponentially. The concern around 'cartohypnosis' could only emerge from a shifting context where the space of the Second World War gave way to a new kind of fear of proximity that enemies were closer than ever. Over the course of the twentieth century, cartographers were ensnared between the idealistic possibilities of the 'one world' (and the scientific cooperation and peace it could bring) and the realist need for maps to strategically contain hostile spaces. Geographer Neil Smith has pointed out that, as the century unfolded, a crucial reconception of space took place, concurrent with new perspectives from the air, where absolute geography (seeing spaces as a preexisting identity – that space 'is') shifted to a relational geography where distance is relative and space is constituted socially (Smith, 2003). New, interdependent spatial relationships meant that transportation fluidly connected capital and communication networks. Geographers no longer had to travel the land in order to describe its contours; increasingly sophisticated technology challenged such expertise.

That notion of a relational geography was evident, at least in fits and starts, in one of the very first modern 'geopoliticians', Sir Halford Mackinder. S.W. Boggs was just one example of many who revived and expanded the precepts of Mackinder, the eminent British geographer, for the new air-age. Mackinder wrote during the turn of the century where land power still reigned, declaring, in the footsteps of commentators like Fredrick Jackson Turner, that the globe now housed a 'closed political system . . . of world-wide scope' where:

> [e]very explosion of social forces, instead of being dissipated in a surrounding circuit of unknown space and barbaric chaos, will be sharply re-echoed from the far side of the globe, and weak elements in the political and economic organism of the world will be shattered in consequence.
>
> *(Mackinder, 1904: 422)*

For Mackinder, the new century was about the wide-open possibilities of a 'closed world', where reverberations of the periphery could now rattle the world centers. Proclaiming the 'heartland' of the Russian Empire as a strategic 'pivot area' for the world's balance of power, Mackinder foreshadowed the Soviet Union's dominance in the area. Mackinder's most famous map was an oval-framed appropriation of the Mercator projection – his 1904 lecture and ensuing paper for the Royal Geographical Society cohered around a map of the 'pivot area' in the center with an 'inner crescent' comprising both continent and ocean, and an 'outer crescent', which was entirely oceanic. Thematic mapping was still a relatively new concept in 1904, and Mackinder's innovation was activating politics and strategy as the thematic 'data' for his map. His simplistic map recognized the coming together of time and space in a closed world. But the power of his theory came in the form of forecasting a threatened Britain in decline, and more importantly, subordinating European history to what he called 'Asiatic' history. In other words, Mackinder was prefiguring a social, economic, and political shift in the twentieth century towards a globalized world, all on the flat page of the map (Smith, 2003). He was subversively making geography and specifically geopolitics as a central historical, political, and even scientific concern for the foreseeable future. This was the vision of the map as, in Jerry Brotton's terms, a 'giant imperial chessboard', an influential function for twentieth-century geopolitical cartography featuring innovative and abstract concepts of 'pivots' and 'crescents' that, despite new sophistications and technologies, would continue to capture the geographic imagination of cartographers and politicians (Brotton, 2013: 365). Such abstractions, while on Mackinder's part were intended to keep a balance of peace, set forth a legacy of vying for resources in a shrinking world of perpetual warfare, helping to blur the lines in the twentieth century between peace and war.

That notion of global war also marked the cartographic and geopolitical legacy of the contemporaneous American naval historian, Alfred T. Mahan. If Mackinder's world encompassed the centrality of land power, Mahan influentially seized upon the importance of sea power in the future of foreign policy and strategic wartime dominance. Mahan, according to Guntram Herb, 'illustrated the control of the Gulf of Mexico and the Caribbean through a triangle that linked the major maritime choke points with the proposed canal across the Central American isthmus' (Herb, 2015: 539). Mahan's maps and treatises were published in popular outlets like

*The Atlantic Monthly* and *Harper's*, where he argued that the United States was a burgeoning world power that needed to protect itself on the seas. Like Mackinder, though, Mahan's world was based on Mercator, and these visions were soon to be supplanted by wartime imaginaries of the world that would revolutionize the practice and experience of cartography.

It has often been said that the 'short' twentieth century did not begin until the First World War; in some ways, the cartographic history and geopolitics of the twentieth century did not begin until the Paris Peace Conference of 1919 that dealt with the war's wreckage. As Geoffrey J. Martin has written, 'The value of maps had been recognized prior to the Paris Peace Conference of 1919, yet at Paris the map suddenly became everything' (Martin, 2015: 1053). While Mackinder's and Mahan's geopolitical mapping had outlined a kind of program for cartographic power politics in the twentieth century, the First World War left the imprint of true globalism in practice and sheer ubiquity of maps to diplomatic leaders and elites. In terms of mapping's effect on the tensions between war and peace, the First World War also affirmed cartography as a 'carving knife' between self-determining nations, not just in terms of state boundaries, but of race, ethnicity, and language as well. The thematic capacities of mapping during the close of the First World War and after were dramatically expanding. In fact, it was because of the Paris Peace Conference that the State Department started its own geographic division and began what would eventually be Boggs' office – there was a dramatic increase in the needs for all kinds of thematic and political maps in the executive branch, and a fully staffed unit was soon required. Mapping in the US had heretofore been conducted by the individual defense agencies and domestically by the United States Geological Society. However, the fact that a central agency was needed for research and cataloguing on the political aspects of geography and cartography revealed the immense importance that mapping had assumed on a global level.

It was after the First World War when Mackinder wrote the famous dictum, 'Who rules East Europe commands the Heartland; Who rules the Heartland commands the World-Island; Who rules the World-Island commands the World' (Mackinder, 1919: 106). It was after the Second World War that S.W. Boggs playfully established the corollary that 'He who would solve world problems must understand them; He who would understand world problems must visualize them; and He who would visualize world problems should study them on the spherical surface of a globe' (Boggs, 1954: 910). Much happened between the two wars to transform and complicate world space on the map. The Mercator world of strategists like Halford Mackinder was giving way to a host of new worlds that could account for truly global relations, and new maps to draw such worlds. The National Geographic Society in the US adopted the Van der Grinten projection in 1922, which still greatly distorted the areas closer to the poles (hence still enlarging regions like the Soviet Union) but more innovatively projected the world onto a globe-simulating sphere (Snyder, 1993). John Paul Goode's influential Goode Homolosine projection was introduced in 1923 and popularized by Rand McNally in *Goode's Atlas* – a publication, according to Susan Schulten, whose 'very existence served as an indictment of the American cartographic tradition that Rand McNally had itself helped to create' (Schulten, 2001: 186). Goode's disorienting projection interrupted the world into four quadrants like a peeled orange held together at the top, decentered the United States from the focus, and fixed the land distortion of the Mercator with its use of equal-area. S.W. Boggs' own eumorphic projection of 1929 was not as widely used as Goode's offering, but it refined even further the depiction of spatial distributions of global phenomena across the world (Boggs, 1929). In the early 1940s, Boggs would also try to refurnish the halls of government with the Miller projection, O.M. Miller's attempt to compromise the recognizability and clarity of the Mercator with better fidelity to area (Miller, 1942). Each of these projections moved fluidly through popular, academic, and government contexts, making the contingent and situational nature of maps more prominent in post-First World War discourse.

But it was the so-called 'air-age globalists' of the 1930s to the 1950s who most strongly emphasized the sphericity of the Earth, through their innovative use of new and long-forgotten projections and their employment of novel perspectives (Ristow, 1957). In both perspective and projection, one of the most prominent innovators was Richard Edes Harrison, *Fortune* magazine's artist and cartographer-in-residence from the mid-1930s to the early 1950s (and prominent at *Time* and *Life* magazines as well). Harrison trained as a graphic artist for advertisements for products companies, and fell into map-making during a temporary post at *Fortune*. He quickly gained notoriety for his novel viewpoint of world spatial relationships. His earliest map, for example, employed a 'vulture's-eye' view of the Italo-Ethiopian conflict in the mid-1930s – a rolling, spherical, pilot-like perspective of the landscape rolled out before the viewer. Harrison popularized little-known projections like the polar azimuthal projections, which centered on the increasingly strategic ground of the Arctic and made Europe closer than ever, and the orthographic projection, which most closely replicated the sphericity of the Earth as a globe on the flat page. Amidst these innovations, one of Harrison's most important contributions was the emphasis that war strategy was something that the everyday citizen could simulate alongside political elites, thus broadening the sense of civic participation, or at least consent, during the war (Schulten, 2001). President Roosevelt followed suit with this line of thinking, working off his own freely floating glass globes (handy for seeing the world strategically from any perspective), and encouraging his fireside audiences to 'look' at their maps as they followed along on the latest war update (Roosevelt, 1992).

Harrison was also a talented and intelligent writer in his own right, and his graphics were often accompanied by cartographic treatises on the need for flexible maps (Harrison, 1944). In many ways, Harrison became a minor popular celebrity, often asked for his views on the state-of-the-war and maps' role for the military, and often publishing his reviews of the latest atlases in the popular literature of the day, like the *Saturday Review*. In fact, Richard Edes Harrison corresponded with S.W. Boggs as a consultant for several State Department projects during the 1940s, especially to bring in the new, flexible mapping perspectives, evidencing the fluidity of government, academic, and popular cartographic cultures. While his output lessened after the end of the Second World War, his perspectives were remarkably easy to adapt to the burgeoning Cold War, emphasizing the US's proximity to the Soviet Union. He continued to produce such visions into the 1950s.

A host of other diverse actors picked up Harrison's mantle. Academics such as George Renner at Columbia and Erwin Raisz at Harvard entered air-age maps into the popular culture of the Second World War alongside newspaper and magazine graphic artists like Harrison. Renner was a representative case of the very complexity of the new air-age perspectives – he was a central figure in appropriating the tenets of Mackinder for new world war realities, and he worked with Rand McNally to popularize new air-age perspectives for textbooks and classrooms, but he also fell prey to controversial 'civilizational' arguments that came uncomfortably close to the kind of German *geopolitik* that he was arguing against (DeBres, 1986; Schulten, 2001). Raisz's work embodied the interconnectedness of nations and regions in the age of aviation, and popularized the fusion of maps with statistical diagrams through 'cartograms' (Schulten, 2007: 202). A newspaper artist like *The Los Angeles Times*' Charles Owens uniquely used Southern Californian geography and the precepts of Harrison-style perspective novelties and projections to create an imaginative cartography of the wartime Pacific (Cosgrove and della Dora, 2005). The Associated Press artists drew scores of maps to accompany their stories in the Newsfeatures series, often mixing cartographic images into infographics and cartoon-type designs. The popular monthly magazine artists like Robert M. Chapin, Jr., William Rowley, Richard Erdoes, and Vincent Puglisi, among others, also turned into important 'armchair' strategists during the Second World War and especially into the transitional Cold War (Henrikson, 1975). Even *Time* portrait artist Boris Artzybasheff brought maps into his distinctive work, also 'moonlighting' as a consultant with the State Department. By some measures considered propaganda, by other measures considered iconic or 'suggestive', or in Walter Ristow's more neutral terms, 'journalistic cartography', these maps often reduced geopolitical concepts to particular metaphors and bold, easily understood arguments (Ristow, 1957). Industrial factories, tools, storms and weather, clocks, and animals were just a few of the visual icons integrated into popular cartography to instantiate political messages. The stylistic flair of these maps expanded cartography's popular reach during the Second World War and into the Cold War. Chapin's signature techniques at *Time*, for example, impressed his boss Henry Luce at *Time/Life*, calling Chapin's airbrush a 'sort of highpower atomizer with which he sprays paint over his maps in an infinite number of shadings' and featuring a 'library of celluloid stencils – bomb splashes, flags, jeeps, sinking ships' (Barney, 2015: 104). The ubiquitous 'red' splashed all over the pages of the wartime and postwar *Time/Life* maps was also a signature, able to capture the increasing fears of expansions, interventions, and infiltrations of the global world.

Wartime multimedia discourses also demonstrated the new fluidity of maps in popular culture. Newsreels extensively employed cartography to show America's new and important strategic relationships. As Susan Schulten points out, Mackinder's famous law was taken up by discourses such as Frank Capra's *Why We Fight*, as it quoted the Heartland doctrine onscreen – but rather than using the old Mercator projection, Capra displayed a polar-centered map to show America's closeness to its European expansionist enemies (Schulten, 2001). Throughout the course of the war, feature films also mobilized maps, often of the more propagandistic variety. Walt Disney's *Victory Through Air Power* (1943) was the archetypal air-age artifact. The best-selling author, pilot, and air executive, the Russian-born Alexander De Seversky, was invited to offer his thesis about the superiority of US air power for winning the Second World War, couched between innovative cartoon histories of air power's beginnings and evolution. This kind of geopolitics was a long way from Mackinder and Mahan. Throughout the film, animated sequences would use graphic maps to show the increasing closeness of the world and the strategic need for vigilance against the US's enemies on multiple fronts, while Seversky would lecture during the live segments in front of sophisticated wall-sized interactive maps on novel air-age projections and spinning globes.

While the US public engaged with these popular geographers and cartographers, many of these practitioners' academic colleagues were being conscripted into the business of intelligence and top-secret research with the Office of Strategic Services (OSS) (Wilson, 1949). The OSS brought together an extraordinarily interdisciplinary group of scientists and humanists, among government bureaucrats, with the diverse skills needed to fight a truly global war (Barnes, 2006). Leaders in geography and cartography such as Richard Hartshorne and Arthur Robinson worked in the OSS, where more than 8,000 new maps were produced and over two million maps collected and analysed (Clarke, 2015: 1667).

Importantly, much of the era's theorizing about maps, both behind closed doors and in the public, worried aloud about the pernicious influence of propaganda cartography and the ideological use of geopolitics by Nazi Germany (Speier, 1941). The Nazi geographer Karl Haushofer has perhaps been overestimated in terms of both his influence on Hitler's apparatus and to the Allied enemies, but certainly much of the geographical discourse of the period mentions Haushofer by name, including Boggs. The notion of a 'cartohypnosis', for example, was premised on the idea that Axis mapping during the Second World War was an ideological practice, while the Allies conceived that they were doing more scientific work. The OSS geographers during the war and after (when they splintered into academic and government positions) continually worried about the inadequacy of their expertise and training, as they believed that their German counterparts had better schooling and techniques (Wilson, 1949). And, inside the wartime State Department, a memo by analyst Herbert Block offered a map sketching Haushofer's theory that the United States and the USSR were empires looking to expand to South America and Southeast Asia, while railing against Hitler's 'geo-mania' and declaring that 'German geopolitics is not a science; it is a slimy cluster of wishful thinking, political scheming and mendacious propaganda, interspersed with scientific facts' (Barney, 2015: 67).

Block's map was an almost eerie foreshadowing of a new world system. The split between what was seen as 'propaganda' and scientific mapping marked much of the cartography that would begin to arise during the Cold War, as the enemy in the world bi-polar system would fast become the Soviet Union (Pickles, 1992). Cartography and geography itself underwent significant transformations during the transition to the Cold War (Farish, 2010). The intelligence-gathering and research functions of agencies like the OSS (now morphing into the CIA) helped ease a 'quantitative revolution' that was also shared in other social-scientific disciplines (Barnes, 2008). OSS specialists like Arthur Robinson would lead the quantitative direction of postwar mapping, establishing a sophisticated scientific information-communication theory for cartography, and even pioneering new projections to match his philosophy (eventually the Robinson projection was picked up by *National Geographic* in the late 1980s) (Robinson and Petchenik, 1976). In the process, the business of mapping became much more of a data-amassing and scientific affair. As early as 1945, the Cold War fight for global spatial information began. A *Life* article from 1956 made a hero out of army geodesist Floyd Hough for stealing an enormous treasure of German maps in April of 1945, including some key cartographic images of Russian areas, mere minutes before the area he was pillaging was claimed for the Russian zone (Dille, 1958). John Cloud has written that the Hough finds 'would change the course of the Cold War', as it helped to revolutionize the science of cartography and geodesy (Cloud, 2002: 266). Concurrently, the Arctic became a Cold War focus of strategic concern as the US and the Soviet Union vied over the area's scientific and military/strategic potential, and maps followed suit with more sophisticated polar-centric viewpoints (Chaturvedi, 1996). Furthermore, a 1956 geodetic breakthrough by academic and government collaborative efforts revealed the actual circumference of the Earth to be significantly shorter than expected, confirming the Earth's ideological shrinkage through scientific means (Warner, 2000). During this time, government-military-academic collaborations also produced the Heezen-Tharp map of the ocean floor, a significant demonstration of the fluid lines between peace and war mapping in the Cold War. While academics hailed the breakthrough of an accurate map of the sea-floor, commercial outfits like Bell Labs eyed its use for transatlantic cable, while the military sought to benefit from the map as a defense weapon (Doel *et al.*, 2006). At the same time, the map led an eventful 'rhetorical life' as it also captured the geographic imagination of the public, circulating as it was through *National Geographic* and other outlets. With the Heezen-Tharp map, yet another new cartographic perspective was opened.

That circulatory power indicates that a significant part of maps' functions during the long Cold War was their very usability and employability as 'informational weapons', brought into diplomatic and public exchanges as provocative evidence for intervention and antagonism. For example, US Congressman O.K. Armstrong badgered Soviet Foreign Minister, Andrei Gromyko, at the 1951 San Francisco Peace Treaty conference with a map of the reported Gulag slave labor camps. This map turned out to be a highly orchestrated collaboration between anti-communist journalists, the State Department, the CIA, and crucially, the American Federation of Labor. The map incident itself was widely covered in the press, and the map was spread into

Cold War Europe and requested over Voice of America extensively (Young, 1958). The Gulag map confirmed that cartography had an incendiary power of 'place' in the Cold War to establish authentic knowledge of what was clandestine and provoke ideological battles.

With such projects, the Cold War significantly blurred the lines between 'war cartographies' and 'peace cartographies'. As Isaiah Bowman, Franklin Roosevelt's chief geographic consultant, himself forged in the Paris Peace Conference's foundry of new world spatial relationships, warned in 1948 that:

> It is so much easier to deal in simplicities and broad generalizations, or search for salvation in a single idea, or think ourselves secure because of a recent military victory. All must realize that the boundary between war and peace can be smudged with resulting disorder and danger. Events and ideas are now involved in a vast turbulence which statesmen must direct into strong and positive currents.
>
> *(Bowman, 1948: 142)*

For cartography, this 'smudge' meant the hailing of maps into wartime preparation under the guise of often peaceful uses – the idea that mapping was a central part of a perpetual mobilization (Crampton, 2010). An important example is how thematic mapping was used to amass large amounts of data about the so-called Third World, as the 'South' saw an astounding sequence of decolonizing independence movements where nations opted for self-determination in both peaceful and violent terms. The major Cold War powers were scrambling to document the immense amount of information about economic, political, and social issues that was emerging from these areas. Ostensibly a 'peacetime' cartographic effort, the *Atlas of Disease* project, through the American Geographical Society, also fluidly moved into the spaces of the Cold War, as the work was being marshaled for other purposes of national security and funded by defense agencies to chart problem areas for interventions by US troops and advisers. While nominally a scientific endeavor, the work was sponsored by the Office of Naval Research and supported by pharmaceutical companies like Pfizer, who were looking for new global markets for their medicine. The cartography of the *Atlas of Disease* itself comprised impressive, scientifically detailed map sheets on cholera and tropical skin diseases, even starvation. The maps remain important and complex examples of both a progressive and socially responsible, data-rich quantitative mapping of new global relationships *and* a paternalistic and racialized, quasi-colonial depiction of a Third World that was perpetually 'becoming' (Barney, 2014).

The idealism of this kind of scientific internationalism became increasingly difficult to manage – the International Map of the World (IMW), for example, was a massively ambitious collaboration between nations to map themselves at a common 1:1,000,000 scale. The project dated back as far as the 1890s, and much progress had been made by many nations across the world – but after the Second World War, the political climate had changed and that kind of open exchange, despite the efforts of actors like S.W. Boggs, was no longer tenable. Some of the functions of the IMW project were taken up by the United Nations. Certainly, the UN set up cartographic initiatives and conferences that brought different regions together to share cartographic methods and data, but the success of these conferences was limited by Cold War security realities, and the tense unwillingness to share cartographic data that might violate the sovereignty of nations became an issue (Heffernan, 2002). Participants in these cartographic exchanges were also hamstrung by the choice to adopt Soviet or US cartographic methods, as that choice had much larger symbolic implications beyond maps (Gardiner, 1961).

Still, the nonaligned nations often reclaimed cartography for themselves, attempting to avoid the binary positioning of the superpowers. For example, one of the original Organization of the

Petroleum Exporting Countries (OPEC) emblems from the early 1970s dispensed with showing the Northern hemisphere at all, enclosing the so-called Third World within an oval made of the constituent nations' flags, and conspicuously leaving out the US and Russia (Henrikson, 1979: 175–177). Likewise, the graphic designs of the mostly tri-annual nonaligned summits in Belgrade (1961), Cairo (1964), Zambia (1970), Algeria (1973), Sri Lanka (1976), and Havana (1979) often featured cartographic motifs that emphasized both the interconnected globality and independence of the self-determining Third World movement.

If air-age globalism had shrunk the world dramatically and altered the cartographic perspective of a host of audiences, the world of the satellite would revolutionize that perspective even further. The airplane changed the measurement standard to minutes instead of miles, but the advent of missile technology revised that standard to mere seconds. Thus, the major Cold War powers sought to put 'eyes in the sky' to photograph and map these networks of missile bases popping up across the world. Aerial photography had impressively begun during the First World War, where it made leaps and bounds in sophistication in a few short years. While lagging a bit during peacetime, on the US side Pearl Harbor activated the coordination of sophisticated aerial photography for intelligence work. The OSS had already affirmed the extensive utility of cartography in the intelligence community, and now that kind of knowledge branched into a host of agencies, including chiefly the CIA. In the process, the premium on the perspective from 'above' went from its novelty status in popular culture into the deepest classified spaces of governments (Cloud, 2002).

During the Cold War, the lack of spatial knowledge of the Eastern Bloc posed significant intelligence problems – hence the establishment of the CIA's highly classified Corona program, an initiative that saw an immense growth in the US ability to capture high-quality remote-sensing imagery, and became what Keith Clarke has called a 'game-changer as far as intelligence operations and mapping were concerned' (Clarke, 2015: 1668). After a few notable failures, Corona's success was fairly swift upon its establishment in 1960 – a single day in orbit over Soviet territory produced more images than had been collected in twenty-four U-2 flights in four years (Obermeyer, 2015: 1228). Until the project's demise in 1972, Corona reeled over 2.1 million feet of film in tremendously clear resolution (Campbell and Campbell, 2015: 1294). With the Corona images staying classified until the Clinton administration's order in 1995, the project's top-secret nature brings up the debate of public access to cartographic information, an ongoing and heated exchange since cartography became a function of intelligence in the first place. The cartography of the Second World War and the postwar often saw this multilayered combination of public, private, and classified uses of maps commingling together. The quantitative revolution in cartography saw defense professionals, academics, and even private companies developing GIS innovations as early as the 1960s, with GPS to follow in the 1970s. Especially at the classified level, as satellites took the cartographic eye higher and higher and computerized technologies came into the picture, maps were increasingly seen on digital screens. The iconic image of men (the gendered implications are important) sitting at a large computer screen tracking maps can be seen, for example, in Strategic Air Command (SAC) training films of the 1960s. As SAC managed the event of a nuclear retaliation, maps were ubiquitous in dark control rooms (parodied and dramatized in the era's films like *Dr. Strangelove* and *Fail Safe*). The perspectives of overhead cartography being on a flat page were now giving way to customizable and fluid spatial images, even if computerized maps started primitively in fits and starts. Maps, then, were fast becoming part of the Cold War's doctrines of 'mutually assured destruction', where cartographic expertise was needed to organize and spatialize potential targets through abstract images.

At the same time, the increasingly technologized Cold War brought maps more and more into the world of diplomacy and global politics. America's UN Ambassador Henry Cabot

Lodge entered maps prominently into a notorious skirmish with the Soviet Union at the tail end of the Eisenhower administration. With more and more sophisticated surveillance technologies, Americans were able to produce a map that charted the circumstances of an RB-47 plane being shot down by the USSR and Lodge conscripted the evidential authenticity of mapping into a blistering speech indicting his Cold War enemies (US Department of State, 1960). The U-2 air photography produced detailed images that gave US agencies the power to make maps of areas previously shrouded by the iron curtain. When Adlai Stevenson stood in front of the UN railing against the Soviet missile sites being built in Cuba, he was drawing on such state-of-the-art satellite photo and mapping technology to indict and inflame Cold War tensions (Clarke, 2015: 1668).

Mapping was also activated into the highest echelons of US foreign policy and the foray into domestic and international public opinion through the presidency itself. Using more traditional maps than the technologized images of Lodge and Stevenson, John F. Kennedy advocated action in Laos in 1961 by taking to the televised airwaves wielding a map of communist influence. Richard Nixon would call for even more drastic interventionist measures in Cambodia in his infamous April 30, 1970 speech, using his finger to circle on a map where an 'incursion' by US troops was needed to roll back communist gains. Four days later, four students lay dead from National Guard gunfire after protesting the move into Cambodia. The Mackinder legacy allowed for maps to become abstract chessboards, but such abstractions obscured the staggering effects of people living on these actual, material landscapes. The clean lines of a geopolitical map housed prodigious empty space for powers like the US to fill in with its own interventionary ideologies. The Cold War was a perfect fit for the spatial logics of maps, but cartography could often belie that there were myriad hot wars on the ground below.

The most public manifestation of the kinds of overhead views available from on high came from the cameras of Apollo astronauts – what were essentially photos of the Earth became themselves enormously influential maps of the world that transfixed a generation. The Earthrise photos mapped the Earth as the now-iconic blue marble, adopted discursively in global culture, according to Denis Cosgrove, as both a 'one world' of realist loneliness and a 'whole Earth' of ecological, idealist harmony (Cosgrove, 1994). As a conflict like the Vietnam War and other 'Third World' satellite Cold Wars raged inside the marble, the calm of the Apollo's map of the Earth was both reassuringly hopeful and devastatingly ironic.

By the end of the 1970s and into the 1980s, a so-called 'Second Cold War' had begun in earnest, with declarations by the US that the doctrines of mutually assured destruction should be fully renewed. For example, with an 'Evil Empire' supposedly ready to strike, the US indicted the Soviet Union through virulent cartographic propaganda – Defense Secretary Caspar Weinberger's pamphlets like *Soviet Military Power* were laced with maps showing a quantified world of deterrence through massive missile armament, along with projections that the Soviet Union was overpowering US weapon systems. The Soviets responded with treatises like *Disarmament: Who's Against?* showing maps of a bellicose and imperial United States bombarding the rest of the world with imposing arrows representing missiles (US Department of Defense, 1981; USSR Ministry of Defense, 1983). The Warsaw Pact also produced many maps simulating the potential onset of nuclear war over Europe. An ominous arrow-filled 1970 map forecasts the logistics of a Greece and Turkey invasion, while another encircles Denmark and Northern Europe. The secret Warsaw Pact exercise 'Seven Days Over the River Rhine' from 1979 used cartography extensively to chart, complete with red mushroom clouds strewn about the continent, an all-too probable nuclear clash between Cold War powers. And while this chapter does not generally focus on topographic mapping, it is worth noting the Cold War implications of the massive and secret Soviet mapping project that mapped the entire world at detailed scales right down to streets and buildings – a surveying initiative of unprecedented magnitude (Miller, 2015). Sir Halford Mackinder likely could not have foreseen how his geopolitical chessboard of a world would turn into cartography of the globe as a massive scientific apparatus of surveillance, military targeting, and weapon counting. And like his predecessors, Ronald Reagan would argue for global intervention in places like Latin America through the medium of the map, except now his were sophisticated news-style animations rather than the poster board and easel set-up of old. Counter-maps, in turn, appeared from the very rebel groups that Reagan was indicting. For example, a widely distributed (and English-translated) propaganda map supporting the left-wing FMLN coalition in El Salvador referred to Reagan's policies as 'imperialistic', and instead spatialized into cartography what the FMLN saw as an 'inevitable revolution' ('El Salvador: The Inevitable Revolution', 2015).

The doom-laden world of armaments was further challenged by a robust international peace movement, and maps were able to provide a central medium for these arguments of disarmament. The Pluto Project was a radical group of British social justice advocates who produced an innovative and influential group of atlases, under titles like *The War Atlas* and *The State of the World* (Kidron and Segal, 1981; Kidron and Smith, 1983). These atlases essentially parodied the world of Weinberger and mutually assured destruction by showing a Cold War world system that could annihilate itself a million times over. Blackly comic titles such as 'Funny Money', 'Slumland', and 'The Nuclear Club' accompanied icons of death and overwhelming color contrasts to rouse readers away from the staid 'scientific' mapping of the Cold War and to graphically display the effects of state violence (Barney, 2009). William Bunge, a renegade geographer who came out of that very same scientific and quantitative geographic tradition, abandoned his lucrative academic career to join the civil rights movement and eventually the nuclear disarmament movement – producing a provocative set of maps under the title *The Nuclear War Atlas*. Even while Bunge was a pioneer of the kind of cartography that would be used for the digital revolution, his maps were deliberately drawn as crude and defiant, accusing the Cold War powers of guaranteeing the destruction of humanity. Bringing the changing global geopolitical perspectives of the twentieth century full circle, he used Harrison-style perspectives and polar projections to outline a bleak world, including face icons melting from blasts and red swathes of nuclear fallout across regions of the world (Bunge, 1988). The German historian Arno Peters' projection, a 1973 equal-area map, reached its saturation level in the 1980s as the Cold War was both lighting up again and fading away. A self-conscious indictment of cartography's complicity in systems of inequality, Peters' map was adopted by many development and social justice organizations as it, like the air-age geographers' projections sought decades before, 'corrected' the Mercator world of a bloated Northern hemisphere. On the Peters' projection, the world was elongated to correct the spatial relationship between African, South American, and Asian areas and the Northern hemisphere. The projection incensed geographic and cartographic disciplinarians because Peters and his supporters billed it as the 'one' projection to right the world's wrongs rather than as one useful projection among many, but the projection also usefully called into question the role of ideology in mapping – and after an ideological century of cartography, the debate was an important one as the walls of the Cold War tumbled and a new set of globalizing ideologies came into the lines of the map (Vujakovic, 2003).

These critical challenges at the end of the century to cartographic power signal a kind of full circle. Back in 1947, with his fears of 'cartohypnosis', S.W. Boggs was both surveying the tumultuous half-century of world wars marked by often exciting but dangerous cartographic perspectives, and also looking forward cautiously at a future where the map was ubiquitous but not always accompanied by the critical sensibility he believed was needed to handle maps. Boggs certainly captured the complexity of the changing perspectives of maps and the tremendous growth of Mackinder's geopolitics, but he may have missed that cartography could not be a black-and-white enterprise of 'good' and 'bad' maps. The changing geopolitical perspectives presented by cartography in the twentieth century showed a remarkably fluid capacity to shape the world, not just to reflect it. From the Heartland to Apollo and beyond, the world somehow shrunk and enlarged at the same time, creating inescapable tensions in the lines of the century's geopolitical maps. For better or for worse, modern cartography promoted the idea that the ground and the space below could be ordered, classified, and therefore altered to fit the ideologies of the elites, the publics, and the challengers using them.

# References

Barnes, T.J. (2006) "Geographical Intelligence: American Geographers and Research and Analysis in the Office of Strategic Service, 1941–45" *Journal of Historical Geography* 32 pp.149–168.

Barnes, T.J. (2008) "Geography's Underworld: The Military-Industrial Complex, Mathematical Modeling and the Qualitative Revolution" *Geoforum* 39 pp.3–16.

Barney, T. (2009) "Power Lines: The Rhetoric of Maps as Social Change in the Post-Cold War Landscape" *Quarterly Journal of Speech* 95 pp.412–434.

Barney, T. (2014) "Diagnosing the Third World: The 'Map Doctor' and the Spatialized Discourses of Disease and Development in the Cold War" *Quarterly Journal of Speech* 100 pp.1–30.

Barney, T. (2015) *Mapping the Cold War: Cartography and the Framing of America's International Power* Chapel Hill, NC: University of North Carolina Press.

Boggs, S.W. (1929) "A New Equal-Area Projection for World Maps" *The Geographical Journal* 73 pp.241–245.

Boggs, S.W. (1947) "Cartohypnosis" *Scientific Monthly* June pp.469–476.

Boggs, S.W. (1954) "Global Relations of the United States" *Department of State Bulletin* 30 pp.903–912.

Bowman, I. (1948) "The Geographical Situation of the United States in Relation to World Policies" *The Geographical Journal* 112 pp.129–142.

Brotton, J. (2013) *A History of the World in 12 Maps* New York: Viking.

Bunge, W. (1988) *The Nuclear War Atlas* New York: Blackwell.

Campbell, J.B. and Campbell, J. (2015) "Remote Sensing" in Monmonier, M. (Ed.) *The History of Cartography, Volume Six: Cartography in the Twentieth Century* Part 2 Chicago, IL: University of Chicago Press, pp.1273–1304.

Chaturvedi, S. (1996) *The Polar Regions: A Political Geography* Chichester, UK: John Wiley & Sons.

Clarke, K.C. (2015) "U.S. Intelligence Community, Mapping by the," in Monmonier, M. (Ed.) *The History of Cartography, Volume Six: Cartography in the Twentieth Century* Part 2 Chicago, IL: University of Chicago Press, pp.1666–1672.

Cloud, J. (2002) "American Cartographic Transformations During the Cold War" *Cartography and Geographic Information Science* 29 pp.261–282.

Cosgrove, D.E. (1994) "Contested Global Visions: One-World, Whole-Earth, and the Apollo Space Photographs" *Annals of the Association of American Geographers* 84 (2) pp.270–294.

Cosgrove, D. E., and della Dora, V. (2005) "Mapping Global War: Los Angeles, the Pacific, and Charles Owens's Pictorial Cartography" *Annals of the Association of American Geographers* 95 pp.373–390.

Crampton, J.W. (2010) *Mapping: A Critical Introduction to Cartography and GIS* Malden, MA: Wiley-Blackwell.

DeBres, K. (1986) "Political Geographers of the Past IV: George Renner and the Great Map Scandal of 1942" *Political Geography Quarterly* 5: 385–394.

Dille, J. (1958) "The Missile-Era Race to Chart the Earth" *Life* 12 May, pp.132–148.

Doel, R. E., Levin, T. J. and Marker, M. K. (2006) "Extending Modern Cartography to the Ocean Depths: Military Patronage, Cold War Priorities, and the Heezen-Tharp Mapping Project, 1952–1959" *Journal of Historical Geography* 32 pp.605–626.

"El Salvador: The Inevitable Revolution" (2015) Persuasive Maps: PJ Mode Collection, Cornell University Library Available at: *https://digital.library.cornell.edu/catalog/ss:3293985* (Accessed: 4 January 2016).

Farish, M. (2010) *The Contours of America's Cold War* Minneapolis, MN: University of Minnesota.

Gardiner, R.A. (1961) "A Re-Appraisal of the International Map of the World (IMW) on the Millionth Scale" *International Yearbook on Cartography* 1 pp.31–47.

Harrison, R.E. (1944) *Look at the World: The Fortune Atlas for World Strategy* New York: Alfred A. Knopf.

Heffernan, M. (2002) "The Politics of the Maps in the Early Twentieth Century" *Cartography and Geographic Information Science* 29 pp.207–226.

Henrikson, A.K. (1975) "The Map as an 'Idea': The Role of Cartographic Imagery During the Second World War" *American Cartographer* 2 pp.19–53.

Henrikson, A.K. (1979) "All the World's a Map" *The Wilson Quarterly* 3 pp.164–177.

Herb, G.H. (2015) "Geopolitics and Cartography" in Monmonier, M. (Ed.) *The History of Cartography, Volume Six: Cartography in the Twentieth Century* Part 1 Chicago, IL: University of Chicago Press, pp.539–548.

Kidron, M., and Segal, R. (1981) *The State of the World Atlas* New York: Simon & Schuster.

Kidron, M., and Smith, D. (1983) *The War Atlas: Armed Conflict – Armed Peace* New York: Simon & Schuster.

Mackinder, H.J. (1904) "The Geographical Pivot of History" *The Geographical Journal* 23 pp.421–437.

Mackinder, H.J. (1919/1942) *Democratic Ideals and Reality: A Study in the Politics of Reconstruction* New York: Henry Holt.

Martin, G. J. (2015) "Paris Peace Conference" in Monmonier, M. (Ed.) *The History of Cartography, Volume Six: Cartography in the Twentieth Century* Part 2 Chicago, IL: University of Chicago Press, pp.1049–1053.

Miller, G. (2015) "Inside the Secret World of Russia's Cold War Mapmakers" *Wired* July Available at: *www.wired.com/2015/07/secret-cold-war-maps/* (Accessed: 4 January 2016).

Miller, O.M. (1942) "Notes on Cylindrical World Map Projections" *Geographical Review* 32 pp.424–430.

Obermeyer, N. J. (2015) "Public Access to Cartographic Information," in Monmonier, M. (Ed.) *The History of Cartography, Volume Six: Cartography in the Twentieth Century* Part 2 Chicago, IL: University of Chicago Press, pp.1227–1231.

Pickles, J. (1992) "Text, Hermeneutics and Propaganda Maps" in Barnes, T.J. and Duncan, J.S. (Eds) *Writing Worlds: Discourse, Text & Metaphor in the Representation of Landscape* London: Routledge, pp. 193–230.

Ristow, W.W. (1944) "Air Age Geography: A Critical Appraisal and Bibliography" *Journal of Geography* 43 pp.331–343.

Ristow, W.W. (1957) "Journalistic Cartography" *Surveying and Mapping* 17 pp.369–390.

Robinson, A.H. and Petchenik, B.B. (1976) *The Nature of Maps* Chicago, IL: University of Chicago Press.

Roosevelt, F.D. (1992) "Fighting Defeatism: February 23, 1942" in Buhite, R.D. and Levy, D.W. (Eds) *FDR's Fireside Chats* Norman, OK: University of Oklahoma Press, pp.206–218.

Schulten, S. (2001) *The Geographical Imagination in America, 1880–1950* Chicago, IL: University of Chicago Press.

Schulten, S. (2007) "Mapping American History" in Akerman, J.R. and Karrow Jr, R.W. (Eds) *Maps: Finding Our Place in the World* Chicago, IL: University of Chicago Press, pp.159–205.

Smith, N. (2003) *American Empire: Roosevelt's Geographer and the Prelude to Globalization* Berkeley, CA: University of California Press.

Snyder, J.P. (1993) *Flattening the Earth: Two Thousand Years of Map Projections* Chicago, IL: University of Chicago Press.

Speier, H. (1941) "Magic Geography" *Social Research* 8 pp.310–330.

US Department of Defense (1981) *Soviet Military Power* Washington, DC: Government Printing Office.

US Department of State (1960) "Security Council Rejects Soviet Complaint Against U.S. in RB-47 Incident" *Department of State Bulletin* 43 pp.235–244.

USSR Ministry of Defense (1983) *Disarmament: Who's Against?* Moscow: Military Publishing House.

Vujakovic, P. (2003) "Damn or Be Damned: Arno Peters and the Struggle for the New Cartography" *The Cartographic Journal* 40 pp.61–67.

Warner, D.J. (2000) "From Tallahassee to Timbuktu: Cold War Efforts to Measure Intercontinental Distances" *Historical Studies in the Physical and Biological Sciences* 30 pp.393–415.

Wilson, L.S. (1949) "Lessons From the Experience of the Map Information Section, OSS" *Geographical Review* 39 pp.298–310.

Young, W.R. (1958) "Gulag – Slavery, Inc.: The Use of an Illustrated Map in Printed Propaganda" in Daugherty, W.E. (Ed.) *Psychological Warfare Casebook* Baltimore, MD: Johns Hopkins University Press, pp.597–601.

# Part III

# Measuring the Earth

## From geodesy to GPS

# 14

# Modelling the world

*Miljenko Lapaine*

Cartography is the art, science, and technology of making and using maps. Scientific and technological developments over the last few decades have brought new ways to create and use maps. The Internet has spread maps to all corners of our planet and allows everyone to use maps, on a desktop or mobile device, while everyone seems to be able to create maps, since both data and tools are widely available. However, while the design of communicative maps is still an art, it also requires the science of cartography (Kraak, 2015). Science underpins the cartographic project, and one of the most important aspects of this is how the world, as a three-dimensional entity, is modelled. This is the realm of geodesy – the science of mapping the Earth.

## Geodesy: determining the Earth's shape and size

The Greek name for geodesy was first γεωμετρία, and later γεωδαισία. The first term originates from the word γη, i.e. the Earth, and μετρείν, 'to measure'. Hence, a literal translation of γεωμετρία is 'land surveying'. According to Neidhardt (1950), geometry, the science of geometric shapes, first arose from geodesy. Geometry concerns not only theoretical speculation and reflection, but purely practical tasks and problems. The notion of the triangle or rectangle as geometric shapes, their properties, and the discovery of formulae to calculate areas and so on, were required by purely practical, geodetic tasks that had previously been solved empirically. Later, the properties of every triangle, every rectangle and so on were devised and the science of geometry was born.

Some confusion over the terms *geometry* and *geodesy* has continued even to the present. A geodetic expert is still called a *land surveyor* (De Graeve and Smith, 2010). Some German authors used to call geodesy *Praktische Geometrie* (practical geometry), unlike theoretical geometry taught in high schools. This makes it sound as if geometry was a science, and geodesy only a practical activity, but geodesy has developed as an independent branch of science and has equal standing among other sciences. In French and Italian, the expression *topography* is used for practical geodesy, and *geodesy* for 'higher' or pure geodesy. There are many modern definitions of geodesy, with Vaníček and Krakiwsky (1986) describing the science as 'the discipline that deals with the measurement and representation of the Earth, including its gravity field, in a three-dimensional, time-varying space'.

According to the definition of the surveyor as adopted by the General Assembly 23 May 2004 of the International Federation of Surveyors (FIG, 2004), a surveyor (French *géomètre*, German *Vermessungsingenieur*) is a professional person with the academic qualifications and technical expertise to conduct one, or more, of the following activities:

- to determine, measure and represent land, three-dimensional objects, point-fields and trajectories;
- to assemble and interpret land and geographically related information;
- to use that information for the planning and efficient administration of the land, the sea and any structures thereon; and
- to conduct research into the above practices and to develop them.

Geodesists are very proud of their predecessors. About 2,330 years ago, Eratosthenes of Cyrene (c.276–195 BC) measured the length of the meridian arc between Alexandria and Syene (Aswan) and determined the radius of the Earth. Egyptian land surveyors used to recover plot boundaries every time the Nile flooded. In the age of great geographic discoveries, Mercator made a huge contribution to the safety of ocean navigation by inventing the conformal cylindrical map projection in 1569. The last great contributions to a better understanding of our planet were made through geodesy in the late nineteenth and early decades of the twentieth century, when the last 'white spots' disappeared from the maps of South and North America, Africa and Asia. At that time, the highest peak on the Earth was named after a geodesist, the British Surveyor-General of India, Sir George Everest (1790–1866).

## Modern geodesy

*Geodesy* is a branch of applied mathematics and Earth sciences, the scientific discipline and technology that deals with the measurement and representation of the Earth, including its gravitational field, in three-dimensional, time-varying space. Geodesy can be divided into applied, physical, geometric, and satellite geodesy and is associated with the fields of photogrammetry, remote sensing, and cartography.

*Applied geodesy* is the part of geodesy covering land surveying, engineering geodesy, and management of spatial information. Lower geodesy is an older term for land surveying, when the Earth's curvature was not taken into account. Engineering surveying is the part of geodesy dealing with designing, measuring, and supervising constructions and other objects, e.g. roads, tunnels, and bridges.

*Physical geodesy* is the part of geodesy dealing with the Earth's gravity field and its effect on geodetic measurements. *Geometrical geodesy* is about determining the Earth's shape and size geometrically, and precisely determining the location of its parts. This can be accomplished by means of different geodetic surveying methods, such as triangulation and trilateration. Higher geodesy is an older term for the geodetic surveying of large regions, when the Earth's curvature was taken into account. The main goal of physical geodesy is determining the geoid, the level surface where the gravity field potential is constant. In combination with measuring the arcs of meridians and parallels, the dimensions of the Earth's ellipsoid can be determined.

*Satellite geodesy* is a part of geodesy in which satellites are used for measurements. It belongs to the wider discipline of space geodesy, which includes Very Long Baseline Interferometry (VLBI) and lunar laser ranging. Astronomical geodesy is traditionally about the astronomical determination of positions, and is not a common part of satellite geodesy. Measuring techniques in satellite geodesy include geodetic measurements from global navigation satellite systems (e.g. GPS, GLONASS, Galileo); distance determination to the satellite (Satellite Laser Ranging or SLR); land applications of interferometric synthetic aperture radar (InSAR); satellite radar altimetry (mostly at sea), e.g. from satellites Seasat, Geosat, TOPEX/Poseidon, ERS-1, ERS-2, Jason-1, Envisat; land satellite radar altimetry, e.g. ICESat; satellite determination of the Earth's gravity field, e.g. CHAMP; satellite gradiometry (measurement of gravity gradient from orbits), e.g. GOCE; and tracking satellites from other satellites, e.g. GRACE.

A *satellite navigation system* (see Chapter 18) is a system of satellites that provides autonomous geo-spatial positioning with global coverage. The locations of small electronic receivers are determined (longitude, latitude, and altitude) to within a few metres, using time signals transmitted along a line-of-sight by radio from satellites. The receivers calculate the precise time as well as their positions. A satellite navigation system with global coverage may be termed a global navigation satellite system or GNSS. These include those belonging to the United States (NAVSTAR Global Positioning System) and Russia (GLONASS). China is in the process of expanding its regional Beidou navigation system into a global compass navigation system (by 2020). The European Union's Galileo positioning system is a GNSS in the initial deployment phase, scheduled to be fully operational by 2020 at the earliest. France, India, and Japan are in the process of developing regional navigation systems. Global coverage for each system is generally achieved by a satellite constellation of 20–30 medium Earth orbit satellites distributed among several orbital planes. The actual systems vary, but use orbital inclinations of >50° and orbital periods of roughly 12 hours, at an altitude of about 20,000 kilometres.

*Photogrammetry* (see Chapter 16) is the science of acquiring reliable quantitative information on physical objects and the environment by using aerial photography, which can be traditional film or digital photography, and may use visible light or other wavelengths of electromagnetic radiation. *Remote sensing* is a method of collecting and interpreting data on objects from a distance. The method is characterized by the fact that the measuring device is not in contact with the object surveyed. Its most frequent application is from aerial or space platforms.

*Cartography* is the discipline dealing with the art, science, and technology of making and using maps. Maps include the study of spherical or ellipsoidal transformation from the Earth's surface model to a two-dimensional representation, requiring the use of the following concepts: ellipsoid, datum, and coordinate systems.

## Geodetic coordinate systems

Coordinates, from the Latin *co* (with) and *ordinatus* (ordained, defined), are numbers whose designation defines a point on a line, a plane, a surface, or in space. The first coordinates to enter the system of use were astronomic and geographic – longitude and latitude – to determine a point in the heavens or on the surface of the Earth's sphere. In the fourteenth century, the French mathematician, physicist, and economist Nicole Oresme (c.1320–1382) applied coordinates in a plane by constructing graphs, designating their horizontal and vertical positions by what are called today the abscissa and ordinate. Coordinates began to be applied more systematically in the seventeenth century to resolve geometric problems in a plane. The French philosopher and mathematician René Descartes (1596–1650), in Latin *Cartesius*, is credited with interpreting the significance of coordinates, which allowed the systematic transformation of geometric tasks into the language of mathematical analysis, and vice versa. In addition to the coordinates of points, the coordinates of lines, planes, and other geometric objects are also considered. Chapter 15 covers the topic of map projections in more detail.

## The Cartesian system of coordinates

This is a rectilinear system of coordinates in Euclidean space. In the plane, the *general Cartesian coordinate system (affine coordinate system)* consists of a given point O, which is known as the origin, and an ordered pair of non-collinear vectors *i* and *j*, which are called the base vectors (Figure 14.1). The straight lines which pass through the origin in the direction of the base vectors are called the coordinating axes of the Cartesian coordinate system. The first axis, determined by vector *i*, is called the abscissa axis (or *x* axis), and the second is the ordinate axis (or *y* axis). The Cartesian coordinate system is designated by *Oij* or *Oxy*. In the Cartesian coordinate system, the *Cartesian coordinates of point M* are the ordered pair of numbers (*x*, *y*), which are the coefficients of the depiction of vector *OM* in the base *i, j*, hence:

$$\overrightarrow{OM} = x\vec{i} + y\vec{j}.$$

Number *x* is called the abscissa, and number *y* the ordinate of point *M*. If (*x*, *y*) are the coordinates of point *M*, this is written as *M(x,y)*, identifying the point with its ordered pair of coordinates.

The Cartesian coordinate system is called rectangular if the base vectors are orthonormal, i.e. perpendicular to each other and of equal length. In this case, vectors *i* and *j* are called orths. In the rectangular Cartesian coordinate system, the coordinates (*x*, *y*) of point *M* are equal to the following: *x* – to the scalar orthogonal projection of vector *OM* on the abscissa axis, and *y* – to the scalar orthogonal projection of vector *OM* on the ordinate axis. The left and right coordinate systems in the plane are distinguished (Figure 14.2). Occasionally, the oblique-angled Cartesian coordinate system is used, which differs from the rectangular system in that the angle between the base vectors is not a right angle.

In mathematics, we usually use the right Cartesian coordinate system in the plane, as shown in Figure 14.2b. However, in other areas, the left coordinate system is applied. In mathematical cartography, the left coordinate system is used, as in Figure 14.2a, with the proviso that the positive direction of axis *x* points north, and the positive direction of axis *y* points east: or that seen in Figure 14.2d, where axis *x* points south, and axis *y* points west.

By analogy, the *general Cartesian coordinate system (affine coordinate system)* is defined in space (Figure 14.3), by providing point O, the origin, and the base vectors – a triple of

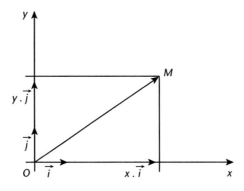

*Figure 14.1* Rectangular Cartesian coordinate system in a plane

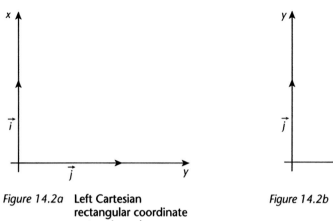

Figure 14.2a  Left Cartesian rectangular coordinate system in a plane

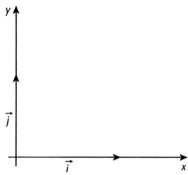

Figure 14.2b  Right Cartesian rectangular coordinate system in a plane

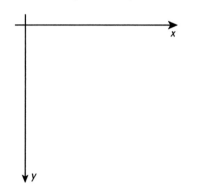

Figure 14.2c  Left Cartesian rectangular coordinate system for a monitor screen

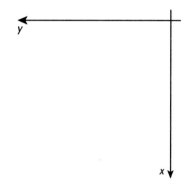

Figure 14.2d  Old coordinate system used in some European countries (left)

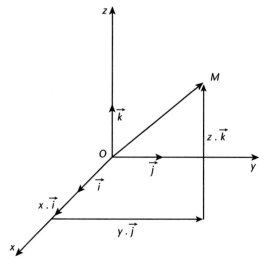

Figure 14.3  Cartesian rectangular system in space

non-complementary vectors $\vec{i}$, $\vec{j}$, and $\vec{k}$. In order to acquire the coordinates $x$, $y$, $z$ of point $M$, vector $\overrightarrow{OM}$ is shown using the form:

$$\overrightarrow{OM} = x\,\vec{i} + y\,\vec{j} + z\,\vec{k}.$$

The position of a single spatial rectangular Cartesian coordinate system in relation to another rectangular Cartesian coordinate system with a common origin and the same orientation can be defined by means of three Euler angles, after L. Euler (1707–1783).

The Cartesian coordinate system was named after the French mathematician and philosopher René Descartes (1596–1650), who in *Geometria* (1673) used the oblique coordinate system with coordinates that could only be positive numbers. However, in the 1659–1661 edition of *Geometria*, the work of the Dutch mathematician J. Hudde (1633–1704) was presented, which allowed negative values for the coordinates for the first time. The Cartesian spatial coordinate system was introduced by the French mathematician and physicist Philippe de La Hire (1640–1718); he was the first to use the word origin (*origine*), from which we have the designation O. The designations $x$, $y$, and $z$ were not immediately proposed for Cartesian coordinates, but had become common by the eighteenth century. The modern terms for the left and right coordinate systems were introduced by the English physicist J.C. Maxwell (1831–1879).

## Curvilinear coordinates

Curvilinear coordinates were first applied by the Swiss mathematician Jacob Bernoulli (1654–1705), and were named by the German philosopher-idealist, mathematician, physicist, inventor, lawyer, historian, and linguist G.W. Leibnitz (1646–1716). Curvilinear coordinates in a plane are linked to the name of a German mathematician who provided essential contributions to astronomy and geodesy, C.F. Gauss (1777–1855). The term 'curvilinear coordinates' was first introduced by the French mathematician and engineer G. Lamé (1795–1870).

## Polar coordinates

In addition to using Cartesian coordinates in a plane, the position of a point can be described using the polar coordinates $\rho$ and $\varphi$ (Figure 14.4), which are connected to the rectangular coordinates $x$ and $y$ by the formulae:

$$x = \rho \cos\varphi, \quad y = \rho \sin\varphi,$$

where $0 \le \rho < \infty$, $0 \le \varphi < 2\pi$. The coordinate lines are concentric circles ($\rho = $ const.) and rays ($\varphi = $ const.). To each point in the plane $Oxy$ (apart from point $O$, for which $\rho = 0$, and $\varphi$ is not defined) the pair of numbers $(\rho, \varphi)$ corresponds, and vice versa. The distance $\rho$ of point $M$ from the pole is called the *polar radius*, and angle $\varphi$ is called the *polar angle*.

The ancient Greek mathematician Dinostratus (fourth century BC) noted the polar coordinates in implicit form when he investigated quadratrices. The English physicist and mathematician Isaac Newton (1643–1727) used polar coordinates and provided formulae for their connections with rectangular coordinates. Polar coordinates appeared in an almost contemporary form in the work of Jacob Bernouilli. The term 'polar coordinates' appeared in the nineteenth century.

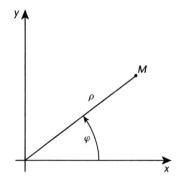

*Figure 14.4*  Polar coordinate system in a plane

In mathematical cartography, polar coordinates are applied in azimuthal and conical projections. In normal or polar azimuthal projections, the polar coordinates in the plane are $\rho$ and $\lambda$, and their connection with rectangular Cartesian coordinates is:

$$x = \rho \cos \lambda, \quad y = \rho \sin \lambda,$$

where $\rho \geq 0$, $-\pi \leq \lambda \leq \pi$. At the same time, if we want to obtain a natural view of the upper hemisphere from above, the coordinate system must be set as in Figure 14.5a. However, this will give a view of the lower hemisphere from within, i.e. a mirror image of the distribution of land and sea in relation to the one with which we are familiar. Thus, in this case, the coordinate system must be set up as in Figure 14.5b. There is of course another possibility, according to which the coordinate system set-up remains the same all the time, but then the equation for the projection must be altered appropriately.

For a normal aspect conical projection, the polar coordinates in the plane are $\rho$ and $\delta$, and their connection with rectangular Cartesian coordinates is:

$$x = q - \rho \cos \delta, \quad y = \rho \sin \delta, \delta = k \lambda,$$

where $\rho \geq 0$, $-\pi \leq \lambda \leq \pi$, and $q$ and $k$ are constants of the projection. At the same time, if we wish to obtain a natural view of the upper hemisphere, the coordinate system must be set up as in Figure 14.6.

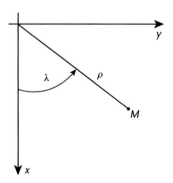

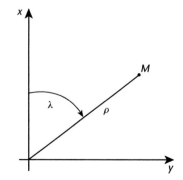

*Figure 14.5a*  Coordinate system set-up for azimuthal projection of the upper hemisphere

*Figure 14.5b*  Coordinate system set-up for azimuthal projection of the lower hemisphere

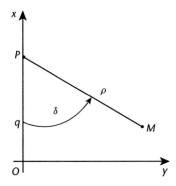

*Figure 14.6*   Polar coordinate system for a normal aspect conical projection

## Geographic coordinates on a sphere

The formulae which link spherical coordinates with rectangular coordinates were given by the French mathematician and physicist J.L. Lagrange (1736–1813), and the term *spherical coordinates* was proposed by the German mathematician and teacher H.R. Baltzer (1818–1887). The sphere is often taken for a model of the Earth's surface. The equation for a sphere with its centre in the point of origin of the rectangular Cartesian system *Oxyz* and with radius *R* is:

$$x^2 + y^2 + z^2 = R^2.$$

This kind of sphere is called the *Earth's sphere*. The point with the coordinates $(0, 0, R)$ is called the *North Pole*, and the point with the coordinates $(0, 0, -R)$ the *South Pole*. A circle on the sphere equidistant from the poles is called the *Equator* and it divides the sphere into two *hemispheres*. The straight line which passes through both the poles is known as the *axis of the Earth's sphere*, and the plane where the Equator is situated as the *equatorial plane*. The angle which encloses the normal (at the same time, the radius-vector) of an arbitrary point *M* on the Earth's sphere and the equatorial plane is called *geographic latitude* and noted with the symbol φ. All points on the Earth's sphere which have the same latitude lie along a circle called a *parallel*.

The semicircles on the terrestrial sphere which join the North and South Poles are called *meridians*. One is designated the *zero* or *prime meridian*. This is usually the meridian which lies in the plane $y = 0$. The *geographic longitude* of arbitrary point *M* on the terrestrial sphere is noted with the symbol λ, and this is the angle between the meridian which passes through point *M* and the prime meridian. Accordingly, all points which lie on the same meridian have the same longitude. Latitude is measured in intervals of $-\pi/2 \le \varphi \le \pi/2$, and longitude (usually) in intervals of $-\pi \le \lambda \le \pi$. The geographic coordinates φ and λ create a curvilinear coordinate system on the sphere. The coordinate curves are the parallels and meridians.

A geographic coordinate system can also be interpreted as the restriction of the spherical spatial coordinate system of radius *R*. On the basis of this explanation (and see Figure 14.7), the connections of the rectangular Cartesian spatial coordinates *x*, *y*, and *z* of arbitrary point *M* on the sphere can be derived, along with their geographic coordinates φ, λ:

$$x = R\cos\varphi\cos\lambda, \quad y = R\cos\varphi\sin\lambda, \quad z = R\sin\varphi.$$

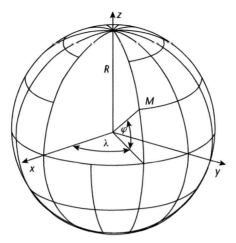

*Figure 14.7* Geographic coordinates on the Earth's sphere

## Geodetic coordinates on the rotational ellipsoid

Apart from the sphere, another model frequently used for the Earth's surface is the rotational ellipsoid. The equation for the rotational ellipsoid with its centre in the point of origin of the rectangular Cartesian system $Oxyz$ and the semi-axes $a$ and $b$ is as follows:

$$\frac{x^2}{a^2} + \frac{y^2}{a^2} + \frac{z^2}{b^2} = 1.$$

The point with the coordinates $(0, 0, b)$ is called the *North Pole*, and the point with the coordinates $(0, 0, -b)$, the *South Pole*. The circle on the ellipsoid which is equidistant from the poles is called the *Equator*, and it divided the ellipsoid into two parts. The straight line which passes through both poles is called the *axis of the rotational ellipsoid*, and the plane on which the Equator is located is the *equatorial plane*. The angle which encloses the normal (but not the radius-vector) of an arbitrary point $M$ on the terrestrial sphere and the equatorial plane is called *geodetic latitude* and noted with the letter $\varphi$. All points on the rotational ellipsoid which have the same latitude lie along a circle called a *parallel*.

The semiellipses on the ellipsoid which connect the North and South Poles are called meridians. One is designated the *zero* or *prime meridian*. This is usually the meridian which lies in the plane $y = 0$. The *geodetic longitude* of arbitrary point $M$ on the rotational ellipsoid is noted with the symbol $\lambda$, and this is the angle between the meridian which passes through point $M$ and the prime meridian. Accordingly, all points which lie on the same meridian have the same longitude. Latitude is measured in intervals of $-\pi/2 \leq \varphi \leq \pi/2$, and longitude (usually) in intervals of $-\pi \leq \lambda \leq \pi$. The geographic coordinates $\varphi$ and $\lambda$ create a curvilinear coordinate system on the rotational ellipsoid. The coordinate curves are the parallels and meridians.

Based on the above explanation (and see Figure 14.8), the connection of the rectangular Cartesian spatial coordinates $x$, $y$, $z$ of arbitrary point $M$ on the rotational ellipsoid can be derived, along with the geodetic coordinates $\varphi$, $\lambda$:

$$x = N \cos\varphi \cos\lambda, \quad y = N \cos\varphi \sin\lambda, \quad z = N(1 - e^2) \sin\varphi,$$

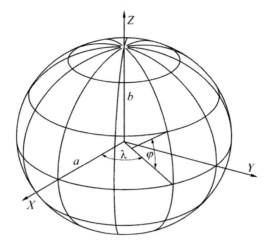

*Figure 14.8*  Geodetic coordinates on the rotational ellipsoid

where:

$$N = \frac{a}{\sqrt{1 - e^2 \sin^2 \varphi}}, \quad e^2 = \frac{a^2 - b^2}{a^2}.$$

A *geodetic datum* or *geodetic system* is a coordinate system and set of reference points used to locate places on the Earth. In other words, a geodetic datum should define the relation of geodetic coordinates to the Earth (Krakiwsky and Wells, 1971). The geodetic coordinates $\varphi$, $\lambda$ and height $h$ above the ellipsoid may be transformed to an Earth-centred Cartesian three-dimensional system using the following equations (Vaníček, 2001):

$$X = (N + h)\cos\varphi\cos\lambda, \quad Y = (N + h)\cos\varphi\sin\lambda, \quad Z = \left[N(1 - e^2) + h\right]\sin\varphi.$$

## Rotational ellipsoid and the Earth's sphere

The physical surface of the Earth divides our planet from its atmosphere. It is very irregular. In order to represent even a small part of the Earth's surface in a plane, a simpler surface is required on which the horizontal relations of points are represented, while altitude is written or drawn for certain points.

When the entire Earth is considered, the surface should be as close to the actual physical surface as possible. Since seas and oceans make up ~70 per cent of the Earth's total surface area, the ideal surface of motionless sea stretched over continents has been taken as the reference surface defining the Earth's shape. However, the surface is not regular due to the distribution of masses of various densities, making it difficult to use in calculations. A rotational ellipsoid approximates the Earth's surface relatively well. Such an ellipsoid is very close to a sphere with a radius of ~6,400 km (the minor semi-axis of the ellipsoid is ~6,400 km, while the major semi-axis is only 21 km longer). In solving problems in geodesy, remote sensing, navigation, and cartography, we consider the Earth to be a rotational ellipsoid or a sphere.

### Elements of the Earth's ellipsoid

The Earth's ellipsoid can be understood by rotating an ellipse (Figure 14.9) around its shorter axis, which we assume corresponds to the Earth's axis. In considering the properties of a rotational ellipsoid, it is sufficient to know the elements of a meridian ellipse, which is rotated to produce the Earth's ellipsoid. We will label the ellipse's semi-major axis $a$, and its semi-minor axis $b$. The dimensions of the ellipsoid are often given with major semi-axis $a$ and flattening $f$, which is defined as:

$$f = \frac{a-b}{a}.$$

In geodetic and cartographic calculations, the first and second numerical eccentricities are frequently encountered, and are defined as:

$$e = \frac{\sqrt{a^2 - b^2}}{a}, \quad e' = \frac{\sqrt{a^2 - b^2}}{b}.$$

If we wish to represent a large part of the Earth, a continent or even the whole world, the flattening of the Earth can be neglected. In this case, we can use a geographic coordinate system instead of a geodetic coordinate system.

A spherical coordinate system can be obtained as a special case of the ellipsoidal coordinate system, assuming that flattening equals zero, $f = 0$, or equivalently stating that the second eccentricity equals zero, $e = 0$. Sometimes, in geodetic and cartographic practice, it is necessary to transform Cartesian three-dimensional coordinates to spherical or even ellipsoidal coordinates. Furthermore, it is sometimes necessary to make a transformation from one three-dimensional coordinate system to another. The appropriate methods or equations exist, but the reader should consult the available literature.

Numerous geodetic, astronomic, gravimetric, and satellite measurements are conducted in order to determine the numerical values of the Earth's ellipsoid elements. Many scholars and institutions have been interested in determining the dimensions of the Earth's ellipsoid over the centuries (Defense Mapping Agency, 1984). Table 14.1 contains the dimensions of the Earth's ellipsoid which have been most frequently used in geodetic and cartographic practice.

In geodesy, the Earth's ellipsoid is often identified with the Earth's spheroid. A spheroid is a geometrical solid similar in shape to a sphere. Therefore, the Earth's spheroid is considered to be a rotational ellipsoid with slight flattening.

### The Earth's ellipsoid

The *Earth's ellipsoid* is the ellipsoid which best represents the Earth as a planet. However, it is a mathematical model of the Earth which can be defined in several ways. All aim for the greatest

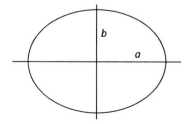

*Figure 14.9* **Ellipse**

*Table 14.1* Dimensions of the Earth's ellipsoid

| Name | Year | a | b | 1/f |
|------|------|---|---|-----|
| Everest | 1830 | 6377276.345 | 6356075.415 | 300.801726 |
| Bessel | 1841 | 6377397.155 | 6356078.963 | 299.152813 |
| Clarke | 1866 | 6378206.400 | 6356583.800 | 294.978698 |
| Clarke | 1880 | 6378249.145 | 6356514.967 | 293.466308 |
| Hayford International | 1909, 1924 | 6378388 | 6356911.946 | 297 |
| Krasovskiy | 1940 | 6378245 | 6356863.019 | 298.3 |
| GRS 1967 | 1967 | 6378160 | 6356774.516 | 298.247166 |
| GRS 1980 | 1980 | 6378137 | 6356752.3141 | 298.257222 |
| WGS 72 | | 6378135 | 6356750.520 | 298.26 |
| WGS 84 | 1987 | 6378137 | 6356752.314 | 298.257223563 |

possible congruence between the ellipsoid and the Earth. This ellipsoid is oriented absolutely in space, i.e. its equatorial plane corresponds to the Earth's equatorial plane, and its minor axis corresponds to the central position of the Earth's rotational axis (the central position for a particular surface). Available data and mathematical formulae have been used to derive numerous parameters of the general ellipsoid, while striving to achieve absolute orientation. This was not successful until observations of Earth's artificial satellites became possible.

The ellipsoid to which geodetic measurements are reduced, and on which they are analysed, is called a *reference ellipsoid*. Therefore, a general ellipsoid can be a reference ellipsoid if geodetic measurements are reduced to it and analysed. The Earth's general ellipsoid was not used as a reference ellipsoid in the past, mostly because it could not be oriented in an adequate way. In addition, it was not possible to calculate the difference between the potentials of the geoid and a general ellipsoid in at least one point. It was therefore not possible to calculate the distances between those two surfaces and reduce measurements to the ellipsoid. This is why each country or group of countries determined its own reference ellipsoid with the most suitable dimensions and orientation for their area. The consequence was discontinuity and incoherence in the geodetic networks and maps of different countries. The economic–political and military integration of countries which began after the Second World War consolidated geodetic work within their areas and transferred it onto a specially selected and oriented ellipsoid.

Contemporary satellite techniques and information technology have enabled accurate determination of the Earth's centre of inertia, the position of its rotational axis, and its dimensions and shape. Thus, the general Earth ellipsoid could be oriented and unique global triangulation developed. In addition, new methods could be employed to obtain a great number of high accuracy measurements, which were then processed by computers. Therefore, it is possible to calculate increasingly accurate parameters of the Earth's shape, dimensions, and gravity field. After they are compared and analysed, the International Association of Geodesy (IAG) adopts them periodically and recommends them as reference parameters. For example, in 1979, the 17th Congress of IAG (Moritz, 1980) recommended the Geodetic Reference System 1980 (GRS 80). The basic parameters of the corresponding ellipsoid were: $a = 6\ 378\ 137 \pm 2$ m, $1/f = (298\ 257 \pm 1) \times 10^{-3}$.

## WGS 84 Coordinate System

The World Geodetic System 1984 (WGS 84) was developed in the USA as a replacement for WGS 72. This was initiated by the need to provide more precise geodetic and gravimetric data for the navigation and weapon systems of the Department of Defense. The new system represents Earth modelling from geometrical, geodetic, and gravimetric perspectives, using the data, techniques, and technology at the disposal of the Defense Mapping Agency at the beginning of 1984 (Defense Mapping Agency, 1984).

The WGS 84 coordinate system is a conventional system obtained by modifying the Navy Navigation Satellite System Doppler Reference Frame (NSWC 9Z-2) in the origin and scale, including rotation, which superposes the reference meridian with the prime meridian of the Bureau International de l'Heure (BIH). Therefore, the origin of the WGS 84 coordinate system is in the Earth's centre of mass. Its $x$ axis is parallel to the direction of the Conventional Terrestrial Pole (CTP) for pole movement; as defined by BIH, its $x$ axis is the intersection of the WGS 84 plane of the reference meridian and the CTP Equator plane, and the reference meridian is parallel to the prime meridian defined by BIH. The $y$ axis completes the right-oriented orthogonal system linked to the Earth. The origin and axes of the WGS 84 coordinate system are also used as the geometric centre and $x$, $y$, and $z$ axes of the WGS 84 ellipsoid. The $z$ axis of the WGS 84 coordinate system is the rotational axis of the WGS 84 ellipsoid.

Geodetic applications usually include three different Earth surfaces or shapes. In addition to the Earth's natural or physical surface, there are the geometrical or mathematical reference surface (ellipsoid) and the equipotential surface (geoid). In determining the WGS 84 ellipsoid and its parameters, the procedures of the International Union of Geodesy and Geophysics (IUGG) are accepted, the body which established and accepted the Geodetic Reference System 1980 (GRS 80). The WGS 84 ellipsoid's major semi-axis equals $a = 6\ 378\ 137 \pm 2$ m. The value is equal to that of the GRS 80 ellipsoid and is 2 metres longer than that of the WGS 72 ellipsoid. The major semi-axis value is based on an estimate of data obtained between 1976 and 1979 from laser, Doppler, radar altimeter and combined measurements.

## Land surveying

As measurements relate to human activity, the history of measurement has the same origins as the history of humankind. There is clear evidence of a high level of measuring activity around the rivers Euphrates and Tigris, where the oldest races go back 6,000 years. Excavations have proved that there was a 160-km long irrigation system functioning in the area even then (Benčić, 1990: 548; Irrigation Museum, 2016). The system was maintained successfully for about 2,000 years, but today, everything is covered in sand. Such an enterprise could be characterized as grandiose for the time, but also for the present, as it could not have been accomplished without sound surveying. It becomes obvious when we compare this irrigation system with the Roman water supply system used about 4,000 years later. The Roman system was about 400 kilometres long and comprised 19 aqueducts. It was maintained by a full-time crew of 700 people, enabling 600 to 1,000 litres of water a day per citizen to be delivered to the city (Benčić, 1990: 548).

Egyptian surveyors were called 'cord stretchers', after the cords they used to measure distances. In the past, geodetic measurements were only used to determine the mutual locations of points on the Earth's surface, and for gathering data on the types of surfaces, or buildings erected on them. The results of such measurements were maps, plans, or location sketches. With the development of technical means for measuring and processing data, particularly photogrammetry and electronic computers, the conditions were met for measuring not only plots of land, but also other objects, and for producing geodetic models, i.e. simulations of various modelling processes. A geodetic model may be, for instance, a topographic map, a cadastral-topographic plan, a perspective depiction of terrain, an orthogonal projection of a building, or an orthogonal projection of a sculpture. The basic property of a geodetic model is that it links data on a surveyed object with its position in space. Modern computing resources allow data on geodetic measurements to be stored in memories and shown in the desired form. So, for example, it is possible to depict a surveyed object in any projection and at various scales.

Basic measurements during a geodetic survey serve to define the position of points. Five basic types of measurements can be distinguished: measuring horizontal angles, measuring vertical angles, measuring horizontal distance, measuring vertical distance, and measuring spatial coordinates. In practice, distance is often measured along a slanting surface, but is then recalculated in a horizontal plane. A horizontal plane is a plane which at a certain point makes the bubble in a spirit level rest in the highest position, and a vertical plane is at that point perpendicular to the established horizontal plane. Accordingly, any point has only one horizontal plane, but an infinite number of vertical planes. The horizontal distance between two points is the distance between the orthogonal projections of those points on the horizontal plane. The vertical distance (difference in height) between two points is the distance, in the vertical plane which passes through both objects, between one of them and the horizontal plane which passes through the other point.

A geodetic land survey is the recording, processing, and systematization of measuring and descriptive data with certain contents on a piece of land and objects on it, in order to produce plans and maps. Plans and maps are used in the spatial organization and use of land, for keeping land records in cadastres and land registries; for establishing and keeping other spatial records, in the design of hydrotechnical objects, roads, and other communal infrastructure; for geological, geophysical, and other scientific research; and for other agrarian and technical requirements. The purpose for which a piece of land is surveyed and data on it used for various purposes demands special procedures when producing plans and maps, so geodetic measurements are regulated by legal provisions and ordinances, so as to preserve uniformity and continuity in measurement data.

A geodetic land survey includes: establishing and determining the network of permanent geodetic points (in the past); detailed recording of the terrain (the plot and objects on it); and the production of plans and maps. Detailed measurement of the terrain (the plot and the objects on it) means collecting measurement data regarding horizontal and height depiction, names, and characteristics of the terrain. These data are then processed numerically and graphically, resulting in geodetic plans and maps.

## Methods of measuring plots

By applying basic types of measuring, the mutual positions of points are always determined according to two or more given points whose positions are already known or considered known. The orthogonal or polar method is usually applied, using an arc intersection, forward intersection and backward intersection. Along with these methods for determining the mutual positions of the sought points in the horizontal plane, the procedure of levelling is used to determine their height relations.

When the photogrammetric method is applied, a stereo pair is first oriented with the aid of the known points, in order to establish the spatial relations which existed at the time of recording. In this way, the relative position of all points can be determined in reference to the known points by adducing markers of the photogrammetric instrument at each point, and registering the position of each marker. Registration can be performed graphically or numerically.

A particular feature of geodetic measurement is a host of control measures applied to identify errors and eliminate them; in other words, to assess the accuracy of geodetic measurements and geodetic models. In these, and in many other ways, geodesy contributes to our understanding of the Earth and underpins the creation of accurate and reliable maps today.

# References

Benčić, D. (1990) *Geodetski instrumenti* (Geodetic Instruments, in Croatian) Zagreb, Croatia: Školska knjiga.

Defense Mapping Agency (1984) *Geodesy for the Layman* Report No.TR 80-003 Aerospace Center, St. Louis Air Force Station.

De Graeve, J. and Smith, J. (2010) *History of Surveying* (FIG Publication No.50) Copenhagen, Denmark: International Federation of Surveyors (FIG) and International Institution for the History of Surveying and Measurement.

FIG (International Federation of Surveyors) (2004) "Definition of the Functions of the Surveyor, as adopted by the General Assembly 23 May 2004" Available at: *www.fig.net/about/general/definition/index. asp* (Accessed: 14 November 2016).

Irrigation Museum (2016) "Irrigation Timeline" Available at: *www.irrigationmuseum.org* (Accessed: 14 November 2016).

Kraak, M.-J. (2015) "Cartography Today" (Keynote) *EuroCarto 2015 Vienna, 10–12 November 2015* Available at: *http://eurocarto.org/keynote/* (Accessed: 15 May 2016).

Krakiwsky, E.J. and Wells, D.E. (1971) *Coordinate Systems in Geodesy* (Lecture Notes 16) Department of Geodesy and Geomatics Engineering, University of New Brunswick, Canada.

Moritz, H. (1980) "Geodetic Reference System 1980" *Bulletin Géodésique* 54 (3) pp.1128–1133. Republished with corrections in Moritz, H. (2000) "Geodetic Reference System 1980" *Journal of Geodesy* 74 (1) pp.128–162.

Neidhardt, N. (1950) „Naziv struke" *Geodetski List* 1–3 p.72.

Vaníček, P. (2001) *An Online Tutorial in Geodesy* Geodesy Group, University of New Brunswick, Canada. Available at: *http://einstein.gge.unb.ca* (Accessed: 25 May 2016).

Vaníček, P. and Krakiwsky, E.J. (1986) *Geodesy: The Concepts* Amsterdam: North-Holland.

# 15

# Understanding map projections

*E. Lynn Usery*

## Introduction

It has probably never been more important in the history of cartography than now that people understand how maps work. With increasing globalization, for example, world maps provide a key format for the transmission of information, but are often poorly used. Examples of poor understanding and use of projections and the resultant maps are many; for instance, the use of rectangular world maps in the United Kingdom press to show Chinese and Korean missile ranges as circles, something which can only be achieved on equidistant projections and then only from one launch point (Vujakovic, 2014).

Map projection is the transformation of geographic data from ellipsoidal or spherical coordinates to a plane coordinate system. This transformation is required to bring data from various systems into a common reference framework. It is not possible to preserve all angles, sizes, directions, and other spatial relationships that exist in the ellipsoidal or spherical data, thus, map projection always involves distortion, not error, of some characteristics to preserve others. Generally, a trade-off is made between preservation of angles and areas, leading to conformal and equal area map projections. In a conformal map projection, angular relationships are preserved while equal area projections preserve sizes of areas in the transformation. Alternatives are to preserve specific properties such as direction from a point, distances, or a compromise in which all types of distortion are minimized.

## Reference ellipsoid and datum

Map projection proceeds from the definition of a reference *ellipsoid* (spheroid or sphere) defined by specification of the lengths of the semi-major and semi-minor axes of the ellipsoid. The mathematical representation of the ellipsoid uses $a$ for the semi-major axis and $b$ for the semi-minor axis and thus a flattening factor, $f$, can be defined as:

$$f = (a-b)/a$$

Note: Any use of trade, product, or firm names in this chapter is for descriptive purposes only and does not imply endorsement by the U.S. Government.

This factor is the measure of the oblateness of the ellipsoid, and with an approximate factor of 1/298, it is not apparent to the naked eye in satellite views of the Earth from space. We define the first eccentricity from the flattening factor as:

$$e = (2f - f^2)^{1/2}$$

We define geodetic coordinates of latitude, $\varphi$, and longitude, $\lambda$, and height above the ellipsoid as:

$$X = (v + h) \cos \varphi \cos \lambda$$

$$Y = (v + h) \cos \varphi \sin \lambda$$

$$Z = \{1 - e^2) \, v + h\} \sin \varphi$$

where

$$v = a/(1 - e^2 \sin^2 \varphi)^{1/2}$$

For mathematical derivation of geodetic coordinates and transformations with Cartesian coordinates, see Mugnier (2004).

Common ellipsoids in use today are the Clarke 1866, used in the United States for maps prior to about 1980; the Global Reference System 1980 (GRS 80), the basis of the North American Datum of 1983; and the World Geodetic System 1984 (WGS 84), which is the basis of coordinates from the Global Positioning System (GPS). The WGS 84 and GRS 80 ellipsoids, which were established by satellite positioning methods, are geocentric (referenced to the centre of mass of the Earth), and provide a reasonable fit to the entire Earth. The differences between WGS 84 and GRS 80 coordinate values are minor and only relevant for high order geodetic control information. For a listing of some more common ellipsoids and their characteristics, see Snyder (1987).

A *datum* is the basis of a coordinate system and defines the initial point from which all other coordinates are defined. Datums can be global, if referenced to the ellipsoid, or local in origin. A horizontal datum, such as WGS 84, allows specification of latitude and longitude, or *x* and *y* Cartesian coordinates, relative to the initial point.

## Coordinate systems

The geographical reference system (GRS) is the name of the ellipsoidal coordinate system of latitude and longitude. With the datum and projection defined, we can then design a plane coordinate system, commonly expressed as *x* and *y* coordinates, sometimes referred to as Eastings and Northings. The *x* and *y* coordinates are defined on a grid system of selected units of measure from the initial point on the datum. A right-handed Cartesian system defines *x* increasing to the right and *y* increasing to the top. Two Cartesian systems in common use are the U.S. State Plane Coordinate System used in the United States and the Universal Transverse Mercator Coordinate System, a worldwide system projected in zones of six degrees of longitude.

## Map projections

A map projection is defined as a systematic transformation of ellipsoidal coordinates of latitude and longitude to a plane coordinate representation. Mathematically,

$$x = f_1\ (\varphi, \lambda)$$

$$y = f_2\ (\varphi, \lambda)$$

The transformation is developed from a 'generating globe', which is a reduced scale model of the Earth as either a sphere, an ellipsoid, or an 'aposphere' (Figure 15.1).

Map projection transformation always results in distortion with only a single point, circle, or one or two lines maintaining true scale from the generating globe. Whereas distortion always

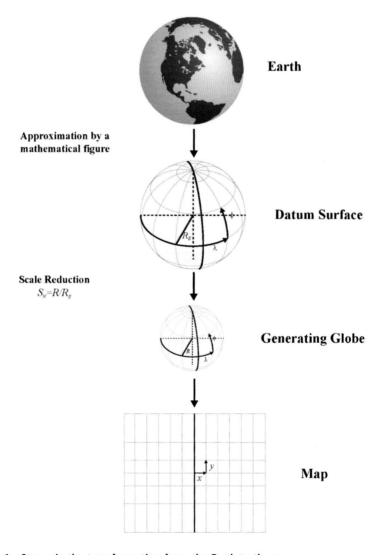

*Figure 15.1* Stages in the transformation from the Earth to the map

results from the transformation, the mathematics can be developed to preserve specific characteristics such as angular or areal relationships. Pearson II (1990), Bugayevskiy and Snyder (1995), Iliffe (2000), Yang *et al.*, 2000, and Canters (2002) provide complete documentation of map projection theory. A history of map projection development is available in Snyder (1993) and a documentation of characteristics of various projections is provided in Snyder and Voxland (1989). Snyder (1987) provides a description of many different projections, and the forward and inverse transformation equations with worked numerical examples.

Geometry, shape, special properties, projection parameters, and nomenclature provide the basis for classifying map projections (Canters, 2002). The purpose of such classification is to group projections with similar properties in common classes. Often the classification is based on the shape of the graticule (grid lines of latitude and longitude) that results from the projection transformation. A geometric classification uses cylindrical, conical, and azimuthal, for the look of the resultant projection, but this classification is incomplete because many projections do not fit these classes. The geometric system is usually expanded to include pseudocylindrical and pseudoconical. Maurer (1935), cited in Canters (2002), developed a hierarchical system including five levels mainly based on the appearance of the meridians and parallels. Lee (1944), Goussinsly (1951), and Maling (1968, 1992) provide classifications of map projections based on special properties. Tobler (1962) used parametric classification and generated four groups. Thus, there are a variety of classification methods and schemes for map projections.

## Some commonly used projections and their characteristics

Selecting an appropriate projection for a specific application is dependent on the purpose of the map, the type of data to be projected, the area of the world to be projected, and the scale of the final map. Advice on selection is available from a variety of print and Web sources, including Finn *et al.* (2004), USGS (2006), and Lapaine and Usery (2017).

Detailed characteristics of a selected set of map projections, organized by the property preserved, are presented after Table 15.1. For each projection, the name and creator are identified and specific characteristics including properties preserved, shapes of meridians and parallels, lines or points of true scale, and the extent of the Earth that can be projected are identified, and for specific cases, characteristics that make the projection unique are identified. Table 15.1 provides a summary of the characteristics of the projections included in the discussion. A graphic representation produced with the Geocart software (Mapthematics, 2015) is included for each projection. This graphic also includes Tissot's Indicatrix (Tissot, 1881; Canters, 2002), which is a set of distortion circles plotted on the graphic. The circles allow a quick assessment of areal or angular distortion for the projections.

## Conformal projections

### *Mercator*

Probably the most well-known projection is a cylindrical conformal projection developed by Gerardus Mercator in 1569. The projection was developed to show loxodromes or rhumb lines, which are lines of constant bearing, as straight lines making it possible to navigate a constant course based on drawing a rhumb line on the chart. Meridians are shown as equally spaced parallel lines. Parallels are shown as unequally spaced straight parallel lines, perpendicular to the meridians and closest near the Equator. In the tangent case, meaning a single line of contact between the projection and the generating globe, scale is true along the Equator.

Table 15.1 Characteristics of projections described

| Projection | Type | Retains | Parallels | Meridians | Pole | True scale | Other |
|---|---|---|---|---|---|---|---|
| **Mercator** | Cyl | Angles | Straight, parallel | Straight, parallel | NA | Along Equator | Straight rhumbs |
| **Transverse Mercator** | Cyl | Angles | Complex curves | Complex curves | NA | Along a meridian | Used for large scale |
| **Lambert Conformal Conic** | Con | Angles | Concentric arcs | Straight, converge to a point | Point | Along standard parallels | Good for east-west areas |
| **Stereographic** | Az | Angles | Circles centred on pole | Straight intersect at pole | Point | Circles on projection centre | Used for hemisphere maps |
| **Space Oblique Mercator** | Cyl | Angles | Complex curves | Complex curves account for motion | NA | Along ground track | Designed for satellite data |
| **Lambert Cylindrical Equal Area** | Cyl | Area | Unequally spaced, perpendicular to parallels | Straight, equally spaced, parallel | Lines equal in length to Equator | Equator | Spacing of parallels maintains areas |
| **Albers Equal Area** | Con | Area | Unequally spaced circles | Straight, equally spaced, centred on pole | Circular arcs | Along standard parallels | Spacing of parallels maintains areas |
| **Lambert Azimuthal Equal Area** | Az | Area | Unequally spaced circles centred on pole | Straight, equally spaced, centred on pole | Point | Standard parallel | Used in polar aspect for atlases |
| **Mollweide** | PCyl | Area | Unequally spaced straight lines | Equally spaced semi-ellipses | Point | Parallels 40°44' north-south | Central meridian is ½ length of Equator |
| **Sinusoidal** | PCyl | Area | Equally spaced straight lines | Sine curves | Point | Along central meridian and every parallel | Meridian is ½ length of Equator |
| **Gnomonic** | Az | NA | Unequally spaced circles centred on pole | Equally spaced straight lines intersect at pole | Point | Point of intersection of central parallel and central meridian | Great circles are straight lines |
| **Polyconic** | Con | NA | Non-concentric circular arcs spaced at true distances | Complex curves except central meridian is straight | Point | Along central meridian and each parallel | Easily constructed from tables of coordinates |
| **Orthographic** | Az | NA | Unequally spaced, centred on pole | Equally spaced straight lines | Point | At centre and along any circle circumference | Prospective view from infinite distance |
| **Robinson** | PCyl | NA | Equally spaced Equator to 38° N and S, decreasing beyond | Central meridian straight, others elliptical arcs concave | Line 0.53 length of Equator | No point free of distortion | Designed for look of the world |
| **Van der Grinten I** | PCyl | NA | Circular arcs concave toward pole | Central straight, others circular | Point | Along Equator only | Shows world in a circle |
| **Goode Homolosine** | PCyl | Area | Straight lines | Sine curves | Point | Along Equator | Combined Mollweide and sinusoidal |

Note. Type: Cyl = Cylindrical; Con = Conformal; Az = Azimuthal; PCyl = Pseudocylindrical.

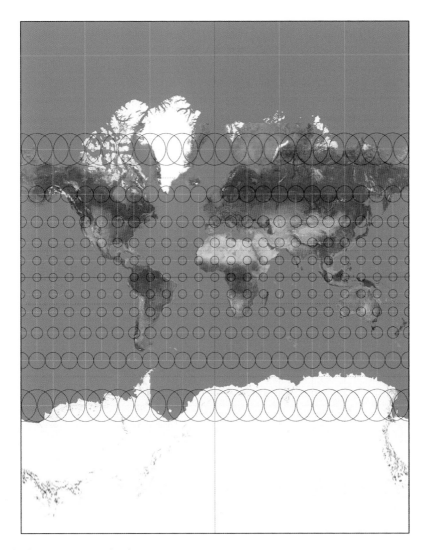

*Figure 15.2*  Mercator projection

In the secant case, meaning contact along two parallel lines, scale is true along two parallels equidistant from the Equator. The North and South poles cannot be shown and significant size distortion occurs in the higher latitudes as shown by the circle sizes in Figure 15.2. Greenland appears to be larger than Africa, when in fact Africa is almost 14 times the area of Greenland. Note the perfect 90° intersections of the parallels and meridians, a characteristic of cylindrical projections in the normal aspect. The Mercator projection, a standard for marine charts, was defined for navigational charts and is best used for navigation purposes.

## Transverse Mercator

The Mercator projection in the transverse aspect is a conformal projection with the line of constant scale along a meridian rather than the Equator. Straight lines are used for the central meridian, each meridian 90° from the central meridian, and the Equator. Other meridians and

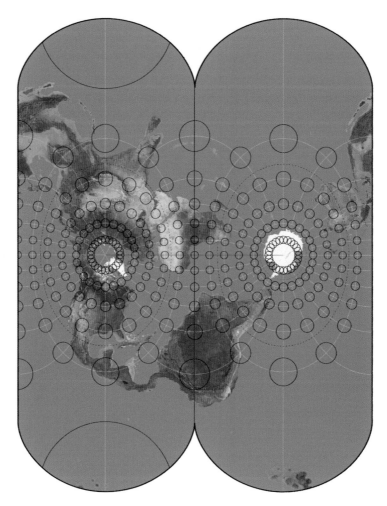

*Figure 15.3*　Transverse Mercator projection

parallels are complex curves and the poles are points along the central meridian. Scale is true along the central meridian or in the secant case, along two meridians equidistant from and parallel to the central meridian. The Transverse Mercator Projection, also known as the Gauss-Krüger projection, is commonly used for large scale, small area presentations, with many of the world's topographic maps from 1:24,000 scale to 1:250,000 scale based on this projection. The Universal Transverse Mercator (UTM) coordinate system and many of the State Plane Coordinate systems for states with an elongated north-south axis in the Unites States are based on this projection. Figure 15.3 shows the transverse Mercator projection in a global view. The projection is commonly only used for projecting narrow strips of the Earth centred along a meridian as is done to create the six-degree projection zones of the UTM coordinate system.

## Lambert Conformal Conic

The Lambert Conformal Conic (LCC) projection shows meridians as equally spaced straight lines converging at a common point, usually one of the poles. Concentric circular arcs centred

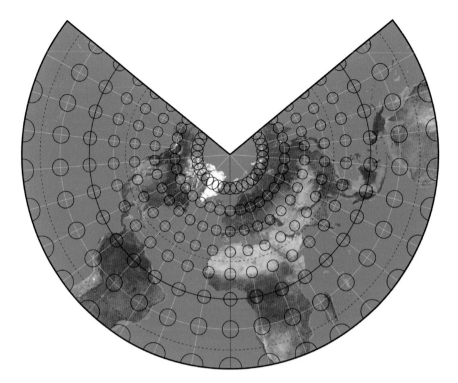

*Figure 15.4* Lambert Conformal Conic projection

on the pole of convergence of the meridians form unequally spaced parallels. Angles between the meridians on the globe are larger than the corresponding angles on the projection. Parallel spacing increases away from the pole. The pole nearest the standard parallel is represented as a point and the other pole cannot be shown. Scale is constant, not true, along any given parallel, but scale is true along the standard parallel or the two standard parallels in the secant case. Angles are maintained at the expense of area as shown by the perfect circles of different sizes in Figure 15.4. The LCC projection finds its best use for mapping of large east-west oriented regions in middle latitudes, such as China, Russia, and the 48 contiguous states of the United States. The U.S. State Plane Coordinate system uses this projection for states with an elongated east-west axis, such as Tennessee. The U.S. Geological Survey uses it for state based maps at 1:500,000 scale and for maps of the entire U.S. contiguous 48 states, when preservation of angles is important.

## Stereographic

The Egyptians and Greeks had developed the stereographic projection by the second century BC. The stereographic projection is a perspective azimuthal projection that is conformal, that is, it preserves angles. The polar, Equatorial, and oblique aspects result in a different appearance of the graticule. Projecting from one pole to a plane tangent at the other pole creates the polar aspect of this projection. Parallels are unequally spaced circles centred on the pole, which is represented as a point, and the spacing of the parallels increases away from the pole. Meridians are equally spaced straight lines intersecting at the pole with true angles between them. The projection is commonly used only for a hemisphere, but can be used to show most of the other hemisphere

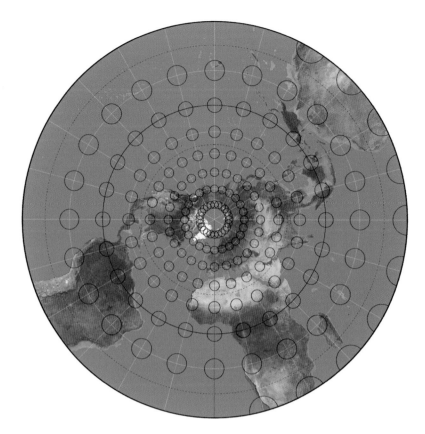

*Figure 15.5* Stereographic projection

at an accelerating scale (Figure 15.5). Scale is only true where the central parallel crosses the central meridian or along a circle concentric about the projection centre. Scale is constant along any circle with the same centre as the projection. The stereographic projection in the polar aspect is a projection of choice for topographic maps of the polar regions. The Universal Polar Stereographic is the sister projection to the transverse Mercator for the polar regions in the UTM coordinate system for military mapping. The oblique ellipsoidal form of this projection is used in Canada, Romania, Poland, Netherlands, and other nations (Thompson *et al.*, 1977). The projection is normally used only in the secant case where the scale factor is less than 1.0, and is generally chosen for regions that are roughly circular in shape. The Equatorial aspect of the stereographic projection is used for east and west hemisphere maps.

## Space Oblique Mercator

Alden P. Colvocoresses in 1973 conceived the Space Oblique Mercator (SOM) projection and John P. Snyder developed it mathematically in 1977. The SOM projection was designed to map the ground track of a satellite and maintain conformality, or angular relationships, and scale is true along the ground track of a satellite. To account for the motion in time of the satellite, meridians are complex curves with varying intervals. The scale is true along the ground track, but varies about 0.01 percent within the normal sensing range of the satellite. The projection is conformal to within a few parts per million for the sensing range and is used for satellite images from Landsat and other platforms.

## Web Mercator

The advent of the World Wide Web and the development of tiled data services led to the implementation of a modified version of the Mercator projection to support indexing and seamless transitions while enlarging the map on display. This modification, commonly referred to as the Web Mercator Projection, uses the equations for the spherical Mercator projection, but uses the semi-major axis length of the WGS 84 ellipsoid as the radius of the sphere, and coordinates from WGS 84. This combination of characteristics from the spherical Mercator (the equations) and the WGS 84 ellipsoid has drawn criticism from geodesists and cartographers, but persists as the dominant projection used for Web maps (Zinn, 2010; Battersby *et al.*, 2014). These characteristics make the projection not truly conformal as is the spherical Mercator. The graphic for the Web Mercator looks the same as for the Mercator projection in Figure 15.2.

## Equal area projections

### Lambert Cylindrical Equal Area

Johann Heinrich Lambert designed the cylindrical equal area projection, first presented in 1772, which became the basis for many other similar equal area projections including the Gall Orthographic (re-identified by Peters, Behrmann, and Trystan-Edwards). To create the variants of Lambert's original projection, one simply makes the projection secant at two small circles (parallels). The similar projections listed above use different parallels as the lines of constant scale. Lambert's cylindrical equal area projection has meridians that are 0.32 as long as the Equator and equally spaced straight parallel lines. Parallel lines of latitude are unequally spaced, perpendicular to the meridians (product of cylindrical projection in normal aspect), and furthest apart near the Equator. The pole lines are straight lines and equal in length to the Equator. Equal areas are maintained by changing the spacing of the parallels, which causes significant shape distortion. The distortion is greater in high latitudes near the poles as shown by the ellipses in Figure 15.6. Although this projection is little used, it is a standard to describe map projection principles in textbooks and also serves as a prototype for other projections as described earlier.

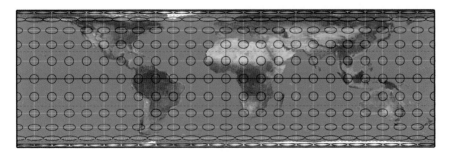

*Figure 15.6*  Lambert Cylindrical Equal Area projection

## Albers Equal Area

Heinrich Christian Albers presented his conical equal area projection in 1805. It has meridians as equally spaced straight lines converging at a common point, which is normally beyond the pole. Poles are circular arcs and the scale is true along one or two standard parallels. Angles between the meridians on the globe are greater than the angles on the projection. Parallels are centred on the point of convergence of the meridians, and are unequally spaced concentric circular arcs with the spacing of the parallels decreasing away from the point of convergence. The criterion required for equal areas is that the scale factor at any given point along a meridian is the reciprocal of the scale factor along the parallel, which is true for this projection. Figure 15.7 shows the projection graticule and Indicatrix in which circles maintain size but change in shape to ellipses away from the standard parallel. The projection has one (tangent case) or two (secant case) standard parallels free of angular and scale distortion. Similar to the LCC projection, the Albers equal area projection is commonly used for maps to show areas of east-west extent in mid latitudes. It is used in applications where preservation of area is important. It is the projection of choice for showing the 48 contiguous states with true relative areas and is the projection on which the U.S. National Atlas is based.

## Lambert Azimuthal Equal Area

The Azimuthal Equal Area projection was developed by Lambert in 1772. It is a non-perspective azimuthal equal area projection and in the polar aspect, meridians are equally spaced straight lines intersecting at the central pole with true angles between them. Parallels are centred on the pole, a point, and are unequally spaced circles. The spacing of the parallels decreases away from the pole. When the projection is used for the entire Earth, the opposite pole appears as a bounding circle with a radius 1.41 times that of the Equator. Scale is true at the centre in all directions and decreases rapidly with distance from the centre. A global view centered on the origin of the GRS

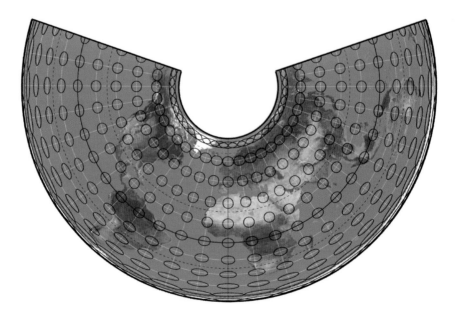

*Figure 15.7*  Albers Equal Area projection

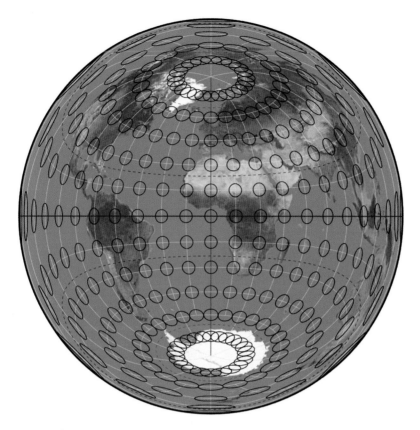

*Figure 15.8* Lambert Azimuthal Equal Area projection

coordinates is projected on the Lambert Azimuthal Equal Area in Figure 15.8 with distortion circles showing the equal area preservation and the shape distortion into ellipses. The projection is often used in the polar aspect for atlases of the polar regions and the Equatorial aspect is used for the east and west hemisphere maps. It works well for areas that are roughly circular in shape.

## *Mollweide*

This pseudocylindrical equal area projection was developed by Carl Mollweide in 1805. The projection forms an ellipse for the entire globe with the central meridian as a straight line half as long as the Equator. Meridians that are 90° east and west of the central meridian form a circle, with other meridians as equally spaced semi-ellipses that curve concave toward the central meridian and intersect the poles. Perpendicular to the central meridian are parallels that are unequally spaced straight parallel lines farthest apart near the Equator with spacing changing gradually. Scale is true along latitudes 40°44' north and south and constant along any given latitude, although not true. The poles are shown as points. Figure 15.9 shows the entire globe projected to the Mollweide and centred on the Greenwich meridian. The circles indicate distortion of shape and preservation of area since all are the same size and they become ellipses toward the poles. The Mollweide has been occasionally used for world maps, particularly thematic maps where preservation of area is important. Goode (1925) combined it with the sinusoidal projection to create the homolosine, a popular interrupted projection.

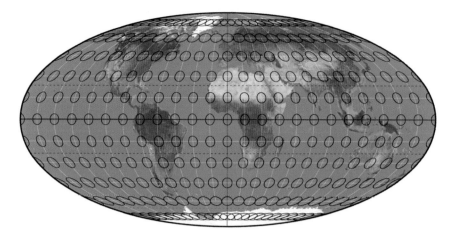

*Figure 15.9*   Mollweide projection

## *Sinusoidal*

The equal area sinusoidal, one of the oldest pseudocylindrical projections, was developed in the sixteenth century and used by various cartographers in atlases. Later users know this projection as the Sanson-Flamsteed. The central meridian, which is half as long as the Equator, is a straight line. Other meridians are equally spaced sinusoidal curves, hence the name of the projection, intersecting at the poles, and curve concave toward the central meridian. Perpendicular to the central meridian are the parallels which are equally spaced straight lines. The poles are shown as points and the scale is true along the central meridian and along every parallel. The sinusoidal projection preserves area, but distorts angles (Figure 15.10), with the greatest distortion occurring near outer meridians and in high latitudes while the Equator is relatively free of distortion. The sinusoidal has been used for world maps and for maps of South America and Africa. It was combined with the Mollweide by Goode (1925) to create the homolosine projection.

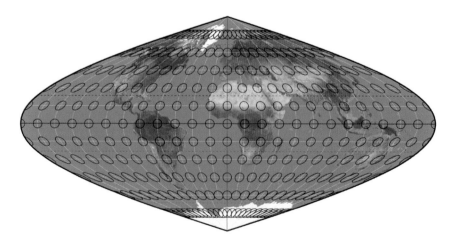

*Figure 15.10*   Sinusoidal projection

## Some projections preserving special properties

### *Gnomonic*

The gnomonic projection maps great circles on the generating globe to straight lines on the map projection. The Greek Thales possibly developed the gnomonic projection around 580 BC. The gnomonic is a perspective azimuthal projection that is neither conformal nor equal area, but it has the unique characteristic that any great circle, including all meridians and the Equator, appears as a straight line. The point of projection is at the centre of the Earth. The name gnomonic, adopted in the nineteenth century, is derived from gnomon, since the meridians radiate from the pole, or are spaced on the Equatorial aspect, exactly as the corresponding hour markings on a sundial (Snyder and Voxland, 1989). The graticule appearance changes with the aspect as in other azimuthal projections. In the polar aspect, meridians have true angles between them and are equally spaced straight lines intersecting at the pole. The spacing of the parallels, which are unequally spaced circles centred on the pole, increases from the pole. Only one hemisphere can be shown and the Equator and opposite hemisphere cannot be shown because of the perspective projection from the centre of the Earth. Scale is true only where the central parallel crosses the central meridian and increases rapidly with distance from the centre. Distortion circles (Figure 15.11) show that the projection is neither conformal nor equal area. Its usage results from the special feature of great circles as straight lines and thus assists navigators and aviators in determining the shortest and most appropriate courses.

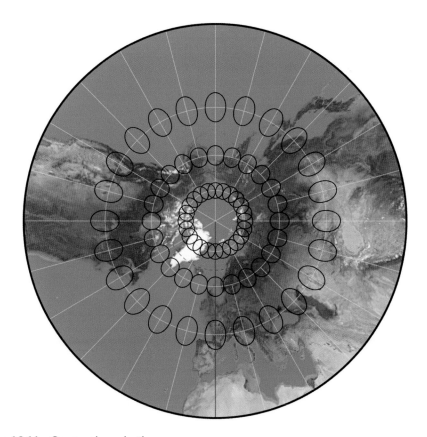

*Figure 15.11*   Gnomonic projection

## Polyconic

The polyconic projection has the unique property that it is easily constructed from tables of rectangular coordinates. Ferdinand Rudolph Hassler of the U.S. Coast and Geodetic Survey developed the polyconic projection for plane table and alidade coastal mapping. The ease of construction from simple tables while in the field led to the polyconic becoming a standard for the U.S. Geological Survey quadrangle maps. The projection uses one cone along each parallel and thus uses many cones for the total projection, hence the name 'poly' conic. Meridians are complex curves except for the central meridian which is a straight line. Parallels are non-concentric circular arcs spaced at true distances along the central meridian, with the Equator being the only parallel that is a straight line. Scale is true along the central meridian and along each parallel, and the projection is free of distortion only along the central meridian. Significant distortion occurs if the range is extended far to the east and west. The projection is aphylactic meaning that it preserves neither area nor shape. Because of its ease of construction from tables of rectangular coordinates, it was the only projection used by the USGS for topographic maps until the 1950s. The tables may be used from any polyconic projection on the same ellipsoid by applying the proper scale and central meridian. Thus, for each quadrangle map the same tables could be used. These quadrangle maps for the same ellipsoid and for the same central meridian at the same scale will fit exactly from north to south and they also fit exactly east to west, but cannot be mosaicked in both directions simultaneously. Figure 15.12 presents a graphic illustration of the projection, although the projection is not recommended for regional or global maps since better projections are available.

## Orthographic

The orthographic projection is essentially a perspective projection of the globe onto a tangent plane from an infinite distance (orthogonally). It is commonly used for pictorial views of the

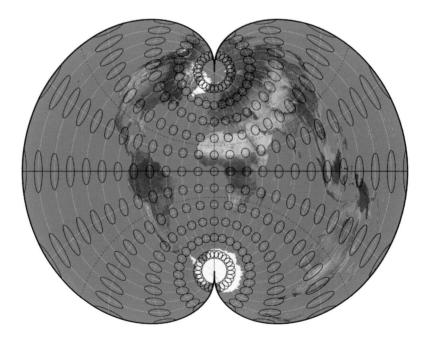

*Figure 15.12*  Polyconic projection

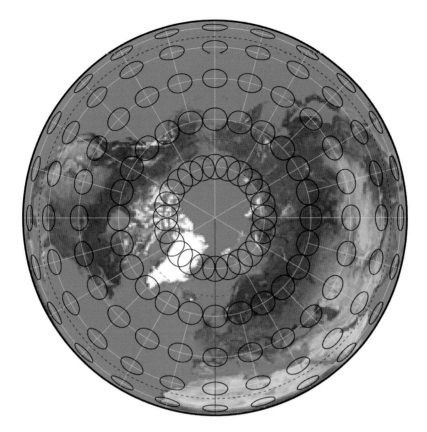

*Figure 15.13*   **Orthographic projection**

Earth as if seen from space. Developed by the Egyptians and Greeks by the second century BC, the orthographic projection is a perspective azimuthal (planar or zenithal) projection that is neither conformal nor equal area. It is used in polar, Equatorial, and oblique aspects resulting in a view of an entire hemisphere of the Earth. In the polar aspect, shown in Figure 15.13, meridians are equally spaced straight lines with true angles intersecting the central pole. Parallels are centred on the pole, which is a point, and are unequally spaced circles with spacing decreasing away from the pole. Other aspects are described in Snyder and Voxland (1989). Scale is true at the centre and along any circle circumference (parallels in the polar aspect) with its centre at the projection centre. Scale decreases radially with distance from the centre. The globe-like look of the projection and its distortion circles are shown in Figure 15.13.

## Projections developed to look correct

### *Robinson*

At the request of Rand McNally and Company in 1963, Arthur Robinson developed a projection using graphical and visual means to make the world 'look right', that is preserved the look of the globe as much as possible. The Robinson projection is a pseudocylindrical projection that is neither conformal nor equal area that uses a set of tabular coordinates rather than mathematical formulae. The central meridian is a straight line that is 0.51 the length of the Equator, while

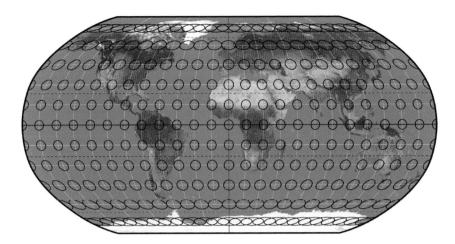

*Figure 15.14* Robinson projection

other meridians resemble elliptical arcs and concave toward the central meridian but are equally spaced. Equally spaced straight parallel lines between 38° north and south form the parallels, with space decreasing beyond these latitudes. The projection has true scale along the 38° latitudes north and south and is constant along any given latitude. The pole lines are 0.53 the length of the Equator. There is no point completely free of distortion and both size and shape change as shown by the circles in Figure 15.14. The Robinson projection was used for world maps by Rand McNally in their *Goode's Atlas*, and the National Geographic Society adopted it for world maps for a time during the 1990s.

### Van der Grinten

In 1898, Alphons J. Van der Grinten of Chicago presented the Van der Grinten projection (also called Van der Grinten 1), which is a polyconic projection that is neither conformal nor equal area. Parallels are circular arcs, concave toward the nearest pole (poles are points), with the Equator as a straight-line exception. The projection has a straight central meridian, while other meridians are circular and equally spaced along the Equator, concave toward the central meridian. Scale is true along the Equator only and it rapidly increases with distance from the Equator. The projection encloses the entire world in a circle and has significant distortion near the poles (Figure 15.15). The Van der Grinten I has been used for world maps by the U.S. Department of Agriculture, the U.S. Geological Survey, and the National Geographic Society.

### Combining map projections

The combination of two map projections for different parts of the Earth was initially developed by Goode in 1823 with the homolosine projection. Goode combined the sinusoidal projection for the area of the Earth from the Equator, zero degrees, to 41°44'11.8"N/S and the Mollweide projection from 41°44'11.8"N/S to each pole. Figure 15.16 shows the resulting homolosine projection without interruption, but the kink in the projection outer boundary shows the point of joining of the two different projections. The homolosine projection is equal area since both the input projections are equal area. Goode also developed

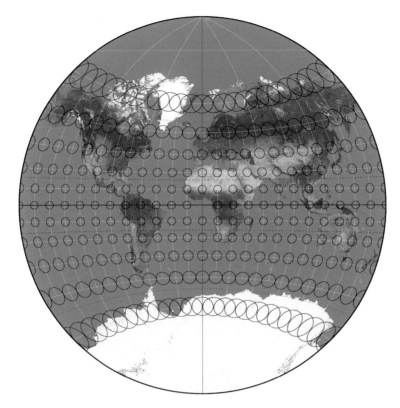

*Figure 15.15*   Van der Grinten projection

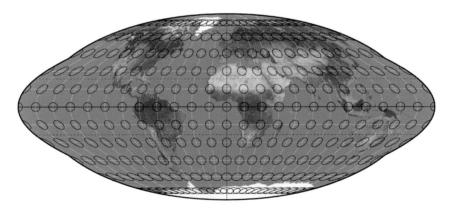

*Figure 15.16*   Goode homolosine projection

an interrupted version as shown in Figure 15.17. Interruption is a standard feature for global projections and is supported in most projection software packages.

With the current availability of digital computer technology, methods have been developed to rapidly combine projections using a graphical user interface (Jenny *et al.*, 2010; Jenny and Šavrič, 2017). The Flex Projector (Jenny *et al.*, 2008; Jenny and Patterson, 2013) software

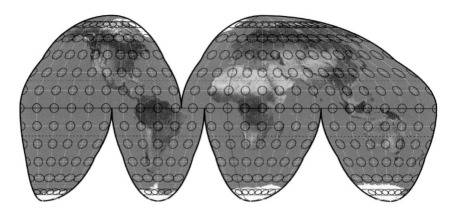

*Figure 15.17*   Interrupted Goode homolosine projection

extracts geometric characteristics from two source projections and builds tables of values, then allows the user to mix the table values, and create a new projection from the mixed table values. The tables encode (1) the horizontal length of parallels, (2) the vertical distance of parallels from the Equator, (3) the distribution of meridians, and (4) the bending of parallels. The ability to rapidly and easily combine projections in an adaptive fashion, e.g. based on latitude or scale, supports the increased use of maps in the digital environment of the World Wide Web today. Tools such as the Flex Projector make it easy for lay persons to create new map projections.

## Sources of tools and tutorials for map projections

There are many tools and tutorials for map projection design, development, and use online in both open source and commercial offerings. Flex Projector (www.flexprojector.com), mentioned earlier, is an application for creating custom world map projections. It includes a graphical user interface allowing the user to create a new projection by adjusting the length of the pole line and the Equator. The software uses cylindric, pseudocylindric, and polyconic projections as starting points for user adjustment to desired graphical look.

Proj.4 (*http://trac.osgeo.org/proj*) is a map projections code library. The library can be used with the Geospatial Data Abstraction Library (GDAL), which is based on C++ and operates on raster and vector geospatial data formats. GDAL is released under an X/MIT style open source license by the Open Source Geospatial Foundation. The software supports more than 100 projections and handles datum shifts with either a three- or seven-parameter transformation.

The National Aeronautics and Space Administration (NASA) has the Global Map Projector (G.Projector) available for free download. This software transforms a base equirectangular map into more than 100 different map projections. The program is cross-platform and executes with Windows Mac and Linux operating systems.

Commercial offerings include packages specifically designed for map projection or coordinate transformation tasks, such as Mapthematics' Geocart and Blue Marble's Geographic Calculator, and software for geographic information systems, image processing, and other geospatial processing which usually includes map projection capability. For example, Esri's ArcGIS, ERDAS Imagine, Excelis ENVI, Global Mapper, and other geospatial software include the ability to map to hundreds of different projections. Open source geospatial software systems, such as Quantum GIS, also include projection capabilities.

A variety of tools is available for learning about map projections and selecting an appropriate projection for a specific map project. The USGS developed and makes available the Decision Support System (DSS) for map projection selection. This system is based on recommendations in Snyder (1987) and based on input user information of data type (raster or vector), area and location of the portion of the world to be projected (region, continent, hemisphere, or globe), and the property to be preserved in the transformation, determines an appropriate projection selection. The user also has the option of tracing the logic to select the projection in the tutorial mode of the DSS. This logic process is presented in Finn *et al.* (2017). Additional sources of information on selecting appropriate map projections are contained in Jenny *et al.* (2017). Snyder (1987) provides a comprehensive workbook of map projections used by the U.S. Geological Survey with complete forward and inverse mathematical formulations and worked examples of transforming from spherical and/or ellipsoidal coordinates to map projection space and the inverse. Snyder and Voxland (1989) provide an album of map projections including verbal and graphic descriptions of many projections, and Snyder (1993) includes 2,000 years of flattening the Earth with map projections. A general guide to map projections is provided by Lapaine and Usery (2017) in a book on map projections that includes historical map projection selection documents developed by the American Cartographic Association in the 1980s as well as several articles by current projection experts.

## Conclusions

Map projection is an essential task today in a world flooded with geospatial data. Geospatial processing software includes capabilities for projecting data to hundreds of different projections. The choice of the appropriate projection for a specific dataset in a specific area remains a challenge, but the availability of decision support tools and educational material on map projection meets this challenge. The design of map projections has moved from the graphical and mathematical approach with tables of coordinates to interactive graphical user interfaces for map projection creation and on the fly calculation as users change map scale and orientations. The need for understanding map projections is far greater now than in the past because of the availability of geospatial data and the ubiquity of maps in today's world.

## References

Battersby, S.E., Finn, M.P., Usery, E.L. and Yamamoto, K.H. (2014) "Implications of Web Mercator and its use in Online Mapping" *Cartographica* 49 (2) pp. 85-101.

Bugayevskiy, L.M. and J.P. Snyder (1995) *Map Projections: A Reference Manual* London: Taylor and Francis.

Canters, F. (2002) *Small-Scale Map Projection Design* London: Taylor and Francis.

Finn, M.P., Usery, E.L., Posch, S.T. and Seong, J.C. (2004) "A Decision Support System for Map Projections of Small Scale Data" *U.S. Geological Survey Scientific Investigation Report 2004–5297.*

Finn, M.P., Usery, E.L., Woodard, L.N. and Yamamoto, K.H. (2017) "The Logic of Selecting an Appropriate Map Projection in a Decision Support System (DSS)" in Lapaine, M. and Usery, E. L. (Eds) *Choosing a Map Projection* New York: Springer.

Goode, J.P. (1925) "The Homolosine Projection: A New Device for Portraying the Earth's Entire Surface" *Annals of the Association of American Geographers* 15 pp.119–125.

Goussinsly, B. (1951) "On the Classification of Map Projections" *Empire Survey Review* 11 pp. 75–79.

Iliffe, J.C. (2000) *Datums and Map Projections* Caithness, UK: Whittles Publishing.

Jenny, B. and Šavrič, B. (2017) "Combining World Map Projections" in Lapaine, M. and Usery, E.L. (Eds) *Choosing a Map Projection* New York: Springer.

Jenny, B. and Patterson, T. (2013) "Blending World Map Projections" *Cartography and Geographic Information Science* 40 (4) pp.289–296.

Jenny, B., Patterson, T. and Hurni, L. (2008) "Flex Projector: Interactive Software for Designing World Map Projections" *Cartographic Perspectives* 59 pp.12–27.

Jenny, B., Patterson, T. and Hurni L. (2010) Graphical Design of World Map Projections *International Journal of Geographic Information Science* 24 (11) pp.1687–1702.

Jenny, B., Šavrič, B., Arnold, N.D., Marston, B.E., Preppernau, C.A. (2017) "A Guide to Selecting Map Projections for World and Hemisphere Maps" in Lapaine, M. and Usery, E.L. (Eds) *Choosing a Map Projection* New York: Springer.

Lapaine, M. and Usery, E.L. (Eds) (2017) *Choosing a Map Projection* New York: Springer.

Lee, L.P. (1944) "The Nomenclature and Classification of Map Projections" *Empire Survey Review* 8 pp.142–152.

Maling, D.H. (1968) "The Terminology of Map Projections" *International Yearbook of Cartography* 8 pp.11–65.

Maling, D.H. (1992) *Coordinate Systems and Map Projections* (2nd ed.) Oxford, UK: Pergamon Press.

Mapthematics (2015) "Mapthematics Geocart 3" Available at: *www.mapthematics.com/* (Accessed: 22 October 2015).

Maurer, H. (1935) "Ebene Kugelbilder, Ein Linnésches System der Kartenenturfe" *Petermanns Mitteilungen Erganzungsheft* No.221.

Mugnier, C.J. (2004) "Object Space Coordinate Systems" in American Society of Photogrammetry and Remote Sensing *Manual of Photogrammetry* (5th ed.) Falls Church, VA: American Society of Photogrammetry and Remote Sensing.

Pearson II, F. (1990) *Map Projections: Theory and Applications* Boca Raton, FL: CRC Press.

Snyder, J.P. (1987) *Map Projection: A Working Manual* U.S. Geological Survey Professional Paper 1395 Washington, DC: U.S. Government Printing Office.

Snyder, J.P. (1993) *Flattening the Earth: Two Thousand Years of Map Projections* Chicago: University of Chicago Press.

Snyder, J.P. and Voxland, P.M. (1989) *An Album of Map Projections,* U.S. Geological Survey Professional Paper 1453 Washington, DC: U.S. Government Printing Office.

Thompson, D.B., Mephan, M.P. and Steeves, R.R. (1977) "The Stereographic Double Projection" *University of New Brunswick Technical Report* No.46.

Tissot, N.A. (1881) *Mémoire sur la représentation des surfaces et les projections des cartes* Paris: Villars.

Tobler, W.R. (1962) "A Classification of Map Projections" *Annals of the Association of American Geographers* 52 pp.167–175.

USGS (2006) "Cartographic Research" U.S. Geological Survey, Rolla, MO Available at: *http://carto-research.er.usgs.gov/* (Accessed: August 2006).

Vujakovic, P. (2014) "The State as a 'Power Container': The Role of News Media Cartography in Contemporary Geopolitical Discourse" *The Cartographic Journal* 51 (1) pp.11–24.

Yang, Y., Snyder, J.P. and Tobler, W.R. (2000) *Map Projection Transformation Principles and Applications* London: Taylor and Francis.

Zinn, N. (2010) "Web Mercator: Non-Conformal, Non-Mercator" Available at: *www.hydrometronics.com/downloads/Web%20Mercator%20-%20Non-Conformal,%20Non-Mercator%20(notes).pdf* (Accessed: 30 October 2016).

# 16

# Photogrammetry and remote sensing

*Stuart Granshaw*

## Introduction

Topographic mapping requires the measurement of the Earth's surface features and may result in map series such as those produced by many nations at scales such as 1:25,000 and 1:50,000 (where map generalization is required) or plans at, say 1:1,000 or 1:5,000 (requiring no generalization and perhaps used as cadastral maps for land ownership). Traditionally, the measurements required for such topographic maps were made by land survey techniques (see Chapter 11) and employing map projections (normally a conformal projection such as a Transverse Mercator – see Chapter 15). Such topographic surveys require two elements. First a control survey must be undertaken, which prior to the 1950s was conducted by triangulation using theodolites, but since that time has involved electromagnetic distance measurement and, more recently, global navigation satellite systems (GNSS) such as GPS (see Chapter 18). The second element of a topographic survey is the detail survey. The control survey provides a skeletal framework of points from which the map detail can be surveyed. Historically this was done by simple methods such as plane tabling or chain surveying, but today can be carried out by modern survey instruments such as total stations. However, mapping any extended area by land survey techniques is time and labour intensive and, consequently, expensive. Such methods are also not easy to implement in difficult terrain, such as in mountainous, tundra or desert areas. It was appreciated towards the end of the nineteenth century that photographs taken from the air (e.g. from a balloon) could be used to derive a topographic map. The invention of the aircraft at the beginning of the twentieth century allowed more systematic aerial photographs to be taken. These were originally used in mountainous terrain, especially in southern Germany and Switzerland, to derive topographic map detail (both planimetric information such as roads and rivers, and height information such as spot heights and contours). Although both world wars provided some impetus for mapping from aerial photographs, with the development of specialized *plotting instruments* during the 1920s, it was only after the Second World War that aerial photography was widely adopted over much of the world as the most efficient method of providing topographic map detail over large areas that was sufficiently accurate to meet national map accuracy standards. The technological discipline to achieve this is called *photogrammetry* (*photos* = light; *gramma* = drawn or written; *metron* = to measure). Thus an early definition of photogrammetry was 'the science or art of obtaining reliable measurements by means of photography' (McGlone, 2013).

223

In 1960 the first meteorological satellite, TIROS-1, was launched. The images produced were clearly not aerial, neither were they strictly *photographs*, as the instruments used extended observations from beyond the visible spectrum used in most photography into the infrared and microwave areas of the electromagnetic spectrum (Figure 16.1). The term *remote sensing* was introduced to provide a more accurate description of these *images* (itself a more generic term than photographs), although aerial photography provided much of the material used in this new discipline. In 1972 the US National Aeronautics and Space Administration (NASA) launched Landsat 1, the first of many Earth observation (EO) satellites. Campbell and Wynne (2011) note three contributions made by Landsat:

(a) the routine acquisition of *multispectral* data of large regions of the Earth's surface;
(b) the rapid extension of computer *digital image analysis* (rather than the film prints from aerial photography);
(c) a model for the development of other EO satellites, so that today there are many satellites from various countries used for mapping purposes (Tables 16.1 and 16.2).

Most EO satellites are polar-orbiting, so their orbits are roughly north-south and successive orbits are over different land areas. However, some meteorological satellites are in a *geostationary* orbit (alongside most communications and television satellites), which is a specific orbit 36,000 km above the Equator where the satellite remains over the same place (on the Equator).

Lillesand *et al.* (2015) define remote sensing as 'the science and art of obtaining information about an object, area, or phenomenon through the analysis of data acquired by a device not in contact with the object, area, or phenomenon under investigation'.

One may, justifiably, be confused about the difference between photogrammetry and remote sensing. Today they both use aerial photography and satellite imagery; additionally, both utilize visible and non-visible parts of the electromagnetic spectrum (such as microwave radar radiation). They also both rely on digital computer methods. As well as mapping, both disciplines encompass non-mapping applications (such as engineering or medicine), although mapping remains their primary focus. They are certainly related, as demonstrated by their national and international bodies, such as the International Society for Photogrammetry and Remote Sensing (ISPRS). Remote sensing relates mainly to the nature of features and their characteristics, whereas photogrammetry is concerned with the geometric attributes of features (where they are and their size).

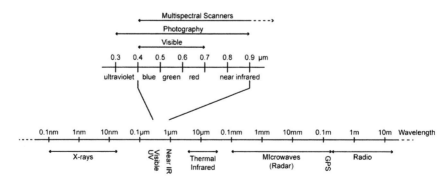

*Figure 16.1*   Part of the electromagnetic spectrum. 1000 nm = 1 µm; 1000 µm = 1 mm. Short wavelengths such as X-rays are almost completely blocked by the Earth's atmosphere. Long wavelengths do not have sufficient resolution. Most photogrammetric and remote sensing systems use the optical (visible and near infrared), thermal-infrared and microwave portions of the spectrum

*Table 16.1* High-resolution satellites used in photogrammetry and remote sensing. PAN is panchromatic; MSS is multispectral scanner; GSD is ground sample distance. For lower-resolution satellites see Table 16.2

| Satellite | Country | PAN GSD (m) | MSS GSD (m) | MSS bands | Revisit (days) |
|---|---|---|---|---|---|
| ALOS (PRISM-2) | Japan | 0.8 | 5 | 4 | 14 |
| Cartosat | India | 0.8 | – | – | 4 |
| EROS | Israel | 0.7 | – | – | 4 |
| FormoSat (−1, −2) | Taiwan | 2.0 | 8 | 4 | – |
| GeoEye | USA | 0.46 | 1.7 | 4 | – |
| IKONOS | USA | 1.0 | 4 | 4 | 3 |
| KOMPSAT 2-3A | S. Korea | 0.7 | 2.5 | 4 | 28 |
| OrbView | USA | 1.0 | 4.0 | 4 | 3 |
| Pleiades | France | 0.7 | 2.8 | 4 | 26 |
| QuickBird | USA | 0.6 | 2.4 | 4 | – |
| Resurs | Russia | 1.0 | 4 | 5 | – |
| SPOT 6-7 | France | 1.5 | 6 | 4 | 26 |
| THEOS | Thailand | 2.0 | 15 | 4 | – |
| WorldView | USA | 0.3 | 1.3 | 8 | 1 to 4 |
| ZY-3 | China | 2.1 | 5.8 | 4 | – |

*Table 16.2* Primary examples of polar-orbiting, lower-resolution optical, lidar and radar satellites, together with geostationary meteorological satellites. For higher-resolution satellites see Table 16.1

| Optical satellites | Country | Lidar satellites | Country | Radar satellites | Country | Geostationary meteorological | Country |
|---|---|---|---|---|---|---|---|
| Aqua | USA | ICESat | USA | ALOS (PALSAR) | Japan | Elektro | Russia |
| ASTER | USA/Japan | LIST | USA | COSMO-SkyMed | Italy | Fengyun | China |
| IRS | India | Sentinel-3 | Europe | Envisat | Europe | GOES | USA |
| Landsat | USA | | | ERS | Europe | INSAT | India |
| NOAA | USA | | | KOMPSAT 5 | S. Korea | Meteosat | Europe |
| RapidEye | USA | | | PAZ | Spain | MTSAT/Himawari | Japan |
| Sentinel-2 | Europe | | | RADARSAT | Canada | | |
| SPOT 1-5 | France | | | Sentinel-1 | Europe | | |
| Terra | USA | | | TanDEM | Germany | | |
| | | | | TerraSAR-X | Germany | | |

In terms of cartography, photogrammetry is primarily concerned with topographic mapping (where accurate positions are paramount, especially for large–scale mapping) whereas remote sensing is more concerned with certain environmental aspects of thematic mapping (such as vegetation distributions or climatic data) where change over time may also be important. Both provide digital data that can be input into geographical information systems (see Chapter 17).

## Platforms

Although both photogrammetry and remote sensing can use terrestrial sensors (e.g. on a tripod), for most mapping applications the sensor is raised well above the ground using the following platforms:

(1) *Aircraft*. Traditionally used to provide a stable vehicle to house a large- or medium-format camera or other sensor. Can take a series of images (overlapping for photogrammetric applications) in a systematic fashion at a large scale over a specific area to be mapped.

(2) *Unmanned aerial vehicles* (UAVs). Sometimes called unpiloted aerial vehicle and also unmanned/unpiloted aerial system; colloquially a *drone*. A smaller and less expensive, but more unstable, platform that can only accommodate light-weight sensors, yet increasingly used for specialized mapping over small areas.

(3) *Satellites*. This is the platform commonly associated with remote sensing, although high-resolution satellite imagery is also widely used in modern photogrammetry. Provides systematic and often worldwide coverage on a regular basis.

(4) *Spacecraft*. Less systematic and of much shorter duration than satellites, yet some mapping applications exist (e.g. Shuttle Radar Topography Mission).

## Sensor systems

Both modern photogrammetry and remote sensing use a variety of sensors to derive their topographic and thematic mapping products (Figure 16.1). These can be divided into two categories: *passive* sensors that rely on naturally available energy and *active* systems that use an artificial energy source. Note that many systems are classified by how many *spectral bands* they use. For example, a black-and-white (greyscale) photograph uses a single (*panchromatic*) range of wavelengths (usually covering the visible spectrum) whereas a colour photograph uses three spectral bands (blue, green and red).

## Passive systems

(1) Panchromatic (black-and-white) images from a film or digital camera, or a pushbroom scanner (one visible spectral band) (Figure 16.2a).

(2) Colour photograph (Figure 16.2c) from a digital camera or pushbroom scanner (three visible spectral bands) (Figure 16.2c).

(3) Colour infrared (IR) photograph (a camera can detect near (reflected) infrared wavelengths longer than the human eye can see). The red colour in Figure 16.12 represents near infrared.

(4) Multispectral scanner (MSS) images. For example, the early Landsat satellites used four spectral bands (blue, green, red, near IR) and the later Thematic Mapper (TM) had seven bands. Landsat 8 (launched in 2013) has a nine-band Operational Land Imager (Figure 16.14). The colour picture in Figure 16.2(d) uses three of the eight bands on the WorldView-2 satellite.

(5) Thermal camera. Uses a single band in the thermal (not near) infrared (detects heat).

(6) Hyperspectral scanner. Captures many (hundreds) of very narrow spectral bands.

Note that passive systems (1) to (4) collectively are often termed *optical imagery* as they use the visible or near-visible parts of the electromagnetic spectrum and rely on sunlight as their energy source. In contrast, thermal cameras (5) rely on the Earth's own emitted energy.

## Active systems

(7) Radar (radio detection and ranging). The most common system is synthetic aperture radar (SAR) using microwaves transmitted from either an aircraft or a satellite. (Figures 16.3 and 16.4).

(8) Lidar (light detection and ranging). Aerial or satellite laser scanning systems (Figure 16.2b).

(a)

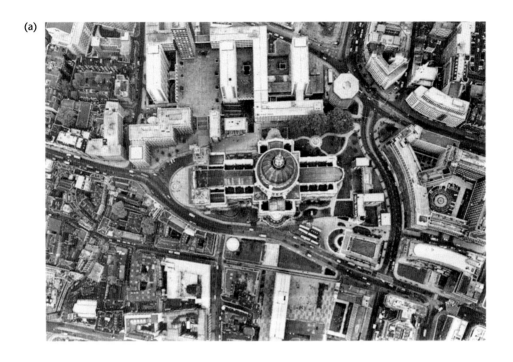

(b)

*Figure 16.2    (continued)*

*(continued)*

(c)

(d)

*Figure 16.2*  Images of St Paul's Cathedral, London. (a) A panchromatic film aerial photograph taken in 1975 (Courtesy Hunting Surveys Limited); (b) A 50 cm GSD lidar image taken in 2007 (Courtesy Blom UK); (c) A 4 cm GSD digital colour photograph taken in 2007 (Courtesy Blom UK); (d) A 46 cm GSD WorldView-2 satellite image in 2013 (Courtesy DigitalGlobe). Note the height displacement of the cathedral dome in (a), (c) and (d) which is not present in the lidar image (b)

## Cameras

Most digital cameras used in photogrammetry and remote sensing are not standard digital single-lens reflex (DSLR) or compact cameras that an amateur photographer might use (although high-end *prosumer* versions may be utilized in UAVs). Rather, specialist large- or medium-format cameras are employed. There are two basic types. First, *frame-array* cameras, which use a rectangular grid of pixels, much as in standard DSLRs and compact cameras, but with many more pixels and improved lenses. Second, *linear array* cameras where only a line (or sometimes a few lines) of detectors image a line perpendicular to the fight path of an aircraft or UAV, or the orbital track of a satellite; these are also termed *pushbroom* cameras (think of the bristles of a brush represent the detectors and the image is formed by the forward movement of the aircraft or satellite, much like a broom cleaning).

## Multispectral and hyperspectral scanners

These are not strictly cameras, but rather detectors that scan from side to side (as in the early Landsat satellites) or in a pushbroom fashion. The term *multispectral* relates to the fact that they do not form a single image of the ground, but rather take several images or *bands* (also called *channels*) of the same areas using different parts of the electromagnetic spectrum (Figure 16.1). For example, Landsat TM had the following seven bands: blue, green and red visible bands; two near-infrared bands; a mid-infrared band; and a thermal-infrared band. Figure 16.2(d) is formed from the blue, green and red channels of the multispectral scanner aboard the WorldView-2 satellite. Multispectral scanners are widely used in remote sensing and are discussed in more detail later. Hyperspectral scanners take the concept of multispectral scanners one step further, using hundreds of spectral bands.

## Thermal cameras

The thermal infrared, which depends on emitted (heat) energy, should be carefully distinguished from the near infrared (just beyond the visible spectrum and, like it, depending on reflected solar energy). The general public may have seen thermal cameras used by the police or rescuers, detecting the body heat of fleeing criminals at night or trapped survivors after an earthquake. Their poor resolution makes them unsuitable for photogrammetry, but they have specific applications in remote sensing where the thermal properties of ground features provide a very different 'view' to a camera or multispectral imagery, especially if heat is a component characteristic.

## Radar

Radar is an acronym for *radio detection and ranging*. It is an *active* microwave system so that the sensor on the aircraft or satellite generates its own signal (as well as detecting the reflected signal) and *ranging* means the return distance of the signal is measured (using the time delay between the transmitted and received signals). A radar image (Figures 16.3 and 16.4) is very different to an aerial photograph, both in the way it looks and under the conditions in which it can be taken. It can be seen from Figure 16.1 that microwaves have wavelengths of between 3 and 30 cm (somewhat shorter than those used in radio transmissions). These wavelengths are much longer than either visible light (0.4 to 0.7 $\mu$m) or the thermal infrared (around 10 $\mu$m) and, at the upper end, are similar to the carrier wavelengths (20–25 cm) used by GPS satellites. This has several consequences. First, the unfamiliar look of a radar image is due to what appears to be 'smooth' or 'rough'. For example, a dry lake-bed which may appear as rough in an aerial photograph may be smooth relative to a microwave signal of 10 cm. In Figures 16.3 and 16.4, water areas are dark because calm water 'bounces' the radar signal away from the sensor, whereas buildings in the urban area return a strong signal (light areas) because they act as corner reflectors. Second, the particles in the Earth's atmosphere that will either scatter or totally block visible light (such as water droplets in a cloud) are much smaller than a radar's wavelength and generally do not block the signal. This means that radar imagery (like GPS signals) can be taken in almost all weather conditions, which is a huge bonus in severely cloud-covered parts of the world. In fact, the first systematic radar survey was carried out in the Darien province of Panama in 1967 (almost permanent cloud cover meant it had never previously been comprehensively photographed or mapped). Figure 16.4 depicts modern ALOS satellite imagery of this region where North and South America meet. In 1971 half-a-million square kilometres of Venezuela were mapped in a similar fashion and Project Radam (Radar of the Amazon) began in Brazil in the same year. In 1978 the Seasat radar satellite was launched (sea roughness is easily differentiated on radar imagery) followed by other radar satellites (see Table 16.2).

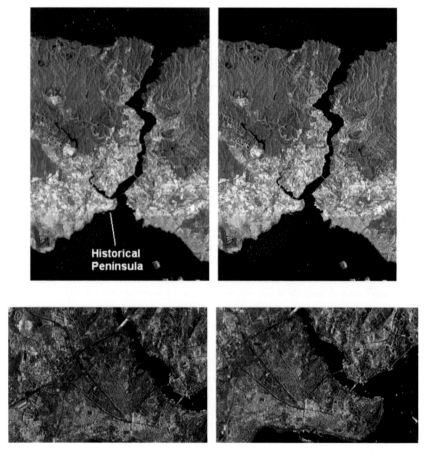

*Figure 16.3*   Radar images. Top: a pair of synthetic aperture radar images taken in different orbits of the TerraSAR-X satellite over the Bosporus, Turkey. The Black Sea is at the top and the Sea of Marmara at the bottom. Bottom: enlargements of the Historical Peninsula of Istanbul. From Sefercik and Yastikli (2014)

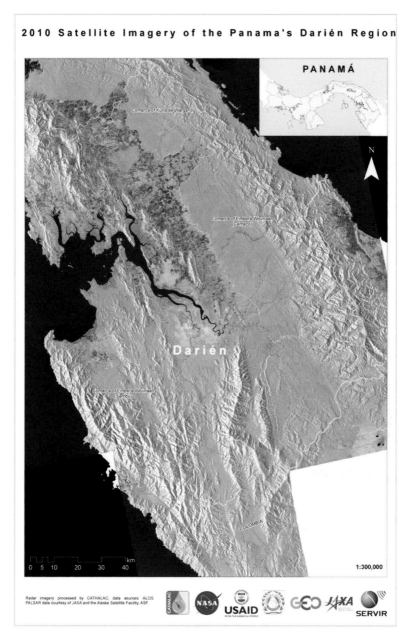

*Figure 16.4* **ALOS PALSAR satellite radar image of Darien Province, eastern Panama. Photography of this region is extremely difficult due to almost permanent cloud cover, but radar penetrates cloud and light rain. Courtesy NASA ASF and JAXA**

Most imaging radar systems are *side-looking airborne radar*; this means the radar pulse is transmitted obliquely to the ground which allows discrimination between features on the same scan line due to their differing distances from the sensor. Almost always they also use SAR. This is because to obtain a sufficiently high resolution to be of use for mapping, the physical antenna would need to be, say, 100 m long. This is impractical, but complex processing simulates a long antenna from a short physical one, hence the term *synthetic*. Two or more SAR images from different orbits can be used in the process of *interferometric* SAR (InSAR) to form a digital elevation model (DEM) of the topography of the ground. Although optical imagery provides greater resolution, the all-weather capabilities of radar systems make them attractive for mapping in many regions.

## *Lidar*

Lidar stands for *light detection and ranging* and is used (like radar) as a word in its own right (although *LiDAR* is also common). Also like radar, it is an *active* system so that the sensor on the aircraft or satellite generates its own energy (as well as detecting the reflected signal) and *ranging* means the return distance of the signal is measured. Unlike radar which uses relatively long wavelength microwave energy, lidar (like photography and some bands of multispectral scanners) uses visible light. However, lidar is based on lasers which use *coherent* light, which is very 'pure' and is composed of a very narrow band of wavelengths (unlike the broad range of blue to red light that the human eye (and panchromatic images) can detect (e.g. a rainbow). Although there are some satellite systems (Table 16.2), most lidar systems operate from aircraft and Figure 16.2(b) provides an example. *Direct georeferencing* using a GNSS for position and an inertial measurement unit (IMU) for attitude means that the *pose* of the lidar unit is known. The lidar scanner transmits over 300,000 laser pulses per second via a rotating mirror across the swath width below the aircraft. Several signals can be reflected from the ground. The *first-return pulse* is reflected from the first surface contacted, such as the top of a building or the upper part of a tree, and can be used to form a *digital surface model* (DSM). The *last-return pulse* is generally reflected from the last solid surface that the lidar penetrates (in other words the actual ground surface) and can be used to form a DEM. However, modern systems usually use the *full waveform* to provide more information than the first- or last-return pulses. The time the laser signal takes is precisely measured and converted to a distance. Combined with the GNSS/IMU data, this provides full 3D coordinates of the ground feature and thus can be used to create a map. For detailed information on lidar, see Vosselman and Mass (2010) and Renslow (2012).

## *Resolution*

This normally refers to *spatial resolution*, in other words the degree of detail that can be detected on the ground (called the *ground sample distance* – GSD) One major difference between photogrammetry and remote sensing is that the former requires a high resolution (small GSD) to 'see' the detail (often linear [vector] features, such as tracks or streams) that are required on a topographic map, whereas remote sensing is frequently concerned with larger area (raster) data (such as rock formations or vegetation coverage) where a lower resolution (higher GSD) is acceptable. The panchromatic high-resolution satellite imagery in Table 16.1 generally has a GSD of 2 m or smaller. A large-scale digital aerial photograph, such as that in Figure 16.2(c), may have a GSD of 4 cm. The original Landsat MSS sensor had a GSD of just 80 m whereas the WorldView-2 image in Figure 16.2(d) has a GSD of 46 cm.

## Photogrammetric mapping

In its mapping applications, photogrammetry is concerned with identifying the features to be shown on a topographic map (roads, railways, rivers, buildings, forests, field boundaries) from

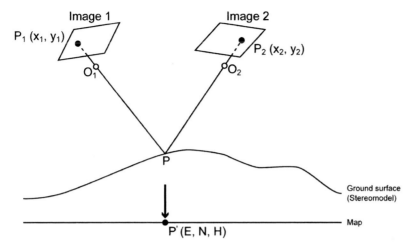

*Figure 16.5*  **Principle of intersection in photogrammetry. At the time of photography, light rays from the ground surface at P passed through the camera lens at O$_1$ and O$_2$ to form image points p$_1$ and p$_2$. Photogrammetry reconstructs the position and orientation of the two images and reprojects in the reverse direction to form P on a stereomodel and hence map point P'**

aerial or satellite imagery and determining the 3D positions of these features (plus purely height information such as spot heights and contours). Mikhail *et al.* (2001) and Wolf *et al.* (2014) describe in detail the various photogrammetric processes. It can be seen in Figure 16.2(a) that features cannot be simply traced off an aerial photograph and called an accurate map. For example, the tops and bottoms of buildings are in different positions, whereas on a topographic map these should be in the same location. Active systems such as lidar and radar determine these 3D positions using angles and distances (range) to the ground features. Passive systems, such as cameras, use 2D images, and 3D information about the ground surface can only be determined using two (or more) overlapping images of the same ground area (*stereo-images*) which use intersection from these two images to determine 3D positions (Figure 16.5).

The area of ground to be mapped is termed a *block*, and can rarely be mapped from just two overlapping photographs. Instead the area is divided into several overlapping strips, and each component strip contains a series of overlapping individual photographs (Figure 16.6). The overlap between adjacent strips is termed the *lateral overlap* (*sidelap* in the US); it is commonly about 20 per cent for standard topographic mapping from an aircraft, but may increase to 60–80 per cent for UAVs to avoid gaps from this more unstable platform. The overlap between successive photographs within a strip is termed the *forward overlap* (*endlap* in the US); it must be more than 50 per cent to ensure that each ground point appears on at least two photographs and is normally 60 per cent for standard topographic mapping; however, it may increase to 80 per cent in urban areas where buildings may *occlude* (obscure) adjacent ground features. Note that aircraft normally make tight turns at the end of each strip so that adjacent strips are exposed sequentially. This also sometimes happens with UAVs (the so-called lawnmower pattern), but such tight turns can be problematic for many UAVs and so a Zamboni pattern may be adopted (named after the inventor of resurfacing vehicles for ice-skating rinks) – see Figure 16.6.

Aerial blocks will therefore consist of a series of overlapping photographs. These photographs may be *near* verticals, although in urban areas both near-vertical and oblique photographs are increasingly taken with cameras such as that shown in Figure 16.7. In either case their location

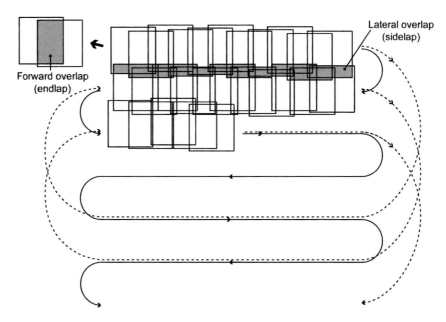

*Figure 16.6* **A block of photography comprises several strips of overlapping photographs. Solid: standard (lawnmower) flight lines adopted by aircraft and some UAVs. Pecked: Zamboni pattern adopted by some UAVs**

and rotation (their pose) will be unknown and will be different for each individual photograph. To determine the pose of each photograph, two methods can be adopted:

(1) The traditional approach is to choose common points in the overlap between each forward overlap (in other words they are common to three successive photographs) called *pass points*, and also in the common lateral overlap, called *tie points* (to tie adjacent strips together). A few of these points, particularly around the outside of the block, will have their geographical coordinates determined by land survey methods (often GNSS). By using a method called *aerial triangulation*, normally a so-called *bundle block adjustment*, the position and orientation (pose) of each photograph can be determined.

(2) Increasingly, *direct sensor orientation* (*direct georeferencing*) is being used. This is where a GNSS is mounted in the aircraft to measure the location of each image exposure, and an IMU, part of an *inertial navigation system* (INS), uses three accelerometers and three gyroscopes to determine the exact direction in which the camera was pointing. For this method to be successful, the GNSS, INS and camera must be carefully related to one another in terms of time and positional offsets.

In contrast to aircraft and UAVs, satellites use a very different system due to their regular orbits. A satellite will image a *swath* which is a band of land either side of the ground line of the satellite's orbit. The swath is generally composed of individual lines – or groups of lines – of detectors (a linear array of pixels) at right angles to the orbital track in the pushbroom fashion. One difference between remote sensing and photogrammetry is that the latter must 'see' each ground feature from two different positions. Originally this was achieved using images from different orbits, but obtaining cloud-free images might cause these images to be separated by several months with the potential for concomitant changes

*Figure 16.7* Leica RCD30 Oblique aerial camera. The central lens takes a conventional nadir (near-vertical) photograph whilst the other four cameras take intentionally tilted forward, backward, left and right images. Courtesy Leica Geosystems

in the ground features. More recently, satellites designed for topographic mapping (e.g. the Chinese ZY-3 satellite) can take forward and backward images from the same orbit.

## Acquiring topographic map detail by photogrammetry

Once the pose of each aerial or satellite image has been determined, planimetric and height detail can then be determined. Note that it is the forward overlap area common to two adjacent photographs that is plotted as a unit. After determining the pose of the two photographs, a 3D model of the ground can be simulated; this is called the *stereomodel* (*stereos* means 'solid'; the ground surface in Figure 16.5 can be interpreted as a stereomodel). This requires the left eye to only see the left photograph and the right eye to only see the right photograph, just as with wearing the special glasses in a cinema showing a 3D film. Up to the mid-1970s, photogramme-try inevitably used film photographs and the topographic map detail was plotted using *analogue*

(a)

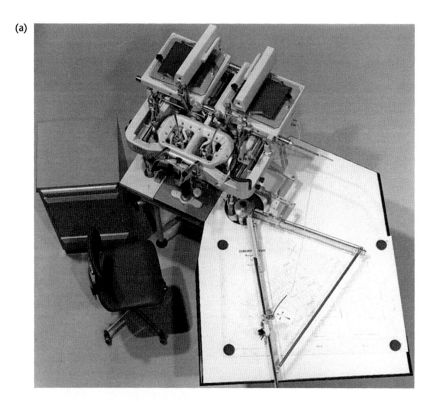

(b)

*Figure 16.8* Photogrammetric plotting. (a) Kern PG2, an example of an analogue photogrammetric stereoplotter, used up to the 1980s. The two images can be seen at the top of the figure. The operator would view the 3D stereomodel through the binocular eyepiece in front of the chair. Two metal space rods simulate the rays from the ground to the two photographs. Map features are traced onto the plotting table to the right. Courtesy Kern Aarau. (b) A Leica photogrammetric workstation using current digital photogrammetry. Courtesy Leica Geosystems

*stereoplotters*, normally in a graphical mode (Figure 16.8a). Analogue stereoplotters used optical and/or mechanical constructions to simulate the intersections of light rays from the camera of two overlapping photographs that formed a stereopair. This meant the aerial cameras had to accurately replicate a pinhole camera with virtually no distortion and they were, consequently, very expensive and used a large format (commonly 230 mm square).

From the 1970s to the 1990s, a hybrid system using *analytical stereoplotters* was employed where computers, rather than optical or mechanical instruments, were used to form the stereo-models, but the photographs were still on film. The output was digital data, which could be input to a computer-assisted cartography system (a development that had begun towards the end of the analogue era). The 1990s saw the advent of the current *digital photogrammetry*, first using scanned film photographs and then using the digital aerial cameras that are used today with pixel arrays of several hundred megapixels. Photogrammetry thus finally achieved the fully digital workflow experienced by remote sensing two decades before.

In the analogue and analytical eras, map detail was plotted by an operator at the stereoplotter (Figure 16.8a) viewing the stereomodel in 3D where a so-called *floating mark* (two dots, one on the left image and one on the right image, fused into a single 3D dot) could be moved around the model by the operator. Planimetric features (such as roads, rivers, buildings, fences and so on) were plotted by tracing such features whilst adjusting the height of the floating mark to ensure it remained on the surface of the model (equivalent to the ground surface). To plot contours, the floating mark was fixed at the particular contour height and moved around the stereomodel by the operator so that it appeared to remain on the model (ground) surface. The movement of the floating mark either produced a graphical plot as in Figure 16.8a or a series of (*X, Y, Z*) coordinates that could form a variety of digital products. Figure 16.8b depicts a digital photogrammetric plotter, where the reconstruction is done by computer.

## Image matching in photogrammetry

Although the derivation of topographic map detail using photogrammetry is generally much quicker and less expensive that using land surveying (except over very small areas), in the analogue and analytical eras it still required the orientation of each stereomodel and the manual plotting of every map feature by an operator. The problem that has been attempted to be solved since the 1950s is to develop an automatic method of reproducing what the human stereoplotter operator was doing. The problem boils down to identifying on the left stereo–image the same point that appears on the right photograph. If this can be achieved, then intersection from the correctly oriented photo pair will provide the ground position (and thus map position) of that point (Figure 16.5). This automated technique is termed *image matching*, and much recent research in photogrammetry and computer vision has concentrated on this area. There is now even free or low-cost software based on this technology that amateur photographers can use to automatically form 3D computer models of objects from a large number of overlapping photographs. Whilst the results in mapping projects can produce dense *point clouds* that can be used, for example, to create DEMs and othophotographs, fully automated extraction of all features to be shown on a topographic map remains elusive. Although some techniques can automatically determine features such as buildings, most still require a human operator to identify the map features.

## Digital elevation models

As their name suggests, DEMs are a computer representation of the topography of the ground surface (i.e. the shape of the ground, but without specific planimetric features being identified).

*Figure 16.9*    A digital elevation model. This DEM in Liaoning Province, China was derived completely automatically from ZY-3 satellite imagery. Different colours represent different heights. After Zhang *et al.* (2014)

They are sometimes also termed a *digital terrain model*. In modern mapping, they play a huge role in deriving certain information (for example, contours can be interpolated) and they can be used to generate orthophotographs. They can be generated automatically using image matching from aerial or satellite optical imagery, from lidar images or using interferometric SAR radar images. Figure 16.9 is a representation of a DEM where different colours represent different heights. However, there is one problem with automated methods, in that they generally produce a DSM, rather than a DEM (see, for example, the lidar image in Figure 16.2b). On bare ground, there is no difference between a DEM and a DSM, but where there is vegetation (such as trees) or man-made infrastructure (such as buildings), aerial and satellite methods generally produce the heights of the DSM, which is the top surface (such as the tops of trees and buildings), rather than the ground surface (DEM) which is generally required in cartography. Some method of removing the heights of such features therefore needs to be adopted (thus transforming the DSM to a DEM).

## Orthophotos and orthophotomaps

One current major problem in mapping from aerial and satellite imagery has already been touched upon. This is the automatic derivation of topographic map features such as roads, rivers, buildings and so on. The human eye and brain find the interpretation of such features relatively easy; deriving automated computer methods to accomplish this task is very difficult and the subject of much current research. One way of automating the mapping process is to provide images with correct map positions but to leave the interpretation to the end user. Essentially the original aerial or satellite image is replaced by a new image which is corrected, totally or in part, for the distortions in the original imagery. Figure 16.10a depicts a truly vertical aerial photograph (so no tilt) of totally flat ground. Such a photograph would have features in their exact map positions, which could be simply traced off and scaled to the relevant map scale. Taking a precisely untilted photograph is virtually impossible, but a tilted photograph can be *rectified* to an equivalent untilted one (Figure 16.10b) with ground control points and has been used in the past for map revision. However, the most realistic situation is a tilted photograph of undulating terrain with, additionally, features such as buildings and trees above the ground surface. Often a DEM is available and if the position and orientation of the photograph are known (e.g. by direct georeferencing), then a computer can automatically reproject the ray from the camera, and where this intersects the DEM provides the map position. In such a differentially rectified image (Figure 16.10c), each image pixel is geometrically transformed individually according to the DEM height, and the image is termed an *orthophotograph* or *orthophoto* (ortho meaning right or correct). One final problem relates to features above the ground surface, such as buildings. An orthophoto created from a DEM will only correct the positions of photographic features on the ground surface, and the sides of buildings will still be visible (Figure 16.2a). A DSM is required to correct for such elevated features (Figures 16.10d and 16.11) to produce a so-called *true orthophoto*.

One might well argue that 'passing the buck' of image interpretation to the user is not cartographically sound. However, annotations can be added to form an *orthophotomap*, and users of Google Maps will have appreciated the aerial or satellite view as a complement to the map view. Indeed, in many remoter parts of the world, such as the Florida Everglades, Kalahari Desert of

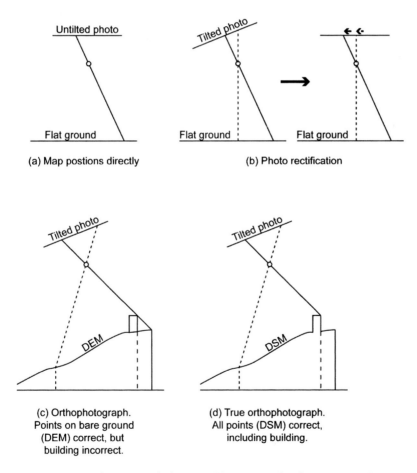

*Figure 16.10* Progressively corrected photo positions. An orthophoto uses a DEM. A true orthophoto uses a DSM

*Figure 16.11*  Digital orthophotograph in southern Germany. Unlike the original image, features are in their correct map positions and can be formed automatically from a DEM or DSM. After Zhang *et al.* (2014)

Botswana, or Antarctica, the fine textures and patterns of vegetation, sand, rock and ice that may be difficult to represent cartographically, can be readily seen on an orthophotomap to the benefit of the user.

## Thematic mapping from remote sensing

We have already seen that remote sensing is concerned with obtaining information about an object with a non-contact device. In cartography, the 'object' is invariably part of the Earth's surface, and the overwhelming applications of remote sensing relate to environmental subjects such as geography, geology, meteorology, oceanography and so on (hence the term *Earth observation* – EO). The non-contact 'device' can be any of the sensors listed above in 'Sensor systems', and they can be elevated above the Earth's surface aboard any of the platforms listed above under 'Platforms'. Thus, although the term 'remote sensing' was only coined in the early 1970s, the exploitation of (especially) photographs taken from aircraft had been used during and since the First World War for the interpretation of ground features and derivation of geographic information as well as for photogrammetric mapping. However, the real impetus for remote sensing as a distinct discipline was the launch of sensors aboard spacecraft and satellites. The practical need for the transfer of data to Earth by digital means was also associated with the rapid development of computer methods to process that data. Today, satellite remote sensing is mature and it is alternative platforms such as UAVs that have joined the forefront of development.

Although the earliest satellites were devoted to meteorology (with some concomitant cartographic applications), it was the launch of Landsat 1 in 1972 that really spurred the application of remote sensing to obtaining environmental data over the Earth's land areas. The primary characteristic of satellites that no other platform listed earlier possesses is the systematic and repetitive coverage of almost all the Earth's surface. In a 24-hour period, Landsat satellites complete 14 orbits. At the Equator, each orbit is just under 1,700 km apart (closer in higher latitudes). However, the imagery covered an area with a width (the so-called *swath width*) of 185 km. The orbits on the next day would be displaced slightly from those on the previous day and these adjacent orbits would begin to fill the gaps in the ground coverage. After 18 days, all the 'gaps' would be filled and the coverage cycle would repeat once again.

Most topographic (photogrammetric) applications of satellite imagery do not really require colour. Colour requires at least three spectral bands, and it will be seen from Table 16.1 that by using panchromatic imagery (one wide spectral band) the GSD can be improved, which (alongside a known geometry) is one of the most important characteristics for topographic photogrammetric mapping. Thus, the 80 m GSD of the early Landsats was almost useless for topographic mapping (where it would be difficult to identify a motorway, never mind a small road). This is because many topographic features are *linear* (and the resulting topographic maps tend to be composed of *vectors*). However, for most environmental applications, including those that can result in certain thematic maps, geometric accuracy and a small GSD are not crucial. If one is mapping the surface geology of an arid region, or the vegetation coverage in a tropical forest, most features cover hundreds of square metres or many square kilometres. These features are *areal* and the resulting maps are composed of *rasters* (just like the pixels that form digital images). Additionally, it is the regional coverage (resulting from the satellite's orbital height) that cannot be readily achieved from an aircraft or a UAV which is important, as well as the repetitive (*temporal*) coverage, which means that changes over time (e.g. in vegetation or climate) can be detected (see Figure 16.14). Note also that the photogrammetric requirement for stereoscopic coverage, where each piece of land appears on at least two images taken from different positions, is also not generally a high priority. What is important is to discriminate between (areal) ground features (e.g. limestone from sandstone, or forest from grasslands), with obvious implications for thematic maps of such features. Distinguishing such features on panchromatic (single band) images can be very difficult, and even if possible would probably rely on skilled and time-consuming manual methods. However, by using multispectral and hyperspectral imagery (that can be taken from satellites, aircraft or UAVs), automatic classification of land types can be undertaken.

## Multispectral and hyperspectral imagery

The early Landsat satellites (Landsat 1 (1972–1978), Landsat 2 (1975–1982) and Landsat 3 (1978–1983)) included multispectral scanner (MSS) and return beam vidicon (RBV) sensors. The RBV was designed to obtain images with known geometric properties, but malfunctions meant that the MSS became the primary sensor (although it was designed with little concern for positional accuracy). The RBV had three spectral bands (designated bands 1 to 3) and the MSS had four spectral bands (bands 4 to 7). These latter four bands were (Figure 16.1): Band 4, visible green (0.5–0.6 μm); Band 5, visible red (0.6–0.7 μm); Band 6, near-infrared (0.7–0.8 μm); and Band 7 (0.8–1.1 μm). Satellites have to orbit beyond the Earth's atmosphere to prevent atmospheric drag pulling the satellite back to the ground. As well as meaning that the *instantaneous field of view* (roughly equivalent to the GSD or pixel size) was 80 m on the ground for these bands, imaging through the entire atmosphere means that the blue wavelengths, which are scattered most by atmospheric particles (hence the blue colour of the sky) are the first to be sacrificed (hence Landsats 1–3 had no visible blue band, although Band 1 of the RBV was blue-green). Many later satellites also had, and continue to have, these multispectral scanners (thus the abbreviation MSS can also be used as a generic abbreviation, as well as for the specific Landsat system). Thus, Landsat 4 (1982–1993) and Landsat 5 (1984–2013) included both the four-band 80 m MSS of its predecessors but also a new TM sensor with seven bands (three in the visible spectrum, three in the near- or mid-infrared spectrum, and one in the thermal infrared with 120 m GSD). Landsat 6 (1993) failed before orbiting but Landsat 7 (1999–) has an eight-band Enhanced Thematic Mapper Plus ETM+ (15 m) sensor and Landsat 8 (2013–) has a nine-band Operational Land Imager (OLI) and a two-band Thermal Infrared Sensor (TIRS). Tables 16.1 and 16.2 provide details of many other remote sensing satellites, and the common characteristic is that many carry (possibly in addition to a high-resolution panchromatic sensor) a multispectral scanner. Note that if both panchromatic and multispectral images exist, the higher-resolution panchromatic image can be used to *pan-sharpen* the multispectral image, providing the benefits of both.

Figures 16.12, 16.13 and 16.14 show examples of Landsat imagery. Although each individual band will result in a black-and-white (greyscale) image, any three bands can be combined to form a *colour composite* and this is what is depicted in these images. Three of the nine bands on Landsat 8 are blue, green and red visible light, and these can be combined to form the fairly natural colour images of Mopti, Mali in Figure 16.14. However, most colour composites of early Landsat (1 to 3) imagery (which has no blue band) conventionally depict the green band in blue, the red band in green and one of the near-infrared bands in red (infrared is invisible to the human eye). These images are therefore called *false-colour* composites, as seen in Figures 16.12 and 16.13. Healthy vegetation strongly reflects the near infrared (but not dying vegetation, hence its development for camouflage detection during the Second World War) and provides the vivid irrigated red areas in Figure 16.12.

Below, the potential of multispectral scanners to derive environmental data is outlined. However, with MSS the number of bands is relatively small and width of the bands relatively wide (e.g. just four bands for Landsat 1, each covering 0.1 μm). Starting in the 1980s, scientists began to develop hyperspectral sensors which took the MSS concept to a new level, with perhaps over 200 bands, but with each component band only 0.01 μm (10 nm) wide. The main application of hyperspectral remote sensing has been to mineral exploration (it is essentially *imaging spectroscopy*), but now includes agriculture, forestry, and air or water pollution. An example is the Aqua satellite which has five different sensors with 3, 5, 12, 15 and 2,300 spectral bands; the first four are multispectral, but the last is clearly hyperspectral.

## *Image classification from multispectral and hyperspectral imagery*

Space precludes a detailed explanation of how thematic data are derived from remote sensing imagery. For more information, see introductory remote sensing texts such as Campbell and Wynne (2011) or Lillesand *et al.* (2015). Essentially, image classification is concerned with assigning the component pixels in an aerial, UAV or satellite image to classes relevant to the particular environmental subject, such as the aforementioned forest, grassland, or limestone, sandstone, and so on. There are a variety of classification methods. General groups include *point* and *neighbourhood* classifiers, or *supervised* and *unsupervised* classifications.

All pixels in a digital image have an intensity or digital number that corresponds with the amount of light or other electromagnetic radiation detected by that pixel. For example, an 8-bit image can have radiometric values from 0 to 255. No energy received by that pixel is represented by 0 and, in a panchromatic image, would normally be depicted as black; the maximum intensity (white) is represented by 255. However, most pixels will have intermediate values (e.g. 184 – light grey; 67 – dark grey). What image classification attempts to do is assign each pixel value to a land class. The difficulty is that with just one spectral band

(a)

(b)

*Figure 16.12*  The growth of Cairo, Egypt. The urban area is shown in dark grey tones either side of the River Nile (in black). (a) 31 August 1972 image from Landsat 1; (b) 3 August 2013 image from Landsat 8. These are false-colour composites where Landsat's multispectral green band is depicted in blue, the red band in green and the near-infrared band in red. Healthy vegetation has a strong infrared reflectance and irrigated areas are thus shown in red. The expansion of the (grey) urban area between 1972 and 2013 is readily apparent, as are new irrigated zones. Courtesy NASA

(a)

(b)

(c)

*Figure 16.13*   Escondida Mine, Atacama Desert, Chile. This open-cast copper mine is the largest in the world. Its expansion is apparent in these three colour composite images. (a) Landsat 1 image on 30 October 1972; (b) Landsat 7 image on 10 November 2000; (c) Landsat 8 image on 12 November 2015

(thus a panchromatic image), a pixel with a value of, say, 93 may represent both tarmac and water. Additionally, different illumination and other differences may mean that same piece of tarmac may register a value of 78 on a second panchromatic image or a 108 on a third. Thus, both manual interpretation and image matching in photogrammetry rely on other characteristics, such as texture, linearity across multiple pixels, or detection of edges (sharp changes in adjacent pixel values). In remote sensing, the various multispectral (and hyperspectral) images can be used to great advantage. Thus, although a particular area of tarmac and water may both have pixel values of 93 in one band, in another band the tarmac may register 82 but the water only 14 (near-infrared radiation, for instance, is absorbed by water to provide very limited reflectance back to the sensor and, incidentally, is useful for coastal mapping of high and low water marks). So, whatever the method of classification, it is the computer comparison of the same pixels in various spectral bands that will lead to a successful image classification and resultant derivation of thematic information. The limited number of spectral bands in multispectral imagery can lead only to broad categorization (such as forest or grassland, not individual tree species) whereas hyperspectral imagery can provide greater differentiation, for example as needed for the identification of different surface minerals.

An unsupervised classification is performed automatically, using 'natural' classes, without any prior knowledge of the area being mapped and with little opportunity for human error.

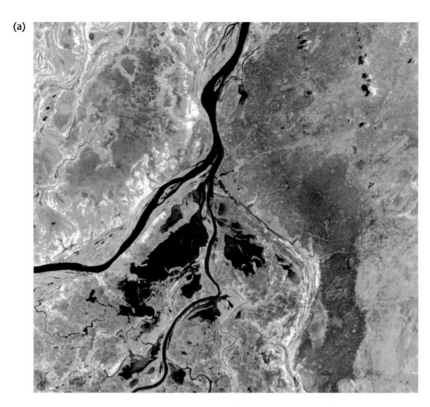

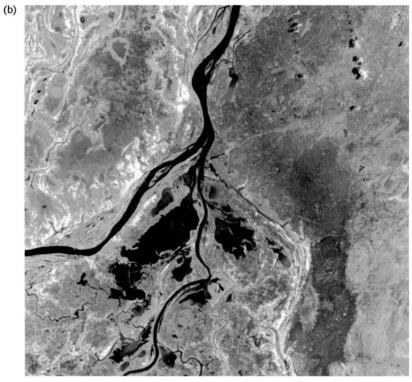

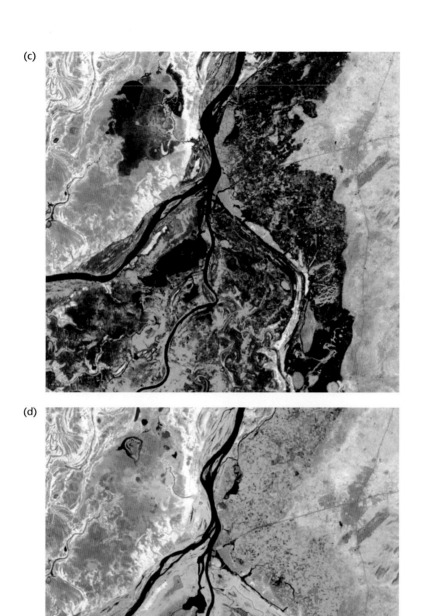

*Figure 16.14* Mopti, Mali from Landsat 8. Area around the confluence of the Bani and Niger rivers (dark blue). Seasonal waxing and waning of vegetation is apparent in these four Landsat 8 colour composite images from 2014. (a) 6 June; (b) 25 August; (c) 28 October; (d) 31 December. Courtesy NASA

(a)

However, it identifies homogeneous classes that may not correspond well with the categories that are required to be mapped. Supervised classification, on the other hand, uses *training areas* where the ground type of the component pixels is known and extends this to much larger (unknown) areas. Therefore, the classes to be mapped are predetermined, but the determination of training areas is expensive and time-consuming and may not be representative of the entire area.

Ideally, the various spectral bands used in image classification would show very different responses, for a given ground feature, from one another. In practice, bands may be highly correlated. Campbell and Wynne (2011) provide an example from Landsat TM where the three visible bands have correlations over 0.9 with one another, whereas their correlation with the thermal-infrared band is under 0.1. *Principal components analysis* is a remote sensing method that derives a series of 'replacement' images, the first showing the greatest differences (low correlations) and the remainder decreasing differences (higher correlations), which can be used as a classification tool. A huge variety of classification techniques exist, each depending on different statistical analyses from simple methods such as *minimum distance classification* (assigning an unknown pixel to the class that is 'closest', in terms of spectral characteristics, in a training area) to *artificial neural networks* (computer programs that simulate human learning through linkages). Figure 16.15 shows an example of land-use classification from remote sensing imagery.

(b)

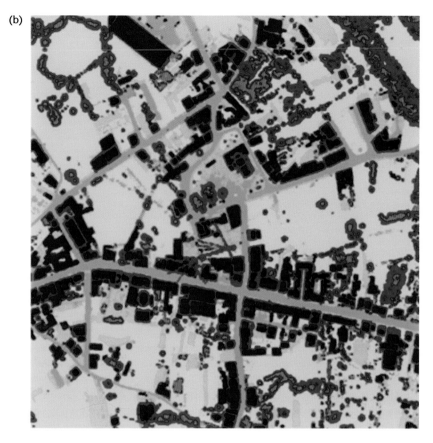

*Figure 16.15*　Image classification. Comparison of (a) an orthophotograph and (b) the resultant land classification. Courtesy S. Bujan

## Cartographic applications of remote sensing imagery

Thematic maps that depict environmental data, such as geological, vegetation and meteorological maps, cover a huge variety of scales. We have already noted that, in general, such applications do not require the very high spatial resolution (small GSD) that photogrammetric topographic mapping demands. In addition, such thematic data may be readily transferred to a topographic or other base map, so that geometric accuracy is not normally an issue. However, a large-scale geological or land-use map derived (at least partially) over part of a country from, say, current 6 m SPOT-6 or -7 MSS imagery is a very different proposition to a vegetation map covering the whole globe from 1.1 km AVHRR imagery. Certain geological maps can be determined using *photogeology* techniques by using stereoscopic aerial and satellite imagery (e.g. detecting lineaments such as fault lines), through *geobotany* (where the underlying geology influences the vegetation seen on remote sensing images) or, in arid areas, the direct mapping of rocks and soil using multispectral classification techniques. Figure 16.16 depicts areas of the globe, primarily the southern hemisphere, where carbon stored in plants has increased during the month of December 1999 (the height of the southern summer).

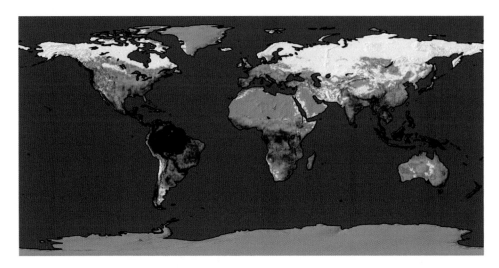

*Figure 16.16*   Changes in stored carbon. Data derived from remote sensing satellites showing increases in net carbon capture by vegetation in December 1999. Green areas represent an increase in carbon stored in living biomass. The increase in summer activity in the southern hemisphere its readily apparent

## References

Campbell, J.B. and Wynne, R.H. (2011) *Introduction to Remote Sensing* (5th ed.) New York: Guilford Press.

Lillesand, T.M., Kiefer, R.W. and Chipman, J. (2015) *Remote Sensing and Image Interpretation* (7th ed.) New York: John Wiley & Sons.

McGlone, J.C. (Ed.) (2013) *Manual of Photogrammetry* (6th ed.) Bethesda, MD: American Society of Photogrammetry and Remote Sensing (ASPRS).

Mikhail, E.M., Bethel, J.S. and McGlone, J.C. (2001) *Introduction to Modern Photogrammetry* New York: Wiley.

Renslow, M. (Ed.) (2012) *Airborne Topographic Lidar Manual* Bethesda, MD: American Society of Photogrammetry and Remote Sensing (ASPRS).

Sefercik, U.G. and Yastikli, N. (2014) "Assessment of Interferometric DEMs from TerraSAR-X Stripmap and Spotlight Stereopairs: Case Study in Istanbul" *Photogrammetric Record* 29 (146) pp.224–240.

Vosselman, G. and Mass, H.-G. (2010) *Airborne and Terrestrial Laser Scanning* Caithness, UK: Whittles Publishing.

Wolf, P.R., Dewitt, B.A. and Wilkinson, B.E. (2014) *Elements of Photogrammetry with Application in GIS* (4th ed.) New York: McGraw-Hill.

Zhang, Y., Wang, B., Zhang, Z., Duan, Y., Zhang, Y., Sun, M. and Ji, S. (2014) "Fully Automatic Generation of Geoinformation Products with Chinese ZY-3 Satellite Imagery" *Photogrammetric Record* 29 (148) pp.383–401.

# 17

# Geographical information systems

*Paul A. Longley and James A. Cheshire*

Other chapters in this book have addressed the role of cartography in helping us to address the importance of location in many of the problems society must solve. Some problems involving location are so routine that we almost fail to notice them under ordinary circumstances, such as the daily question of which route to take to and from work. Others, such as earthquakes or security emergencies, are quite extraordinary occurrences in which many attributes of locations must be known and effectively communicated in order to formulate and implement an effective response. All involve aspects of location, either in the information used to solve them, or in the solutions themselves.

We can think of three bases for classifying geographic problems. First, there is the question of *scale*, or level of geographic detail – with geography conventionally defined as concerned with problems at geographic scales ranging in extent from the architectural (e.g. the distribution of buildings on a university campus) to the global (e.g. the manifestations of global warming in terms of changed vegetation cover). Second, geographic problems can be distinguished on the basis of motivation, or *purpose*. Some problems are strictly practical in nature – they must often be solved as quickly as possible and/or at minimum cost, in order to achieve practical objectives such as saving money, or coping with an emergency. Others are better characterized as driven by human curiosity, such as understanding the effects of glaciation upon Earth surface landforms, where there is no urgent problem that needs to be solved, but rather a wider quest for improved understanding of the way that the world looks and/or works. However, such differences in motivation should not be taken to imply that different tools and methods are necessary to resolve them, however. Third, geographic problems can be distinguished on the basis of their *time scale*. Some decisions are operational and are required for the smooth functioning of an organization, such as how to regulate the spacing of trains on a subway network. Others are tactical and concerned with medium-term decisions, such as how to use advertising hoardings to publicize the opening of a new retail store. Still others are strategic, such as how to develop effective flood defenses in anticipation of rises in sea level.

In each of these instances, location is central to problem solving and geographic *information* is likely to be required about multiple attributes of locations – and often more attributes than can be conveniently or clearly indicated on a paper map. The term *information* can be defined in more or less specific terms. Most generally, information can be treated as devoid of meaning, and therefore as essentially synonymous with data – that is, as raw numbers or text. More specifically, information can be differentiated from data in its degree of selection, organization, or preparation for a particular purpose – most conventional maps display information that is the outcome of some degree of interpretation. As national mapping organizations will testify, information is often costly to collect and assemble, but once digitized it is cheap to reproduce and distribute (e.g. through web mapping). Moreover, different information sources may be used in combination with each other to add still further value, and hence add still further value to cartographic outputs.

## Geographic information (GI) science and systems

GI systems are computer-based systems for storing and processing geographic information about sets of locations. We can think of each atom of GI as comprising three $(x, y, z)$ elements – a coordinate $(x, y)$ pair and a location attribute $(z)$. GI systems are a collection of tools that improve the efficiency and effectiveness of handling information about geographic objects and events. They can be used to carry out many useful tasks including storing vast amounts of GI in databases, conducting analytical operations in a fraction of the time it would take to do it by hand, and automating the process of making useful maps. GI systems process information, but there are limits to the kinds of procedures and practices that can be automated when turning data into information. However, it is important to understand whether this selectivity and preparation for purpose actually adds any value, or whether the results are of sufficient value in geographic applications. Such issues are the realm of GI science, a rapidly developing field concerned with the scientific underpinnings of GI systems. GI science thus provides a framework for learning from the accumulated experience of using GI systems.

GI systems can thus be thought of as providing the technology for problem solving. Longley *et al.* (2015) discuss how the label 'GIS' is used to describe many things, including: a collection of software tools (sometimes bought from a commercial company) to carry out certain well-defined functions; digital representations of various aspects of the geographic world, in the form of datasets; a community of people who use and perhaps advocate the use of these tools for various purposes; and the activity of using a GI system to solve problems or advance science. In all of these senses, the history of GI systems is intimately bound up with that of computer cartography, because a GI system is a computerized tool that can be used as a container of maps in digitized form. It provides an environment for creating and maintaining inventories of geographically distributed features and facilities, which may share the same geographic location.

## GI systems and the origins of modern cartography

As described by Longley *et al.* (2015), there is general consensus that the term 'GI system' was first coined by Roger Tomlinson in the mid-1960s when his Canada Geographic Information System or CGIS was used by the federal and provincial governments to inventory the nation's land resources and their existing and potential uses. At around the same time, the US Bureau of the Census's DIME programme (Dual Independent Map Encoding) created digital records of all US streets, to support automatic referencing and aggregation of 1970 Census records. Recognition of the similarity of the technologies used in these applications led to a major programme at Harvard University's Laboratory for Computer Graphics

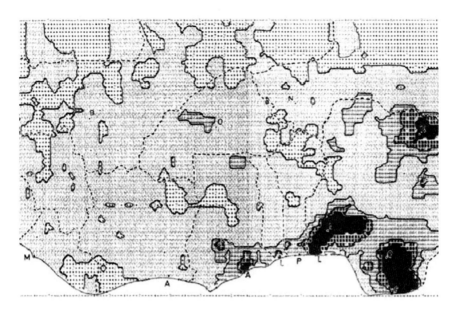

*Figure 17.1*  First generation digital cartography; a lineprinter map by John Adams

and Spatial Analysis and the creation of SYMAP, the first raster system for automated cartography, in 1969 (see Figure 17.1). This in turn led to the development of a general-purpose GI system, the ODYSSEY GIS software of the late 1970s, and the establishment of Environmental Systems Research Institute (Esri) Inc. in 1969.

In the latter half of the 1960s, cartographers and mapping agencies also began to use computers to reduce the costs and shorten the time of map creation. The UK Experimental Cartography Unit (ECU) pioneered high-quality computer mapping in 1968, publishing the world's first computer-made map in a regular series in 1973 (with the British Geological Survey). The ECU also pioneered GI system work in education, post and ZIP codes as geographic references, visual perception of maps, and much else. National mapping agencies, such as the UK's Ordnance Survey, France's Institut Géographique National, and the US Geological Survey and Defense Mapping Agency (now the National Geospatial-Intelligence Agency) began to investigate the use of computers to support the editing of maps, to avoid the expensive and slow process of hand correction and redrafting. The first automated cartography developments occurred in the 1960s, and by the late 1970s most major cartographic agencies were already computerized to some degree. But the magnitude of the task ensured that it was not until 1995 that the first country (the UK) achieved complete digital map coverage in a database. At the same time, military requirements (see Chapter 13) led to the initial development of the Global Positioning System (GPS: Chapter 18). The modern history of GI systems dates from the early 1980s, when the price of sufficiently powerful computers fell below a critical threshold. The history of GI systems and their relationship with modern cartography is inevitably more complex than this very brief review suggests, and the reader is referred to Longley *et al* (2015: chapter 1, especially table 1.4) for some further details.

The use of the World Wide Web to provide access to maps dates from 1993. Web mapping underpins GI system applications, disseminates GI and enables participatory GI systems based around volunteered GI (Goodchild 2007). Internet technology became increasingly portable in the early 2000s, which meant that locationally enabled devices could be used in conjunction with publically available or subscription wireless networks and broadband. This has led to a growth in real-time geographic services such as mapping, routing, and location-based advertising. First generation applications provided only unidirectional flows of data and information from websites to their user bases. Over time, this system has evolved into services that facilitate bidirectional collaboration between users and sites, the outcome of which is that information is collated and made available to others. The two main technologies that stimulated this development were Asynchronous Javascript And XML (AJAX) and Application Programming Interfaces (API). AJAX enabled the development of websites that retain the look and feel of desktop applications. They have improved the usability of web mapping significantly by enabling direct manipulation of map data where user interactions (such as 'click and drag') are visualized instantaneously. Today, the API of web mapping sites such as Google Maps provide basic GI system operations, such as the ability to draw shapes, place points, geocode locations, and display these on top of high-resolution base-map or satellite data.

GI systems are central to the academic discipline of geography, fundamentally because of their facility to bring together representations of what is unique about places on the Earth's surface (the *idiographic* tradition of the subject) with understanding of general physical and social processes that are extant upon them (the *nomothetic* tradition). Recent years have seen the popularization of the term *neogeography* to reflect the widening interest in geographical data beyond 'traditional' geographers and cartographers into domains such as computer science. Neogeographers – and by extension neocartographers (see Chapter 20) – have pioneered developments in web mapping technology and spatial data infrastructures that have greatly enhanced our abilities to assemble, share, and interact with GI online. Allied to this is the increased 'crowd sourcing' by online communities of *volunteered geographic information* (VGI) (see Chapter 35). Neogeography is founded upon the two-way, many-to-many interactions between users and websites that has emerged under Web 2.0, as embodied in projects such as Wikimapia (*www.wikimapia.org*) and OpenStreetMap (*www.openstreetmap. org*). Today, Wikimapia contains user-generated entries for more places than are available in any official list of place names, while OpenStreetMap is well on the way to creating a free-to-use global map database through assimilation of digitized satellite photographs with GPS tracks supplied by volunteers. This has converted many new users to the benefits of creating, sharing, and using GI, often through ad hoc collectives and interest groups. As such, bidirectional interactions between users and web services simultaneously facilitates crowd sourcing of VGI while making basic GI system functions increasingly accessible to an ever-broader community of users. The creation, maintenance, and distribution of databases is no less than a 'wikification of GI systems'. Neogeography has brought GI systems and some use of spatial data 'infrastructures' to the masses. The empowerment of many non-expert users also brings new challenges of ensuring that tools are used efficiently, effectively, and safely, and re-emphasizes that there is no purely technological solution to the effective deployment of GI systems. Where applications are successful in technical terms, they may also raise issues of privacy and confidentiality, not least where diverse data sources are brought together in ways that were not envisaged when they were created.

There is no doubt that these are positive developments that have, in part, been responsible for the huge interest that maps and mapping have enjoyed in recent years. They have, however, generated a new set of challenges for GI systems and the maps that depend on them. The largest of these relates to the quality, provenance, and cataloguing of the data collected, stored, and utilized by volunteers. In the previous era of information generation by a few large organizations,

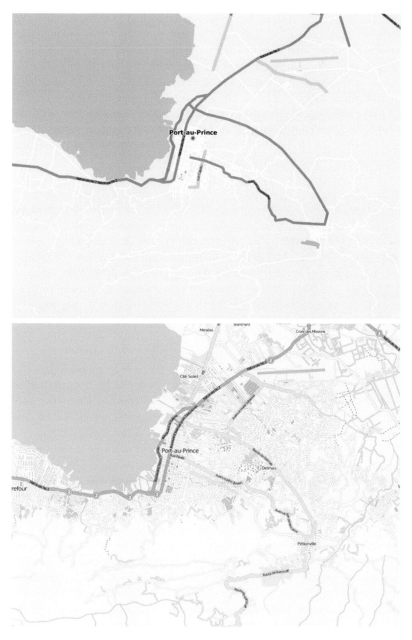

*Figure 17.2* **Map of Port-au-Prince before (top) and after (bottom) the 2010 Haiti earthquake. The bulk of the new data were digitized from satellite imagery by volunteers (Kent, 2010)**

such as Ordnance Survey, all data were structured in the same way and went through a rigorous checking process prior to dissemination. For example, strict rules exist for what constitutes a particular land use or road classification, and all features would be topologically correct. The bottom-up maturation of initiatives such as OpenStreetMap has led to relatively few rules related to such things – leading to inconsistent categorization of feature types or roads that do not connect. The costs of centralized data collection are highest in trying to resolve such issues since they require a large amount of staff resource, both in terms of time and training. It would be unsustainable to expect volunteers to achieve the rigour of Ordnance Survey, but for a large number of applications it is not needed and it comes with two key advantages: zero financial cost – although extra time is often required to clean the data for a specific application – and the expediency with which data are updated. The most powerful example of this was the 2010 Haiti earthquake where hundreds of volunteers were able to create a detailed map of the devastation in a matter of hours to be used in relief efforts (see Figure 17.2 and Kent, 2010). The success of this initiative launched the Humanitarian OpenStreetMap (HOTOSM) movement that has had a major impact in many serious natural disasters since.

GI systems have therefore had to adapt to a new era of large volumes of rapidly generated and less precise data. Leading such developments are the open source GIS platforms that, arguably, are symbiotically linked to the VGI communities – more effective software leads to greater data generation and utilization. QGIS, for example, has a series of dedicated toolkits for the creation and utilization of OpenStreetMap data. ArcGIS too enables users to pull in OpenStreetMap data via its dedicated APIs. In addition to the increased detail and diversity of base mapping, GI systems have had to rapidly develop to reflect the huge volumes of attribute information now generated. Maps are extremely effective at creating a clear picture from a seemingly messy and complex world, and so have become essential to our understanding of 'big data'. GI systems now have to be tightly coupled with powerful database infrastructures, such as PostgreSQL and Oracle Spatial, in order to process and overlay data generated by everything from social media through to real-time transport reporting systems.

These developments are increasingly combined to create highly advanced web mapping interfaces that can be used on a variety of platforms. These can be characterized as a move away from serving image tiles to users towards the use of vector tiles where the web browser effectively performs the GIS operations previously undertaken on servers 'in the cloud'. Such tiles have the advantage of being generated 'on the fly', which facilitates the inclusion of real-time data or rapid updates. This is particularly powerful in the context of mobile phone technology where user locations can be used to feedback information, which in turn is shown on the map. Google Maps' real-time traffic maps and routing is one such example of this. Cartography, like the analytical approaches of GI science, has therefore become increasingly dynamic to adequately represent these new forms of data.

Such developments have led to an increasing emphasis on the need for GIS practitioners and cartographers to acquire the necessary programming skills required to process and analyse large datasets as well as to automate as much of the map production process as possible. Recent years have seen a move away from the graphical user interface and back to the command line as software has sought to keep pace with the advances in data collection fuelled by technological innovation. This has not diminished the importance of 'traditional' maps, such as those from Census data; instead, it is responding to an increasing demand for innovative cartographic representations of new data to be served to a large number of non-specialist users. For example, the UK-based Consumer Data Research Centre has developed its own mapping platform (Figure 17.3) that seeks to address this need. Users can access maps generated from millions of data points depicting a range of data from deprivation through to Internet usage and all have links to the raw data for use in their own analysis. This has proved to be very useful to analysts in local and national government, the commercial, and third sector who lack the budget and skills to produce their own maps from complex data, but who appreciate the value of cartographic representations for providing insight.

*Figure 17.3*   A screen shot from the maps.cdrc.ac.uk website, which has pioneered the mapping of large and complex datasets relating to consumer behaviour

## Concluding comments

The acquisition, creation, maintenance, dissemination, and sale of GI accounts for a large volume of economic activity. Traditional data providers are national mapping agencies such as the UK's Ordnance Survey. In most countries, the funds needed to support national mapping come from sales of products to customers – with the notable exception of the US where federal government policy requires that prices be set at no more than the cost of reproduction of US Geological Survey data. The innovation of free-to-view mapping services, such as Google Maps (maps.google.com) and Microsoft Virtual Earth, with its business model based upon advertising revenues, has had profound implications for the provision of map data. Such mapping services provide mapping, geocoding, and routing services that are used every day by tens of millions of people, and that often provide a precise fit between individuals and offerings of goods and services – not least because location underpins many aspects of consumption behaviour.

Volunteer-driven, open source approaches to online cartography such as OpenStreetMap (*www.openstreetmap.org*) have also revolutionized online cartography with their novel approach. More generally still, the Open Data movements in several countries have spawned new applications, often associated with the open software products described in the previous section. Together, new data and software products have spawned an industry based around the collection and assembly of ancillary data sources, with the creation of geographically enabled Apps for mobile devices a prominent area of activity. The provision of Apps for small, hand-held devices presents a range of important issues for digital cartography, many of which are discussed elsewhere in this volume.

It is clear that cartographers are at the same time facilitators and end users of software that organize, store, analyse, and represent GI. They are therefore drivers of, and subject to, the developments in GI science that continue to redefine geographic concepts and their use in the context of GI systems in the manner outlined in Goodchild's (1992) seminal paper. There are ongoing and shared challenges for both fields, such as coping with new technological innovations and their associated large data volumes. But the field of GI science also brings focus to society's use of GI systems (and, as geographer David Mark has pointed out) the ways in which GI systems are used to represent society. This is particularly prescient to cartography and, in many respects, is where maps become powerful advocates for the value of geographic representation.

## References

Goodchild, M.F. (1992) "Geographical Information Science" *International Journal of Geographical Information Systems* 6 (1) pp.31–45.

Goodchild, M.F. (2007) "Citizens As Seniors: The World of Volunteered Geography" *GeoJournal* 69 (4) pp.211–221.

Kent, A.J. (2010) "Helping Haiti: Some Reflections on Contributing to a Global Disaster Relief Effort" *The Bulletin of the Society of Cartographers* 44 (1,2) pp.39–45.

Longley, P.A., Goodchild, M.F., Maguire, D.J. and Rhind, D.W. (2015) *Geographic Information Systems and Science* (4th ed.) Chichester, UK: John Wiley & Sons.

# 18

# Global positioning systems

*Martin Davis*

Of the plethora of technologies developed during the Cold War, few would proceed to have such a substantial and lasting effect on cartography as the Global Positioning System (GPS). The United States Department of Defense's (DoD) 24-satellite NAVSTAR GPS (NAVigation System with Time and Ranging Global Positioning System), which facilitates the determination of latitude, longitude and altitude at any point on or above Earth at any time of day, began as a solely military venture, though the subsequent willingness of the US Government to make NAVSTAR data available for civilian applications has profoundly impacted both the collection and visualization of spatial data (Dorling and Fairbairn, 1997). The recent development of satellite navigation systems by the Russian Federation (GLONASS), the European Space Agency (Galileo) and China (Beidou) indicates that the technology remains a desirable and necessary part of numerous military and civil applications around the world. Indeed, the acronym 'GPS' commonly refers to any satellite positioning system, although the more generic term is properly 'Global Navigation Satellite System' or (GNSS). This chapter gives a brief outline of how cartographers have used this technology since its inception, the development and organization of the system and the fundamental principles behind its operation. Similar systems, in operation or development, are then outlined.

## GPS and cartography

The availability of GPS has wholly transformed the processes of surveying and mapping in recent years and the technology is now an indispensable tool for professional and amateur cartographers alike. The ever-increasing availability, affordability and portability of GPS receivers has only increased applications of GPS in cartography (Xiao and Zhang, 2002). Between 23 March 1990 and 2 May 2000, with the exception of an 11-month period during the first Gulf war, GPS signals were systematically degraded for civilian users; a policy known as selective availability (SA). This resulted in a two-stream framework within the GPS – a Precise Positioning Service (PPS) being made available for a selected group of authorized (mainly military) users and a degraded Standard Positioning Service (SPS) for civilian users (Sleewaegen, 1999; Xiao and Zhang, 2002; Seeber, 2003). As part of an effort to broaden the civil and commercial applications of a number of technologies developed by the US Government, President Bill Clinton announced the removal of SA, predicting that this action would 'allow new GPS applications to emerge and continue to enhance the lives of people around the world' (Clinton, 2000). Mapping was one such application; the near ten-fold increase in accuracy overnight made GPS considerably more attractive to surveyors and cartographers by providing a new source of medium to large scale data for use in Geographical Information Systems (GIS) (Xiao and Zhang, 2002). SPS and PPS still exist as distinct entities, differentiated by the fact that SPS broadcasts on one frequency, whereas PPS uses two (NCO, 2015).

Military and civilian navigation by land, sea and air undoubtedly constitute the largest application of GPS, and indeed is the system's *raison d'être*. However, in addition to displaying a real-time position on a base map, most commercial GPS receivers now facilitate the recording of point, line and polygon data, which can be converted to most major vector formats. In 2002, Ordnance Survey Director General Vanessa Lawrence acknowledged that GPS had become a vital technology in the creation of Ordnance Survey products, stating that the UK's national mapping agency had 'invested heavily in specialist GPS equipment that is fundamental to [producing] data of the highest precision' (Lawrence, 2002).

Since the removal of SA, GPS has also changed the medium through which many consumers of map products receive geographical data. In-car and smartphone-based satellite navigation has largely replaced the market for printed road maps, with studies indicating that, for some user groups, 'sat-navs' are now by far the preferred means of navigation (Speake, 2015). Research conducted by Ochieng and Sauer (2002) accounts for this surge in popularity, concluding that, since the removal of SA, GPS data in urban road navigation applications reach a 20-metre accuracy level 99 per cent of the time, compared with only 44 per cent of the time prior to the removal of SA. Recreational use of GPS (e.g. walking and cycling) has also grown since the turn of the century, aided by the portability of modern GPS receivers.

Post-SA GPS has also enabled a significant democratization of mapping, notably including the growth of community mapping and online-based open source map platforms, such as OpenStreetMap (OSM), established in 2006. Perkins (2007) highlighted that, for the first time, such initiatives had enabled the long-established paradigm of 'top-down' mapping to be challenged. No longer was cartography exclusively reserved for professional cartographers, as it broadly had been for centuries, but anyone with a GPS device and computer could now collect spatial data and incorporate it into a map, individually or collaboratively, anywhere on Earth. This democratization of mapping has led to a new wave of map types emerging, separate from traditional mapping. Maps, often together with GPS, are being increasingly used in innovative applications as diverse as contemporary artistic projects (Schulz, 2001), travel patterns and spatial mobility (Duncan and Regan, 2015), animal tracking and habitat mapping (Hulbert and French, 2001), offshore survey (NCO, 2015), and infrastructure maintenance (Kumar and Moore, 2002).

## NAVSTAR GPS: principles of operation, structure and organization

The development of the NAVSTAR GPS programme, more simply referred to as GPS, began in 1973, and the system has been fully operational in its current form since 1995. By calculating 'pseudoranges' between a ground receiver and at least four satellites, the exact locations of which are known at all times, the position of the ground receiver can be determined geometrically almost instantaneously. This is achieved by timing how long signals from the satellite take to reach the ground receiver (the difference between the times of transmission and reception). This process is accurate to within approximately 10–15 m in ideal conditions (Seeber, 2003). GPS consists of three segments, each with vital functions. The space and control segments are largely operated and maintained by the United States Air Force (USAF), whereas the user segment now encompasses a broad range of users, no longer confined to the USA.

## Space segment

The space segment of the GPS currently consists of 31 satellites, known as the 'baseline constellation' (NCO, 2015). These orbit the Earth at an inclination of 55° along one of six equally spaced orbital planes (A–F) (Seeber, 2003). Of the 31 satellites, at least 24 are operational 95 per cent of the time, with the 'spare' satellites kept available for use during maintenance, or in the event of a fault (NCO, 2015). In addition, decommissioned NAVSTAR satellites are kept in orbit in case a need to reactivate them emerges (NCO, 2015). GPS satellites orbit the Earth at an altitude of 20,200 km and have a repeat phase of 12 hours (Seeber, 2003; NCO, 2015). The number and configuration of the satellites and their orbits has been determined to ensure that at least four satellites are 'visible' at all times from all points on Earth, ensuring continuous global coverage. Given the importance of accurate timing to the calculation of pseudoranges, perhaps the most important, and costly, elements of GPS satellites are their highly precise, synchronized clocks. GPS satellites also contain several spare clocks for use in the event of the main clock malfunctioning.

## Control segment

The control segment of the GPS consists of a series of ground facilities which control and track the satellites, while monitoring the data being transmitted from them. In order to perform these functions across the entire constellation of satellites, the control segment comprises 12 command and control antennae and 15 monitoring stations distributed across the globe (Figure 18.1); coordinated by a Master Control Station (MCS) at Schriever USAF Base, near Colorado Springs, USA (which also functions as a further monitoring station) (NCO, 2015). Monitoring stations are responsible for observing the transmissions and orbits of satellites passing overhead. Pseudoranges between monitoring stations and satellites, together with meteorological data, are transmitted to the MCS in real time (Seeber, 2003). Originally, a total of six monitoring stations were used, all operated by the USAF (three in the USA – Hawaii, Cape Canaveral and Schriever MCS – in addition to Ascension Island, Diego Garcia and Kwajalein). In 2008, as part of the Legacy Accuracy Improvement Initiative (L-AII),

*Figure 18.1* The control segment (adapted from NCO, 2015)

ten further monitoring stations, operated by the National Geospatial-Intelligence Agency (NGA), were brought into use (in Alaska and Washington, DC, Argentina, Australia, Bahrain, Ecuador, New Zealand, South Africa, South Korea and the UK). Vandenberg USAF Base in California is equipped to function as an alternate MCS if necessary (NCO, 2015).

## User segment

Although applications of GPS have been dealt with in more detail elsewhere, it should not be overlooked that the user segment is the largest of the three components of the GPS, both in terms of the number of people involved and the variety of applications of GPS data, especially since the removal of SA. As GPS receivers are incorporated into an ever-increasing variety of commercial devices (e.g. smartphones, smart watches (see Figure 18.2) and Unmanned Aerial Vehicles (UAVs)), the growth of the user segment is likely to continue. The falling price of receivers accounts for much of this growth (Lechner and Baumann, 2000). Early receivers, produced during the mid-1980s, cost in excess of US$100,000. However, prices fell swiftly as mass-production of receivers began; the first receiver costing under US$1,000 appeared on the market in 1992 and the $100 milestone was reached in 1997 (Kumar and Moore, 2002). Entry-level commercial receivers, including in-car navigation units, are now available at less than US$50.

*Figure 18.2*  A 2015 *Apple* watch, which is GPS-enabled when connected to a compatible smartphone (*Macworld*, May 2015)

## Major sources of error

Inherent in several elements of GPS data collection is scope for varying degrees and types of error to be introduced. As Seeber (2003) highlights, the GPS is reliant on modelling – most notably the modelling of a reference system for the Earth (WGS 84) and modelling the behaviour of transmissions between space-borne satellites and receivers on the ground. Any inaccuracies or uncompensated fluctuations in these physical frameworks will lead to the inaccuracy of such models, which will in turn inevitably lead to the introduction of errors in GPS data.

The first of these error sources is the Earth's atmosphere. The degradation and delay of signals as they pass through the ionosphere and troposphere can interfere with the resulting pseudoranges which need to be highly accurate in order to provide accurate position data (Jin, 2004). The second is Dilution of Precision (DOP) (see Figure 18.3); a phenomenon through which the precision of data is reduced due to the geometric arrangement of visible satellites from a receiver at a given time. DOP is reduced when there is a greater angle between visible satellites in the sky (Dussault et al., 2001). Different indexes exist to measure the effect of DOP on different aspects of accuracy; Horizontal Dilution of Precision (HDOP), Vertical Dilution of Precision (VDOP), Position Dilution of Precision (PDOP) (a combination of HDOP and VDOP) and Time Dilution of Precision (TDOP) (Seeber, 2003). Geometric Dilution of Precision (GDOP) incorporates both satellite location and time.

Given the importance of accurate timing to the measurement of pseudoranges, clock errors are another potential source of error. Lack of synchronization between the satellite clocks and receivers, even by the smallest margins, can lead to substantial errors in positioning. According to Seeber (2003), an error of 1µs in a satellite clock will result in a pseudorange error of 300 m. It for this reason that the close observation of 'GPS time' in the control segment is vital.

Although a lack of visible satellites is perhaps the most fundamental cause of poor accuracy of GPS data, the reflection or delay of signals by objects in close proximity to the receiver (e.g. trees, buildings, terrain) can disrupt pseudorange measurement, and therefore overall data accuracy, regardless of the number of satellites visible.

## Improving accuracy

Differential GPS (DGPS) refers to a means of improving the accuracy of data by correcting or compensating for it based on measurable discrepancies between the true location of a receiver and the location of the same receiver as stated by the GPS. This is achieved by measuring the error in the GPS position of a receiver at a reference station, the true position of which is precisely and accurately known. The positions of other nearby GPS receivers can then be corrected based on this known error, often providing accuracy to within 1 m (Kumar and Moore, 2002).

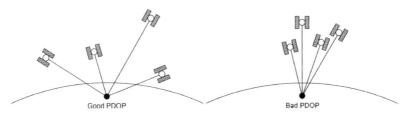

Good PDOP    Bad PDOP

*Figure 18.3*   **Example graphical representations of good and bad Position Dilution of Precision (PDOP) (adapted from Seeber, 2003)**

The increase in accuracy provided by DGPS makes it a more viable means of positioning in some applications where precision is vital (Lechner and Baumann, 2000).

In some parts of the world, commercial DGPS reference stations have been established, giving users the opportunity to obtain correction data. Examples of such stations can be found in Germany (Satellite Positioning Service of the German Topographical Survey Administration [SAPOS]), The Netherlands (Eurofix, with coverage across Europe) and the UK (General Lighthouse Authority) (Lechner and Baumann, 2000).

A Satellite-Based Augmentation System (SBAS) is a means of improving GPS accuracy within a large region by incorporating DGPS data from multiple reference stations with clock error and ionospheric corrections. The corrected data is then transmitted to receivers across the region via a geostationary satellite. The SBAS for the USA is known as the Wide Area Augmentation System (WAAS) and was primarily developed to improve GPS accuracy in the aviation industry (Lechner and Baumann, 2000). Similar systems are in operation elsewhere in the world, notably in Europe (European Geostationary Navigation Overlay Service – EGNOS) and Japan (Multi-Functional Satellite Augmentation System – MSAS).

## The development of GPS

The technology which would become vital to the modern GPS perhaps has its origins in 1948 when the US Army Signal Corps successfully transmitted radar waves to the moon and detected their deflection from the Earth's surface – indicating for the first time that microwave transmission in space was possible (Kumar and Moore, 2002). The first artificial satellite, *Sputnik 1*, was launched on 4 October 1957 by the USSR, prompting interest in space technology in the US and the start of the 'space race'. Scientists studying the satellite at Johns Hopkins University, Baltimore, Maryland, observed a variance in the frequency of the 'beeps' transmitted by *Sputnik 1* as it orbited overhead. They reasoned that if the orbit of the satellite was known, as well as the exact degree to which the signal frequency varied during each pass, the position of the receiver on Earth could be determined (Ramsey, 1984).

It was not until 31 January 1958 that the first US satellite, *Explorer 1*, was launched. *Explorer 1* collected pioneering data regarding conditions outside of the Earth's atmosphere, although it was not used for terrestrial positioning. The first system used for this purpose was *TRANSIT*, launched in 1964. Used to aid the navigation of US submarines, this primitive system had several major limitations; given that it only consisted of a single satellite, position data could only be obtained every 35–40 minutes. Furthermore, the submarine needed to be stationary in order to obtain this information as there was no way of determining how long a transmission between the satellite and submarine had taken and it was therefore not possible to accurately compensate for any movement. This issue was rectified in 1967 with the launch of the *TIMATION I* satellite. *TIMATION I* had an atomic clock on board, making it possible to determine the length of time signals had taken to reach the receiver on Earth. Nevertheless, due to the continuing dependence on one satellite, positional accuracy remained low (Kumar and Moore, 2002).

The development of the *TRANSIT* system continued throughout the 1970s, with multiple satellites becoming operational before the end of the programme in the late 1980s. Use of *TRANSIT* data was not restricted to military users. Of the 10,000 receivers active in 1984, approximately 90 per cent were operated by civilian users, notably in the shipping industry (Ramsey, 1984; Gooding, 1992). However, *TRANSIT* was never a GPS as its constellation was too small to achieve constant, global coverage (Ramsey, 1984). From its inception, the NAVSTAR GPS programme was intended to be the first system to achieve this, while also providing altitude and velocity information, making it suitable for airborne navigation (Ramsey, 1984; Ford, 1985).

Early development of the NAVSTAR GPS programme began in 1973, separate from previous projects (Kumar and Moore, 2002). The first four satellites were launched in 1978 and the ensuing years saw significant expansion of the NAVSTAR constellation at a cost of over US$2 billion (Ramsey, 1984). The system became available to private users in 1995 (Lechner and Baumann, 2000).

## Alternative and future GNSS

### GLONASS (Russian Federation)

After NAVSTAR GPS, the largest GNSS in operation is GLONASS (*GLObalnaya NAvigatsionnaya Sputnikovaya Sistema*), operated by the military of the Russian Federation. GLONASS operates by the same principles as NAVSTAR GPS, though it uses a smaller constellation of 27 satellites, organized in 3 orbital planes at a wider inclination of 64.8° and slightly lower altitude of 19,100 km (Lechner and Baumann, 2000; Federal Space Agency, 2016a). As with GPS, GLONASS operates two levels of service for different users; Channel of Standard Accuracy (CSA) and Channel of High Accuracy (CHA). CSA data has been freely available to civilian users since 1996 and offers 60 m horizontal accuracy and 75 m vertical accuracy 99.7 per cent of the time (Lechner and Baumann, 2000). GLONASS uses a Russian geodetic framework (PZ-90) and operates on Moscow time (GMT+3). Greater coverage, and potentially accuracy, can be achieved by using a receiver which combines both GPS and GLONASS data (Melgard *et al.*, 2009; Pan *et al.*, 2014).

The USSR launched its first prototype navigation satellite, *Cosmos 192*, on 23 November 1967. As with its early American counterparts, the positional accuracy of data from *Cosmos 192* was low; after a software upgrade in 1969, average horizontal error remained as high as 100 m if measured over a five-day period (NASA, 2014). The Soviet Union embarked on the development of its first multi-satellite navigation system in 1979. *Cicada* consisted of four low-level satellites which were used to gather positional data for maritime applications. The first test flights for the GLONASS programme took place in October 1982, although the newly formed Russian Federation did not begin operational testing of the system until 1993 (Federal Space Agency, 2016b). An operational constellation of 24 satellites was in place by 1995 (Lechner and Baumann, 2000). The command centre, or Information Analytical Centre, of GLONASS is in Korolyov, near Moscow.

### Galileo (European Space Agency)

The European Space Agency's (ESA) desire to build a European, civilian GNSS has its roots in collaborative discussions which took place between the ESA and the European Union throughout the 1990s. The first product of these discussions was the aforementioned EGNOS augmentation system, launched in 2009, which improves GPS accuracy across Europe. The second was Galileo, a new GNSS, independent of existing systems.

Still in its infancy, relative to GPS and GLONASS, ESA launched its first test satellites, *GIOVE-A* and *GIOVE-B*, in 2005 and 2008 respectively. This was followed by the launch of the first operational Galileo satellite on 21 October 2011 (ESA, 2015). Twelve operational Galileo satellites have been launched to date, with 30 proposed in total. The constellation will be spread across three orbital planes with an inclination of 56°. The programme is partly funded by the European Union and is operated from two ground control centres in Fucino, Italy and Oberpfaffenhofen, Germany.

## Beidou (China)

Between 30 October 2000 and 24 May 2003, China launched three navigation satellites as part of its *Beidou* programme, a regional initiative intended to facilitate civilian positioning applications in Chinese territory (Forden, 2004). Almost immediately afterwards, development began on the *Beidou-2* programme, intended to be the first GNSS independently operated by China. By the end of 2012, 16 *Beidou-2* satellites had been launched, with a fully operational, global system expected by 2020 (Li *et al.*, 2014).

## Regional satellite positioning systems

The Indian Space Research Organization (ISRO) is currently developing a satellite navigation system, the Indian Regional Navigation Satellite System (IRNSS) (ISRO, 2016). The first IRNSS satellite was launched in 2013 and five of the seven proposed satellites are now in orbit.

The Japan Aerospace Exploration Agency (JAXA) is also in the early stages of developing a regional satellite navigation system, designed to complement NAVSTAR GPS, rather than replace it. The first and currently only satellite of the Quasi-Zenith Satellite System (QZSS) was launched on 11 September 2010. The development of three further satellites has been approved by the Japanese Government, with JAXA planning to have a seven-satellite constellation in operation sometime after 2023 (QZSS, 2016).

## References

Clinton, B. (2000) *Statement by the President Regarding the United States' Decision to Stop Degrading Global Positioning System Accuracy* Washington, DC: Office of the Press Secretary, The White House.

Dorling, D. and Fairbairn, D. (1997) *Mapping: Ways of Representing the World* Harlow, UK: Longman.

Duncan, D.T. and Regan, S.D. (2015) "Mapping Multi-Day GPS Data: A Cartographic Study in NYC" *Journal of Maps* 11 pp.1–3.

Dussault, C., Courtois, R., Ouellet, J. and Huot, J. (2001) "Influence of Satellite Geometry and Differential Correction on GPS Location Accuracy" *Wildlife Society Bulletin* 29 (1) pp.171–179.

ESA (2015) *The Story So Far* Available at: *www.esa.int/Our_Activities/Navigation/The_story_so_far* (Accessed: 29 January 2016).

Federal Space Agency (2016a) *GLONASS Constellation Status* Available at: *www.glonass-iac.ru/en/GLONASS/* (Accessed: 29 January 2016).

Federal Space Agency (2016b) *GLONASS History* Available at: *www.glonass-iac.ru/en/guide/index.php* (Accessed: 29 January 2016).

Ford, T. (1985) "Satellite Based Navigation" *Aircraft Engineering* 57 (10) pp.6–15.

Forden, G. (2004) "The Military Capabilities and Implications of China's Indigenous Satellite-Based Navigation System" *Science and Global Security* 12 (3) pp.219–248.

Gooding, N.R.L. (1992) "Navstar GPS: Charting Aspects" *Journal of Navigation* 45 (3) pp.344–351.

Hulbert, I.A.R. and French, J. (2001) "The Accuracy of GPS for Wildlife Telemetry and Habitat Mapping" *Journal of Applied Ecology* 38 (4) pp.869–878.

ISRO (2016) *IRNSS Programme* Available at: *www.isro.gov.in/irnss-programme* (Accessed: 29 January 2016).

Jin, S. (2004) "A Method to Establish GPS Grid Ionospheric Correct Model" *Proceedings of the Pan Ocean Remote Sensing Conference (PORSEC)* pp.1–7.

Kumar, S. and Moore, K.B. (2002) "The Evolution of Global Positioning System (GPS) Technology" *Journal of Science Education and Technology* 11 (1) pp.59–80.

Lawrence, V. (2002) "Mapping Out a Digital Future for Ordnance Survey" *The Cartographic Journal* 39 (1) pp.77–80.

Lechner, W. and Baumann, S. (2000) "Global Navigation Satellite Systems" *Computers and Electronics in Agriculture* 25 pp.67–85.

Li, M., Qu, L., Zhao, Q., Guo, J., Su, X. and Li, X. (2014) "Precise Point Positioning with the BeiDou Navigation Satellite System" *Sensors* 14 pp.927–943.

Melgard, T., Vigen, E., De Jong, K., Lapucha, D., Visser, H. and Oerpen, O. (2009) "G2: The First Real-Time GPS and GLONASS Precise Orbit and Clock Service" *Proceedings of the 22nd International Technical Meeting of The Satellite Division of the Institute of Navigation (ION GNSS 2009)* pp.1885–1891.

NASA (2014) *Cosmos 192* Available at: *http://nssdc.gsfc.nasa.gov/nmc/spacecraftDisplay.do? id=1967-116A* (Accessed: 29 January 2016).

NCO (2015) *The Global Positioning System* Available at: *www.gps.gov/systems/gps/* (Accessed: 12 January 2016).

Ochieng, W.Y. and Sauer K. (2002) "Urban Road Transport Navigation: Performance of the Global Positioning System After Selective Availability" *Transportation Research Part C* 10 pp.171–187.

Pan, L., Cai, C., Santerre, R. and Zhu, J. (2014) "Combined GPS/GLONASS Precise Point Positioning with Fixed GPS Ambiguities" *Sensors* 14 (9) pp.17530–17547.

Perkins, C. (2007) "Community Mapping" *The Cartographic Journal* 44 (2) pp.127–137.

QZSS (2016) *Quasi-Zenith Satellite System* Available at: *http://qzss.go.jp/en/overview/services/sv02_why.html* (Accessed: 29 January 2016).

Ramsey, W.E. (1984) "Navigation Satellites: Their Future Potential" *Philosophical Transactions of the Royal Society of London* A 312 (1519) pp. 67–73.

Schulz, D. (2001) "The Conquest of Space: On the Prevalence of Maps in Contemporary Art" *Henry Moore Institute Essays on Sculpture* 35 pp.1–8.

Seeber, G. (2003) *Satellite geodesy* (2nd ed.) Berlin/New York: Walter de Gruyter.

Sleewaegen, J.M. (1999) "GPS Selective Availability Error Contains a Small Component with a Period of 3 Seconds: Influence on the Phase Measurement Noise" *Geophysical Research Letters* 26 (13) pp.1925–1928.

Speake, J. (2015) "'I've Got My Sat Nav, It's Alright': Users' Attitudes Towards, and Engagements with, Technologies of Navigation" *The Cartographic Journal* 52 (4) pp.345–355.

Xiao, B. and Zhang, K. (2002) "Handheld GPS and Mobile Mapping" *Cartography* 31 (1) pp.99–107.

# 19

# Mobile mapping

*Martin Davis*

Mobile mapping has proved to be a significant advance in the field of cartographic data collection methods since its inception, largely due to the speed and efficiency it offers relative to traditional surveying techniques. In the broadest sense of the term, mobile mapping refers to any form of spatial data acquisition which takes place from a moving platform on land, in the air or below ground. Utilizing and combining technologies associated with digital imaging and georeferencing, mobile mapping today provides cartographers with a means of rapidly gathering large quantities of GIS-ready data for use in an array of applications. The swift growth of both of these technologies in recent decades has facilitated the streamlining of mobile mapping processes, while increasing their versatility (Schwarz and El-Sheimy, 2004). The resulting data are often of a much larger scale than those of more established remote sensing technologies, given the closer proximity of sensor and target.

Not only can spatial data be derived from mobile mapping but valuable attribute data about ground features can also be measured or observed in order to further enrich output datasets (Li, 1997). Some recent technologies offer real-time mapping capabilities, in which large-scale spatial databases can be automatically updated, according to the features being observed (Ou *et al.*, 2013). Such real-time mapping is vital in a growing number of applications which require the data to be used almost immediately after it has been collected, such as emergency response and military applications (Li, 1997). This chapter introduces the origins and principles of mobile mapping before exploring the development of a variety of these applications.

## Origins and development of mobile mapping technology

The origins of mobile mapping can be traced to Ohio State University (USA) and the University of Calgary (Canada) in the late 1980s, where it was the focus of small groups of researchers shortly after data from the Global Positioning System (GPS) became available to civilian users (Li, 1997; Karimi and Grejner-Brzezinska, 2004; Li and Chapman, 2005). John Bossler, of the former institution, offered the first definition of the term in 1995 when he referred to:

[a] technique used to gather geographical information, such as natural landmarks and the location of roads, from a moving vehicle. The technology has been around for decades, but recent advances in computers and satellites have made mobile mapping easier, cheaper and more accurate.

*(Li and Chapman, 2005: 375)*

Ohio State University hosted the first International Symposium on Mobile Mapping Technology between 24 and 26 May 1995, which prompted wider interest in the fledgling technology and the adoption of the term across a growing body of literature (Tao and Li, 2007). The symposium is now a regular event, providing a platform for recent research and an opportunity to showcase the latest technological developments in the field.

From the outset, mobile mapping has been concerned with the fusion of positioning and remote sensing technologies, both of which developed significantly throughout the second half of the twentieth century. The majority of technologies presented at the inaugural symposium featured a GPS receiver, an Inertial Measurement Unit (IMU) and a number of video cameras (Li and Chapman, 2005). Bossler's statement alludes to an early driving force behind the bringing together of these technologies, the need for transport infrastructure mapping (Niu *et al.*, 2006; Tao and Li, 2007). The earliest mobile mapping platforms were used along highways and railway corridors, with large-scale data subsequently being extracted from the resulting georeferenced images (Li and Chapman, 2005). Traditionally, aerial photogrammetry and ground survey would be the sources of such data, although aerial images are rarely of sufficient spatial resolution to adequately map small features such as kerbs, road markings, centrelines and manholes. Although this issue is overcome by manual ground survey, the time and costs associated with this method make mobile mapping an attractive alternative (Li, 1997; Rau *et al.*, 2011).

## Principal mobile mapping technologies

### Data acquisition sensors

The nature of the sensor used for data acquisition depends on the nature of the mobile platform, the accuracy required and the intended application of the data (El-Sheimy, 2005). Systems using multiple imaging sensors are becoming more common, diversifying the data being collected and making them suitable for a broader range of tasks (Karimi and Grejner-Brzezinska, 2004). Early mobile mapping used analogue, film-based sensors – commonly used for aerial photography. Digital imaging, in which an electronic sensor gathers data regarding the intensity of electromagnetic radiation incident upon it, has become increasingly important to mobile mapping concurrent with the development of this technology more broadly. As the quality of these electronic sensors, or charge coupled devices (CCDs), has improved over time, the scope and potential accuracy of mobile mapping has improved likewise. Despite being able to capture multiple, colour frames in quick succession, the relatively low spatial resolution provided by early CCDs restricted their use to ground mobile mapping applications, where the distance between the sensor and the target was minimal. As the achievable resolution has improved, airborne applications have become more practical, while the falling financial cost of digital imaging sensors has also improved accessibility (Schwarz and El-Sheimy, 2004).

The development of multispectral digital sensors has also broadened the scope of mobile mapping considerably, allowing radiation from both visible and non-visible portions of the electromagnetic spectrum to be detected. As with satellite remote sensing, this technology allows

*Figure 19.1*   Lidar image of fields near Cricklade, Wiltshire, UK

*Source: https://historicengland.org.uk/research/approaches/research-methods/airborne-remote-sensing/lidar/.*

data acquisition pertinent to non-visible environmental characteristics, such as mineral composition and vegetation health. Sensors using longer wavelengths (e.g. radar and microwave) have the added advantage of being able to penetrate cloud, thus facilitating operation independent of weather conditions.

In addition, laser scanning and ranging technology is commonly used in mobile mapping, providing a means of rapidly measuring distances between the sensor and surrounding features, as well as the angle at which these measurements took place. Lidar (Light Detection and Ranging) typically employs this principle from airborne scanners, allowing the construction of dense point clouds which can be used to model and visualize the surveyed area (Figure 19.1). Lidar data provides the fine detail in terms of terrain and surface features required by localized applications, including environmental management and archaeology.

## Positioning and georeferencing

To determine the locations of objects with precision and accuracy, the position and orientation of the sensor must be known, in addition to any geometric distortions introduced by imaging equipment (e.g. optical lens distortions and camera calibration) (El-Sheimy, 2005). These position and orientation (kinematic modelling) data are three-dimensional, incorporating parameters on three axes ($x$, $y$ and $z$ values) (Li, 1997). After data are collected, they are georeferenced. Georeferencing refers to the process of assigning this positional and orientation information to a datum (or feature), relative to a modelled global framework. Naturally, the accuracy of the data to be mapped depends heavily on the accuracy of the positional data for features, the moving platform, the imaging sensors and the relative positions (horizontal and vertical) of each of these elements in the world, or a representation of it (Li, 1997; Rau *et al.*, 2011; Madeira *et al.*, 2012). In early applications, the sensor orientation (i.e. pitch, yaw and roll or rotation) was determined using traditional photogrammetric block adjustment, and georeferencing relied on matching images to a series of Ground Control Points. However, subsequent improvement in GPS accuracy has enabled 'direct georeferencing' by which orientation data can be gathered without this process (Li, 1997; El-Sheimy, 2005; Li and Chapman, 2005).

Commonly, the georeferencing process is undertaken using a combination of GPS data (see Chapter 18), or its international equivalents, and an IMU, a system which detects movement using a series of gyroscopes and accelerometers. While both technologies are theoretically capable of acquiring both positional and orientation data, GPS data are commonly used for the former, while IMUs frequently offer greater accuracy in terms of orientation (Schwarz and El-Sheimy, 2004). Naturally, more sophisticated IMUs and GPS receivers come at a high, sometimes prohibitive, cost and are normally the most expensive components of a mobile mapping system (Rau *et al.*, 2011). The accuracy of this process can be improved by using differential GPS (DGPS, see Chapter 18) (Li, 1997). A homogenous time system also needs to be shared by each of these hardware elements, to ensure that the positional data are synchronized. The GPS time system outlined in Chapter 18 often provides a convenient and precise basis for this (El-Sheimy, 2005). A more detailed, technical introduction to these concepts is provided by El-Sheimy (2005).

Theoretically, real-time georeferencing is possible, as geometric calibrations can be completed before the data collection process, and GPS data are transmitted in real time. Nevertheless, bypassing post-processing in this way makes the resulting data accuracy vulnerable to the variability of GPS satellite visibility and the impact of this on positional data quality (El-Sheimy, 2005). Research regarding the use of GPS Quality of Service (QoS) data has highlighted means of establishing the optimum day and time for GPS data collection and therefore, in principle, the time at which mobile mapping data can be georeferenced with the greatest accuracy (Karimi and Grejner-Brzezinska, 2004). Early research has indicated that in indoor applications of mobile mapping technology, where no GPS signal is available, a 'total station' (an electronic theodolite and distance measurer) may be used as an alternative positioning tool (Keller and Sternberg, 2013).

## Mobile platforms

Both the data acquisition sensors and the positioning and orientation equipment need to be mounted on a moving platform, manned or unmanned, in order to move around the survey area. This platform can take the form of road or rail vehicles, aeroplanes, helicopters, boats or, more recently, backpack-mounted frames that can be carried by humans. The ongoing development of such backpack systems has the potential to increase the accessibility of mobile mapping technology to low-budget or non-specialist users due to its significantly lower cost and complexity relative to more established mobile mapping platforms (Ellum and El-Sheimy, 2000; El-Sheimy, 2005). Airborne platforms tend to be the costliest, given the higher quality of sensor demanded by the large distance from the target, as well as the inherent costs associated with aviation (Madeira *et al.*, 2012). Nevertheless, in some applications, these costs are offset by the ability to survey large areas at speed, making aircraft the most suitable platforms in these instances.

## Accuracy

As with most methods of data acquisition, appraisals of mobile mapping technology naturally tend to focus on the accuracy of the resulting data. Given the very large-scale uses of mobile mapping data, acceptable error levels are relatively small. The overall accuracy of a mobile mapping system depends on the accuracy of its constituent parts (Li, 1997). Data acquisition sensors with a high spatial resolution and effective positioning and orientation systems are both necessary elements of the ideal mobile mapping system.

In land-based, digital mobile mapping applications, with GPS data available, a Root Mean Square Error (RMSE) of 3–5 cm has been suggested as an achievable positional accuracy standard (Schwarz and El-Sheimy, 2004). Longer outages of GPS will result in greater degradation of accuracy levels, although the nature and quality of the IMU being used is also significant. Scherzinger (2002) conducted a mobile mapping survey in central Tokyo where, due to the nature of the urban environment, a GPS signal was only available for 50 per cent of the data acquisition period. Where GPS data were available, typically only one or two satellites were visible and multipath reflections of signals from surrounding buildings introduced further error. Nevertheless, accuracy to within at least 25 cm was achieved for 90 per cent of the survey by correcting the data using inertial positioning and orientation systems.

## Applications of mobile mapping technology

Mobile mapping is commonly applied in contexts requiring the construction or revision of large-scale map databases (Hwang et al., 2013; Ou et al., 2013). The most common of these applications are in industries pertinent to infrastructure and utilities (Li, 1997; Karimi and Grejner-Brzezinska, 2004). Highway mapping was among the earliest applications of the technology, including the compilation of traffic sign inventories, road network databases and assessments of road surface condition. As well as assisting the maintenance of roads, large-scale highway models are also vital to traffic monitoring systems and advanced driver guidance and navigation systems (Jin et al., 2012), and road networks also often form the bases for other networks, including sewerage channels, water pipelines and electrical and communications cables (Rau et al., 2011). Frequently, mobile mapping data collected on roads can be used for more than one of these purposes, adding to the efficiency of the process (Li, 1997). Mapping from a moving platform also negates the need to close sections of roads for mapping to be carried out. Given the abundance and importance of roads in all parts of the world, the development of methods for mapping these features has been a prominent area of research and development within mobile mapping technology (Rau et al., 2011). The ease of using a moving platform on a road network (typically a car or van) has also facilitated widespread use of mobile mapping in this application.

Considering the speed at which mobile mapping allows surveying to take place, roads may be surveyed and re-surveyed at regular intervals at a feasible cost (Li, 1997). Highway applications of mobile mapping may require some manual digitization of images, such as the tracing of centre lines and kerbs, although automatic and semi-automatic methods of feature extraction in this context continue to improve, such as those involving 3D stereo plotting and laser scanning, and have been the focus of considerable research throughout the field's existence (Li, 1997; Tao et al., 1998; Jin et al., 2012), though the technology required to do this is still in relatively early stages of development (Ou et al., 2013). In road surface inspections, surface damage can be measured automatically using laser scanners and visualized spatially to create a Digital Road Surface Model (DRSM). In mountainous areas in which it is proposed that a road be constructed or widened, laser mobile mapping data can also be used to estimate the volume of material to be excavated by sampling a series of points across the existing surface (Wang et al., 2013).

The use of mobile mapping technology in infrastructure mapping is not restricted to road transport. Overhead power lines can also be mapped using mobile mapping imagery, provided

*Figure 19.2*   Images from a car-mounted mobile mapping system showing damage after the 2010 Haiti earthquake (left to right: collapsed building, restricted road access, temporary shelter)

*Source*: Ajmar *et al.*, 2013.

that the imaging sensors are oriented appropriately. Automated feature extraction is perhaps more reliable in this application, as the contrast between the overhead cables and the sky beyond makes it easier to distinguish the desired features from the background of the image (Li, 1997).

Mobile mapping technology provides a faster and less labour-intensive means of mapping underground areas than traditional surveying techniques. This is particularly useful in the mining industry, where mine development, maintenance and emergency response rely heavily on the availability of accurate spatial data (Zlot and Bosse, 2014). In such an environment, the use of an adapted laser scanner to create a 3D point cloud provides accurate data and eliminates the need for ambient light.

In the aftermath of the earthquake in Haiti in 2010, some saw the potential for mobile mapping to be a fast means of gathering up-to-date spatial data, which is invaluable in the emergency response, aid distribution and damage assessment phases of such catastrophes. In Haiti, this data was remotely acquired through the digitization of aerial and satellite imagery shared via open source web platforms. Although this provided large quantities of useful data in a short time period, Ajmar *et al.* (2013) highlight the issues associated with solely relying on vertical and oblique views and suggest that a means of ground truthing would significantly promote data quality. A series of car-mountable webcams and a GPS receiver provide a low-cost means of undertaking this where other methods of survey would be too time-consuming in such emergency situations (Figure 19.2).

*Figure 19.3*   A selection of land-based mobile platforms used by Google for the collection of *Street View* imagery

*Source*: *www.google.com/maps/streetview/understand*.

273

Physical applications of mobile mapping technology are also extant, including the tracking of riverside topographic change using mobile laser scanners. Although the technology provides a useful means of tracing erosion and morphological change, accuracy is reduced when vegetation is present (Vaaja *et al.*, 2011).

Since 2007, Google has collected *Street View* imagery; a series of navigable, panoramic views from roads and paths around the world, collected from various mobile platforms (see Figure 19.3). Linked to the Google Maps web mapping service and freely available to the public, the imagery can provide a useful navigation aid and, since past imagery was made available in 2014, a historical resource.

## Conclusion

Since the late 1980s, mobile mapping technology has developed rapidly, both in terms of its capabilities and its cost and accessibility. Mobile mapping provides a means of rapidly gathering large quantities of highly accurate spatial and attribute data for use in a number of applications in which large-scale mapping holds a vital function. Given the obvious pairing of moving platforms and transport networks, these features have been the predominant focus of mobile mapping research to date. However, as more versatile mobile mapping systems become more feasible and affordable, indoor and portable (e.g. backpack-mounted) systems that are currently in development have the potential to broaden applications of the technology well beyond this initial sphere of use. Decreasing costs and complexity, together with broader democratization trends in cartography, provide scope for widespread utilization of mobile mapping principles, perhaps utilizing open source platforms, and have already facilitated new ways of representing geographic space to large consumer groups, as the *Google Street View* project perhaps best illustrates (Dodge *et al.*, 2011). Ever-faster data collection and georeferencing processes, alongside advancing automatic feature extraction methods, provide the modern cartographer with an unprecedentedly swift surveying method, forging opportunities for the field to contribute to rapid-response and emergency applications in ways that would have seemed impossible half a century ago.

## References

Ajmar, A., Balbo, S., Boccardo, P., Tonolo, F.G., Piras, M. and Princic, J. (2013) "A Low-Cost Mobile Mapping System (LCMMS) For Field Data Acquisition: A Potential Use to Validate Aerial/Satellite Building Damage Assessment" *International Journal of Digital Earth* 6 (2) pp.103–123.

Dodge, M., Kitchin, R. and Perkins, C. (2011) 'Editors' Overview' in "Mobile Mapping: An Emerging Technology for Spatial Data Acquisition" in Dodge, M., Kitchin, R. and Perkins, C. (Eds) *The Map Reader: Theories of Mapping Practice and Cartographic Representation* Chichester, UK: Wiley, p.170.

Ellum, C. and El-Sheimy, N. (2000) "The Development of a Backpack Mobile Mapping System" *International Archives of Photogrammetry and Remote Sensing* 33 (B2) pp.184–191.

El-Sheimy, N. (2005) "An Overview of Mobile Mapping Systems" *Proceedings of the FIG Working Week 2005 and GSDI-8* pp.1–24. Available at: *www.fig.net/resources/proceedings/fig_proceedings/ cairo/papers/ ts_17/ts17_03_elsheimy.pdf* (Accessed: 15 March 2016).

Hwang, J., Yun, H., Jeong, T., Suh, Y. and Huang, H. (2013) "Frequent Unscheduled Updates of the National Base Map Using the Land-Based Mobile Mapping System" *Remote Sensing* 5 pp.2513–2533.

Jin, H., Feng, Y. and Li, M. (2012) "Towards an Automatic System for Road Lane Marking Extraction in Large-Scale Aerial Images Acquired over Rural Areas by Hierarchical Image Analysis and Gabor Filter" *International Journal of Remote Sensing* 33 (9) pp.2747–2769.

Karimi, H.A. and Grejner-Brzezinska, D.A. (2004) "GQMAP: Improving Performance and Productivity of Mobile Mapping Systems through GPS Quality of Service" *Cartography and Geographic Information Science* 31 (3) pp.167–177.

Keller, F. and Sternberg, H. (2013) "Multi-Sensor Platform for Indoor Mobile Mapping: System Calibration and Using a Total Station for Indoor Applications" *Remote Sensing* 5 pp.5805–5824.

Li, R. (1997) "Mobile Mapping: An Emerging Technology for Spatial Data Acquisition" *Photogrammetric Engineering and Remote Sensing* 63 (9) pp.1085–1092.

Li, J. and Chapman, M.A. (2005) "Introduction to MMS Special Issue" *Journal of the American Society for Photogrammetry and Remote Sensing* 71 (4) pp.375–376.

Madeira, S., Gonçalves, J.A. and Bastos, L. (2012) "Sensor Integration in a Low Cost Land Mobile Mapping System" *Sensors* 12 pp.2935–2953.

Niu, X., Hassan, T., Ellum, C. and El-Sheimy, N. (2006) "Directly Georeferencing Terrestrial Imagery Using MEMS-based INS/GNSS Integrated Systems" *Shaping the Change XXIII FIG Congress Munich, Germany*, October 2006. Available at: *www.fig.net/resources/proceedings/fig_proceedings /fig2006/papers/ ps05_07/ps05_07_03_niu_etal_0619.pdf* (Accessed: 15 March 2016).

Ou, J., Qiao, G., Bao, F., Wang, W., Di, K. and Li, R. (2013) "A New Method for Automatic Large Scale Map Updating Using Mobile Mapping Imagery" *The Photogrammetric Record* 28 (143) pp.240–260.

Rau, J., Habib, A.F., Kersting, A.P., Chiang, K., Bang, K., Tseng, Y. and Li, Y. (2011) "Direct Sensor Orientation of a Land-Based Mobile Mapping System" *Sensors* 11 pp.7243–7261.

Scherzinger, B.M. (2002) "Inertially Aided RTK Performance Evaluation" Available at: *http://applanix. com/media/downloads/articles_papers/POSLV_2002_09_InertiallyAided.pdf* (Accessed: 15 March 2016).

Schwarz, K.P. and El-Sheimy, N. (2004) "Mobile Mapping Systems: State of the Art and Future Trends" *Proceedings of the XXth ISPRS Congress, Istanbul, Turkey*, July 2004 Available at: *www.isprs.org/proceedings/ XXXV/congress/comm5/papers/652.pdf* (Accessed: 15 March 2016).

Tao, C., Li, R. and Chapman, M.A. (1998) "Automatic Reconstruction of Road Centrelines from Mobile Mapping Image Sequences" *Photogrammetric Engineering and Remote Sensing* 64 (7) pp.709–716.

Tao, C.V. and Li, J. (2007) "Foreword" in Tao, C.V. and Li, J. (Eds) *Advances in Mobile Mapping Technology* London: Taylor and Francis, pp.xi–xiv.

Vaaja, M., Hppä, J., Kukko, A., Kaartinen, H., Hppä, H. and Alho, P. (2011) "Mapping Topography Changes and Elevation Accuracies Using a Mobile Laser Scanner" *Remote Sensing* 3 pp.587–600.

Wang, J., González-Jorge, H., Lindenbergh, R., Arias-Sámchez, P. and Menenti, M. (2013) "Automatic Estimation of Excavation Volume from Laser Mobile Mapping Data for Mountain Road Widening" *Remote Sensing* 5 pp.4629–4651.

Zlot, R. and Bosse, M. (2014) "Efficient Large-Scale Three-Dimensional Mobile Mapping for Underground Mines" *Journal of Field Robotics* 31 (5) pp.758–779.

# 20

# Neocartography and OpenStreetMap

*Steve Chilton*

Massive changes in the cartographic landscape took place in the first decade of the new millennium. Changes, both technological and social, that may have seemed small in themselves but that had far-reaching consequences. A mapping conference programme in 2005 proved to be a classic example of prescience on behalf of the programme organizers, in that it identified and highlighted many of these changes in just one session of the programme.

## The beginnings

In September 2005, the Society of Cartographers' Annual Summer School was held at the University of Cambridge. It had always covered new developments in cartography, and that year included a session on the theme of 'public access to maps and geospatial data'. As part of the session Steve Coast, described in the conference report as 'a freelance cartographer' (a title he is unlikely to have bestowed on himself), gave a presentation about OpenStreetMap (OSM). Apart from his previous talk to a computer conference, this was the first time the project had been exposed to the mainstream of cartography.

OSM was started in London by Steve Coast in August 2004 because of his frustration with the restrictive copyright that pertained to the Ordnance Survey (UK's National Mapping Agency) maps and data. In his view, this was restricting him from producing a local map, and at a macro level stifling innovation by not making the map data available to users without excessive royalty payments. So, with a consumer-grade GPS device, he started collecting GPS tracks around his local area of central London, and writing some reasonably unsophisticated software to display this data. He soon realized, after talking to people of a similar frame of mind (he was a research student at University College London at the time), that there could be much more to this than he had perhaps originally thought. Shortly after presenting his idea at the EuroFoo conference, he started getting others interested and contributing data in like mode. The mushrooming interest in the feasability of actually mapping a significant geographical area (now spreading rapidly from its UK source) is shown by the fact that within 16 months there were 1,000 registered OSMers. By now, coders galore were being attracted to this novel project, and editors, map display tools and other significant refinements to allow scaling of the work were being developed. Exactly three years on, in August 2007, there were 10,000 registered participants. In March 2009, another milestone was reached when the figure of 100,000 participants was reached – an amazing increase in uptake (it reached 2 million in 2015).

The other five 'access to data/maps' sessions in that 2005 conference can now be seen to be pointers to other developments that were either already happening or were soon to became part of the new data landscape. Access to geospatial data and the degree to which it should be made freely available was becoming one of the most debated topics among cartographers and allied professionals. These conference sessions brought together six professionals with differing views about the 'free' availability of data for mapping purposes and were aimed at exploding some of the myths surrounding the extent to which data is 'free' across the world.

The first paper, from Roger Longthorn (a policy advisor), discussed existing data access policies and the issues involved with data that is freely available and that for which a fee is payable. The question was posed: 'Should governments be making their data more widely available so that a wider user base can drive the geodata market?' The point was made that adding value to any free data would imply that the data would effectively no longer be free, and that with mapping agency budgets being in most cases unprotected, freely available data may not always be best since the maintenance of the data is subject to continuing funding streams which may not be guaranteed. The presentation also questioned whether it is ethical to charge for geospatial information when not all taxpayers either use or benefit from it, or whether it is more appropriate for those who benefit from the data to pay for it. It was claimed that the UK had some of the best mapping and most highly digitized databases, partly because the primary data is not dependent on annual budgets from the UK's Treasury. This was all anticipating the coming of significant data releases. In 2010 it was consolidated with the creation of the Open Government Licence and the creation of the data.gov.uk website.

The next paper, from Peter Cridland (London Borough of Barnet), then demonstrated how the borough was using interactive second generation web mapping, originally aimed at property professionals, to provide a vast array of information for the public on local facilities such as libraries, medical surgeries and council offices as well as live departure boards for bus routes. The system was developed out of the planning department, where it was piloted to allow officers of the council to retrieve information about both current and historical planning applications in Barnet. The system was seen as a good mechanism for delivering information to the public under the Freedom of Information Act and this was backed up by positive feedback from both internal and external users, as well as the monitoring of around 10,000 data requests per day. Barnet may have been ahead of the game, as such provision varied greatly across the public sector. A survey by the government in 2004 noted that 'government offices were collecting data, some 49 per cent were engaged in data sharing', and 'responses indicated confusion about metadata standards generally . . . 31 per cent of respondents were using other (i.e. un-recommended) metadata standards, while 27 per cent had invented their own and 20 per cent weren't using metadata at all' (Cabinet Office, 2004).

Next up was Jo Walsh, who had recently co-authored *Mapping Hacks*, 'a collection of one hundred simple techniques available to developers and power users who want to draw digital maps' (Erle *et al.*, 2005). She was a campaigner for open geodata, free software and civic information, and began by stating that 'Who controls the description of the world, controls the things in that description'. She used examples from Google Maps mashups to illustrate how collaborative citizen-oriented cartography with a mix of maps, photographs and information has contributed to an open source boom in GIS and geospatial information in the United States. Claiming that this citizen-oriented cartographic revolution depends on freely available geographic information as well as the software source code, Walsh asked why those in the UK had not been able to build these maps in the way that they are flourishing in the USA.

She closed her paper by claiming that open geodata will be a real bonus for the economy and will lead to more engaged citizens.

In March 2008 an authoritative independent economic study, commissioned by the UK government and published along with the Budget, was released. In looking at the government sectors most dependent on selling data and taking conservative and pessimistic scenarios throughout, the study – 'Models of Public Sector Information via Trading Funds' (Newbery *et al.*, 2008) – concluded that the benefits of giving away government data outweighed the loss of income from licence fees from the current practice of 'cost recovery' by more than £160m for the largest six 'trading funds' alone. It also rebutted claims that a move to 'free data' would damage the work done by agencies such as the Met Office and Ordnance Survey.

The Treasury and Cabinet Office commissioned the study from a team of economists at the University of Cambridge in response to a government review, the 'Power of Information'. That called on the government to engage with user-generated websites, in particular by ending restrictive policies of charging outsiders to re-use official data. It urged the government to make data available at the 'marginal cost' of supplying an extra copy (for digital information, zero), so long as there was no evidence that this model would damage trading funds which rely on selling information licences.

Another paper in the set was given by Ed Parsons, the then Chief Technology Officer at the Ordnance Survey (OS). He described how most of what the OS did was the collection, manipulating and maintaining of the UK's national digital database. The database at the time contained over five million individual features. Contrary to common belief among the public, the OS received at the time no funding from the taxpayer, with 60 per cent of its revenue coming from the private sector. In 2004–5 the organization returned a £9.2m profit on turnover of almost £115m and paid back £800,000 to the Treasury in dividends.

Parsons claimed that quite clearly free geodata does not exist, since no matter how it is collected effort is involved and resources must be spent both in collecting it and in ongoing actions if that data is to be successfully maintained. In the case of the OS, Parsons revealed that the collection and maintenance of the national digital database cost around £106m each year. At the time, this involved an extremely expensive infrastructure with aerial survey costs and 300 surveyors on the ground, all of which allowed OS to represent within the database 99.6 per cent of significant real-world features within six months of their completion.

Parsons completed his presentation by comparing the record of the UK's national mapping agency with the mapping products available in the USA, a market widely perceived to provide free geodata. The United States Geological Survey (USGS) at that time spent only £70m on database maintenance activities and had to fight every year for the release of federal funds to allow that to continue. By contrast, in the UK, the OS budget is financed from its commercial activities and funding is therefore not dependent on politicians. This led him to the conclusion that there was indeed no such thing as a 'free lunch' in geodata.

In 2007, Parsons moved to work for Google, as Geospatial Technologist. In October 2004 the Where 2 Technologies company and its web-based map offering had been acquired by Google, where it was transformed into the web application Google Maps. At the same time, Google acquired Keyhole, a geospatial data visualization company, for its Earth Viewer application, which was badged as the highly successful Google Earth application in 2005, while other aspects of its core technology were integrated into Google Maps. From this time, until OSM was more fully developed, Google became many people's web map of choice.

Another paper in the conference session was given by Richard Fairhurst, a freelance cartographer and the then editor of the website of British Waterways. He began by comparing basic street mapping available online where the quality of the mapping is secondary to the functionality of the maps, with more modern mapping such as Google Maps, which make widespread use of Java scripts to provide users with a more intuitive experience. He went on to illustrate how the British Waterway's website has been rebuilt around mapping from the OS using Flash to add functionality to the mapping. He showed how the site was rebuilt using Ming, free software that allows the development of Flash code without purchasing proprietary products. The lesson to take away was that anyone could post their mapping on the web cheaply and easily, and non-cartographers could post their own geographic information on existing maps. Fairhurst had introduced the topic and the real possibilities of what became known as mashups. Fairhurst was one of the OSM contributors who also brought excellent coding skills to the project, his development work on the Potlatch editor being just one example. If you look at both coding and mapping aspects of OSM, you will find a small number of individuals such as him contributing a very high percentage of the input. This has been categorized by describing the shape of the OSM project as having a 'long tail' (Lin, 2015).

Steve Coast's earlier contribution to the conference gave the background to how OSM aimed to generate free geodata for people to use, claiming that people want free data; they want cheaper, more timely and accurate vector-based maps without having to pay for licences. OSM hoped to achieve this using data collected by volunteers using GPS receivers and submitted via the OSM website. Applets were then used to edit the vector data and Wikipedia-like technology was used to back up everything to protect the integrity of the data. At that time, street mapping of the central London area was mostly complete, since this was where most data is currently available – with a courier company agreeing to supply 300,000 GPS track points each week. OSM was beginning to attract programmers, as well as volunteers, for the collection of data, postcode acquisition and ground truthing the data.

## A developing project

From the humble beginnings described above, the OSM project has developed way beyond what even Steve Coast probably originally envisaged. OSM is often compared to Wikipedia, which is held up as the best known and most successful example of crowdsourcing on a global scale (incidentally that term itself is derided by Wikipedia co-founder Jimmy Wales), and has certainly changed the way many people approach what might loosely be termed 'fact finding'. There were certainly those who thought that OSM would never even get off the ground. But it did, and the OSM project has had a huge impact in changing the geodata landscape, particularly in places like the UK. As can be seen, the coverage that has been achieved, its accuracy, its availability and global impact are all changing the way individuals and organizations are thinking about the collection, pricing and (re)use of geodata.

The genesis of the project was further expanded on in the book *OpenStreetMap: Using and Enhancing the Free Map of the World* (Ramm *et al.*, 2011). The authors explained how the GPS traces were collated with additional information, such as street names, and then uploaded to a central database. Many difficulties were anticipated, such as the GPS devices not being accurate enough, the people's styles not being consistent, the technical methods inadequate, and there being no way to guarantee even a minimum level of data quality. In the early days, and in most areas, OSM mapping started from a blank sheet. But sometimes it was possible to import a base dataset from existing sources. The Netherlands, for example, benefited from having an almost complete dataset donated by Dutch company, AND. In the US, mapping of the road network was given a head start by importing the freely available TIGER dataset, and Canada, Germany and many other countries also had a number of completed import projects.

The parallels with Wikipedia are worth looking into more deeply. For example, Wikipedia has deliberately avoided creating rigid strictures such as 'articles about composers must contain at least their date of birth, date of death, and a list of works'. This principle is mirrored in OSM, where users can upload data in the form they find most appropriate. For both projects, however, time has brought some consensus about 'standard operating procedures', thus creating some degree of structure. Both projects also have a detailed change history, so that anybody can see who made which changes and when. However, the structure of the data that is collected in OSM is also very different as it is much more interconnected than Wikipedia, which is fundamentally a collection of individual items.

Mashups have already been mentioned, but they do have problems of their own. The service provider for the map component (for example, Google) could in theory curtail their service at any time. Equally, they could embed advertising into their maps, or start to charge a fee (this did happen with the Google Maps API – Application Programming Interface – in October 2011, over a certain usage level). They could even shut down the service completely (which happened with Yahoo maps in June 2015), leaving stranded all mashups that were based on their service. But above all, mashups are unsatisfactory from a cartographic perspective because they can never really change the look and feel of the map. Mashups have been categorized as just 'push pins on a map', that you cannot do anything interesting with, like change the appearance of a road depending on how many accidents happened there. In contrast to nearly all other data providers, OSM offers full access to its geodata. Everyone is free to take it and create whatever type of map (or other artefact) they desire with it.

Right from the beginning, the OSM project decided to use its own data formats and custom software that best served the OSM purpose. OSM has three main *entities* in its model: Node, Way and Relation. For any of these, you can apply tags to identify their attributes. For OSM, it was very important that software and data formats could be used by people without prior GIS experience, and the technology needed to support the free-style attributes for objects required in an open, international project. It was virtually impossible to create a pre-defined data structure for OSM data and to expect everyone to use the same categories and the same level of detail. Allowing this flexibility was incompatible with the more rigid, top-down structures found in traditional GIS approaches, but was taken anyway.

At a basic level, OSM comprises a large, shared database containing data that comes from lots of different sources and that can be used for a number of different purposes. It was important for OSM that the road data forms a network which can be operated on automatically, for example to derive driving directions. But right from the beginning, connectivity with other systems was built in to the OSM development. So, OSM data could be converted to shapefiles or imported into PostGIS databases, formats that are very common in the GIS world. The reverse of that process could be used for importing data into OSM. OSM maps could also be used in the GIS world through a Web Map Service (WMS). There were soon commercial and non-commercial third-party organizations offering such WMS servers.

## The knock-on effect

It was not long before the OSM data had become so well developed that the same cartography conference just two years on included a presentation entitled 'OpenStreetMap: Real data, real uses', which gave examples of the value and variety of uses the author had found. OSM was becoming a serious player. It was no coincidence that by 2008 the major commercial map data providers had recognized the power of the crowd and now had community data collection and/ or feedback systems in place, for example, Google (Map Maker), Tele Atlas (Map Insight) and NAVTEK (Map Reporter). However, data collected via these routes remains copyright of the commercial companies involved, and is not open data.

Then in late 2009 there was a development that surprised many, not least some at the OS, when the UK government announced that it intended 'to make [some] Ordnance Survey maps free for use online by any organization – including commercial ones – at resolutions more detailed than commercial 1:25,000 Landranger maps' (*The Guardian*, 2009). This led to the opening of a consultation on the plan, which was announced by Gordon Brown (the Deputy Prime Minister), at a seminar on making data public – set in the wider context of public service reform, under the 'Smarter Government' umbrella. Sir Tim Berners-Lee, the inventor of the world wide web, had been recruited by the Prime Minister that year to help open up government data. Berners-Lee said that 'the revised terms for use of OS maps would also remove the *derived data* problem, under which OS claims full copyright on any intellectual property that is created with reference to an OS map' (*The Guardian*, 2009). OS immediately hinted that the government needed to back the theory with appropriate funding if the OS was to retain its position as supplier of high quality geodata.

While this was going on, many examples of the OSM crowdsourced data being taken in interesting and innovative directions appeared – often due to its very availability to researchers, hackers, government institutions and commercial companies. Just some of them will suffice to illustrate the point: specialist maps for cycling, routing applications, skiing, topography and for maritime use. There were also many and varied instances of the OSM map tiles being used as part of high profile websites. They do not come much more notable than the US White House and the German Supreme Court (Chilton, 2009).

One aspect that has troubled many about the OSM phenomenon has been the accuracy and completeness of the data, given its crowdsourced nature, and its lack of formal quality controls. Fairly early in the life of the OSM project, research on both these aspects was being carried out and already was suggesting its use could be justified for most potential users. Haklay (2010) compared OSM and OS datasets and concluded that volunteered geographic information 'can reach very good spatial data quality'. Research in this area has subsequently become very fertile ground. The work of Neis and Zielstra (2014) and the OSM wiki pages (OpenStreetMap, 2015) both identify much of this work and the subsequent conclusions.

## Neocartography

The developments outlined here have created an environment that has allowed the collection of geodata to move from National Mapping Agencies and major commercial data providers to what is now sometimes called User Generated Content (UGC) providers – or crowdsourced data collectors. These changes have been made possible because of: the unscrambling of satellite signals, cheaper domestic GPS units, the availability of satellite imagery in the public domain, and the availability and ease of use of open source development tools. In parallel with this has been the movement to release government and other analytical data rather than having it contained inside silos, as was previously the case. With these changes a new cartography (or neocartography as it is sometimes called) has emerged. It has been such a significant movement that the International Cartographic Association (ICA), whose mission is 'to promote the disciplines and professions of cartography and GIScience in an international context', established a Commission on Neocartography in 2011.

The ICA Commission on Neocartography's (2011) mission and aims stated that:

> Many examples of new and innovative mapping are being produced outside the normal orbit of existing cartographers or map producers. The term neocartographers is being used to describe map makers who may not have come from traditional mapping backgrounds,

and are frequently using open data and open source mapping tools. Another difference is in the blurring of boundaries between map producers and map consumers. The availability of data and tools allows neocartographers to make their own maps, show what they want, and often be the intended audience as well – that is to say they may make the maps for themselves, just because they can. There is a real need for a discipline to be established to study this essentially undisciplined field of neocartography.

Although some commentators dislike both the term neocartography and the concept behind it, there are characteristics of some mapping and mappers today that distinguish this 'movement' and give it credence. One aspect is the 'broadcast' nature of much of today's cartography. Many maps are developed first (or often just) for display on the web and for dissemination via the web and via social media. This can give much greater exposure to maps than traditional/printed maps might have, but also can provide disproportional prominence to less worthy maps that can 'go viral' by multiple mentions by indiscriminate social media participants. Another characteristic is the contrast between paper maps which can get out of date before they are published and that of new mappings which are actually never finished, i.e. are ever-evolving experiments. The plus side of that situation is, of course, that maps can be dynamic and take advantage of change by perhaps taking on board constantly changing data and visualizing the most recent data (possibly 'live'). Thus, some of the best examples of the new cartography are those using an Application Programme Interface (API) feed.

One characteristic of some of the new mappers is their tendency to want to 'get the data out there' as soon as possible. This urge can often produce an outcome that has not been sufficiently thought through, and time not taken to consider the best method of visualizing the data, data classification and symbology, and an appropriate colour palette. In this situation, a cry that is often heard is 'just because the data exists, it doesn't mean it should be mapped'. Having given those caveats, there are numerous examples of new (i.e. untrained) cartographers who spend time absorbing the lessons of their own work and also somehow absorbing some or all of the principles of cartography that trained cartographers would have learnt in their formal training. In conclusion, there are more maps being produced now than ever before, and there are many good and many bad (and many just average) maps being produced by these so-called neocartographers.

## Meanwhile, in the real world

In some ways the OSM project became a victim of its own success. The increasingly large volume of volunteer data contributors wanted to map everything – down to bus stops, individual trees and grit bins. Not only that, but there were strong demands to include everything recorded in the database on the map. With many in the project not being aware of the cartographic principles of scale and selectivity, it was difficult to control the feature bloat that started to happen on the public-facing OSM map. Thus was born multiple map outputs on the OSM main website. So, some project members developed the *Osmarender* version of the map to show 'all' features, which at least served the function of showing mappers what had been mapped, or not. Meanwhile Steve Chilton led a small team, including initially Richard Fairhurst and Artem Pavlenko, in creating the *mapnik* OSM map style, which had a tightly integrated symbol and feature set which was designed to show data appropriately at the various scales (commonly called zoom levels in OSM parlance). An attempt was made to incorporate a slightly washed out colour ramp that hopefully gave a 'designed' feel to the resulting map(s), particularly recognizing that in many cases the generated map tiles were increasingly being used as backgrounds to map mashups.

In around 2007, several companies started building on the crowdsourced data by supplying OSM-based services, support and consulting. These included Cloudmade, geofabrik, Mapnik Consulting and Itoworld. This showed a serious belief in the value of the data and the business that could be built around it. Later, two companies started making inroads and may now be considered to offer 'go to' mapping toolsets. Mapbox is an 'open source mapping platform for custom designed maps', which has been available since 2010, and cartoDB is a 'Software as a Service cloud computing platform that provides GIS and web mapping tools for display in a web browser', which was developed in 2011.

Currently the OSM website has five variants of the map/data as exemplars, but this changes as new ideas are developed and chosen or rejected. Alternative map tilesets from the same OSM data are served by the likes of Mapquest and Stamen Design. Recently, consideration has been given to producing country-based styles, as many users were questioning the UK-centric nature of the symbology employed (blue for motorways, for instance). Geofabrik sponsored the development of a 'German style' map in 2011. Other countries are considering following suit as this has been seen by many as a means of increasing OSM's impact even further globally. To happen though, this needs stylesheet design skills, a considerable amount of time and good server availability – all in a voluntary project.

## Humanitarian work

Perhaps the most socially significant impact of OSM has been its use in disaster responses, such as after major earthquakes. On 12 January 2010, a magnitude 7.0 earthquake struck the island of Hispaniola in the Caribbean Sea, with an epicentre 25 km west of Port-au-Prince, Haiti. Official estimates suggest that around 230,000 people were killed, as in less than a minute the event levelled approximately 20 per cent of the buildings in greater Port-au-Prince and left a million homeless.

The OSM community, in partnership with CrisisCommons, geared up in the days following the earthquake to update the basic open source base maps of Haiti, providing the most detailed mapping of locations of road networks and critical infrastructure (Kent, 2014). Google, DigitalGlobe and GeoEye worked together to release high quality satellite imagery within 24 hours of the disaster and OSM contributors began digitizing the imagery, as Keegan (2010) explains:

> When the earthquake happened it was a signal for OSM members around the globe to start downloading satellite images (either freely available or donated by Yahoo) and then to start tracing the outlines of streets on top so a map emerged. Volunteers on the ground in Haiti, often using Garmin GPS locators, added vital local information – such as which roads were passable, where the hospitals were situated, where refugee camps were, or walls, pharmacies, hedges and so forth – so rescue workers had an invaluable tool. The result is a new, detailed map that is updated frequently, unlike most commercial maps.

In 2010 Humanitarian OpenStreetMap Team (HOT) (*http://hotosm.org/*) was formed. HOT creates and provides up-to-date maps, which are a critical resource when relief organizations are responding to disasters or political crises. HOT does a range of work though mapping projects (e.g. in Nepal in 2015), community development, technical projects (like OpenAerialMap and learnOSM) and partnerships, such as Missing Maps. The project both proves the value of OSM data and adds more value to that data by its mapping and tool provision, such as the Export Tool and the Tasking Manager.

# References

Cabinet Office, (2004) *Geographic Information: An Analysis of Interoperability and Information Sharing in the United Kingdom* London: HMSO.

Chilton, S. (2009) "Crowdsourcing is Radically Changing the Geodata Landscape: Case Study of OpenStreetMap" Proceedings of the 24th International Cartographic Conference, Santiago, Chile, 15–21 November 2009 Available at: *http://icaci.org/files/documents/ICC_proceedings/ICC2009/html/nonref/22_6.pdf* (Accessed: 20 July 2015).

Erle, S., Gibson, R. and Walsh, J. (2005) *Mapping Hacks: Tips and Tools for Electronic Cartography* Sebastopol, CA: O'Reilly Media.

The Guardian (2009) "OS Mapping Data: A New Landscape Unfolds" 19 November 2009 Available at: *www.theguardian.com/technology/2009/nov/19/ordnance-survey-maps-free-online (Accessed: 15 September 2015)*

Haklay, M. (2010) "How Good is Volunteered Geographical Information? A Comparative Study of OpenStreetMap and Ordnance Survey Datasets" *Environment and Planning B* 37 pp.682–703.

ICA Commission on Neocartography (2011) "Mission and Aims" Available at: *http://neocartography.icaci.org/mission-and-aims/* (Accessed: 29 October 2016).

Keegan, V. (2010) "Meet the Wikipedia of the Mapping World" *Guardian Unlimited* 4 February 2010 Available at: *www.theguardian.com/technology/2010/feb/04/mapping-open-source-victor-keegan* (Accessed: 20 July 2015).

Kent, A.J. (2014) "Helping Haiti: Some Reflections on Contributing to a Global Disaster Relief Effort" in Chilton, S. and Kent, A.J. (Eds) *Cartography: A Reader* Reading, UK: Society of Cartographers, pp.267–273.

Lin, W (2015) "Revealing the Making of OpenStreetMap: A Limited Account" *The Canadian Geographer / Le Géographe canadien* 59 pp.69–81.

Newbery, D., Bently, N. and Pollok, R. (2008) "Models of Public Sector Information Provision via Trading Funds" London: Office of Public Sector Information. Available at: *http://webarchive.national-archives.gov.uk/20090609003228/http://www.berr.gov.uk/files/file45136.pdf* (Accessed: 3 August 2015).

Neis, P. and Zielstra, D. (2014) "Recent Developments and Future Trends in Volunteered Geographic Information Research: The Case of OpenStreetMap" *Future Internet* 6 (1) pp.76–106.

OpenStreetMap (2015) Available at: *http://wiki.OpenStreetMap.org/wiki/Research* (Accessed: 6 August 2015).

Ramm, F., Topf, J. and Chilton, S. (2011) *OpenStreetMap: Using and Enhancing the Free Map of the World* Cambridge, UK: UIT Cambridge.

# Part IV

# Understanding cartographic design

## Principles in practice

# 21

# An introduction to map design

*Giles Darkes*

This chapter examines the process of map-making from the point of view of what makes for good design and effective graphic communication. Its aim is to summarize the essential features of map design, and to introduce the key elements of map creation which are explored in more detail in the other chapters of this part of the book.

There are more maps in the world than at any time in history. Maps are as popular, more widely available, cheaper (or free), and used by a wider cross-section of society in a bigger range of situations than ever. As a by-product of the information age, geographical information is readily available, and a range of software means that more people are using geographical data and are actively involved in the map-making process. The basic output of most Geographic Information Systems (GIS) is a map. However, most of the people creating maps are not cartographers. Historically, the map output of many GIS queries has been of a poor cartographic standard, and as a result has been poor at communicating a message.

The chapter is deliberately not allied to specific software or map-production platforms. The assumption is that almost all maps are now created on computers, but the range of software used to make maps is very wide and mapmakers will no doubt be using a program that they are familiar with and know how to get the best out of. The basic principles discussed here apply, no matter what are the means of production of the map.

Given that the vast majority of information in general and statistical data in particular has some geographical component (i.e. *where* something is, is an important aspect of it), much information lends itself to being mapped. A map can therefore be an efficient way of communicating spatial information in a succinct and straightforward graphic way. However, maps vary greatly in quality. By looking at the sequence of production, and understanding the basic principles behind good map design, the chances of creating an attractive and useful graphic are considerably increased.

## The map-production process

The process of making a map is not always a linear one, but there is a basic sequence which is generally followed. First, the initial planning stage involves consideration of the reason for producing the map, and of the requirements of the end-user. This leads to thinking about basic practical factors: the medium of output (print *versus* screen) and the layout of the map graphic.

287

The next stage involves compilation and analysis of the information to appear on the map; the final stage is the design and production.

## The purpose of a map

Understanding *why* a map is to be produced should drive the design process. It is often the case that maps are commissioned with neither a firm idea of precisely what is needed as the final output nor what the intention of the map is, but by asking a few questions it is easy and worthwhile to find out. The cartographer may need to take the initial brief which has been given and probe more to find out the specific intention. Ask the author what message the maps are intended to convey and how they complement a text. If you are producing a map as a product of GIS analysis, consider how the map will enhance and illustrate the conclusions that the analysis has produced.

## The target audience

Understanding the end-user of a map helps to define the map design, symbology and the level of complexity of the information shown. As a broad principle, maps can be made to fit the level of knowledge of an audience by increasing or decreasing the number of categories of data shown, and varying the level of sophistication, as well as the number of symbols. An atlas designed for children aged 8–11 will use a much-simplified depiction of the world, with fewer and perhaps bolder symbols, compared to a reference atlas aimed at an adult readership, but the geographical information it is showing is the same (see Chapter 28 for more on designing maps for use in schools).

Clearly, once a map has been published, it will be seen by a wide-ranging readership, and a good map will be of use to all of them, but the degree to which the map's information is 'mined' will vary. General-purpose reference maps are intended for a wide audience with varying levels of map-reading skills. National topographic map series are typical of this genre; likewise, Internet map sites have the largest and therefore the most diverse map-using audience, and use a correspondingly small set of map symbols. Other maps are aimed at a specialist audience who have a high level of technical knowledge, and are familiar with a complex set of symbols. Common examples include geological maps and nautical charts, both of which use conventional, complex, and highly stylized symbols. In practice most maps will have a varying audience – in which case you may need to consider the most important intended readers and cater to their needs; other readers make of the map what they will.

## Map viewing conditions

The viewing conditions of a map may have more influence on the design than the intended audience. Looking at a printed paper map when you are sitting down and concentrating on it is a very different process from the quick glance needed at a sat-nav map on a small screen whilst driving. Map design therefore often reflects variant viewing conditions. Compare a reference atlas with a map of ski-slopes; or a street map on a smartphone screen with a planning strategy map; or a nautical chart with a metro map – the necessary difference in designs based on the viewing conditions becomes obvious.

## Medium of reproduction

As a general rule, maps designed for viewing on screen have fewer symbols, larger symbols, fewer and larger map labels, and a smaller range of colours than maps designed for printed output.

This is because the resolution of a computer or smartphone screen is lower than the resolution of printed maps, and the range of colours seen on screen is lower. The term 'resolution' means the smallest object which can be seen as a separate object compared to its neighbours. A printed map can show very fine lines, subtle variations in colours, and small type whereas a computer screen will only show thicker lines and larger type, and it is harder to discern one colour from another.

The distinction between printed and computer-screen maps is now not as evident as it once was, largely because of the ability to zoom in and out of a map displayed on screen. A PDF on screen can look just like a printed map, and zooming in and out of it offers better functionality than looking at its printed equivalent. Free maps available through websites work because they use datasets with differing levels of map content according to the scale of the map seen, and how far you have zoomed in or out – the larger the scale, the more detail is shown. But obvious differences remain, not least in the size of the map viewed. A map seen on a smartphone screen cannot show a wide geographical area in detail, whereas a paper map can. Since we often understand geographical information through the context in which it is seen, maps viewed on a screen 7 × 7cm must necessarily show different content and symbology to a 70 × 70cm printed map.

## Format

Knowing the output size of a printed map is essential. It allows the cartographer to size the map correctly from the beginning, and relate the design directly to the space available, and use symbols and text of an appropriate size. For example, it is unusual to use type which is smaller than 5pt on a map, and if you create a map using 5pt type only to find that the map has been reduced in size when it is printed, the smallest type may now be illegible.

If you know the trimmed page size of a printed product, you will be aware that the actual space available for the map will be smaller, allowing for margins and (in the case of book publishing) the other items that may appear on a page such as running headers, page numbers, and captions or figure numbers. For example, many books are published in octavo format (between 200 and 250 mm tall). Using a true example, the actual trimmed page size of a book is 225 × 150 mm, but the working area allowed by the publisher for illustrations and maps is 169 × 110 mm; only about 55 per cent of the total page area is available to the cartographer.

## Colour versus black and white

A basic decision to be made is whether the map is to be produced in colour or in black and white. Although most maps produced are coloured, a significant minority of maps is still produced in black and white. It is important to know that the map will be monochrome, because most colour maps will not appear correctly if rendered in black and white – colours which are distinct hues in a coloured map may look the same grey when shown in black and white. The choice of fills for areas, and of types of line and point symbols, is quite different in monochrome compared to colour mapping.

## Layout

A 'map' page usually contains a variety of graphic elements, of which the geographic map is just one (albeit the principal) component. The other typical contents of a map include a legend, scale bar, North arrow, title, additional text (such as acknowledgements and copyright statements), diagrams to show adjoining map sheets, insets (a continuation of the map, or a detached region, or a reference diagram to show the location of the main map), grid and graticule numbers, and a neat line (the bounding frame of a map, which itself can show detail such as divisions of degrees, minutes, and seconds). This list is not exhaustive. Clearly, not all these elements will always appear, but the design of the map page will need to accommodate them appropriately. A more detailed consideration of layouts and the component elements is given in Chapter 23.

## Analysing and compiling the map data

The map-production process requires the compilation of the different elements that will appear on the map. A distinction can be made between two broad categories of mapping at this point: between general-purpose reference maps; and specialist or thematic maps which combine a base map with symbols which depict a specific theme or topic such as population density or land use.

### *Compilation*

The process of compiling a map involves the editorial selection of the features to be shown on a map, and design decisions on how best to show them. Any map can show only a selection of features from the complex real world, and so the process of compilation is in many ways more about what to leave out than what to include. A well-compiled map has enough information on it to convey the necessary information to the map reader, but not so much as to make the graphic unreadable. However, maps can contain a lot of information and still work well because good design principles have been followed.

Almost all maps are derived from existing printed map material or from GIS-based mapping, and it is quite exceptional to undertake field work to compile a map from completely original survey. Compiling a map is therefore usually a process of taking a selection of existing base maps, extracting information from them, and then adding or emphasizing information relevant to the map you are creating. Increasingly, maps are created from data gathered and analysed in a GIS, and often the challenge is to take GIS information and present it in a readable and clear way. Creating maps from sources other than GIS usually involves finding printed or digital map sources and selecting information from them. Source maps used will depend on the scale and purpose of the map. National topographic maps series will provide information for large-scale maps, and published atlases and sheet maps are suitable for medium- and small-scale maps. Of course almost all published material is subject to copyright and it is only fair that you seek permission to use the map and pay royalties to the owners of your source material – in the same way, you can reasonably expect others to pay you royalties if they use your published material.

For maps of the world or major regions of it, the choice of map projection you use is important. Given that the process of displaying the roughly spherical Earth on a flat plane unavoidably introduces distortions, it is usually appropriate to select a map projection which is either equal area, or one of a handful of widely used projections which are not strictly equal area but approximately so, and which keep the shape of the world familiar. Examples of equal-area projections to consider for maps of the world are the Mollweide and Eckert IV projections, and compromise projections often used are the Winkel Tripel, Robinson, and van der Grinten (usually clipped 80° North and South to exclude the polar regions which look very distorted). Some of the familiar projections once used, especially the Mercator and other rectangular projections, are not usually appropriate because the distortions in area lead to a map which is misleading, especially for distribution maps. The so-called Geographic Projection, which is the default projection of many GIS programmes, is a rectangular projection and not a very good one for a world map. There is a range of useful projections for continents and regions of the world, including conical projections for areas with a wider East–West extent than North–South, such as Europe or the United Sates. The addition of the graticule (the lines of latitude and longitude) is very helpful for map readers to see how the globe has been portrayed on the map.

As a general rule, when compiling a map, the cartographer should always work by taking information from a larger scale to a smaller scale. This is because a smaller-scale map will have been generalized more than you need (generalization is explained below) and by blowing up a smaller-scale map, you cannot get out of it more detail than the scale of the original map allowed. The typical procedure is to scan a larger-scale map and then to reduce it digitally to the scale you need for the final graphic. Ensure that you include the scale bar on the map so that you can keep track of the reduced scale. Digital map data is also scalable, of course, but ensure that you acquire data of a comparable or slightly larger scale than the output scale.

Thematic maps are created by combining base-map information, which provides context, with symbols which show the theme of the map. For example, a population-density map will need a base map which shows the areas (countries, regions, census enumeration districts, and so on) for which the statistics have been gathered. The areas are then filled with a range of colours or gradations of one colour to convey information on population density. The base maps for thematic maps need enough information to show the setting for the theme but are usually selective and simplified versions of topographic maps, typically with coastlines, major settlements, and principal roads or hydrology shown.

## Data quality

The quality of source data is a thorny issue, because it can be very hard to assess but affects the quality of the map you produce. Source material varies in age, quality, and reliability. The comparison of several maps of an area may reveal important discrepancies in all sorts of aspects of the data. Frustratingly, a map that looks good may contain very poor quality data, as is discussed below. Using up-to-date sources is important, as recent maps should show new names, recently added features, and changes to human features (such as the current size of settlements, reservoirs) and natural features (such as changing lake shorelines). However, natural features such as relief and drainage generally do not change substantially, and if they are shown well on maps which are out of copyright, then it is useful to be able to use them.

The term 'data quality' incorporates a number of concepts including currency, completeness, accuracy, and precision. Currency relates to how up-to-date a map is. Completeness is the idea that a map shows all of the categories that it claims to show; for instance if a map includes a symbol for lighthouses, it is reasonable to assume that all lighthouses within the map area are shown. Map accuracy and precision are two topics on which a great deal has been written, not least in connection with GIS map data, and there are varying definitions of both. Accuracy is a catch-all term which incorporates the concept of completeness, the correct classification of mapped information, and the statistical analysis of errors. Take for example a map of land use; a check can be made that, at a series of sample points across the map, what the map shows as the land use is indeed the use found on the ground. The results of a check will give a statistical likelihood that at any given point the map shows the right category. Precision is a related term, but is usually meant as a mathematical concept, relating to the number of decimal places given in a measurement. Accuracy is, moreover, a relative rather than an absolute concept, and is affected by the available information and by the scale of the map. What is deemed 'accurate' at a small map scale may be hopelessly inaccurate at a large map scale.

Assessing data quality in source material is not easy. Digital data may carry extensive metadata which gives information about the features shown on a digital map, such as its source, when it was last updated, etc., but accessing the metadata may be difficult, and it is anyway often impractical to check it for every feature shown. Printed map material should carry a date, and national mapping agency (NMA) material will carry additional information about dates of survey, revision, and edition. But for many parts of the world, NMA material at medium or large scales is not available, and much of the time you will be using maps from other sources.

In practice, the way that most people judge map accuracy is by its appearance, and herein lies the rub: maps which *look* well presented are judged to be more accurate than maps which look poorly presented; in other words, maps which are well designed and produced look more believable than maps which are not. It is a fact that, in the way we always judge books by their covers, we judge maps by their graphics. The (unspoken) reasoning is that, if the cartographer has taken care in designing a good graphic, then the same care must have been taken in acquiring the information. Alas, it is not true. Unfortunately, a symbol placed on a map assumes an authority which it may not merit. For the most part, as map readers, we have no idea how certain the cartographer was in placing any particular symbol on the map. One symbol which is from a certain source looks identical to one which is speculative.

There are many examples of mistakes which have been perpetuated from one map to another. The English cartographer John Speed (1551/2–1629) found an unnamed village in Wiltshire on a map of Christopher Saxton which he was using as a source for his own map of the county. He made a note to himself in Latin (*quaere*) to check the name of the village, but his engraver thought that the village was called Quaere, and it appeared on Speed's map of Wiltshire in 1611. The name then appeared on nine re-issues of Speed's map (1614–1676) and a host of other maps copied from Speed, until about 1760 when the error was finally removed in Thomas Kitchin and Emanuel Bowen's *Large English Atlas*.

Given the reality of time constraints in producing maps, the best practical recommendation for assessing the accuracy of a printed or digital source is to compare it with as many other sources as possible. Comparison of a coastline on a source map with its appearance on Google Earth will reveal how appropriately it has been generalized. Looking at how the same feature has been shown on different maps may help you to assess the quality of those maps – positional accuracy, completeness, categorization of features (such as road types), and other checks can be made.

## Generalization

Generalization is an essential part of the process of compilation. As maps are a graphic representation of geographic reality, the cartographer has to simplify the complex reality of the world in order to present it in an understandable way. The process of compilation of a map is in many ways more about what to leave out than what to include, and so is generalization.

Generalization is the editorial adjustment of the features and characteristics of the real world that will be shown at the reduced scale of a map. This is broadly a process of simplification, and the level of simplification depends on the scale of the map involved – the smaller the scale of the map, the more generalized it has to be. As an example, if you consider the coastline or the lake shore of an area that you are familiar with, you will know that on the ground it is a hugely complex system of rocks, beaches, pebbles, rocky platforms, etc. which protrude and intrude to make a very unsmooth line forming the boundary between water and land. Compare that same coastline with its depiction on a large-scale map or chart; some of the features will be shown (rocks which are just off the coast may be depicted, for example) but many will be omitted. On a medium-scale map, say 1:25,000, only the principal rock formations will be shown and the minor indentations are omitted. On a small-scale map, say 1:500,000, the same length of coastline may look smooth and homogeneous because there is not the room to show its true nature.

When the cartographer generalizes, it is usually a spontaneous and subjective process, and hence hard for a computer to replicate satisfactorily. It is essentially a visual skill, and combines an understanding of real-world geography with an appreciation of how to convey what is important about it in a graphic form. The rules are not hard and fast, and much of the success of generalization lies in cartographers' understanding of the complex nature of the real world and their ability to translate it into graphic form. The process depends on the scale of the map, the reason for making it, and the intended map user.

Generalization is both necessary and desirable. If you take a map and shrink the view of it photographically or on screen, at some point it becomes impossible to read because the density of the information is too great to permit you to see or read any features. Likewise, if you take the map and blow it up in scale, you do not see more information than was on the map in the first place; it is frustrating to buy a street atlas of a town which advertises that it has a large-scale town-centre plan, only to find that it is the same map which has been enlarged, and presents no more geographical details or additional names. Part of the way that we judge the authority and credibility of a map is to assess its level of generalization to see if it is appropriate to its scale and the nature of the contents. It is however an unconscious process, but a well generalized map will 'feel' more credible than a poorly generalized one.

If the degree of generalization depends on the scale of the map, it follows that the quantity of features and the way that they are shown on a map should be proportional to the scale of the map. However, because the reasons for making maps vary hugely, so do the ways in which maps are generalized, and it is not a formulaic process. Maps made for printed media can show much more detail than maps designed for quick reference on screen, and maps for specialist audiences (geological maps and nautical charts, for instance) will probably have less generalization than maps of similar scales made for a less specialist reader. What is desirable is that the process of generalization is undertaken thoughtfully, with the end view that the map shows what it needs to show in the most appropriate and honest way.

Generalization involves a number of different processes. They often take place simultaneously, and they are interdependent. They include simplification, exaggeration, typification, elimination, aggregation, displacement, and symbolization, which are all explained below. They are often applied at the same time; simplification, exaggeration, and displacement tend to go hand-in-hand. The order in which they are applied is also often variable, but as a guide, the information for a topographic map will be generalized in the order of coastlines or lake shorelines, hydrology, contours, settlements, communications (roads and railways), and boundaries.

## Simplification

As map scales get smaller, so the number of real-world features that can be included, the number of symbols that will fit in, and the amount of information that can be conveyed all reduce, if the map is to remain legible. One common hallmark of less experienced mapmakers is that they will often include too much information on a map, making it less effective in communicating its message. The cartographer must decide what features will be shown, retaining the essential geographical characteristics of the area. For example, a map which includes a coastline with many inlets and creeks should convey this to the reader, even if the depiction of that coastline is quite stylized. In choosing the detail to include and omit, it is often not possible (or desirable) to be consistent or systematic and operate to a rule book, because the importance of features varies locally. For example, if you choose to omit all 'B' roads, you may eliminate a road which is of vital importance to the geography of an area; likewise, a well in a desert probably warrants inclusion on a map of the area, even though its physical footprint is tiny.

Consideration of simplification also applies to the level of contents of a map, and it relates closely to the intended audience for a map. For example, if you are required to produce a map of the geology of an area, you will be aware that the source data you are using may well be a highly technical geological map, showing a complex hierarchy of rock and soil types, with many added technical symbols for fault lines and other geological and geomorphological phenomena. Most people do not have the technical skills to interpret such a map, and so you may be required to simplify the contents – amalgamate categories of information, use simpler explanations in the legend, and so on.

## Exaggeration

Exaggeration is often a consequence of simplification. Having omitted features which are less important, the size and characteristics of the features which *do* appear often have to be exaggerated in order to make them legible, and to convey the essence of their geographical nature. For instance, the width of the symbols for roads is usually exaggerated in order that their relative size and importance is understood. In a popular motoring atlas at a scale of 1:160,000 the motorways are depicted by a line symbol 2 mm in width, which represents a scaled-up road of 320 m width – about 8 to 10 times its true width. It can also be very useful to exaggerate the relationship between different features on a map to make it clearer. For example, a motoring atlas may exaggerate the offset between two roads at a staggered junction to make it clear to a driver glancing at the map that you must turn left before turning right. Likewise, the width of inlets of a coastline may be increased on a map to better convey its nature; or a river may have some of its meander loops removed (simplified) but the curving nature of those that remain may be enhanced (exaggerated).

## Displacement

One of the consequences of exaggeration is that you may have to displace a symbol from its true, scaled position so that it appears in the right place relative to other map features. For example, if you are increasing the symbolic width of a road, then any map feature standing next to the road will have to be slightly moved from its true position to avoid showing it in the middle of the road. Likewise, if a river, road, and railway all occupy a narrow valley floor, it is reasonable to separate the symbols for them so that they do not coalesce into an incomprehensible single line on the map. In the latter case, it may be a good thing to keep the natural feature in its right place and move the man-made features, but there are no hard-and-fast rules.

## Typification

Large-scale, detailed maps have the space to include many individual features on them, but small-scale maps do not. Typification is the process of using a single symbol or small number of symbols to represent the pattern of the individual elements of a phenomenon on the ground. For example, a dot map may use a single dot to represent a real-world phenomenon, one dot for every incident of it. However, most dot maps do not do so; they use a scatter of dots which typify the spread of the feature, and no attempt is made to represent accurately the specific locations of the phenomenon, but rather to show the general pattern. As well as point data, typification can be applied to area and line symbols. A row of separate but closely spaced houses may be shown as a single block, for example.

## Aggregation

Aggregation is the joining together of many small map features into a single map feature suitable for a smaller-scale map. For example, a large-scale map may distinguish between separate, discrete patches of woodland. When aggregated, the small patches of woodland are shown as a single patch on a smaller-scale map. Individual houses and buildings are joined together to indicate built-up streets on a medium-scale map.

## Elimination

One of the consequences of aggregation is that small features may have to be disregarded altogether. To continue the example of woodland, if several adjacent patches of woodland are aggregated to form a single unit, an outlying wood may be removed completely because it is too small to show and too far away from its neighbours to be amalgamated.

## Symbolization

The real world is depicted on maps using graphic objects which are *representative* of the real world. All maps are collections of graphics, and the process of choosing what sort of graphics to use is itself one of generalization. Symbols usually represent classes of object, where classification has been undertaken. For instance, a map which shows places of interest to tourists represents the classification of locations into places which are thought to be of interest to the tourist (and, by implication, those which are not). The locations will vary in size and nature but they all have one thing in common. The symbols for roads represent highways of different width and structure, but grouped by attribute.

It will be seen that all processes of generalization by their nature wilfully introduce distortions of the real world into the process of mapping. Does this matter? Is it not incumbent on the cartographer to tell the truth? Well, yes, within reason. Notwithstanding the debate within cartography of the last forty years as to how far maps have an agenda and how truthful they are, cartographers usually try to be as honest as possible. But a map is an agreed set of lies, enacted in the interests of clear communication. The lies are perpetrated by the cartographer and tolerated by the map reader because they are useful. The cartographer draws a blue motorway, but no blue motorways exist in the world. The map reader accepts the blue motorway because it is clearly distinguishable from the green major road. The cartographer exaggerates the distance between two objects so that they are distinguishable. The map reader is grateful because it is better to know that the two objects are not one, even if their absolute positions are wrong. Generalization is part of the process of lying, but it is done with good intention: that of purveying in a graphic the complex world in a simple and clear way. If done skilfully, a well generalized map will be far more useful than a more 'honest' and less generalized map.

## Basic map design principles

Good design is invisible; poor design stands out. This is as true of maps as it is of graphic design in general. Good design is about communicating a message clearly, quickly, and unambiguously. Well-designed maps carry more authority. Well-designed maps can also carry much more information, and more complex information, than poorly designed maps. When you consider a map such as a general reference topographic map produced by an NMA or a reference atlas, you will quickly be aware of just how *much* information is shown on the map. Humans are remarkably adept at interpreting the collection of graphic devices that is a map and assigning meaning to it. However, the process is made much easier if good design principles are adopted.

There are four basic design principles that should be understood when it comes to making a successful map: legibility, hierarchy, figure-ground relationship, and contrast.

## Legibility

Although it sounds obvious, a map should be legible. Some maps fail because they are not. By legible, what we mean is that all the elements on a map are:

- visible – e.g. the type is large enough to be read;
- distinct – e.g. symbols have enough separation to be seen individually;
- interpretable – e.g. symbols are sufficiently different from one another and simple enough to be correctly identified.

Maps that are illegible may contain too much information, or the information is insufficiently well presented to be clear. Symbols vary greatly in how inherently easy they are to interpret. Lines, for example, are quickly seen and, because of their length compared to other symbols, they stand out. Point symbols may be much harder to interpret, especially if they are small. Familiar symbols, such as conventional symbols (like an H for a hospital), will be more quickly recognized than a symbol specially created for a map, particularly if there are many such unfamiliar symbols.

## Hierarchy

Geographical information is almost always sorted into some sort of hierarchical relationship, and when mapped, the map symbols need to reflect that hierarchy. Hierarchies are of different types: administrative boundaries, for example, reflect local, regional, and national units; roads are classified according to width and importance; statistical maps deliberately rank data sets so that the highest or most significant figures stand out. When we read or talk, the information comes to us serially, i.e. one idea after another, but when we look at maps, the information is received synoptically, i.e. all at once, and so the aim is to construct map graphics in a manner that sorts out the more important from the less important. Map hierarchy is essentially achieved by the separation of different types of information into different visual layers.

Hierarchies are established on maps by varying symbols in size, width, length, colour or intensity of the same colour. Consider, for instance, how roads are depicted on most maps; colour is used to denote different classes of road, but the width of the symbol is also varied to reinforce the difference. Different typefaces are used, at different sizes, to denote different features of varying importance. Important areas can be made to stand out by fading other areas into the background, or by using stronger colours for the symbols in the important area.

In some types of map, we deliberately aim not to have much separation into different layers. A general-purpose topographic map is one such example. In a topographic map, all classes of feature should appear to be of *similar* importance, but differences within classes are shown. So roads are intrinsically no more important than railways and hence the symbols used are of comparable visual weight, but within those two feature classes distinction is made between different types of road, and differing numbers of tracks on a railway line. In contrast, most thematic maps deliberately separate the subject of the map (the theme, which looks more important) from the base-map information (which gives context to the theme, but sits in the background).

## *Figure-ground relationship*

Creating a figure-ground relationship is a follow-on from the need for a hierarchical structure in a map. It is the separation of the important (known as the figural) from the background information (the ground). We naturally tend to see some graphic elements as more important than others. Glance at the front pages of newspapers in your local newsagent, for example, and it is probably the photographs that draw your attention initially. They stand out deliberately from the rest of the layout.

At its most basic, map users must be able to distinguish land from water, one country or administrative unit from another, and human from natural features. What are known as graphic closed forms (in the case of maps, features like islands and urban areas) are more easily seen than open forms, and features that are familiar will be recognized more quickly than unfamiliar features. Size is also an issue, as small features, if they have enough contrast with the background, will be seen as figural, against larger areas which form the ground. And detail also promotes figure, so an area which has more detail and information on it than the surround will be seen as more prominent.

## *Contrast*

The symbols on a map need to have sufficient contrast to be seen at all, and sufficient contrast to be distinguished one from another. Perhaps contrast is therefore the most important graphic factor. Again, it sounds an obvious point, but some maps use insufficiently contrasting symbols with the result that the message is not well conveyed.

Contrast on maps can be achieved by varying size and colour (or the saturation of colour) to ensure that two features are seen, and seen as different. For example, land and sea may both be shown as shades of blue, but if the colours are insufficiently different, there may be confusion as to which is land and which is sea. Small point symbols and thin lines need to be in strong colours to show up. Small symbols need to vary in form and colour more than large symbols.

Much depends on the context in which a symbol appears as to how easily it will be seen. A single point in the middle of a large area will be seen more easily than the same symbol in a cluster of similar but different point symbols. Moreover, too much contrast can make a map look gauche and unbelievable, and you will have seen poor maps which use garish reds and greens which cause visual dazzling. When no hierarchy is intended on a map (e.g. on a land-use map, when one category is no more important than another) the ideal solution is to use colours or fill patterns of sufficient contrast to be seen, but where none has a dominant effect.

## Symbology

Almost all maps consist of four basic elements, three of them graphic and one descriptive. The three graphic elements that appear in varying proportions on all maps are points, lines, and areas. All features of the real world can be rendered by the use of combinations of these three graphic basics as symbols. Although linear features of the geographic world (rivers, roads, theoretical lines of movement, for example) tend to remain as lines on a map whatever its scale, other features can be shown as points or areas depending on how much detail the scale of the mapping will allow. So a town on a large-scale map will be shown as an area (perhaps with point and line features within it) but as a point on a small-scale map.

The fourth element is text, and its presence on a graphic separates a map from a mere geographical image. An aerial photograph will reveal a huge amount about a landscape, but it will not tell you the names of settlements, of rivers or lakes, of countries, road numbers, the value of contours or the heights of mountains, or the position in the world in terms of latitude and longitude. By adding text in the form of labels, the graphic becomes a map.

All four of the elements can be varied in a number of ways, and distinctions applied to the same basic type of symbol allow the map reader to distinguish between different categories of real-world feature. Symbols can be varied by size, by colour, and by shape or form. The topic of map symbols is explored further in Chapter 22.

When creating a map graphic, it is often useful to compile the different elements in the order of areas, lines, points, and finally text. Most map-making software allows for the separation of graphics into different layers, and areas are shown below lines, lines below points, and points below text. For GIS-based maps, map area, line and point symbols often equate with the polygon, line and point data of the GIS. Areas on maps are symbols which may have an outline (such as a blue line for a coast or river bank), but will always have a fill of some sort whether it is a colour or a pattern. Lines, representing long, thin real-world features, are displayed on top of the areas. Lines may be cased (meaning that they have an outline to them) or open, and may be simple (like a road) or complex (like the conventional symbols for railways) in form. Point features are small symbols which relate to a specific point on the map, whether real or imaginary. They may be based on geometrical shapes (circles, squares, or other polygons) or be imitative of the subject they represent (a boat used to denote a yacht station, for example), or conventional, such as the letter S placed on a motorway in a circle to represent a service station.

Text is the final element to be placed on the map, and sits on the highest visual plain, seeming to float above the basic graphic elements. Map labels obscure the geographical information underneath them, but map readers unconsciously 'fill in' the missing information. For example, a contour label will give its value, but it is usually placed in a short missing section of the line. However, we reasonably assume that the line in reality continues through the interrupted section.

Type on maps is a complex subject, but a critical one to good map design. A map that is otherwise well designed can be spoilt by poor placement of labels. Map text uses different fonts (families of typefaces) for different functions (for instance distinguishing natural from human features), but there are other ways in which it is varied, including the size of the labels, the spacing and alignment of the words, and the spacing of the letters within words. The topic is dealt with in more detail in Chapter 25.

## Further reading

Anson, R.W. and Ormeling, F.J. (1996) "Communication design and visualisation" in Anson, R.W. and Ormeling, F.J. (Eds) *Basic Cartography for Students Volume 3* London: Butterworth Heinemann for the International Cartographic Association, pp.71–92.

Brewer, C.A. (2005) *Designing Better Maps: A Guide for GIS Users* Redlands, CA: ESRI Press.

Darkes, G.T. and Spence, M. (2017) *Cartography: An Introduction* (2nd ed.) London: The British Cartographic Society.

Dorling, D. and Fairbairn, D. (1997) *Mapping: Ways of Representing the World* Harlow, UK: Longman.

Keates, J.S. (1989) *Cartographic Design and Production* (2nd ed.) Harlow, UK: Longman Scientific and Technical.

Keates, J.S. (1996) *Understanding Maps* (2nd ed.) Harlow, UK: Addison Wesley Longman.

Krygier, J. and Wood, D. (2011) *Making Maps: A Visual Guide to Map Design for GIS* (2nd ed.) New York and London: The Guilford Press.

Mitchell, T. (2005) *Web Mapping Illustrated* Sebastapol, CA: O'Reilly.

Raisz, E. (1962) *Principles of Cartography* New York: McGraw-Hill.

Robinson, A.H., Morrison, J.L., Muehrcke, P.C., Kimerling, A.J. and Guptill, S.C. (1995) *Elements of Cartography* (6th ed.) New York: John Wiley & Sons.

Tyner, J.A. (1992) *Introduction to Thematic Cartography* Englewood Cliffs, NJ: Prentice Hall.

Tyner, J.A. (2010) *Principles of Map Design* New York and London: Guilford Press.

Wood, C.H. and Keller C.P. (Eds) (1996) *Cartographic Design: Theoretical and practical perspectives* New York: John Wiley.

# 22

# Cartographic aesthetics

*Alexander J. Kent*

> As an end product, a representation of reality, a map may be judged either by its artistic qualities – the fineness of line, harmony of colour and lettering and balanced layout or design, or by its usefulness.
>
> *(Board, 1967: 712)*

Are maps *supposed* to be beautiful? Aesthetics, as a line of enquiry that questions beauty and effect, plays a central, if under-researched, role in map-making (Kent, 2005). Using a language of graphical symbols, cartographers wield the power of aesthetics to affect how people approach a place, or a topic. Whether we are using a map to navigate or judging it for a competition, we tend to react to the general appearance of a map first (Petchenik, 1974: 63) and maps trigger emotional responses that can be positive or negative (Fabrikant *et al.*, 2012). Some have found the gravity of aesthetics within cartography disconcerting (e.g. Robinson, 1952; Woodruff, 2012), others have acknowledged aesthetics as a key area of focus (e.g. Imhof, 1982; Huffman, 2013), while some have suggested explicit guidelines for cartographers to follow (e.g. Karssen, 1980). This chapter explores the value of aesthetics in cartography and suggests ways for cartographers to make the most of the opportunities that maps offer as a means of visual communication. Other chapters in this section cover more specific principles that lead to successful cartographic practice in, for example, colour, lettering, and layout.

## In pursuit of the beautiful

One of the major difficulties associated with understanding aesthetics is summarized in the subjective paradigm, i.e. that 'beauty is in the eye of the beholder'; what is regarded as beautiful (and whether beauty means the same as 'aesthetically pleasing') varies by person, as well as by time, place, and culture. Nevertheless, cartographers have tended to support the view that a knowledge and application of aesthetics where map-making is concerned is synonymous with achieving a result that is beautiful. According to Imhof (1982: 359), beauty is concurrent with the greatest clarity, the greatest power of expression, balance, and simplicity. Certainly, Wright's (1942: 542)

*Table 22.1* Elements of good map design according to practising cartographers from Kent (2013: 42), where values indicate the percentage of respondents mentioning a feature within each group

| Features | Relevance |
| --- | --- |
| Clarity \ not cluttered; Look of map – attractive \ joy \ beautiful to look at and use | 82% |
| Emphasis (linework \ lettering \ colour tints); Accuracy; No ambiguity \ uncomplicated symbols; Lettering style; Correct weight of line information | 50% |
| Size of lettering; Consideration of final user; Simplicity; Innovative \ choice of colour used | 36% |
| Consideration of purpose of map \ fulfilling client's stated purpose | 32% |
| Communicates effectively \ understandable | 32% |
| Balance | 14% |

point, that 'An ugly map, with crude colors, careless line work, and disagreeable, poorly arranged lettering may be intrinsically as accurate as a beautiful map, but it is less likely to inspire confidence' suggests that maps *should* be beautiful if they are going to succeed in gaining the user's trust. So, if cartographers seem obliged to put communication before representation (Keates, 1984: 37–38), it is worth exploring what constitutes 'beautiful' within cartographic praxis.

Following an earlier investigation by Wood and Gilhooly (1996) an online questionnaire was put to 26 practising cartographers by Kent (2013) to examine how they saw the relevance of aesthetics in their work. The features of good map design that were mentioned most often were as follows:

These findings suggest that cartographers are principally concerned with creating a thing of beauty, which is perhaps achieved when aspects of a map's design successfully conform to certain aesthetic ideals, such as clarity. It is tempting to suggest that such ideals are restricted to individual cartographers, who can exercise the most control over map design in their desire to create something of lasting value and worth, and as an expression of their own aesthetic ideals. The significance of aesthetics in the pursuit of cartographic excellence has also long been prevalent in the corporate environments of state mapping organizations, such as Ordnance Survey, the national mapping organization of Great Britain, as noted by Withycombe (1925: 533): 'One of the chief preoccupations of the cartographer should be to reconcile these two aspects; to produce a beautiful map which is also accurate and legible'. Whether it is made by an individual or a corporation, a major objective of the cartographic enterprise is to create a map that is at least as beautiful as it is accurate and useful.

This paradigm is also encouraged in the evaluation of maps. Aesthetic judgments have long played a role in assessing cartographic quality, with map reviewers often using the word 'beautiful' to acknowledge the map-maker's skill in achieving a high standard of professionalism. Cartographic design competitions seek to reward those maps which embrace the realms of the beautiful, demanding a quality that extends beyond mere 'mapping'. Nevertheless, aesthetic ideals cannot exist in isolation. If cartographers are in the business of designing maps for others (even if they are aware of designing them for themselves), their maps have to perform, and so there are further reasons why aesthetics matters.

> Beauty is truth, truth beauty, – that is all
> Ye know on earth, and all ye need to know.
>
> *(John Keats (1795–1821)* Ode on a Grecian Urn *(1819))*

The cartographic association between beauty and truth is nothing new. The geographer Al-Idrisi (1100–1165), wrote 'When the observer looks at these maps and sees these countries explained, he sees a true description and pleasing form' (quoted in Brotton, 2014: 11). A map's aesthetic appeal seems to lend to it a timeless quality and even a sense of validity, sometimes appearing to endorse its authority as a truthful document. In commenting on a topographic map of the Kashmir Valley presented at the Royal Geographical Society in 1859, for example, Colonel George Everest stated: 'The beautiful map behind the chair, which could not be characterised in terms that were too high, was a good proof of the knowledge and skill employed in the survey' (Purdon, 1859: 32). In this case, 'beauty' results from a commitment to achieving correctness in cartography (through the skill and care of execution) that implies a dedication to completing a survey which is accurate and 'correct'. Advancing methods of survey may provide more accurate results, but if the ensuing maps do not conform to aesthetic ideals, they may not retain their value and authority. Thus, the exercising of a cartographer's aesthetic judgment in map-making can generate something of a map's value and lasting worth.

In meeting their function as serving the interests of the user, maps can employ both artistic and scientific means of creation simultaneously. Indeed, manuscript maps such as the seventeenth-century English estate map shown in Figure 22.1, were designed to simultaneously offer an accurate record of land ownership based on precise survey and a spectacle whose decoration drew attention to the status and wealth of the patron.

Krygier (1995: 9) urges cartographers to look to instances where art and science have converged – a principle that was best summarized by Eckert (1908: 347) almost 90 years earlier: 'The ideal is the intimate union of the scientific spirit with artistic execution, and when

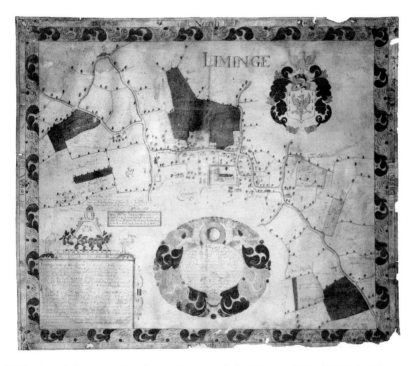

*Figure 22.1* An estate map of Lyminge, Kent, made by Thomas Hill in 1685 for Thomas Bedingfield. Pen, ink, and gold leaf on parchment, 58.5 × 50 cm. (Reproduced courtesy of Rev'd. Peter Ashman and Lyminge Parochial Church Council.)

this is realized it produces those maps which for years remain models of their kind'. Official topographic maps, which present the state's 'good view' of the national landscape, perhaps come closest to this ideal, and many national mapping organizations appear to hold these representations with an esteem and reverence that approaches the sacred. This is especially intriguing, perhaps, when the landscape is not all 'good' – its homogenization through the cartographic eye is also a sanitizing process – beauty can act as a mask.

Cartography is not alone as a discipline that involves art, science, and technology and seeks to meet functional demands with a sense of the aesthetic, as this can also be seen in fields such as architecture. A building may be designed to meet some aesthetic ends but fail to function properly (or worse, it may collapse), or it may be designed to function well but its form may inhibit its use. At its finest, architecture unifies form and function, and an example may be found in the Gothic cathedrals of Europe:

> It has been pointed out many times that everything about a Gothic cathedral, but especially the spire, draws our attention upward, just as the minds and souls of those who worship in it should also be drawn upward. The gigantic nave of the cathedral at Rheims must fill those who stand in it with a sense of how small and fragile they themselves are. The important point is that this is an attitude singularly appropriate for those entering the presence of God.
>
> *(Graham (1997: 145)*

It is for this sort of intimate relation between form and function that cartographers should strive; each symbol on the map must meet its user's need effectively by expressing the object or idea it is intended to symbolize.

Cartography is about communicating space and place through a visual medium and uses a graphical language with infinite possibilities. Maps can be designed to look a particular way: their overall aesthetic is constructed from the microaesthetics of its parts. The precise placement of lettering, the choice of colours that allows distinction yet harmony, the consistency of lineweights, and the visual hierarchy that suggests an order of reading and importance all work together to construct the overall effect. In a competitive geoinformation industry, maps need to sell and they need to function, and if aesthetics varies by time and culture, the surest way forward for cartographers is to pursue a particular aesthetic that supports the message that they need to communicate to the user at a particular time and place.

Yet the language of cartography is broad enough to incorporate designs which are not conventionally beautiful but utilize aesthetics to communicate their theme, from an overall effect to the complex interplay of symbols with a particular use of colour, type and lettering, balance and contrast. This may be seen in the effective design of visitor maps to former concentration camps in Germany, such as Sachsenhausen, whose map employs a harmony of grey, white, and blood red to evoke a particular emotional response to the landscape (see Kent, 2012). Sometimes, maps *need* to be ugly to get their message across. In Figure 22.2, for example, the distortion of continents by the cartogram's algorithm not only serves to draw attention to the countries most affected by disease but lends a grotesque appearance to the map, evoking a sense of revulsion which helps to communicate the message that 'all is *not* right in the world'.

Powerful though it may be, a first impression is not enough. The essence of cartography is generalization, which involves choice; maps speak through symbols – the selective abstraction and emphasis of elements to preserve and communicate the essence of character. People rely on symbols to communicate and whether these are, for example, points, lines, or areas in the grammar of cartographic language, they need to work within the confines of what is recognized and understood.

*Figure 22.2* The prevalence of HIV (Human Immunodeficiency Virus) across the world in 2003, mapped using an equal-area cartogram that re-sizes each territory according to the variable being mapped (in this case, the territory size shows the proportion of all people aged 15–49 with HIV worldwide, living there). Hues indicate regions. From SASI and Newman (2006)

## Symbolizing the aesthetic

If a successful design is an expression of the real and potential needs of the user, the design process is the ordering of that expression through the application of creative imagination and experimentation to meet certain objectives. Maps are symbols: the assortment of individual symbols that are inherent to a map together construct a holistic symbol of its subject. As an inevitable step in the process of cartographic design, symbolization could be defined as the graphical coding of information, or more specifically, using visual variables to represent data summarizations resulting from classification, simplification, and exaggeration (Robinson *et al.*, 1995: 475). To be informationally effective in cartographic representations, it is necessary for symbols to possess clarity of expression to avoid ambiguity; symbols need to be distinct from each other to allow recognition by the map-user. Additionally, symbols need to stand out from their background to be seen clearly, requiring what Robinson *et al.* describe as a 'high visual efficiency' (ibid.: 400), which influences our sense of 'pleasantness' (ibid.: 475). However, if cartographers regard as beautiful those maps which invite attention and further exploration, and this is largely achieved through a sense of balance and harmony (Kent, 2013: 42), it is difficult to establish any universal aesthetic – a particular look – that 'works' for all purposes and audiences of maps.

According to Imus and Loftin (2012: 106), clarity creates visual harmony and when users say that a map is beautiful, they are unconsciously responding to the beauty of clear communication. If clarity itself is aesthetically pleasing and if the goal of all communication is clarity (Crampton, 2001: 237), then it would appear that maps require at least some aesthetic appeal to function. The clearest symbols do not just happen to be aesthetically pleasing; the cartographer's aesthetic principles shape the symbolic representation of a feature towards clarity, thereby enhancing the aesthetic appeal of the map as a whole. If a central goal of the communication models (discussed in Chapter 2) was to promote clarity, they concurrently promoted aesthetics.

But if the view is taken that some features to be mapped themselves possess aesthetic properties, e.g. a waterfall (see Hudson, 2000), then it would follow that these properties, however intangible, should form part of the feature to be symbolized. Complexity encourages exploration, just as simplicity supports decision. Geological maps, which make the intricacies of underlying strata visible, serve as an example and (as Figure 22.3 illustrates) can yield forms that effectively communicate the chaos of nature and invite further exploration.

It is important to remember, however, that the desire to map is not initiated by aesthetics, even if the detailed practice of making maps is; the aesthetic impulse does not direct the inception of a map in the same way as, for example, a photograph (e.g. Figure 22.4). Whereas Ansel Adams (1902–1984) could declare 'Unless I had reacted to the mood of this place with some intensity of feeling, I would have found it a difficult and shallow undertaking to attempt a photograph' (Adams, 1983: 79–80), a cartographer does not instantly react to a scene and decide to map it.

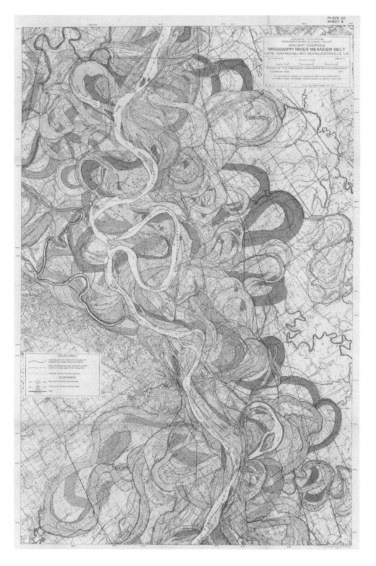

*Figure 22.3*   **Plate 22, Sheet 6 from Fisk's (1944) geological investigation of the alluvial valley of the lower Mississippi River**

*Figure 22.4*  'The Tetons and the Snake River' (1942) by Ansel Adams. Grand Teton National Park, Wyoming. National Archives and Records Administration, Records of the National Park Service. (79-AAG-1). Available at: https://en.wikipedia.org/wiki/Ansel_Adams#/media/File:Adams_The_Tetons_and_the_Snake_River.jpg (Accessed: 1 August 2016)

Topographic maps, for example, are derived from a utilitarian desire to own, manage, and control the land, although the aesthetic impulse influences the symbolization of features, where the cartographer uses his or her imagination for the abstraction and thus (re)construction of a 'good view' of the landscape.

The generalization of detail – the selection of features that construct the map as a whole and the subsequent selection of particular characteristics from those features – is what makes any map an abstract symbol of its subject. As with any map, the smaller the scale, the greater the degree of abstraction needed to ensure clarity, and the more effort required to imagine the sense of place suggested by the symbol. If successful symbolization involves preserving the character of a feature, then its aesthetic properties – whether we consider them to be beautiful or not – may be regarded as integral to the information to be communicated in a map rather than something extraneous to it.

As Tuan (1989: 239) points out:

> In the most general terms, the aesthetic is the human effort to create a *pleasing* world. Although the nature of this world varies from culture to culture, all people want one. The aesthetic impulse is thus also this constructive impulse. A pleasing *world*, rather than isolated sensations and impressions, is the desired gain.

Successful maps should express the aesthetics of features they portray in order to communicate their theme more effectively, whether it is a single theme or a national landscape, but this needs to be done with honesty and authenticity in order to engage the audience in an ethical representation. As suggested above, maps can evoke positive or negative emotional responses which can be triggered according to the aesthetics of the map.

## Towards new aesthetic frontiers

Digital technology and the use of pre-generalized data are now commonplace in map-making and so it is tempting to suggest that cartographers have merely become the beautifiers of geographical information: 'Make me a pretty map'. Users now possess the power to select data, create maps, and to control the way they are presented using methods not previously possible, particularly with the availability and development of open source GIS software such as QGIS. Achieving a consistency of typeface, colour, and line weight, for example, that lends a sense of conformity to the map, no longer lies exclusively within the realm of the expert practitioner. So, if anyone can make a map aesthetically pleasing (Collinson, 1997: 119), why is the user's own aesthetic judgment not to be trusted in constructing these visualizations? And why is there a need for the aesthetic principles of cartographers? Cartographic education (see Chapter 42) aims to lead to the creation of meaningful maps that capture and communicate information accurately and effectively. This depends upon the cartographer's proficiency in cartographic language, and aesthetics are central to its successful expression; the connotations of a particular use of colour, type, projection, and so on, have an immediate influence on the way in which a map is perceived and the degree to which it is trusted. The aesthetic principles of the cartographer therefore determine how a map symbolizes its subject and therefore how the map might appeal to users. While the Swiss maps of the dramatic Alpine landscape were created under the hegemony of objective accuracy, technological progress, and patience, Keates (1984: 42) remarked that no one talking to Eduard Imhof – a significant contributor to their design – could fail to observe his enthusiasm, interest in landscape, enjoyment of terrain, and involvement in landscape painting. As an artist, Imhof was concerned with the creation of the beautiful and sought to include the aesthetics of landscape in these creations. The act of landscape painting can also improve a cartographer's knowledge of the subject, for, according to Wright (1942: 528), 'A topographic map drawn by a man [sic] familiar with geology and physiography is likely to be far more expressive of relief and drainage than one drawn by an untrained observer'.

The degree to which a cartographer achieves the creation of an aesthetically successful map depends upon their experience of, and sensitivity to, the aesthetic character of the feature to be mapped, their awareness of the creative possibilities that cartographic language offers, their aesthetic judgment in evaluating these possibilities, and their practical skill in creating symbols which express the feature within the aesthetic principles of the user (Kent, 2005: 186). As cartographers apply their aesthetic judgment in the imaginative stages of symbol design, they gain experience in knowing what is likely to produce an aesthetically successful result. Since the design of some maps involves the collective experience and judgment of many, these can draw on a wider understanding of the subject.

In effect, the collective development of aesthetic judgment and a command of its expression widen the influence and effectiveness of maps. Trends in aesthetics emerge as best practice is established and replicated; it is no accident that despite their stylistic diversity (Kent and Vujakovic, 2009), the general aesthetic of state topographic maps is constructed from an underlying set of aesthetic values that have more in common than not. Maps are being made, used, and shared more than ever before, particularly through visual media such as the Internet, GIS, and not least computer games, where in-game maps are forming an increasingly important role in providing a sense of realism to support the illusion of the gaming experience (see Chapter 39).

As tools, maps must function to work and aesthetics are part of, and enhance, their function, as the subject or theme of the map should inspire creativity in cartography. More should be made of technology changing the rendering – the cartographic style – of maps depending on where the user is based as well as their preferences, lest international mapping turn into an ever-more-globalized blandscape that appeals to no one (see Kent, 2008).

Maps therefore do not need to be beautiful to be successful; they need to wield a particular aesthetic that supports the effective communication of their theme. Attempts to classify maps into particular styles based on aesthetics have been proposed (e.g. Peterson, 2012), yet each map – or series of maps, such as a national topographic map or atlas – speaks to us with its own 'voice' to some extent. A topographic map series, for example those produced by the national mapping agencies of Switzerland and the Netherlands at 1:50,000 scale, are successful in deriving and expressing a certain aesthetic value from the character of the different landscapes, whether from the awe and splendour of nature or the complex vitality of human civilization. Consequently, both series succeed in their symbolic representations. If the objective has been to create an aesthetically pleasing map, the designs have succeeded by exploiting the aesthetic qualities of the landscape to meet their goal.

The presence or absence of a particular feature in suggesting the overall aesthetic character of the landscape may itself also reveal a more general cultural attitude to the constituents of the environment as a whole. An example of this might be the treatment of wilderness, or more specifically, in relation to whether wilderness supports the theme to be communicated – a 'good view' of the landscape – or not:

> The classical perspective sees most significance in human action and human society. The creation of liveable places and usable spaces is a mark of civilization. Human use confers meaning on space. Outside of society, wilderness is something to be feared, an area of waste and desolation. It is human society which gives meaning and social significance to the world. For the romantics, in contrast, untouched spaces have the greatest significance; they have a purity which human contact seems to sully and degrade. Wilderness for the romantics is a place to be revered, a place of deep spiritual significance and a symbol of an earthly paradise.
>
> *(Short, 1991: 6)*

These attitudes may be expressed through the aesthetic treatment of different phenomena. On a topographic map constructing a view from the 'classical' perspective, for example, the human landscape may attract attention and convey a sense of vitality, order, and civilization through its symbols. It will perhaps form a stark contrast to areas of 'wilderness', which may be represented using 'white' space – a strong connotation indeed for suggesting loneliness, boredom, and desolation – a gap to be filled by 'civilization'. In October 2001, a spot near Ousefleet was named 'the most boring place in Britain' as its square on a 1:50 000 map contains the most 'white' space. Philip Round from the Ordnance Survey remarked 'We're not saying it's the dullest place in Britain. It might be the most fascinating place on earth but on our Landranger maps it has the least amount of information' (BBC, 2001). A 'romantic' perspective, in contrast, may perhaps draw attention to the appeal of natural features through their detailed depiction; artificial features may be subdued, and 'white' space minimized. Indeed, the white background may be replaced by another colour, as in the topographic maps of the Republic of Ireland, where green – as part of a series of hypsometric tints – also seems to suggest the omnipresence of a particular landscape, as opposed to its absence. If these are indeed different ways of interpreting the land – ways of representing a particular landscape,

*Figure 22.5* 'Over the Edge in 3D: Death in Grand Canyon' by Ken Field and Damien Demaj
(2013). Reproduced with permission

a 'good view' – it seems plausible to suggest that more general aesthetic values influence the cartographer's aesthetic judgments. Yet, the relationship between the cartographer's aesthetic values and the aesthetic expectations of the society in which they operate, remains an important and neglected area of study.

Thematic maps, i.e. those whose focus is on communicating a particular theme, can arguably draw on a wider range of aesthetic possibilities. 'Over the Edge in 3D: Death in Grand Canyon', a map designed by Ken Field and Damien Demaj (Figure 22.5), was exhibited at the International Cartographic Conference in Dresden, 2013, and won first prize in the paper maps category. The international jury commented that the entry was an 'Eye-catching and technically innovative product which pushes the limits of what paper maps can do' (ICA, 2016). Here, the landscape is portrayed as hostile – a place of death – and the map utilizes an associated aesthetic to communicate the theme more immediately than a conventional topographic map. The effect is made even more dramatic through the cartographers' intention to demonstrate the effect of chromastereoscopy – the impression of 3D viewing using colour – and the result is an immersion into a sensory experience that is perhaps reminiscent of film director Quentin Tarantino's work, especially through its associations with the visceral, comic-book aesthetic.

As Field (pers. comm.) explains:

> As far as aesthetics is concerned, the print version was deliberately [created] in a cartoon style with strong visuals and a lot of red to match the chromastereoscopic approach. Depth-encoded colour means colours are used in specific ways but I augmented that with saturated red in the top visual plane for the border and other marginalia. For the web version, the likelihood is that more people would see it. It couldn't be as 'gory' and had to deal with the subject a little more respectfully. I'd had one or two complaints about the perceived lack of respect shown by the print version so the web version is toned down and the pure red is reserved for the symbols. The map was to demo chromastereoscopy and not a search or decision to approach map design in a new or different way.

Although maps are versatile media, it is worth contemplating that the negative reaction to the map led to the creation of a more 'rational' choice of symbol (hexagons with numbers, or 'hexbins') for the version created for the Web. Clearly, maps command a certain authority beyond that of words or photographs, and perhaps the user's expectation of a map to exhibit a sense of 'unauthoredness' presents a barrier if cartographers are to realize the full aesthetic potential of cartography. Even though the intention was not (as Field states) to approach map design in a new or different way, that the map gained recognition as an example of excellent cartographic design, and in a category that is usually bound by convention and conservatism, presents something of a milestone in the history of cartographic aesthetics.

## Conclusion

Although attractive things make people feel good (Norman, 2004: 19), and the achievement of a successful cartographic design is recognized through a map's sustained use, the successful application of aesthetics within cartography does not depend on the creation of a beautiful map. There will always be the desire to derive a universal 'style' model to transform the presentation of geographic data, but such a model cannot be entirely valid because the values that construct any set of aesthetic principles will vary with time and culture (Kent, 2005: 186). As more maps are made, used, and shared by more people than ever before, it is possible that new aesthetic directions will emerge as best practices are replicated. It would be more useful for cartographers (and for cartography in general) to undertake further research regarding these developments, and their impact on whether users trust particular aesthetic approaches (see Muehlenhaus, 2012), than to attempt to devise universal principles. As designers, cartographers should continue to refine and exercise their aesthetic judgment if their work is to appeal and therefore individually discover how maps can meet their functional objectives more effectively. If we have become too familiar with maps as a way of bringing order through a singular and authoritative aesthetic, perhaps we are now witnessing the evolution of new cartographic aesthetics which have the scope to reflect the chaos of the world more honestly.

## References

Adams, A. (1983) *Examples: The Making of 40 Photographs* Boston, MA: Little, Brown and Company.

Board, C. (1967) *Maps as Models* in Chorley, R.J. and Haggett, P. (Eds) *Models in Geography* London: Methuen, pp.671–725.

BBC (2001) "The Most Boring Place in Britain" Available at: *http://news.bbc.co.uk/cbbcnews/hi/uk/newsid_1603000/1603136.stm* (Accessed 24 January 2007).

Brotton, J. (2014) *Great Maps* London: Dorling Kindersley.

Collinson, A. (1997) "Virtual Worlds" *The Cartographic Journal* 34 (2) pp.117–124.

Crampton, J.W. (2001) "Maps as Social Constructions: Power, Communication and Visualization" *Progress in Human Geography* 25 (2) pp.235–252.

Eckert, M. (1908) "On the Nature of Maps and Map Logic" (trans. by W. Joerg) *Bulletin of the American Geographical Society* 40 (6) pp.344–351.

Fabrikant, S.I., Christophe, S., Papastefanou, G. and Maggi, S. (2012) "Emotional response to map design aesthetics" *Proceedings (Extended Abstracts) of GIScience 2012*, Columbus, OH, 18–21 September.

Fisk, H.N. (1944) "Geological investigation of the alluvial valley of the lower Mississippi River" Washington, DC: U.S. Department of the Army/Mississippi River Commission.

Graham, G. (1997) *Philosophy of the Arts: An Introduction to Aesthetics* London: Routledge.

Hudson, B.J. (2000) "The Experience of Waterfalls" *Australian Geographical Studies* 38 (1) pp.71–84.

Huffman, D. (2013) "Is Cartography Dead?" Available at: *http://blog.visual.ly/is-cartography-dead/* (Accessed: 17 May 2013).

ICA (2016) "Map of the Month 07/2014: Over the Edge in 3D – Death in Grand Canyon" Available at: *http://icaci.org/map-of-the-month-072014/* (Accessed: 1 August 2016).

Imhof, E. (1982) *Cartographic Relief Presentation* (trans. Steward, H.J.) Berlin: Walter de Gruyter.

Imus, D. and Loftin, P. (2012) "The Beauty of Clear Communication" *Cartographic Perspectives* 73 pp.103–106.

Karssen, A.J. (1980) "The Artistic Elements in Map Design" *The Cartographic Journal* 17 (2) pp.124–127.

Keates, J. S. (1984) "The Cartographic Art" *Cartographica* Monograph 31 pp.37–43.

Kent, A.J. (2005) "Aesthetics: A Lost Cause in Cartographic Theory?" *The Cartographic Journal* 42 (2) pp.182–8.

Kent, A.J. (2008) "Cartographic Blandscapes and the New Noise: Finding the Good View in a Topographical Mashup" *The Bulletin of the Society of Cartographers* 42 (1,2) pp.29–37.

Kent, A.J. (2012) "From a Dry Statement of Facts to a Thing of Beauty: Understanding Aesthetics in the Mapping and Counter-Mapping of Place" *Cartographic Perspectives* 73 pp.39–60.

Kent, A.J. (2013) "Understanding Aesthetics: The Cartographers' Response" *The Bulletin of the Society of Cartographers* 46 (1,2) pp.31–43.

Kent, A.J. and Vujakovic, P. (2009) "Stylistic Diversity in European State 1:50 000 Topographic Maps" *The Cartographic Journal* 46 (3) pp.179–213.

Krygier, J.B. (1995) "Cartography as an Art and a Science?" *The Cartographic Journal* 32 (1) pp.3–10.

Norman, D.A. (2004) *Emotional Design: Why We Love (or Hate) Everyday Things* New York: Basic Books.

Muehlenhaus, I.A. (2012) "If Looks Could Kill: The Impact of Rhetorical Styles in Persuasive Geocommunication" *The Cartographic Journal* 49 (4) pp.361–375.

Petchenik, B.B. (1974) "A Verbal Approach to Characterizing the Look of Maps" *The American Cartographer* 1 (1) pp.63–71.

Peterson, G.N. (2012) *Cartographer's Toolkit: Colors, Typography, Pattern* Fort Collins, CO: PetersonGIS.

Purdon, W.H. (1859) "On the Trigonometrical Survey and Physical Configuration of the Valley of Kashmir" *The Geographical Journal* 4 (1) pp.31–33.

Robinson, A.H. (1952) *The Look of Maps: An Examination of Cartographic Design* Madison: University of Wisconsin Press.

Robinson, A.H., Morrison, J.L., Muehrcke, P.C., Kimerling, A.J. and Guptill, S.C. (1995) *Elements of Cartography* (6th ed.) New York: John Wiley & Sons.

SASI and Newman, M. (2006) "HIV Prevalence" (poster as part of the World Mapper series) Available at: *http://worldmapper.org* (Accessed: 1 August 2016).

Short, J.R. (1991) *Imagined Country* London: Routledge.

Tuan, Y.F. (1989) "Surface Phenomena and Aesthetic Experience" *Annals of the Association of American Geographers* 79 (2) pp.233–241.

Withycombe, J.G. (1925) "Recent Productions of the Ordnance Survey" *The Geographical Journal* 66 (6) pp.533–539.

Wood, M. and Gilhooly, K.J. (1996) "The Practitioner's View? A Pilot Study into Empirical Knowledge About Cartographic Design" in Wood, C.H. and Keller, C.P. (Eds) *Cartographic Design: Theoretical and Practical Perspectives* Chichester, UK: John Wiley & Sons, pp.67–76.

Woodruff, A. (2012) "The Aesthetician and the Cartographer" Available at: *www.axismaps.com/blog/2012/10/the-aesthetician-and-the-cartographer/* (Accessed: 17 May 2013).

Wright, J.K. (1942) "Map Makers are Human: Comments on the Subjective in Maps" *Geographical Review* 32 pp.527–544.

# 23

# Layout, balance, and visual hierarchy in map design

*Christopher Wesson*

Once a subject has been mapped, it must be arranged into a presentable format and carto-graphically finished to continue good and effective communication of the intended story to the map's audience. A lot of the theory around this comes from traditional paper mapping, but the same techniques exist in Geographic Information Systems (GIS) and in the production of maps for web and devices.

## Map layout

This concerns the arrangement of all the different elements of a map before publication. It may include the map body or data frame, title, legend, insets, charts and supplementary figures or graphics, north arrow, scalebar, border and various other elements within the marginalia. These are sometimes referred to as the 'map pieces'.

*Title* should generally include the map's subject, region and any temporal information. It should be easily identifiable as the map's title through its size and positioning. Sub-titles are best in a smaller type size.

*Legend* is a graphic guide that should comprise symbols, including colours, styles and patterns, which are not necessarily familiar to or known by the reader. A well-designed map should involve minimal reference to the legend.

*Insets* allow the cartographer to show areas at a more detailed scale or to reposition features that would have otherwise 'fallen' just off the sheet to make best use of the available page and keep the main mapping at the most appropriate scale. They may also include an overview of where the region is in a wider geographical context, known as a locator map.

*Charts and figures* are common to thematic maps showing geostatistical or geo-numerical data. They should be included only if they add weight to the message of the map. They can be very useful for showing a secondary variable associated with places of interest on the map.

*North arrow* or any other directional indicator is only required if the orientation is not obvious. Walking maps will often show grid north, magnetic north and true north to allow users to set the correct bearing on a compass. Arrows with a clear and suitably long North-South line are easier to use than more decorative ones.

*Scalebar* allows users to make measurements on the map and is a quick reference to size objects. It does not have to represent distance, for example it can represent travel time.

*Border* is a line encasing the edges of the mapped area or a line framing the entire map layout. It is sometimes known as a neatline, a term also used for a finer border line that may form the outer part of a grid and may contain graticule intersections.

Whether considered a layer of the map body or a layout piece in their own right, grids and graticules should be included if their location information is a significant aid to the map's use. Numerous other elements can be added as further map pieces if required, including author, date of creation, map projection, index of places, timelines, explanatory text, data sources, map credits, logos and copyright notices.

Contrary to popular belief, not all the items listed above are always required. Just like the simplicity principle governing the features shown within the body of the map, for the map layout there should be a defensible reason or sufficient need for each item placed. However, a styled map rarely stands alone. The elements or components above are often required for explanation and context, for example to introduce the subject, indicate geographic location, communicate symbology and stipulate orientation. Historically they have also been used for decoration and aesthetic appeal.

Before one can consider a map layout, it is wise to know and understand the formats of final production. The type of media should have already influenced colour settings and fonts by this point, and the critical factor now becomes page size and shape. Other considerations will help determine the need for each of the auxiliary map components, for example a map to be published within a written document will usually be captioned and hence may not need a title or explanation. A map's layout often requires some experimentation, perhaps even 'trial and error'

*Figure 23.1*   Achieving a desirable layout often involves trial and error either on a computer or by preliminary sketching

(Figure 23.1). In the past one might have sketched the layout in advance, which remains a useful planning technique, but now in modern GIS, graphics packages and other publication software, it is very easy to move and resize each component of the map. Successful map layout brings successful reading of a map. A map reader or viewer will often only notice and pay attention to the map layout if it is unsuccessful or obtrusive.

For folded paper maps, it is wise to create some guidelines of where the sheet folds will be. The cartographer should try to limit the number of foldlines that are crossed. A legend for example is often easier to use if kept to one or two panels of the map, and looks good if centrally positioned within that panel of the map sheet. Sometimes the same approach of creating an overlying grid can help design the layout of any map, and aligned components can also aid the map reader. It is thought that map pieces aligned in grid-like symmetry create a view that is easier for human cognition with the divided space creating more obvious sight lines and hierarchy. Earlier we said that the map reader will not notice layout; there is a slight exception with gridded layouts

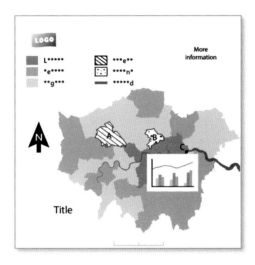

*Figure 23.2*   An example of poor map layout

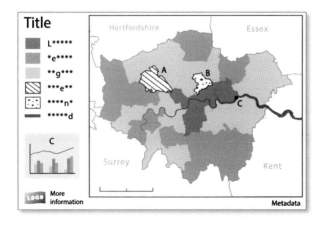

*Figure 23.3*   An example of better map layout

where the viewer will often understand the suggested order in which to view the elements. Grids have been commonplace in art and design for a long time, for instance the famous 'rule of thirds'. Today they are still used extensively in both old and new media alike.

The process of how the elements are afforded different prominence is what is known as *composition*. In map composition, what the cartographer is seeking to achieve is good gestalt (see next section), balance and symmetry through an effective visual hierarchy. There is no one correct answer as there are likely to be many different layout options that satisfy. Like so much of cartography, the layout a cartographer finally agrees upon should ultimately be determined by the purpose, message and audience of the map (Figures 23.2 and 23.3).

## Gestalt

Gestalt is a German word meaning shape or form. In cartography, it is used as a term for how people view multiple elements of the map. It brings together harmony, contrast and visual balance to aid the final composition. What is happening to the composition as a whole cannot be seen when working on or even viewing an individual component of the map, yet the components are controlled by the overall map layout: in their design, for example, in restricting their size and shape, in dictating their style or colour scheme, and so on; and for the map reader by governing the association between the parts. A cartographer must consider the graphic representation of the individual elements but also consider the effect on the whole composition. When elements of a map are presented together in the final layout, the human brain and eye are almost incapable of viewing an element of the map without absorbing its surroundings. Human cognition is based on 'clumping' and the eye seeks the eight gestalt laws of unit, segregation, unification, closure, continuity, proximity, similarity and prägnanz. These can all aid the user's ability to read, process and understand a map. (For more details on map perception and cognitive psychology, see Chapter 3.)

*Unit* is a unique entity which ends with itself, it may refer to a single feature or a group of features. Carvalho and Moura (2009) give the example of map orientation, often represented by three north arrows (magnetic, planed and true). The overall element could be considered a unit, equally so could each arrow. Either way they form standalone, unique elements of the map.

*Segregation* is the ability of the map reader to perceive different groups of features within an area of the map. This can be aided by the cartographic design with good use of colour, line weights, symbolization and contrast.

*Unification* refers to the impact of an object due to its visual appearance and the groupings that occur as a result. Grouping is generally aided by proximity and similarity and has a strong association with the visual hierarchy.

*Closure* is where an enclosed set of features, or even those that are perceived by the human eye as closed due to our brain filling in any small gaps, form a distinct group or parcel within the map.

*Continuity* allows the reader to make a smooth transition across the map and generally occurs where groups of features such as roads and rivers follow a similar direction. Continuity can also be achieved on continuous data such as height, for example by using good colour ramps. In composition, it is the flow between the map pieces.

*Proximity* is that objects located close to one another tend to form groups to the viewer. Equally, items that are vaguely close and of the same form, colour, weight, etc. will also form such groups.

*Similarity* refers to the natural human process of grouping items that are visually alike, for example of a similar colour, size or shape.

*Prägnanz*, also known as the law of good gestalt, literally translates as succinctness and is the idea that people simplify what they see, filtering out anything complex or unfamiliar, and what is left are groups of objects that form orderly patterns.

## Balance

Balance is very much a principle borrowed from the art world. Uneven weight in a composition creates informality, whereas balance builds trust; a well-balanced map is perceived as logical. In cartography, balance refers to the visual stability of the arrangement of the different map components within the map layout and its visual impact. When balance is poor, map readers may be confused, may only get part of the message, and may be put off. When balance is achieved, map readers will focus on the content of the map and take all of it in. Balance is therefore achieved by positioning features in relative equilibrium. Placing objects at equal distances around a midpoint or from a medial axis would not work on its own because not all elements yield the same visual power. This 'power' is often referred to as the visual weight or intensity of the map's components. Size, colour, pattern and contrast are all contributing factors toward an object's visual weight. Small, dark objects can be considered heavy or visually intense and they can balance larger, lighter ones.

Consider each map element to be a person sat on a see-saw (Figure 23.4). Each element will have a different visual weight, just like each person on the see-saw will have a different mass and therefore exert a different downward force. The further that downward force is from the pivot (fulcrum) the larger the turning force (moment). To achieve balance the total 'moment' applied to either side of the see-saw needs to be equal. In mechanics, the mathematical equation is that a moment is equal to the force multiplied by its distance from the fulcrum. So, in Figure 23.4, where the see-saw is balanced, the moment to the left of the fulcrum must be equal to that to the right,

$$A \times g \times d_1 = (B \times g \times d_2) + (C \times g \times d_3),$$

where $g$ is the acceleration of a mass due to gravity. Dividing this out as a common factor leaves the balanced equation:

$$A \times d_1 = (B \times d_2) + (C \times d_3)$$

The see-saw is balanced. In designing a map layout, the cartographer is looking for a similar equilibrium around a medial axis in order to achieve visual balance (Figures 23.5 and 23.6).

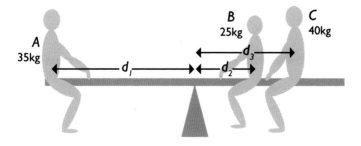

*Figure 23.4* **A see-saw in equilibrium**

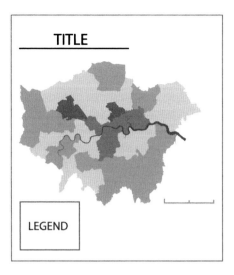

*Figure 23.5*   **An example of poor visual balance**

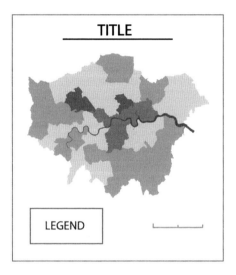

*Figure 23.6*   **An example of better visual balance, with symmetry**

The most important elements should usually appear in more prominent positions and occupy a larger area of the 'paper'. In the West, as we read from top-left to bottom-right, it is generally accepted that the more important elements be placed further toward the top-left and less important elements toward the bottom-right. There are no rules on where the individual map pieces should be placed, but each element should consider the above in addition to the specifics of the particular map. A title is usually placed centrally at the top of the page and a legend on either left or right side; however, the detail of the map may justify placing such elements in a different position, for example at the bottom of the sheet or covering a less important area of the map body. Consider Figure 23.7: the places of interest are all located around the river that runs from the top-left to the bottom-right of the page. It therefore makes sense to place overlying

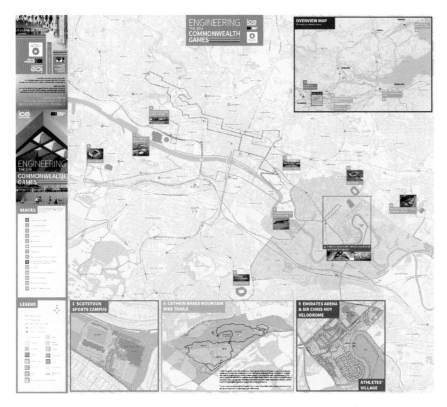

*Figure 23.7* 'Engineering the 2014 Commonwealth Games'
© Crown copyright and/or database right 2014 OS

map elements such as inset boxes and a legend in the positions of least interest, to the bottom-left or top-right of the page.

Scale is often defined in the title or legend, but a scalebar may be in its own box and often toward the bottom of the page. Acknowledgements, data sources and copyright are more often than not placed in the bottom corner. The arrangement should still lead the viewer to the critical information and there should only be one primary, clear message. Drawing too much attention to any of the supplementary map pieces such as a centrally placed figure, an overly ornate north arrow or frame, or an oversized legend will detract from the story of the map body.

For the map itself to be figural, it should be positioned roughly centrally, but the focal centre of the map should be positioned at the height of the optical or visual centre of the map frame, which lies approximately 5 per cent of the total height above the geometric centre of the frame. Again, this is due to the workings of the human brain and eye. Placing the mapping centred on the optical centre will give it prominence and give the map the most balance. The principle of an optical centre extends to influence all the other map pieces. The elements of the map should be arranged around the optical centre of the page. As with many printed documents, it is common to leave a bigger margin at the bottom than at the top of the page (Figure 23.8).

A map can have symmetrical balance by the elements of a map layout being arranged such that the body is central and the other components balance one another. Generally, a map is more likely to be balanced in the vertical axis than the horizontal due to the optical centre and the more important components tending to be nearer the top than the bottom of the page.

317

*Figure 23.8*　**The British Isles placed at the optical centre of the frame (black outline) versus placement at the geometric centre (grey fill)**

A map sheet with landscape orientation, as in Figure 23.3, can often be balanced in the horizontal axis. A balanced map layout is likely to be perceived as traditional, simple and conforming. It gives us the sense of order that we tend to seek.

It is also possible to design a more modern and creative layout design using an asymmetrical balance. Imagine the page once again as a balanced board or see-saw. In symmetrical balance, there is an equal moment either side of the axis between left and right (or top or bottom) of the page. In asymmetrical balance, there is still a balance of moments either side of an axis – so an even distribution of elements considering their areal size and their visual intensity – but this axis is likely to be along the shorter dimension of the page and off-centre, or on a diagonal across the page, rather than straight down the middle.

## Visual hierarchy

Visual hierarchy in map composition decides which of the map's components to promote and emphasize over others in order to best communicate the overall map. The subject and story of a map affects the importance of the features and components within it. Some elements are more important to portraying the map's message than others. This may be called the 'intellectual hierarchy'. Once the intellectual hierarchy has been calculated and understood, the cartographer can consider the appearance of each feature relative to its importance to the map.

The result is that components of the map are effectively assigned to different visual levels – a visual hierarchy. The aim of creating a clear visual hierarchy is to draw attention to certain elements of the map and push those of less importance further down the visual plane. If certain features are less important, they may still be required; if not, then they should be removed. This helps the user differentiate between map features or between map pieces and helps them comprehend the map's message effectively.

Contrast between elements of a map layout, just like the contrast between feature groups within the body of the map, is achieved by a mixture of methods, the result of which is visual hierarchy:

- Colour (of element, text, frame, background);
- Extrusion, shading and 3D effects;
- Figure isolation (white space, framing, vignettes, feathering);
- Figure-ground;
- Graphical effects (drop shadows, outer glows), halos, masks, gaps;
- Orientation;
- Overlap (including layer or draw order);
- Proximity;
- Shape (of element or background);
- Size or aspect;
- Text (font, weight, style);
- Texture or pattern;
- Weight (of element or frame).

Elements that contrast the most with the background or neighbouring elements will stand out at the fore of the visual hierarchy. The concept of *figure-ground* is key as it helps the user to distinguish between the main focus of the map (figure) and that which is background or contextual information (ground). The human eye distinguishes objects against whatever their background, and is one of the first things people will do when looking at any composition. This effect allows us to create visual depth in maps. It is a relationship that can be stable or unstable. Usually sufficient contrast is the key factor, but, as Figures 23.9 and 23.10 show, when contrast is equal, the shape and size of the object and background determine the stability of their relationship.

*Figure 23.9*   The instability of the figure-ground relationship. (Rubin's Vase is based on 'Rubin 2' by John Smithson, 2007, at English Wikipedia.)

319

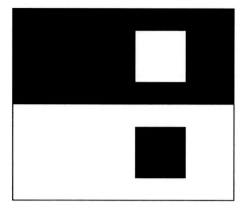

*Figure 23.10*  **The stability of the figure-ground relationship.**

The black chalice on the white background can also be seen as two opposing white faces on the black background, demonstrating that despite maximum contrast, this composition is unstable. Conversely, Figure 23.10 illustrates stability, as it is the small object (regardless of black or white) that assumes the role of figure against a far larger background.

For the cartography of the body of the map, figure-ground is something of an over-simplistic notion. Most maps have many more than two differently styled objects. Each of these features or feature groups has its own level of importance to the map and hence its own level within the visual hierarchy. So, in reality, the contrast between the absolute figure and absolute ground needs to be sufficient to support the variation in contrast at each level in between. It would make sense for the difference in contrast to be proportional to the difference in importance, but few maps are this scientific and sometimes features demand a value, for example a colour. In the UK, motorways are commonly recognized nowadays as blue, something that can remove them from their correct position in any mathematically formulated hierarchy. The same notion of creating different levels of hierarchy between a figure and a ground applies to the map layout just as it does to the map body. The aim is to guide the viewer's eye through the story of your map in the order that you intend. It is a visual narrative which will ultimately bring clarity to the subject matter and its story.

An aesthetically pleasing map will draw a viewer's initial attention, after which the visual levels take over as outlined later in Table 23.1. The first port of call will be the thematic information, the subject of the map and the main information or message that the cartographer wishes to communicate. Next we direct the viewer to the remainder of labels and symbols to start to build relationships and understanding. Then the base map and topography come to the fore to add context and surroundings to the information. After this, the reader can be offered further, supplementary information, such as more in-depth storytelling, figures and so on. Finally, the map journey is completed by providing a useable scale reference and revealing the sources of this information.

As well as maps, take a look at posters, newspapers and websites. Many of the contrast variables listed earlier in this section will become apparent, in particular, a range of type sizes coupled with variations in the amount of white space. Each element has a different visual weight and the different elements of the composition are balanced by their framing, layout and visual hierarchy (Figures 23.11 and 23.12).

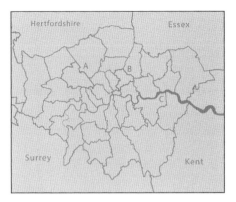

*Figure 23.11*  **An example of poor visual hierarchy**

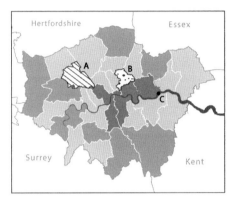

*Figure 23.12*  **An example of better visual hierarchy**

## Web, mobile and new media

Traditional map layout theory is based upon a fixed page size at a set scale. For on-screen displays, the rules around balance and hierarchy remain but there are many other considerations: screen dimensions and resolutions vary greatly, web mapping often has multiple if not continuous scale(s) and new-media cartography is often interactive. Map elements can be minimized or hidden, further explanation often appears as pop-ups or scrolling text. However, a simplified view of the levels of a visual hierarchy is likely to remain very similar to that of paper maps (as outlined in Table 23.1 below).

The screen-size problem can be overcome by deciding on, and designing for, the smallest screen size that you are willing to let the user view your map with. Working with this gives a defined extent similar to a traditional paper size and achieves a balance between a visually pleasing map and one that functions correctly, which are both required for successful communication in new media and hence good cartography.

As in any web or device design, the viewer's screen resolution is often assumed from data on the most frequently used resolutions in the present day. Resolution in pixels per inch also

Table 23.1 Basic visual hierarchy of any map, from Visual Level 1 as the highest, most prominent elements, through to Visual Level 5, as the least prominent. This table was inspired by Dent (1998) and Muehlenhaus (2014)

| Visual level | Elements |
|---|---|
| 1 | Thematic overlay (titles, symbols, legend, key information) |
| 2 | Other (base) map symbols and labels |
| 3 | Base map (topography and other features of map body) |
| 4 | Annotation, locator maps, figures |
| 5 | Other useful elements (scalebar, grid, graticules, frame, tool tips, credits and acknowledgements) |

affects the display size of map tiles or raster images. Developers can help manage such things by determining the user's screen size, resolution and device type on load. Cartographic developers must therefore consider a layout accordingly and consider creating a fluid layout or a responsive design rather than a traditional static layout.

Fluid layouts are based on proportionally laying out a 'page' so that elements take up the same percentage of space on different sized screens. Responsive design is where the layout changes based upon the screen size and/or type. Choosing between the two may be based on considerations such as the pros and cons of generalization, common layout and usability. With either approach, the balance of the map must be retained for the map to be successful. Internet mapping is discussed in more detail in Chapter 27.

## Critiquing

Finally, once the layout is complete, it is normal for the cartographer to undergo a last step of map critiquing. Though not strictly a part of the map layout process, it is best carried out at this stage in a map's production. Some texts go to great length to list every possible consideration, but in short, the aim is to check each component with the context of the overall to make sure the map works well. The steps can be summarized by the checklist below:

✓ Check the subject and message are obvious.
✓ Check that the map can be read by others, that it tells the right story and possesses the required appearance.
✓ Check data sources, projection, scale and generalization are appropriate.
✓ Check map layout, balance and visual hierarchy are as effective as they should be.

## Further reading

Foote, K.E. and Crum, S. (1995) "Basic Elements of Map Composition. Cartographic Communication, Section 4" *The Geographer's Craft Project, Department of Geography, The University of Colorado at Boulder* Available at: *www.colorado.edu/geography/gcraft/notes/cartocom/section4.html* (Accessed: 30 October 2016).

Huffman, D. (2011) "Assembly-Line Map Elements" (Cartastrophe: Mistakes Were Made) Available at: *https://cartastrophe.wordpress.com/2011/09/13/assembly-line-map-elements/* (Accessed: 15 December 2015).

Krygier, J. and Wood, D. (2005) *Making Maps: A Visual Guide to Map Design for GIS* New York: Guilford Press.

Land Trust GIS (2015) "Design a Great Map Layout" Available at: *www.landtrustgis.org/technology/advanced/design/* (Accessed: 23 December 2015).

National Geographic Society (2005) "Xpeditions Lesson Plan: Mapmaking Guide (6–8)" Available at: *www. nationalgeographic.com/xpeditions/lessons/09/g68/cartographyguidestudent.pdf* (Accessed: 15 December 2015).

Ordnance Survey (2015) "Cartographic Design Principles" Available at: *www.os.uk/resources/carto-design/ carto-design-principles.html* (Accessed: 30 November 2015).

Priscilla's Blog (2012) "Grids in Web Design: Online Newspapers" Available at: *www.webdesignstuff.co.uk/ wp101/2012/01/25/grids-in-web-design-online-newspapers-2/* (Accessed: 23 December 2015).

Robinson, A.H., Morrison, J.L., Muehrcke, P.C., Kimerling, A.J. and Guptill, S.C. (1995) *Elements of Cartography* (6th ed.) Chichester, UK: John Wiley & Sons.

Smashing Magazine (2015) "Design Principles" Available at: *www.smashingmagazine.com/tag/design-principles/* (Accessed: 4 December 2015).

## References

Carvalho, G.A. and Mourão Moura, A.C. (2009) "Applying Gestalt Theories and Graphical Semiology as Visual Reading Systems Supporting Thematic Cartography" *Proceedings of the 24th International Cartographic Conference, Santiago, Chile.* Available at: *http://icaci.org/files/documents/ICC_proceedings/ ICC2009/html/refer/20_4.pdf* (Accessed: 30 October 2016).

Dent, B. (1998) *Cartography: Thematic Map Design* (5th ed.) Dubuque, IA: William C. Brown.

Muehlenhaus, I. (2014) *Web Cartography: Map Design for Interactive and Mobile Devices* Boca Raton, FL: CRC Press.

# 24

# Colour in cartography

*Mary Spence*

---

> In visual perception a color is almost never seen as it really is – as it physically
> is. This fact makes color the most relative medium in art. In order to use color
> effectively it is necessary to recognize that color deceives continually.
>
> *(Josef Albers (1963))*

What is true in the world of art is equally applicable in the world of map design. The way we perceive colours is subjective, and our individual preferences evoke different aesthetic and emotional responses. Perception of colour is greatly affected by context – small areas will be more difficult to identify than larger areas. Different surround colours affect the appearance of a colour, making two identical swatches appear to be quite different or two different colours to look alike. Other factors such as ambient illumination or colour vision deficiency should also be taken into consideration in the selection process.

The choice of colour can determine how well a map communicates its message – good use of colour can enhance the reader's understanding whereas bad colour choice can easily confuse or even give totally the wrong impression. The graphic variables of shape, size, orientation and colour are used in map design to symbolize the information shown. Colour can be applied to points, lines, areas and text. It can be used to illustrate quantitative or qualitative data. It is arguably the most significant factor in dictating the look and feel of a map as well as its ability to tell the correct story efficiently and effectively.

It is important to remember that colour is relational – a single colour does not exist in a vacuum but is viewed in relation to the other colours present. This fact, together with the knowledge that 'color deceives', makes the complexities of ensuring the best choice of colour seem like a daunting task. So, how to choose the 'right ones' for a particular map with millions of colours available? The process begins with an understanding of basic colour theory and how colours interact. Then the fundamental map design principles of legibility (making sure it can be seen), contrast (adding emphasis), figure-ground (providing the focus) and visual hierarchy (giving it structure) can be called upon to help determine colour choices that work in each application.

## Colour perception

Colour is the perceivable characteristic of light; the part of light that is reflected by the object viewed. There are three properties – hue (red, yellow, green and so on), saturation (brightness, dullness) and value (lightness, darkness) – that enable us to define different aspects of colour in a common language. Arranging colours in a logical sequence is a useful tool in understanding them in relation to each other.

Traditional artists' colour theory dating from the seventeenth century uses the primary colours of red, yellow and blue – primary because they are not mixtures of other colours. The secondary colours are formed by mixing two primary colours and the tertiary colours by mixing a primary and a secondary colour. A colour wheel (Figure 24.1) contains the primary, secondary and tertiary colours arranged in spectral order. Envisioning where colours fall on the wheel is helpful in the process of understanding the visual effects of colour combinations when selecting harmonious arrangements or colours of equal importance. Complementary colours that are directly opposite each other on the wheel when used together give extreme contrast, but the resulting image can be difficult to look at. More harmonious and balanced effects can be achieved by using colours that are equally spaced on the wheel.

The process of adding colour to maps introduces two further colour wheels based on different primaries. Before preparing a map, it is essential to establish whether it will be delivered online or in print since they use different colour models. A vast array of colours can be produced by combining three primary colours either by additive or subtractive processes.

The colours that appear on computer screens (Figure 24.2) are mixtures of red, green and blue light – RGB. These are called the additive primaries – additive because colour is being added to a black background. The additive secondary colours of cyan, magenta and yellow are the primary colours of the subtractive CMYK model, i.e. cyan, magenta, yellow and black. Digital RGB values range from 0 to 255 (black is 0, 0, 0 and white is 255, 255, 255), which creates theoretically more than 16.7 million colours. However, not all of those combinations can be distinguished apart and, furthermore, monitors are capable of displaying only a limited gamut (range) of the visible spectrum.

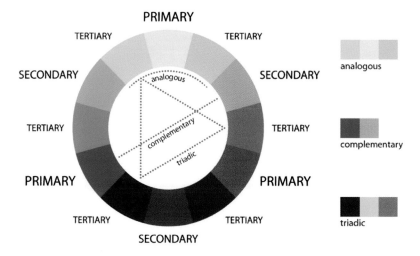

*Figure 24.1* **A traditional colour wheel**

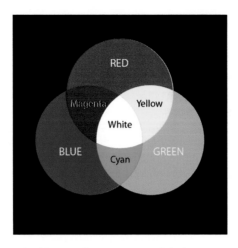

*Figure 24.2*   **RGB circles and additive colour mixing with light**

Another model used in GIS (see Chapter 17) describes colour in terms of its hue, saturation and value – HSV – and is a mathematical transformation of the RGB model. Hue specifies the perceived colour such as red or green and is expressed in angles from 0° to 360° (starting and ending on red). Saturation specifies the intensity or how vivid the colour appears (the amount of white added to a colour) ranging from 0 per cent (no colour) to 100 per cent (intense colour). Value specifies the brightness of the colour; how light or dark it is ranging from dark or low value, 0 per cent (black) to light or high value, 100 per cent (white). The HSV colour space (see Figure 24.3) is similar to traditional colour theory which can make refining colours more intuitive than in RGB. However, it is difficult in HSV to define a set of colours in a systematic manner since colours of the same value do not have the same perceived value. For example, fully saturated yellow and blue have the same value but the blue appears darker.

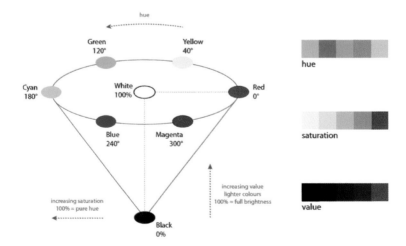

*Figure 24.3*   **HSV cone**

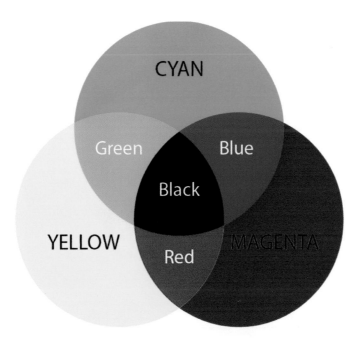

*Figure 24.4*  CMY circles and subtractive colour mixing with pigment

If maps are to be printed, they are most likely to use the four process colours, i.e. cyan, magenta, yellow and black, which combine to provide a wide range of colours. Cyan, magenta and yellow are called the subtractive primaries – subtractive because the addition of more colour (in this case, the ink) subtracts the amount of light reflected from the surface (in this case, the paper). The subtractive secondary colours of red, green and blue are the primary colours of the additive RGB model. With a range of 0 per cent to 100 per cent for each primary, thousands of colours can be created by mixing different amounts of each ink. The overprinting of 100 per cent cyan, magenta and yellow inks theoretically produces black, but a black (K) ink is added for greater depth, hence the term CMYK.

The Pantone Matching System (PMS) is also used in printing – a standardized colour reproduction system of spot colours (i.e. individual colours generated for a single print run) which are created from 13 base pigments plus black. Although the Pantone colours that are readily employed in design work can be applied successfully to mapping, CMYK printing is by far the most cost effective. If corporate colours are required on a map, they can be translated into CMYK, but an exact match may not be possible. To achieve greater colour control, publishers may introduce spot colours as well as CMYK for more prestigious products (Figure 24.5).

If a map is being prepared for electronic as well as paper delivery, careful attention must be given to how the colours translate from RGB to CMYK and vice versa – the colours will not be an exact match, but so long as the relationships and overall balance are good the result will still be effective. No single device is capable of producing all visible colours. You can make bright saturated colours in RGB, but the range becomes more limited when printing in CMYK. Similarly, you will not be able to reproduce the wide variety of Pantone colours on

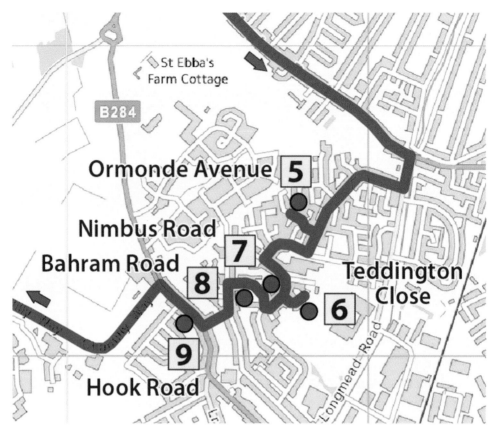

*Figure 24.5* A route map incorporating Pantone colours converted to RGB. Pantone 2415 (the purple) becomes 162, 25, 132 (R, G, B) and is used for the lettering with a paler version for the route

a computer screen. Whatever device the map is viewed upon, whether tablet, smart phone or paper, the fundamental guidelines of what colours work together to convey the right message remain the same.

## Colour interactions

The way an isolated colour appears can become very different once it is surrounded by other colours. This interaction of adjacent colours is called simultaneous contrast and is at its most extreme with complementary colours – where the perceived colour of an area tends to take on a hue opposite to that of the surrounding area. For example, strong green makes a grey appear to take on a reddish hue, and a strong red a greenish hue (Figure 24.6).

The colour wheel described earlier can be divided into warm and cool colours; warm colours (the reds, oranges and yellows) are more vibrant, whereas cool colours (the greens, blues and purples) create a sense of calm. The apparent temperature of the colour affects how close an

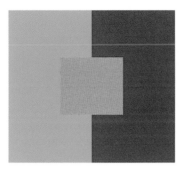

*Figure 24.6* Simultaneous contrast. Note how the effect is emphasized by the introduction of a black line between the red and green

item seems to be; cooler colours recede into the page, and warmer colours, especially yellow, advance towards us (Figure 24.7).

The visual weight of colours influences the hierarchy, balance and harmony of map elements. For example, pure red is considered 'heavier' than blue, which in turn is heavier than green, then orange, with yellow the 'lightest'. In order to achieve equal weights for colours, it is necessary to use relative proportions of each (Figure 24.8). For example, to imply the same importance, a blue needs to be slightly stronger than a red to appear the same. Darker colours appear as stronger or heavier whereas paler colours suggest less importance (Figure 24.9). The effect of this contrast is useful in creating a balanced map using strong colours for the main topic over a softer range of background colours.

*Figure 24.7* The red appears to advance in front of the blue, but note how the yellow advances most of all

*Figure 24.8*   Relative visual weight of colours, from red (heaviest) to yellow (lightest)

Creating colour swatches is helpful in illustrating the different effects caused by the inter-action of colours, but the impact of such effects on map detail can be much more complex. Recognizing these effects is the first step towards selecting colours for linework and symbols that remain legible when they appear over a variety of different backgrounds. Unintentional effects caused by the interaction of contrasting colours can be visually distracting and may even diminish the readability of the map.

## Colour resources

There are many online resources to help with the selection of colours, some of which are specifically created for cartographic applications, for example:

- ColorBrewer 2.0 at *http://colorbrewer2.org* – especially useful for thematic mapping offering colour schemes for different classes of data;
- Adobe Color at *https://color.adobe.com/create/color-wheel/* – a colour scheme generator which includes analogous, complementary and other schemes plus a vast array of existing themes to inspire; and
- Paletton at *http://paletton.com* – a tool for creating colour combinations that work well together including simulations for colour vision deficiency.

## Qualitative maps

Colours selected for qualitative information should not show a hierarchy but be of equal visual importance. Their job is simply to distinguish one category of data from another. However, colour on its own can be a limiting factor with some elements, and other variables such as shape, orientation and form can be introduced to expand the range of possibilities.

*Figure 24.9*   Areas surrounded by lighter colours appear darker, while areas on a darker background appear lighter

## Points

In its simplest form colour can be applied to qualitative point symbols of the same shape where contrast is the major element in differentiating one type from the other. Orientation can be altered to create more variety, but it is helpful if both colour and orientation are used together. Smaller symbols need more intense colours to show up when viewed against a complex back-drop of map detail. Picture symbols which represent a stylized representation of a feature are easier to identify and simple to remember. It can be helpful to users if classes of information are grouped by colour, e.g. heritage information in brown, tourist information in purple and trail information in red (for example, see Figure 24.10).

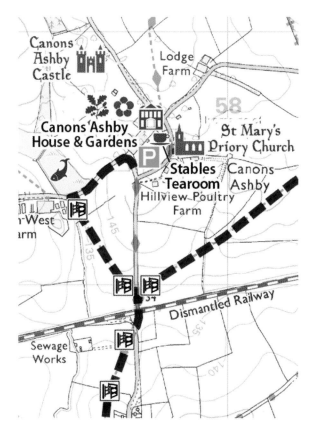

*Figure 24.10*   An extract from *South Northants Heritage Trail* showing walk route and features of interest highlighted over Ordnance Survey 1:25,000 map background (© Global Mapping 2016 and © Crown Copyright 2016 Ordnance Survey. All rights reserved.). Note the following: (1) colour-themed groups of symbols – walk information in red, heritage information in brown, tourist information in purple; (2) features texts darker than symbols so that they appear to be the same strength; (3) solid white type halos to lift feature names above the background detail; (4) walk route transparent so as not to obliterate the OS base detail; (5) rail line added in heritage brown to highlight the dismantled railway; and (6) OS map set to 75 per cent transparency to help the added features to sit above the base

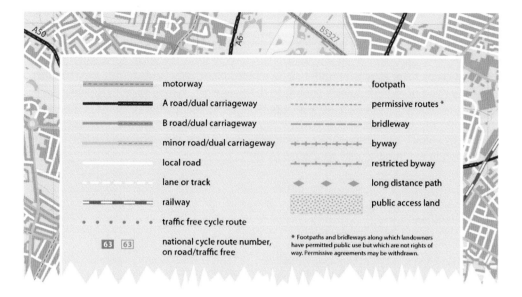

*Figure 24.11*  **The legend from *Visitors' Map of Charnwood Forest Regional Park* (© Global Mapping 2016). Note the following: (1) variety of line widths, colours and styles; (2) conventional (UK) colouration for roads with increasing widths for more important roads and additional pecked centre lines to indicate dual carriageways; and (3) colour coordinated road numbers on the map**

## Lines

On its own, colour would yield a particularly narrow range of different lines, e.g. blue for drainage, red for roads, brown for contours and such like. Introducing other variables, such as weight of line, change in form (e.g. pecked) or multiple lines, can greatly expand that range. For example, cased (outlined) roads of different widths with different colour fills can portray a complex set of road categories with a hierarchy that aids interpretation (see Figure 24.11).

## Areas

The maximum number of colours that can be easily differentiated from legend to map is about 12, but by grouping similar classes together and introducing texture or pattern to the areas, more categories can be identified. Symbols and direct labelling of features can expand the range even further. A notable example of this approach is a detailed geological map where there can be in the order of 50–100 different categories to be identified. Grouping colours by theme is helpful, but eight different greens representing the Mesozoic era would be impossible to distinguish without some additional visual clues. Although equal visual importance is essential in qualitative maps (Figures 24.12 and 24.13), smaller areas may need to use a stronger colour just to be noticeable amongst the larger areas. Balance is the key.

| | Retail | | Retail |
| | Industrial/business | | Industrial/business |
| | Education | | Education |
| | Health | | Health |
| | Park/open space | | Park/open space |
| | Woodland | | Woodland |

*Figure 24.12*   Land use legend with equal value colours. Note there are two options for picking out the building shapes using colour as well as keyline: either the same simple black overprint for all buildings is used, or a separate darker shade in each category which yields a brighter effect

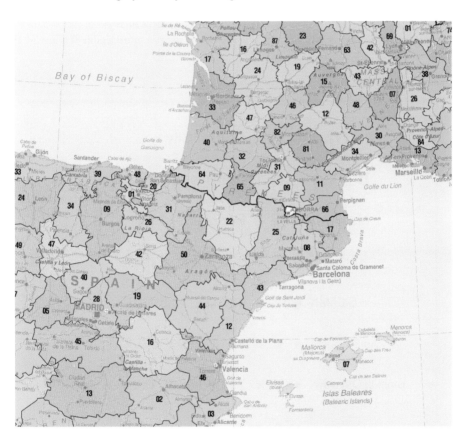

*Figure 24.13*   Relative visual weight of colours, from red (heaviest) to yellow (lightest) in an extract from the Europe Postcodes desk mat (© Global Mapping and data © XYZ Maps 2016). Note the following: (1) the use of five different shades of each colour within countries; (2) the pale grey linework, symbols and type detail allowing postcode information to take precedence yet retaining useful base references; and (3) the colour-matched postcode boundaries and numbers

## Quantitative maps

The purpose of colour in a quantitative map is to illustrate how much of something there is compared with something else. Its job is to make the greater elements appear more important. Ramps of colour can be used to indicate a progression of numeric values and create a visual hierarchy where dark is perceived as more. For example, increases in population density might be represented by the increased saturation of a single colour, or temperature differences shown using a diverging sequence of colours from blue (for cold) to red (for warm).

### Points

With quantitative point symbols, the effect of size enables colour to play a greater role – the larger the symbol, the more colour is seen. Graduated symbols (range of values per symbol size) or proportional symbols (that are individually sized according to specific data values) can use a single colour, bringing smaller symbols to the fore with outlines to separate them (Figure 24.14). Alternatively, adopting a range of increasing saturation as symbols increase in size would offer additional emphasis to the progression, making the whole picture more legible (Figure 24.15).

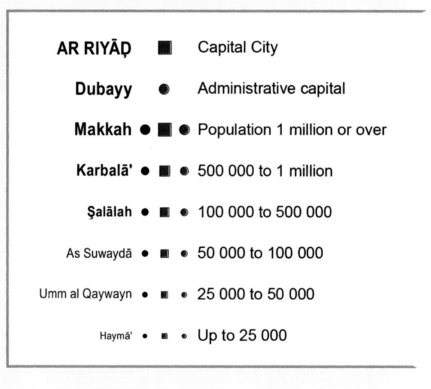

*Figure 24.14* A legend showing settlements graded by population (© Global Mapping and data © XYZ Maps 2016). Note the increase in type size and weight that supplements the small increments in symbol sizes, combining to ease identification

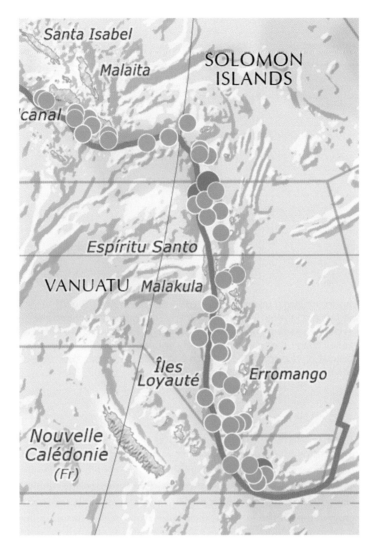

*Figure 24.15* An extract from *The Dynamic World*, featuring earthquakes with magnitudes 6.0–7.4 and >7.5 (© Global Mapping and data © XYZ Maps 2016). Note the following: (1) white outlines enabling the clear depiction of so many overlapping symbols while lifting them above the Natural Earth background; and (2) the combination of larger size and darker green to aid the visibility of the >7.5 magnitude symbols, which would be in danger of being hidden by the profusion of the smaller ones

## Lines

Examples of quantitative line symbols include stream flow, bird migration, commodity exports/imports and the like, where the simplest variable is the thickness of line in a single colour – a wider line signifying a larger number. This may be perfectly adequate for simple datasets, but introducing a range of colours as well as varying the weight of line would improve the interpretation of more complex patterns and enable different amounts of a wide variety of information to be shown clearly (Figure 24.16).

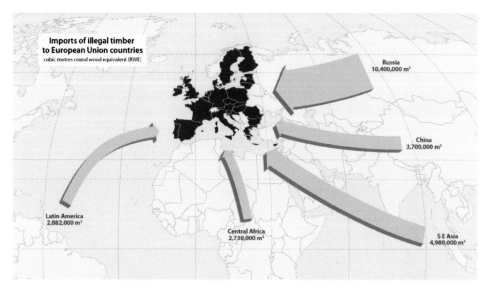

*Figure 24.16*  Imports of illegal timber to the European Union (data © XYZ Maps). Note the following: (1) proportional line widths annotated with volume figures; (2) darker shading (reflecting the EU flag colour), which lifts the EU countries yet still leaves the boundaries clearly visible; and (3) the complementary contrast of blue and yellow which enhances the simple depiction of minimal data over a neutral background

## Areas

Quantitative datasets can have sequential or diverging arrangements. The number of colours that can be easily identified within a single colour range is surprisingly small; while anything more than five or six classes can be easily distinguished in the legend, the differences are not so apparent when isolated and scattered around the map. This is particularly significant in choropleth mapping (data coloured by the ratio of value to the geographical area) where the colours do not sit in a progression on the map and their appearance can be further influenced by the adjacent colours. However, close colour bands can work well for isopleth maps (data shown as a continuous distribution) with gradual change over the map and where the colours appear next to each other as they do in the legend. Labelling isolines can further enhance the interpretation of the data presented (Figures 24.17 and 24.18).

## Text

The legibility of text on a map is determined by the contrast between the lettering and the background detail. Coloured lettering can be used to make certain text stand out or draw attention to a particular feature. It can be an attractive element in the naming of features on a map, e.g. blue for water features, green for primary route numbers, red for places of interest and so on. However, if a product is being published in several languages, only the black plate would be changed to economize the printing, meaning that all lettering would have to be in black (Figure 24.19).

Coloured type works best on a plain or simple background but becomes less legible when the base map is full of multi-coloured detail. Stronger colours and larger fonts would

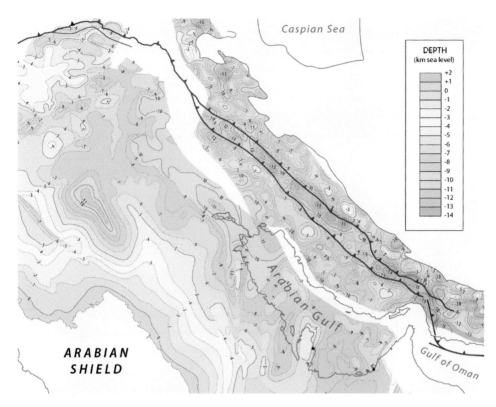

*Figure 24.17*  Extract from *The Geological Evolution of Saudi Arabia* illustrating the depth of the Arabian Platform (© Global Mapping 2007). Note the close colour bands with annotated isolines

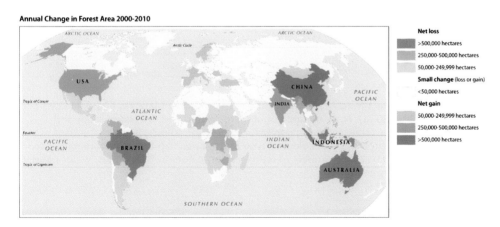

*Figure 24.18*  'Forest Trends' map from *Environmental World: Supplementary Information* (© Global Mapping and data © XYZ Maps 2011). Note the following: (1) the diverging colour scheme illustrating loss and gain; (2) the colour association of red with 'bad' and green with 'good'; and (3) the red-green colourway adapted to be suitable for colour vision deficiency (CVD)

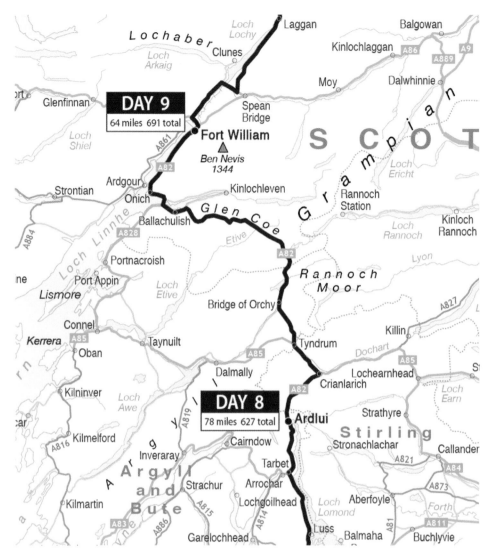

*Figure 24.19* An extract from *Le Jog* cycle route map (© Global Mapping and data © XYZ Maps 2016). Note the cycle route highlighted with overnight stops and distance tables in strong contrast red over the softer colours of the base map, and the use of coloured type in the base map to link feature names to linework – road numbers to roads, region names to boundaries, water names to rivers and lochs

improve legibility, while type halos can be employed to lift the names above the background. Straightforward white halos work well when the base colour is white, but when coloured text sits on different coloured areas it can be beneficial to apply the background area colour to the halo for an effect that is much less stark than white. On a multi-coloured base, such as a Natural Earth rendition, the application of transparency to the white halos can make them less obtrusive by allowing some of the background colour to show through (Figures 24.20 and 24.21).

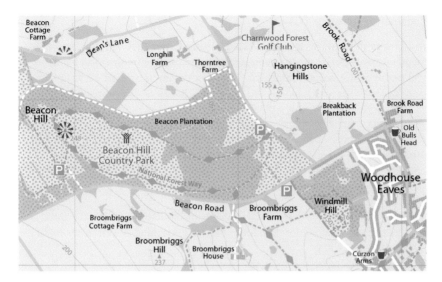

*Figure 24.20*    An extract from *Visitors' Map of Charnwood Forest Regional Park* (© Global Mapping and data © XYZ Maps 2015). Note the halos for the purple (features of interest), orange (contour values) and green (long distance paths) type applied to the green spot-pattern open-access areas, taking on the colour behind them

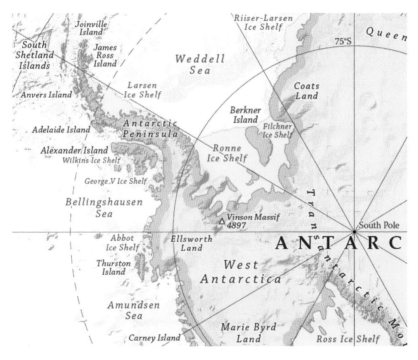

*Figure 24.21*    An extract from an antique-style world map (© Global Mapping and data © XYZ Maps). Note the following: (1) dark brown text for land features and dark blue for sea and ice detail to reflect the antique feel of the map; (2) white text halos for land features set to 50 per cent transparency; and (3) text halos for sea features in the sea colour and ice features in the ice colour

## Colour vision deficiency (CVD)

With 1 in 12 males and 1 in 200 females affected to some extent by CVD, it is important to consider what adjustments can be made to colours to ensure that the information being presented is meaningful to the widest possible audience. The most common types of CVD are:

- protanopia – a red deficiency, i.e. a reduced sensitivity to red light, where, for example, blues and purples can be confused because the viewer cannot 'see' the red element in the colour purple;
- deuteranopia – a green deficiency, i.e. a reduced sensitivity to green light, where colours with an element of green in them are confused; and
- tritanopia – a blue deficiency, i.e. a reduced sensitivity to blue light, where viewers have a tendency to confuse blues and greens.

Protanopia and deuteranopia are known as red-green 'blindness' and are the most common, but blue deficiency and total colour 'blindness' (where everything is seen in shades of grey) are extremely rare (Figure 24.22).

There are a number of online tools that test the viability of colour choice by simulating colour-impaired vision, such as:

- Color Oracle – *http://colororacle.org*;
- Coblis – *http://color-blindness.com/coblis-color-blindness-simulator/*;
- Vischeck – *http://vischeck.com*;
- Adobe Photoshop has a proof setup option to view the image in protanopia- and deuteranopia-type simulations.

Some colour pickers, such as Paletton (*http://paletton.com*), will also test the colours selected for a range of deficiencies, from reduced sensitivity to total colour 'blindness'.

Although these tools are convenient and easy to use, it would be impossible to actually measure the results of the simulation and translate them back, but they are invaluable as a guide to illustrate the potential problems. It is important to recognize where colour confusion is likely to arise to start

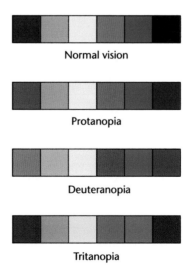

Normal vision

Protanopia

Deuteranopia

Tritanopia

*Figure 24.22*  Simulation of different colour vision deficiencies

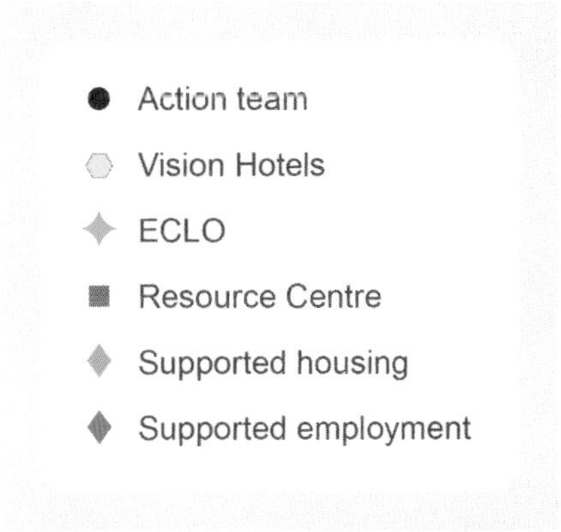

● Action team

◌ Vision Hotels

✦ ECLO

▪ Resource Centre

◆ Supported housing

◆ Supported employment

*Figure 24.23*  The legend from a resources map for the visually impaired. Note the different symbol shapes and colours and 'supported' features in the same shape and the yellow symbol outlined in black to aid visibility against the white background

with, and to be aware of the steps necessary to minimize this without detracting from the overall aesthetics of the map. The inclusion of other variables, such as shape, size, texture, pattern and annotation, all help the user to distinguish between the different features (Figure 24.23).

## Examples of Colour in Cartography

Figures 24.24 to 24.28 from *The Environmental World* present a case study to illustrate the cartographic application of colour.

*Figure 24.24*  An extract from *The Environmental World* (© Global Mapping and data © XYZ Maps 2016). Note the following: (1) transparency is used as an aid to legibility; (2) a transparency of 50 per cent has been applied to white type halos so as not to obliterate the Natural Earth image behind the names and text panels; (3) the colour associations used in the labelling, e.g. the charcoal-coloured lettering for background names and black for highlighted features; and (5) the colour associations of the symbols, e.g. deforestation (green and brown), coral bleaching (coral), flooding (blue), factories and pollution (black), bold town spots in red and black for the most polluted cities, and red and yellow warning triangles with red lettering for the areas suffering extreme environmental damage

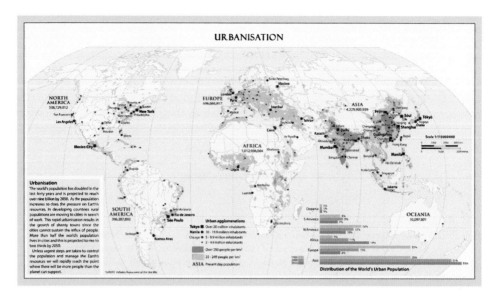

*Figure 24.25* An extract from *The Environmental World* showing the Urbanization inset (© Global Mapping and data © XYZ Maps 2016). Note the following use of colour in the frames and outlines: (1) the use of dark red for the frame (corporate colour), which is not so 'severe' as black; (2) the map surround is pale grey which is softer than white; (3) a blue graticule is used as the map edge; (4) the surround pale grey is used as an inner frame to hold back the map detail; and (6) the text box background colour is themed to suit the map colours

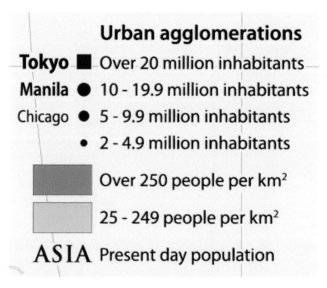

*Figure 24.26* An extract from *The Environmental World* showing the legend from the Urbanization inset (© Global Mapping and data © XYZ Maps 2016). Note that (1) the legend sits in the sea without a box; (2) only the top two categories of population density are shown as the lower categories are not so relevant to the topic; and (3) a coloured type halo (in the same colour as the sea) is used to break the graticule lines

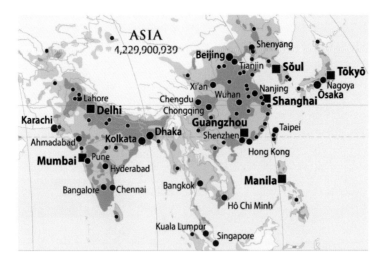

*Figure 24.27* An extract from *The Environmental World* from the Urbanization inset (© Global Mapping and data © XYZ Maps 2016). Note the absence of keylines (boundary lines) for population density and the white outlines applied to the circle symbols, with the smaller symbols overlapping larger ones

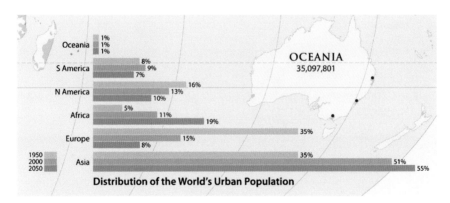

*Figure 24.28* An extract from *The Environmental World* showing a bar chart from the Urbanization inset (© Global Mapping and data © XYZ Maps 2016). Note the following: (1) the use of a complementary colour to the population colours; (2) white outlines to the bars; (3) no scale is required for the bars since the actual percentages are given; and (2) a coloured type halo is used to break the graticule

## Dare to be different

Although colour associations and conventions are a comfortable start to aid the process of choosing colours that work for the data being portrayed, this need not always be the case (see Figure 24.29). Certain products such as poster maps and publicity brochures may merit a different approach. Stepping away from the norm can bring a refreshing look to a map, turning it into an exciting graphic at the same time. So long as the message is clear, there can be no harm in being adventurous (Figures 24.30 and 24.31).

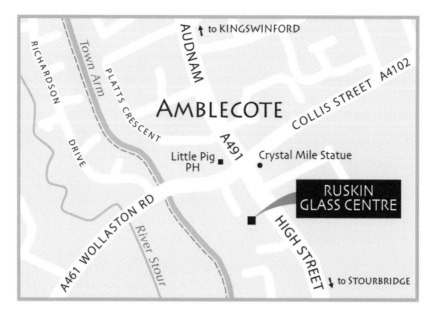

*Figure 24.29*   A map showing the location of premises in a publicity brochure using the
artist's favourite colours

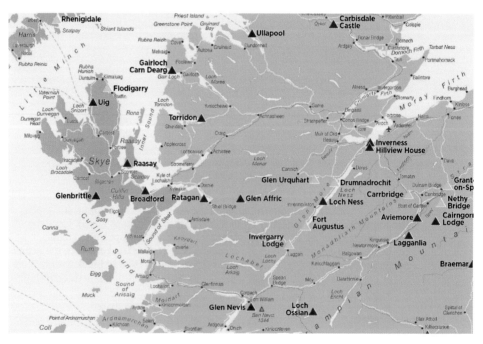

*Figure 24.30*   An extract from a SYHA (Scottish Youth Hostel Association) poster showing youth
hostels (red) and affiliate hostels (yellow) (© Global Mapping and data
© XYZ Maps 2016). Note the following: (1) bold regional colours which necessitate
high-contrast symbols for the hostels; (2) white edges applied to symbols to lift
them above the strong base colours; and (3) the use of grey type for base detail and
black for hostel names, with white halo set to 50 per cent transparency

*Figure 24.31*   An extract from *40°N Planisphere* (© Global Mapping and Mark B Peterson 2016). Note that the night sky lends itself to a black background with the Milky Way highlighted in white shading and the use of bright, high-contrast colours to stand out against the black

## Conclusion

Many things affect how colours appear to the map reader: the device the map is being viewed upon (computer, tablet, smart phone, paper), the viewing conditions (bright sunshine, night viewing, red-light conditions at sea), user abilities (visual impairment, CVD), cultural conventions (purple is associated with wealth and power in many countries, but is the colour of death and mourning in Brazil) and so on. One size does not fit all. There are various tried and tested techniques for achieving clarity and legibility in map design. For example, the layering of information aids interpretation, but, just as a single colour should not be viewed in isolation, neither should only one layer of information be considered at a time. It is essential to be aware of how the various elements of a design fit together – a holistic approach is required.

Colour has a major role to play in data visualization if patterns and relationships are to be depicted correctly, directing attention and emphasizing features. At the same time, the aim should be to produce a harmonious arrangement that provides a sense of order and engages the map reader. Too much colour becomes difficult on the eye, while too little is bland. Balance is required. The ultimate sign of success is a design that works, is imperceptible and looks simple.

## Reference

Albers, J. (1963) *Interaction of Color* New Haven, CT and London: Yale University Press.

# 25

# Lettering and labelling on maps

*Christopher Wesson*

---

Text is a cartographic incongruity. Lettering and labelling a map can be seen as an untidy, destructive, yet often necessary, additional method of communication. It explains the parts of information that simply cannot be expressed easily in any other visual form. Placing text or type is arguably one of the most delicate and specialist skills of the cartographer. As surmised by Imhof (1975), 'Good name position aids map reading considerably and enhances the aesthetics of a map'.

Before lettering a map it is important to consider the purpose of the map and what information the labels are going to portray. Given this, one can begin to understand and deduce the intellectual hierarchy of the features to be labelled and plan the visual hierarchy of the text itself. This allows users to read, search and extrapolate meaning from the map as efficiently as possible by using the properties of the labels as a powerful supplement to the underlying map features. A good hierarchy of text will usually require considerable experimentation, revision and refinement.

*Lettering* is literally the presentation of letters and the process of inscribing with them. Strictly speaking, lettering involves hand–drawn artistry whereas standardized letters are known as type. However in cartography, where both lettering and type may be encountered, 'map lettering' simply refers to the process of choosing a font, preparing names and placing or rendering them in a suitable position.

*Labelling* is describing, naming or classifying an object or feature. Cartographic labelling is map lettering using attributes of, or data associated with, a given geospatial feature, for example its name or a numerical value. From a technical perspective, lettering and labelling combine to form the rules that govern how an item of text, such as a label, will appear.

When labels are put on a map, they often become a figural element of the composition, a source of attraction for the map reader. 'Cartography is a medium of presentation for spatial data and it follows that when such data requires (additional) identification, then that identification becomes an integral part of the map' (Robinson, 1952). Small-scale maps especially are defined by their lettering. Historically, larger-scale maps were multi-purpose or reference materials with many names purposefully subdued so as to not mask too much detail, whereas small-scale maps served a limited audience and hence lettering could be styled to match a specific purpose or theme and really take centre-stage. Today in an age of increased data visualization, this scale–based separation may no longer be true.

While the content of a label represents important information such as a location name, these items of text actually communicate far more: 'Words on a map mean what they say but also mean what they show' (Krygier and Wood, 2011). The labelling in terms of style, form, weight, size and placement all constitute different meaning and so convey extra information to the user.

In their literal function, labels are names of the features to which they are linked by a method of inference, usually position. Their function can be further locative by the use of spaced lettering to indicate extent. The form of the lettering that makes up the label can tie it to one of any number of nominal classes and sub-classes set by the cartographer. By variations of size, case, tone and boldness of lettering, labels can also be ordinal – that is to say that a cartographer can label a feature based not only on its class but also its size or importance. It has previously been suggested (Robinson *et al.*, 1995) that 'the type on a thematic map does not run the gamut of functions that it does on general reference maps', but today the complexity of data to be visualized dictates that even thematic maps can benefit from well-thought-out characteristics of text.

*Style*, or type style, refers to the design physiognomy of the text, i.e. the chosen typeface and its appearance. It affects the overall look of the map, and often in cartography the trend is to avoid indulgence, preferring simple or tried and trusted fonts for the sake of maximum legibility and successful digestion of the map and its information. However, type style can be used as a relatively effortless yet very impactful way to tie a map to a particular theme or genre. With a regular sans serif, non-italic, non-calligraphic typeface, the treasure map in Figure 25.1 would simply lose its integrity and association to the 'Old World' theme, a world of exploration and pirates.

The definition of style used to be dominated by serifs and sans serifs. Serifs are small lines used to finish off a main stroke of a letter whereas sans (meaning without) serifs are cleaner letters. Scientific studies on the merits of each have tended to be inconclusive. Add to this the near endless

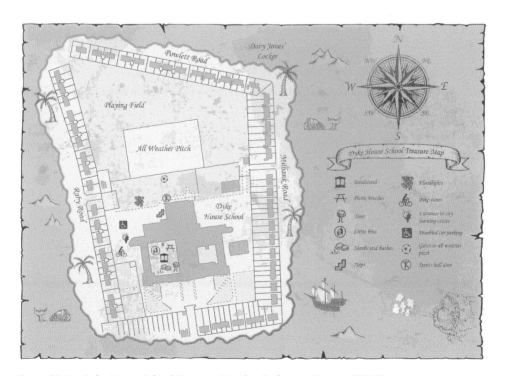

*Figure 25.1    Dyke House School Treasure Map* by Ordnance Survey (2013).

number of typefaces now in existence, and style becomes increasingly a design choice. However, one should still consider the medium for which a typeface was designed, and the subject for which it is to be used. There is no point using a typeface with poor numeral legibility for numbering roads for instance. Some are more legible on-screen, others in print; some work better at small point sizes than others; some possess texture or a clear theme or character (see Figure 25.2).

One should also consider that map letters are not set in a block as they are for literary texts, so the efficiency in perception of words is often in isolation. Furthermore, when maps are viewed on-screen, the monitor or device resolution affects the legibility and perceptibility of text, specifically affecting elements such as size, colour, brightness, contrast and anti-aliasing. It is common to use more than one typeface on a map, but they should work in harmony with the map and with one another. Using too many styles will negatively impact the map's aesthetic and lead to confusion for the map reader. Map publishers may also use typeface to enhance brand recognition and consistency.

Existing published maps offer a way of learning which typefaces suit different map types, but it is good practice to test a shortlist of options to see which works best and to check for any undesirable glyphs (characters). Typefaces and individual fonts can be compared quickly and easily on the websites of most foundries and sellers (as in Figure 25.3). Purchased and installed fonts can be previewed in the context of a generic small-scale map using an online tool called 'TypeBrewer', and 'FontShop' has a plugin to preview its fonts in Adobe Illustrator, but in most modern-day software, for example a Geographic Information System (GIS), it is quick and easy to switch between fonts.

*Form*, or typographic form, describes the physical properties of the text itself. Often in cartography, form is used to associate a label to a specific feature class or grouping. This is achieved by a combination of typographic variables and refers to whether the text is given extra definition such as bold or italic; whether different versions of the typeface (variants called fonts) such as light, condensed, narrow or semi-bold are used; whether type is in upper or lowercase; what the colour of the font is; or the distances between letters.

The traditional convention is for man-made features and general area names to be shown in regular or upright text and for landform, hydrography and other features of the natural environment to be shown in slant or italics. However, with the possible exception of hydrographic text, it is increasingly rare for this approach to be strictly adhered to. Tyner (2010) suggests that for hydrographic text any slant should be in the direction of flow.

*Case*, or letter case, is a way of adding emphasis, with uppercase generally denoting importance (Figure 25.4). In cartography, uppercase, text strings comprised solely of capital letters, is

# Loxley: Robin hood theme

# Corporate A: Mercedes-Benz

# FACE YOUR FEARS: HORROR THEME

*Figure 25.2* A selection of typefaces immediately associable to a theme or brand. All fonts have been rasterised for this book; original fonts are copyrighted and available under licence. Loxley is by Canada Type, Corporate A by URW(++) and Face Your Fears by Hanoded

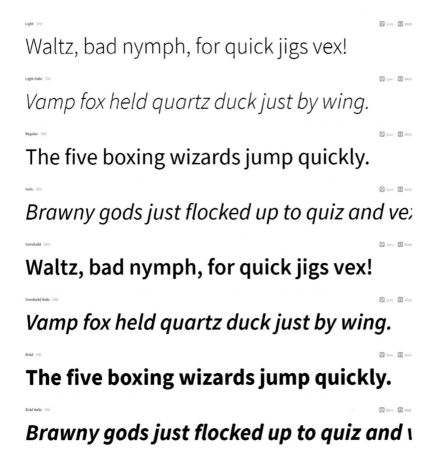

Figure 25.3   Some styles of the open font Source Sans Pro as previewed in Adobe Typekit.
Adobe product screenshot printed with permission from Adobe Systems
Incorporated

traditionally used for large geographic (topographic or oceanographic) areas; for large administrative areas such as counties, countries or regions; as well as at mid to large scales for larger towns and cities, for example postal towns. 'Names requiring considerable separation of the letters are commonly limited to capital (letters)' (Robinson *et al.*, 1995). Uppercase can be used on any feature or theme as a way of attracting attention, although uppercase is also harder to read than mixed, or mixed-use, cases. It is for this reason that many modern maps have broken away from tradition when it comes to more complex text such as the labelling of network features like road names, migrating from a tradition of uppercase to mixed case. Another growing alternative is 'small caps' or text case where the words are all in uppercase but the opening letters are larger than the rest.

Lowercase in its strict form, i.e. all small letters, is rarely used on maps and there are a number of different case types for mixed-use in the English language such as 'sentence case' where only the opening letter of the first word is in capitals unless proper nouns are used, and 'title case' where nearly all words are capitalized, i.e. start with a capital letter. Some software refers to 'proper case'; this is where all words are capitalized, regardless of function and is also known in English as 'start case'.

*Figure 25.4*   Different cases shown in conjunction with other typographic varieties on the Philip's *Modern School Atlas* (98th edition), 2015. © Philip's

The *colour* of text allows us to clearly associate labels with features. In most cultures, blue labels infer hydrography, whereas green is commonly associated with woodland, parkland or any natural habitat. Red implies danger, but if used tactfully can also be utilized to make labels stand out in the visual hierarchy. Shade of colour is also important. When labelling a blue river line, one might be tempted to use the same shade of blue. However, if the map also includes lakes or seas of the same colour then a richer shade of blue would be required in order to be seen on top of those polygons.

Colour can be used as a more subtle variable to size when emphasizing features and creating a visual hierarchy in the map labelling. Black will always be at the top of that hierarchy, but shades of grey can be used where colour is not required. The legibility of a particular colour of lettering is also influenced by the mapping behind it since the legibility is dependent on the visual contrast between the two, with black text on a white background the most legible and most prominent colour combination. This contrast is of special concern for larger labels that in terms of both size and spread might cover several differently coloured features. To date, the 2010s have seen a trend for maps and visualizations on a dark background. In these circumstances, a light or bright colour of text is required to maintain the high contrast between foreground and background. This is known as reverse lettering.

Legibility and impression of size and extent can be further pronounced by the *spacing* of letters. In cartography, this can be seen most commonly on expansive geographic features like mountain ranges. Similarly, it can be used on linear features such as rivers to indicate length. Fine-tuning the spaces between characters is known as kerning and is especially useful for curved labels. Extended name spacing should be reserved for large, important superimposed areas with large text size. In such instances, the spacing should be equidistant and an open face type can mitigate over-dominance (see Figure 25.5). The opposite action of pushing letters closer together is *condensing* which can reduce emphasis as well as be used as a method of reducing labels to fit a particular feature or space.

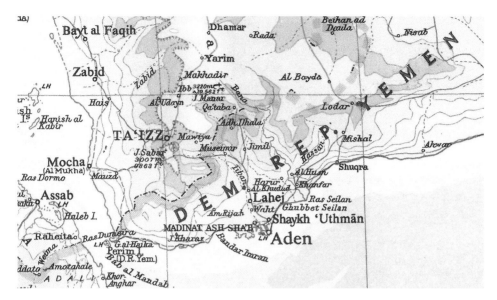

*Figure 25.5* Open face type; actually, the typeface here has horizontal line patterning to create a semi-open effect, to show the Democratic Republic of Yemen without dominating the black text labelling of the settlements of the area. Extract from *Bartholomew World Travel Map: Africa North-East* (1972), used by permission of HarperCollins Publishers

Variations in text size, style, form and perhaps most notably spacing are used extensively by James Wyld in his 1871 map *Adamantia: The Diamond & Gold Fields of South Africa*. Although arguably overused, it is clear to see how the lettering of each label gives an immediate sense of size, extent and importance (Figure 25.6).

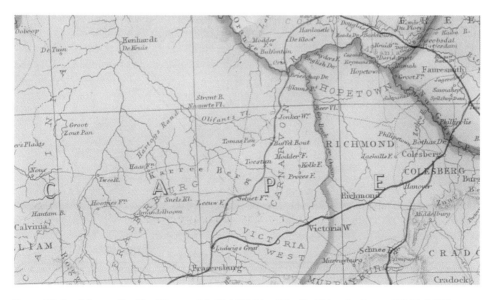

*Figure 25.6  Adamantia: The Diamond & Gold Fields of South Africa*, by James Wyld, 1871

*Weight* and *size* are sometimes considered as elements of style and form. They are both variables that the cartographer has significant control over, they both can be used to show hierarchical importance and/or the size of the associated features, and both influence legibility of the text, which can be negative as well as positive.

Most geospatial software allows text to be 'turned' bold but better results come from using 'true-drawn' or predefined weighted fonts. So, a typeface may have a bold font style as well as a regular one, and possibly many more besides (see Figure 25.3). Generally, the more power that is increased by weight must be proportional to the size of the letter; too strong or too weak and the letters will become hard to read. Many bold fonts are hard to read when used on a map, especially at smaller sizes, due to a lack of white or open space within each ligature, but typefaces are increasingly including a semi-bold style variant. Similarly, light font weights are becoming commonplace but, in cartography, these risk being lost if the map is to be printed.

Size, or type size, is conceptually very obvious. However, the subject of type size is actually quite complex. Size is usually measured in points but this measurement is related to a rectangle, or page, into which the letter was originally drawn for the font. Historically this would have been the size of the cast metal block onto which the letter was embossed for printing; today it is the window into which a digital letter is saved. Also, different typefaces have different lengths of ascenders (the elements extending up from the main bodies of letters) and descenders (the elements extending downwards). So, in short, all typefaces have different sizings: 6 point (pt) text in Arial might not be the same size or height as 6 pt Times New Roman. On average, one point size equates to a letter height of approximately 0.35 mm (see Figure 25.7).

*Placement* may describe the positions of labels relative to one another, their arrangement, and forms a part of the overall map layout. More commonly it describes the positioning of each label usually relative to its associated feature, but sometimes in relation to a given anchor point or pre-scribed cartographic location. It generally concerns the latter part of the 'map lettering' process of positioning and placing although some use the terms interchangeably. Well-placed lettering clearly identifies the feature or phenomenon to which it refers. Map series often make use of a standardized set of positioning rules or guidelines to ensure consistency across the product, and this same uniformity is found in modern automated cartography, for example maps made in a GIS where a certain type of feature or label class will share the same placement rules. Optimum placement in both situations often requires a degree of intervention by a skilled cartographer.

The placed lettering or labels can be associated to point, line or polygon features, each offering different opportunities for conveying information via the placement of their labels. Whereas

*Figure 25.7* Type sizing

points are clearly and simply locative; lines can be labelled to indicate their orientation, length and even shape; and polygons can be labelled to designate the form and extent of their area.

## Point-based

Places or information represented by a point are often labelled around the point. The label should not be broken or spaced and is generally placed horizontal to the grid or at smaller scales placed parallel to the parallels of latitude. Labels should avoid the grid lines. Often the positions around the point are given a ranked preference, based upon cartographic experience, with above and to the right the preferred placement of many (see Figure 25.8). Some texts imply a universal convention in the position hierarchy, but the truth is that many different systems exist. For example, on the OS Explorer Map series of Ordnance Survey (GB), the preferred position is to the right of the point (rather than above and right) (see also Figure 25.9). Some argue that if both feature and label lie on the same horizontal alignment then legibility is decreased due to an optical effect. This is however perhaps more apparent in punctiform labelling of area or polygonal features (see, later, Figures 25.17 and 25.18).

A different system may be chosen specific to a certain type of feature. Alternative systems can also be found in automated cartography. The general consensus, however, is that labelling to the right of a point is far preferable to labelling to the left, and above is better than below due to there being more ascenders than descenders and so less likelihood of the label drifting

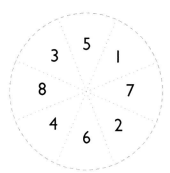

*Figure 25.8* 'Standard' point label placement ranking, as described by Darkes and Spence (2008)

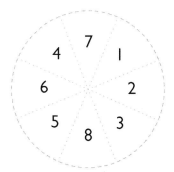

*Figure 25.9* An alternative point placement hierarchy, used for many features in the automated OS VectorMap District raster

away from the feature object. The distance a label should be offset from a feature is purely an optical decision by the cartographer, but there should be a good compromise between 'breathing space' (room for both object and label to be read) and association by proximity (Figures 25.10 and 25.11).

Common and often necessary rule-breakers are public transport network maps. For example, Transport for London and Régie Autonome des Transports Parisiens (RATP) place labels immediately above, below, left or right of a point on lines that are drawn horizontally or vertically, which visually makes most sense (Figure 25.12). Station or stop labels on the diagonal segment of lines are labelled on the perpendicular diagonal to the line: again, this gives good legibility. The Singapore train system map is an intriguing example where the 2015 version of the map is harder to read than previous versions after abandoning these rules. Additionally, in the USA, the labels are often rotated.

Exceptions also occur for certain scenarios and geographic feature types. Labels along the coast are generally shown entirely in the water to improve legibility; although many modern maps use entirely on the land as an acceptable second choice and some use an alternative policy

*Figures 25.10 and 25.11*   Bad (Figure 25.10) and good (Figure 25.11) point label positioning illustrating where point placement ranking is useful either as a guide to the cartographer or to placement software

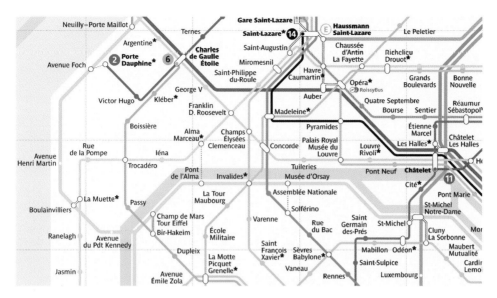

*Figure 25.12* Unique placement rules for public transport maps. Extract taken from 'Paris Metro and RER plan'. © RATP Reproduction Interdite – Avec l'aimable authorization de la RATP

of keeping topographic labels within the land and hydrographic names within the water. The traditional method, as described by Imhof (1975), is that 'names of shore and coastal places should be written neatly on the water surface. Names of places near the shore, but not lying on the shore, should be written completely on the land surface'. Whichever method used, the coastline should not be broken by a label.

*Figure 25.13* Scottish mountain summits near Ben Nevis named and heighted in OS VectorMap District. © Crown copyright and/or database right 2015 OS

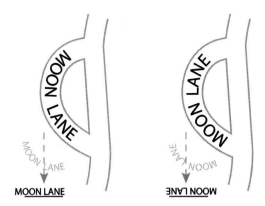

*Figure 25.14*   On the left, correctly placed lettering will fall with gravity to settle in an upright position. Labels should not be positioned such that they would fall upside down when 'let go' (right)

For landform features it is common to have a name and height label referring to the same place, for example a mountain may have a name and a labelled spot or summit height. In such circumstances, the name should be offset enough to allow room for the height to be labelled alongside the corresponding height dot (Figure 25.13). At smaller scales, settlement names may need offsetting in a similar fashion. Note that mountain heights can also be grouped with the name label.

Point labels are more likely to be placed on or directly over the point and/or rotated in situations where the point has been provided solely for the reason of placing text cartographically and does not represent a symbolized feature on the underlying map.

## Line-based

For linear features, labels are usually placed parallel to the feature, or for rivers, may be smoothed to follow the general shape. If space permits they may be placed on top of a styled linear feature, for example, the name of a road or street, but names are better placed above rather than below a line because there are fewer descenders than ascenders in lowercase lettering. Labels placed above or beneath a line should neither touch nor cross the object to which they refer.

For long features, for example, some roads and rivers, labels should be repeated at sensible intervals to make the feature always identifiable to the reader. Preferred label positions are central to the feature and where there is sufficient space to be clearly legible. Horizontal and less meandering parts of a line are also preferable, not just for rivers but for roads and railways too. Names along lines, especially ones that curve, can benefit from being subtly letterspaced. Map type should stand as upright as possible. Lettering should never be placed such that it would 'fall on its back' when let go (Figure 25.14).

Contour lines provide one of the few instances where a cartographer will choose to portray some labels upside down. Contours are generally labelled in what are known as ladders, a sequence of aligned labels at regular intervals from the lowest to the highest contour line with the label always oriented such that the top is uphill and bottom downhill (Figure 25.15).

Terrain brings us features whose names may be represented as points or lines. Mountain passes and valley floors are singular place names but may be aligned to a feature especially at smaller scales and if well-known and identified by a linear feature such as a road, railway, path,

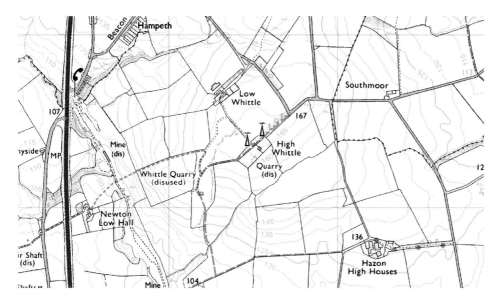

*Figure 25.15* Contour runs or ladders (shown as orange labels) on OS Explorer Map and 1:25 000 Scale Colour Raster. © Crown copyright and/or database right 2015 OS

river or an obvious gap between the mountain slopes or landform rock. The names of the mountains themselves may be point or area-based but are usually a point-based name with the label placed near to the mountain summit (as shown in Figure 25.13). In the past, some cartographers labelled mountains in a ring around the summit. This is no longer fashionable, neither is it very easy to comprehend. Mountain ranges will be spaced in a similar manner to other large areas, i.e. spread centrally across the area of extent.

## Polygon/area-based

Traditionally, area labels were placed, spaced and sized so as to indicate the form and extent of the mapped geographical area to which they refer. However, the authors of similar guidelines contradict this somewhat by stressing that letter-spacing and curved text should be exceptions rather than the norm. So, what most modern mapping has ended up with is a compromise where labels infer the extent of an area without necessarily stretching right across it. The result is less disrupted names, less disjointed letters and hence less confusion, as well as more 'white space', all of which improve map legibility. Sometimes area extents are clearly defined by their feature polygons, such as buildings or lakes; other times the extents are harder to judge, for example seas and geographical regions.

Generally, areal features are labelled inside of the feature polygon. However, there will be some circumstances where it is better to treat the area like a point and place the label at the best position outside of the feature. Examples include building names where the text does not comfortably fit inside the building polygon and island names where the label again cannot fit inside or where the label refers to a group of islands (Figure 25.16). Other labels may require placement outside the area depending on scale and label competition. A final option is to allow the label to overrun the area's extent, but if doing so, place the label in such a way that it is clear which polygon it is labelling. Also, try to avoid optical coincidence (Figures 25.17 and 25.18).

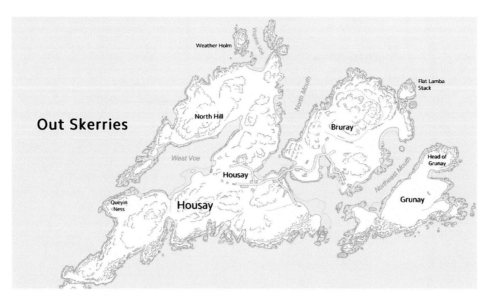

*Figure 25.16*   Island labelling, using OS VectorMap District

*Figure 25.17 and Figure 25.18*   Optical coincidence (Figure 25.17) makes aligned labels harder to read than in the better positioned example (Figure 25.18)

Furthermore, straight, oriented and curved label placements may be used to imply the type and form of features whose boundaries are less clear, for example marine names (Figure 25.19). A bay, channel, stretch of water, tidal inlet, sea and ocean may all be interconnected and identically styled. The placement style of the names can be used as an alternative way of distinguishing between these feature types, for example curved text near the coastline may imply a bay, whereas a straight, oriented name could represent a channel. Most oriented or tilted labels on a map will look more natural when curved.

As well as placing labels centrally and outside, polygons can be labelled around their edge, for example, when naming boundaries. Names or labels of area features should never be repeated at intervals in the manner of long linear features, with the exception of named river polygons, which should be repeated rather than spaced. Adapting labels to the length of ribbon-like areas makes them illegible.

In many cases named extents overlap. For example, larger, higher order areas such as oceans will overlap smaller, lesser order areas such as seas. If there is a choice to be made, then the name most relevant to the scale of the map should generally have priority. There may also be areas of a map cluttered with too many labels. This can often be resolved with the hierarchy previously discussed, but, all things considered equal, the labels should be placed from the centre of the cluster outwards, allowing outer labels to be further displaced without disregard for their feature associations. Generally, large area names can be positioned off-centre to avoid other local labels of any association (point, line or area).

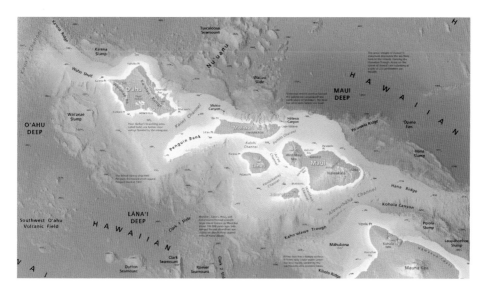

*Figure 25.19*   Extract taken from *Seafloor Map of Hawai'i* by Tom Patterson

## Abbreviations, multi-line text and language

When there is insufficient space for a label to fit, it is possible to abbreviate so long as the map user can easily determine what each abbreviation means. As well as a method for shortening text strings, cartographers use a range of standard abbreviations to describe features on the landscape with minimal disruption in a similar way to the use of symbols. As an example, a list of map abbreviations can be found on the Ordnance Survey website at www.os.uk/resources/maps-and-geographic-resources/map-abbreviations.html.

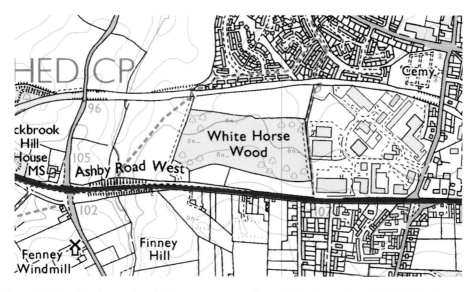

*Figure 25.20*   Good use of multi-line text, extract from OS Explorer Map/1:25,000 Scale Colour Raster. © Crown copyright and/or database right 2015 OS

Sometimes it is best to divide long names into two or more lines of text to make best use of the available space or to be a better representation or fit to a given area (Figure 25.20). British organizations refer to this as double-banking. Using multiple lines for labels can allow more room for other features or other labels to be shown, and can help to ensure that long names do not stray too far from an associated object. It is a common resolution for both the cartographer and for automatic placement engines. With areal labels, especially names, the idea is to show the shape or form of the object. Many lakes or woods, for instance, are of roughly equidistant dimensions in each direction, i.e. are not linear or elongated. Such areas are best labelled with multiple lines to form a similar shape or at least to better fit the polygon.

Alternative names, for example on multilingual maps, or alternative orthographies are usually placed directly beneath and symmetrical to the first or previous label (Figure 25.21). National

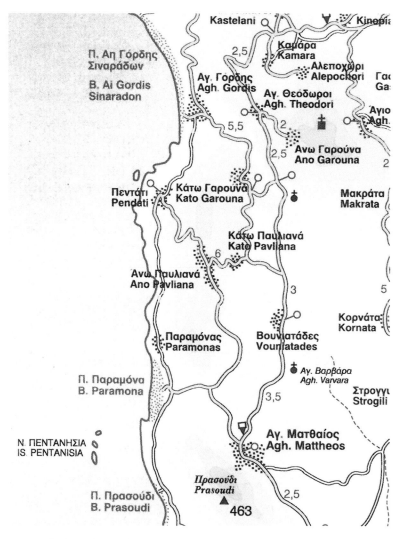

*Figure 25.21*  Multiple orthographies on mapping in Greece to show place names in both Greek and Roman alphabets. Extract taken from sheet 301: Kerkyra by Road Editions (1997)

Land Survey of Finland can show up to five different languages on their map sheets, for example, Inari in Lapland. Cross-border or international maps can either use a common language across the whole map, for example to show all place names in English; or may depict names as endonyms, the geographical names used by the population of that country or region in their own native language.

The spelling of names on maps is often a contentious matter. Historical maps serve as an archive of how toponyms have evolved over time, but even on modern-day mapping, the current spelling of older languages, such as Gaelic, is debatable. Internationally there are different names for the same features such as the Sea of Japan which is known as 'East Sea' by Koreans and a topic of great political debate. The United Nations is developing a multilingual, multiscriptual, georeferenced geographical names database for countries and capitals with other city names provided by national agencies. In the United Kingdom, the Foreign and Commonwealth Office provides a small-scale index of government-approved, geographical names in British English language and Ordnance Survey provides a national names gazetteer.

## Masking

Lettering takes precedent over linework; if there is a conflict then the underling line feature should be masked out unless there is sufficient contrast between the two colours. Masks are holdout shapes that are placed beneath the text but over the conflicting parts of the map to create a buffer in which the lettering can easily be seen and read. There are different methods of masking. Most GIS software uses halos which completely cover the underlying detail, usually with white, although a transparency may be applied (Figure 25.22). There is some merit to this technique, particularly in web map services where the underlying layers may change and in maps with a strong background. A more traditional method still in use, both manually and by software, is a technique known as *selective masking*. This is where a halo-style holdout is applied only to certain layers of the map (Figures 25.23 and 25.24).

*Figure 25.22*   Halos around text on a strongly coloured map of Yellowstone National Park, USA. Labels were added to the painted landscape of Heinrich Berann (1962)

*Figures 25.23 and 25.24*   Selective masking, e.g. in Mercator software, makes names legible without losing the underlying map detail. Images courtesy of 1 Spatial

In selective masking, lines as well as hachures and other patterns should be interrupted when they coincide with labels or dots and tonal contrast is not sufficient. This allows the user to read the labels more clearly without degrading or disrupting the underlying features for which the human eye and brain connects the small gaps automatically. In web- or graphics-based environments, it is usually possible to be more creative and have masks that are stronger nearer to the text and fade away with distance.

## Callouts, charts, symbols and data labelling

Thematic maps are often less densely labelled. If statistical then the output areas may be labelled, depending on scale and usefulness. On choropleth maps, good use of colour often negates the need to label the themed data. Additional text is usually reserved for explanation

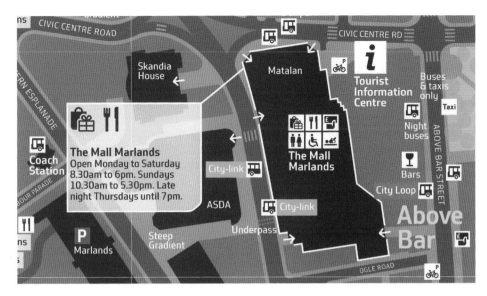

*Figure 25.25*    Extract from Southampton Shopping Map by and courtesy of CityID. Map
produced for Southampton Legible City in 2007 © Southampton City Council.
Southampton Legible City is a Southampton City Council project led by a
project team that has included: Pete Brunskill, Liz Kite, Phil Marshall, Ian
Rothwell and Simon Taylor. City ID developed and designed the project with
specialist support from Andy Gibbins, Dalton Maag, Design Connect, Endpoint,
MK Design/RNIB, Russell Bell and Wood & Wood

or to reinforce a message. Proportional labels can also be used for themed mapping in the same
way as proportional or graduated symbols. There are two applications: one to vary the label size
in conjunction with symbology for example on cluster points; the other to act as the symbology
and display bivariate data. In the latter, the text at each location indicates the value of one dataset
and the size of the lettering denotes the value of another.

Callouts are labels, often in boxes, with leader lines indicating the associated feature. They are a
useful way of labelling or annotating data without obscuring too much detail or when there is no
space on the map to do so, but be warned that too many leader lines in a given space cause confusion for the reader. In many systems, they can also be used to display linked data, including images,
URLs and metadata, and even become interactive. Interactive maps often have callout labels
without leader lines that appear on the 'click' or 'hover' cursor or finger actions. Boxes, whether
considered 'callout' or not, can also be used to highlight features and link them to a data category.
Note how the blue and red boxes in Figure 25.25 each extend from a point location, in this case,
a bus stop. By highlighting the label rather than the feature, the cartographer has added emphasis
and secondary grouping while retaining a shared hierarchical status and primary grouping between
the bus stops and the other point symbols shown. Map symbols can also be considered labels and
arguably in their most efficient, concise form. They label a feature with very little disruption to the
mapping and detail below because they require far less space than a textstring.

## Map sheets, tiles and atlases

Special consideration for name placement is required when creating a set of maps as sheets, tiles
or pages. The cartographer must consider what happens when a label crosses the edge of the

page. For point- and line-based labels, the placement guidelines can be greatly restricted by the edge of the page or sheet. At the right-hand margin, labels may more frequently be placed to the left of a point, and linear labels should be placed central to their visible extent. Important labels may also be shown in the margin.

In the instance of an area name, for example an urban area or a lake, there are two approaches often employed by national mapping agencies. If a reasonable majority of the area extent (polygon) falls on one of the two sheets, then the name is fully placed on that sheet and on the adjoining sheet may be referenced in the margin (as seen in Figure 25.26). If neither sheet has room to display the label, then the name may be duplicated within the margins of each sheet. The other approach is to place the label across the 'join' of the two sheets and on each sheet the name is continued with the missing part placed within the sheet margin, as seen on map sheets from the Federal Office of Topography, swisstopo, in Switzerland. Geographically important names that are near to but do not cross a sheet edge are often also shown in the margin of the adjoining sheet. When creating a map book or atlas, it is wise to insert spacing in the middle of labels that span the page fold or binding and also to prevent labels from ending too close to the bind. In practice this is done by leaving a margin of several millimetres free from labelling at the inner edge of each page.

The word 'tile' in cartography refers to a manageable-sized, usually square-shaped area of mapping that is used for easier distribution and use, and/or as a system of map production. Tiles may be placed together by users to form a larger map or by a cartographer to create a map sheet. Tiles may also be created as part of a cache for web maps. The disadvantage of tiles is that labels are often created as if the tile were an isolated map leading to uneven spacing between and often too frequently repeated labels. This can be overcome by using pre-determined cartographic anchor points for larger areas and in web-mapping by using metatiles, a process of labelling a larger grid of tiles before it is cut up into tile components for raster caching.

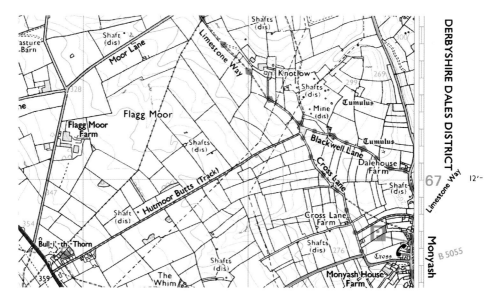

*Figure 25.26*   **Names in the margin of an Ordnance Survey map sheet, OS Explorer Map 24: The Peak District, White Peak area. © Crown copyright and/or database right 2015 OS**

## Automatic label placement

Automatic label placement, commonly known as text placement, can be a time-consuming process in GIS and a difficult problem to solve in automated map-making. Many GIS products have a built-in labelling engine and there are non-platform-specific label placement applications that can be used in a variety of workflows. Web-mapping software has scriptable functions that operate in much the same way. In each of these, labels are placed based upon a set of algorithms normally driven by user parameters. Such rule-based logic uses the same principles and theory of cartographic label placement as described in this chapter, but it is still in its infancy. As well as points, lines and polygons, Environment System Research Institute's *Maplex* engine has specific parameter-driven options for land parcels, rivers, boundaries, streets, street addresses and contours. However, to create advanced cartography akin to traditional topographic paper mapping requires far more algorithms to cover the plethora of feature-specific cartographic placement methods, of which only the most universal ones have been covered in this chapter. This has led some cartographic organizations in Europe to consider creating their own labelling algorithms. As well as the rule-based algorithms, many of the engines employ local optimization algorithms, which is where several different label positions are tested until overlaps are reduced and a 'local optimum' has been found. Each label is moved iteratively and improvements are preserved.

## Systems of orientation

Orienting labels, for example by an attribute value, allows each label on a map to be rotated and aligned according to the data and to a software's settings. Rotation can be set, in some GIS for example, as either geographic – the angle from North in a clockwise direction, or arithmetic – from a horizontal position which equates to East on most maps, and can be clockwise or anti-clockwise depending on the software. Strictly speaking, orientation is a set angle of a feature, map or label in a given view or projection, whereas rotation is the transformation process of getting there. However, in cartography, at least in two-dimensional (2D) cartography, the two terms tend to be used interchangeably.

## 3D

A growing issue for cartography to address is labelling in three-dimensions (3D). Nowadays, 3D visualizations are commonplace with any labels generally floating above features and facing the viewpoint. This is far from ideal and is disruptive to the scene. Alternatives include text laid on or extruded from the associated feature, text standing upright on the associated feature – a method Recce uses for road names (Figure 25.27) – or placing texts as objects in the scene, known as billboards, which are sometimes allowed to rotate on their axes to face the viewer. Obstructions and scene lighting can still cause problems. Showing any information including labels in a 2D plane in front of the scene, for example as a heads-up display, is arguably a better solution, although placement of anchor point and any label offsets can also cause problems in 3D.

   More traditional maps that have aspect or the appearance of 3D, often referred to as 2.5D, are easier to label because the viewing perspective never changes. On the map of the Grand Canyon (Figure 25.28), the features are labelled in a floating 2D plane, yet their nature (path, size, extent, type, position, height, and so on) are inferred in the same fashion as the standard map lettering and labelling principles previously outlined. Such techniques are common in mountain cartography.

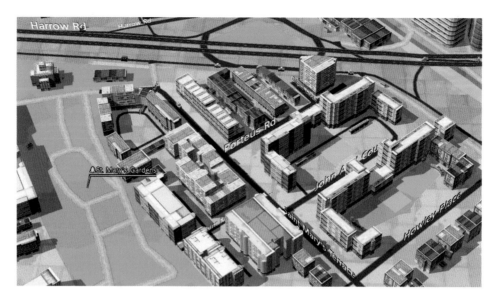

*Figure 25.27* 'Floating' and 'standing' labels in a 3-dimensional scene. Copyright eeGeo Limited www.eegeo.com

*Figure 25.28* Extract of *Exploring Grand Canyon*, a visitor's guide map of the Grand Canyon, USA. National Park Service, U.S. Department of the Interior (2013)

## Summary and final points

Overall, labels should be easy to read, easy to locate and easy to associate – both to the feature to which they refer and to the class to which they belong. Where labels represent geographical names they should immediately indicate extent, connections, importance and feature type. Lettering should be indicative of classification and hierarchy. The amount of geographical

366

extent hidden on a map by lettering and labels is often of concern. Labels can cover up many elements of the landscape and 'prevent the reader from seeing the map as a picture of the Earth' (Raisz, 1938). Thus, labels should disturb and conceal other content of the map as little as possible. Labels should never be upside down, except for contour values which read uphill, as discussed earlier, and building property or house numbers, which can be aligned to face the road or entrance point.

It is not unusual for the guidelines on map lettering and labelling to contradict or oppose one another, therefore every case must be considered individually and the most appropriate principle for the given situation chosen – something that is less achievable in automated cartography. Furthermore, the purpose of labelling a map is to add clarity. Any principles should be overruled by the cartographer in situations where clarity is diminished. Map lettering and label placement can also be used to resolve imbalance in a map or to bring attention to a particular area of a map. Type has apparent weight and oriented text can lead the eye in a particular direction.

Optimal cartographic placement is not always the best approach for labelling, for example in web-mapping where the consistency in placement can sometimes be of more value to the user experience than the best placement possible at each scale of the service. Lettering and labelling should always add to the understanding of a map and never detract from it.

## Further reading

Axis Maps (2009) "Indiemapper: Labelling and Text in Cartography" Available at: *http://indiemapper.com/app/learnmore.php?l=labeling* (Accessed: 14 July 2015).

Bringhurst, R. (2001) *The Elements of Typographic Style* (2nd ed.) Vancouver, BC: Hartley & Marks.

Brownlee, J. (2014) *What's the Difference Between a Font and a Typeface?* Available at: *www.fastcodesign.com/3028971/whats-the-difference-between-a-font-and-a-typeface* (Accessed: 15 July 2015).

Gruver, A. (2014) "Cartography and Visualization: Lesson 3: Labelling and Text" *John A. Dutton e-Education Institute, College of Earth and Mineral Sciences, The Pennsylvania State University* Available at: *www.e-education.psu.edu/geog486/l3.html* (Accessed: 6 August 15).

Peterson, G. (2012) "A Cartographer's Toolkit: Halos Are Evil*" Available at: *www.gretchenpeterson.com/blog/archives/2511* (Accessed: 7 December 2015).

Some helpful entries associated with labelling and lettering have also been developed on Wikipedia and Wiki.GIS.com as follows:

"Automatic Label Placement" *Wikipedia* Available at: *https://en.wikipedia.org/wiki/Automatic_label_placement* (Accessed: 18 July 2015).

"Cartographic Labelling" *Wikipedia* Available at: *https://en.wikipedia.org/wiki/Cartographic_labeling* (Accessed: 13 July 2015).

"Labelling (Map Design)" *wiki.GIS.com* Available at: *http://wiki.gis.com/wiki/index.php/labeling_(map_design)* (Accessed: 30 July 2015).

## References

Darkes, G. and Spence, M. (2008) *Cartography: An Introduction* London: The British Cartographic Society.

Imhof, E. (1975) "Positioning Names on Maps" *The American Cartographer* 2 (2) pp.128–144.

Krygier, J. and Wood, D. (2011) *Making Maps: A Visual Guide to Map Design for GIS* (2nd ed.) New York: Guilford Press.

Raisz, E. (1938) *General Cartography* New York: McGraw-Hill Book Company.

Robinson, A.H. (1952) *The Look of Maps: An Examination of Cartographic Design* University of Wisconsin Press. (Republished in 2010 by ESRI Press.)

Robinson, A.H., Morrison, J.L., Muehrcke, P.C., Kimerling, A.J. and Guptill, S.C. (1995) *Elements of Cartography* (6th ed.) Chichester, UK: John Wiley & Sons.

Tyner, J.A. (2010) *Principles of Map Design* New York: The Guilford Press.

# 26

# Designing maps for print

*Judith Tyner*

The first question that may come to mind for readers of this chapter might be 'Are printed maps obsolete? We have GPS (global positioning systems), WMS (web map servers) and other forms of electronic map that we can view on a computer or smart phone. Does anyone use printed maps?'. The short answer is, yes, printed maps are still widely used. Hikers are advised to have a topographic map and compass in the event of failure of their GPS, such as being in a place where satellite signals cannot be received. Many people print the map and instructions after finding a route online, for example, using Google Maps. A large state map or road atlas is useful for seeing the 'big picture' when planning a trip. Geographic researchers often want a large paper map that can be annotated in the field. Beyond navigation, maps – especially thematic (or special-purpose) maps – are found in books, magazines and newspapers to illustrate and explain spatial information.

Maps have been made in many media – clay, papyrus, stone, cloth and paper. For centuries, maps were hand-drawn, and, if multiple copies were needed, scribes would copy the map by eye. Errors would creep in and each copy was somewhat different from the original. Thus, the invention of printing was a major revolution not just for books, but also for maps because it allowed hundreds of exactly reproducible copies. Through the centuries, printing technology has changed and maps have been printed from woodblocks, metal engravings, lithography, and offset lithography, and change continues by now allowing 3D printing. The changes that have taken place in cartographic production and reproduction, especially, have also been enormous, very likely greater than even the changes brought about by printing, and the rapid pace of change continues.

Since the advent of GIS (see Chapter 17), there has been increased emphasis on exploration, analysis and visualization of data on screen and less on the presentation of data in printed map form. This is especially seen in professional journals, where a GIS map of an analysis is simply reprinted in an article with little or no consideration of how it will look in print, and is largely owing to the rejection of the entire concept of maps as communication. However, in journals, books, and in other printed media, the text is designed to provide information, to 'communicate' information to the reader, and the map should complement this by communicating spatial information.

This chapter will not provide a 'how to' for designing maps for print, but rather raise issues and concerns specific to creating such maps. The references contain a number of sources which cover the basics of map design, including maps for print.

## Map design

To a large extent, designing a map for print and designing a map for online viewing are the same. But there are some differences between the two methods of delivery that must be considered. First, print maps are static. They are not animated, layers cannot be turned on and off, the user cannot click on a symbol and have information about the feature represented 'pop up', and the map reader cannot pan around the map and zoom in on areas of interest. Static maps have a fixed scale. If a feature is too small to be viewed easily, it cannot be enlarged with the click of a mouse. However, the print map is not limited in size to that of the screen of an electronic device, as its size is limited only by the size a printing press can accommodate. Electronic maps are almost always presented in colour, but printed maps may be either colour or black and white, largely owing to publishing considerations. Printed maps have a consistent appearance while electronic maps may be encountered through many different electronic devices and screens, and once a print map is published, it cannot be amended easily as can some online maps.

For any map, whether it is designed to be viewed on a monitor or printed, there are many decisions to be made before beginning the project in order to have a map that communicates its information clearly (see Chapter 21). Designing any kind of map involves planning and asking a series of questions. Most of the questions to be asked before beginning a map apply to both print and electronic maps, but the ones to be emphasized here are those especially important for print. These questions are: What is the purpose of the map? What is the format, i.e. the size and shape of the page? Where will it be displayed or used? How will it be produced? How will it be reproduced? Is it a stand-alone map or one of a series? Will it be reproduced in colour or black and white?

### Purpose

How will the map be used? Obviously, this question applies to both printed and electronic maps, but for printed maps this also involves where it will be viewed. Is it a simple location map, perhaps for tourists; does it illustrate spatial data in an article, such as population distribution or death rates; will the map be in a book or journal to illustrate a story or article; or will it be used as a reference or perhaps be posted on a wall? These various purposes have different requirements.

### Format

The format of the map, that is its size and shape, is governed by several factors. If it is to be printed in a journal or a magazine, there are set specifications for page size and use of colour that are dictated by the publication. Guidelines are available from the publisher or, for some journals, provided in an 'instructions for authors' section of the journal. For professional geography journals, there is sometimes a cartographic editor who can assist with design questions and problems. Because colour printing is still expensive, some journals charge the author for its use. If the map is for a book, the art editor will determine the page format and whether colour is an option, but even there, the author may be asked to help with the costs of colour printing. In recent years, copies of the maps in professional journals may be available online as well as in print, which can complicate the design process, as will be covered later.

### How will the map be produced?

Today, most maps are produced by computer, either by using a GIS, a computer-assisted design and drafting program (CADD), or an illustration program such as Adobe Illustrator or CorelDraw.

It is rare now for a map to be produced by hand, although some cartographers, particularly those who create pictorial maps, do prefer manual methods. Unlike GIS, illustration and CADD software do not allow analysis, calculation or automatic linking of data to location. One can, however, combine GIS with a presentation program. Map production software changes rapidly and the cartographer must be aware of these changes and how they affect the creation of the map. Some software packages that were in common use only five years ago are no longer available.

## Reproduction

Especially important for this chapter is how the map will be reproduced or viewed. The focus here is on maps that are printed, not maps viewed on a monitor or projected on a screen, but not all printing is the same. Will it be reproduced using a laser printer, an inkjet printer or will it be sent to a printing department or commercial printer? Maps produced on laser or inkjet printers are usually for small print runs – dozens of copies, not hundreds or thousands. Books, newspapers and magazines, of course, usually have larger press runs of several thousand or more. The type of printing often governs whether the map can be colour or black and white and how many copies can be printed. For maps to be made by a commercial printing company, discussing the project directly with the owner or operator is helpful.

## Colour or black and white?

Will the map be printed in colour or black and white? Colour is more expensive to print than black and white, and in books, colour is still not usually found on every page, but rather a colour signature or section is printed in colour separately from the rest of the book. Magazines may be printed in full colour, while newspapers still use a mixture of colour and black and white. Professional journals increasingly print in colour, and many have articles and maps that are also viewable online.

Because colour is ubiquitous on electronic maps, we forget that it is not always an option. Cost, as noted above, is a major factor, but so is the type of paper. Colour in newspapers, now widespread, was not always common owing to problems with quality of the printed colours and the registration of colours on newsprint paper. Technology changes rapidly and the topic will not be discussed in detail as the focus of this chapter is on design.

## Designing for black and white

Black and white maps are not simply poor relations; they can be quite effective. A well-designed black and white map can often communicate its message better than a poorly executed colour map. In addition, there has been extensive research on their design, and the focus here will be on area symbolization and type.

It would seem a reasonable method to create a map in colour as shown on the monitor and then to simply print it in black and white. A simple experiment shows the folly of this technique. A series of squares in the spectral sequence – red, orange, yellow, green, blue and violet – will not appear as a gradation in black and white, but rather as a series of random grey tones (Figure 26.1).

On black and white maps, everything is shown in black or shades of grey. This is especially important for choropleth maps that represent quantities with uniform shading by enumeration area.

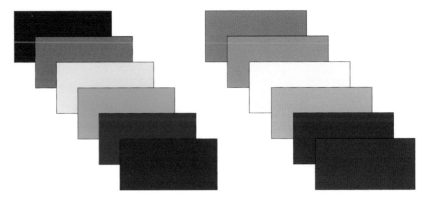

*Figure 26.1*   **A colour map cannot be printed in black and white without loss of information**

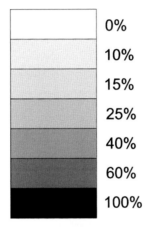

0%

10%

15%

25%

40%

60%

100%

*Figure 26.2*   **No more than six tints should be used or there is loss of information**

Although it is possible to print more than ten shades, the eye cannot perceive more than a limited number, usually considered limited to black, white and five shades or tints usually expressed as per cent black, e.g. 0, 10, 15, 25, 40, 60 and 100 per cent (Figure 26.2). This limits the number of categories or classes when designing choropleth maps.

When using shades of grey on qualitative maps, the cartographer must recognize that greys separated from one another can be hard to distinguish; thus, for such maps the greys must have distinct differences or patterns (Figure 26.3). Again, the number of categories should be limited accordingly.

The use of type is another concern. Size (usually measured in points) is important as well as the background tone or colour. There are 72 points to an inch, and, in general, it is recommended that no smaller than 6 pt type (1/12 inch or 2.1 mm) be used. Smaller type sizes tend to be illegible and even 6 pt is difficult for many to read (Figure 26.4). Type placed over a shaded background can be difficult to read; black on dark grey is essentially illegible. This can be improved by drop shadows or halos around the letters, or by using white lettering on the dark background (Figure 26.5).

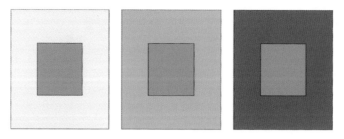

*Figure 26.3* The centre grey square is the same shade, but appears different on different backgrounds

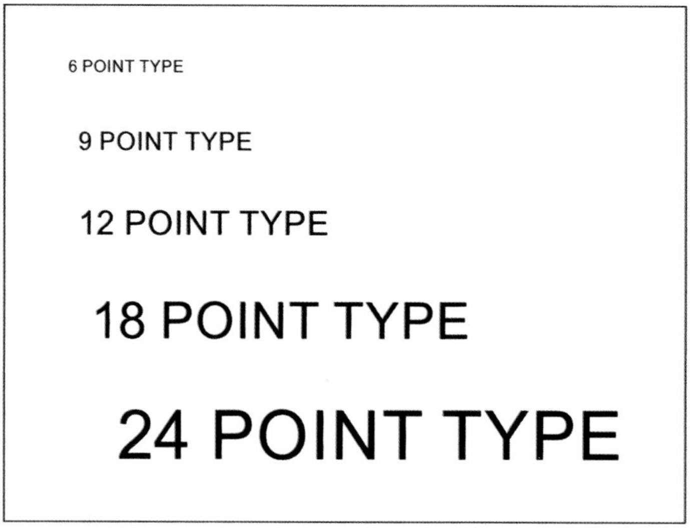

6 POINT TYPE

9 POINT TYPE

12 POINT TYPE

18 POINT TYPE

24 POINT TYPE

*Figure 26.4* Font sizes must be chosen for legibility

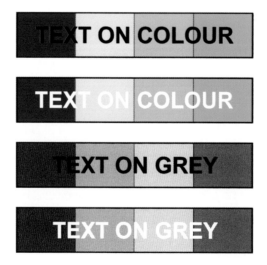

TEXT ON COLOUR

TEXT ON COLOUR

TEXT ON GREY

TEXT ON GREY

*Figure 26.5* Background tints and colours must be considered when using text

## Designing for colour

Often when a map has been designed on a computer monitor and then printed on an inkjet printer, the creator is shocked at the difference in the appearance of the printed map; the colours look 'wrong'. They may blame the printer as being 'poor quality'. The problem is not the quality of the printer or the monitor, but rather the way in which colour is produced on monitors versus paper. Colour on a monitor is created from *light* and three primary light colours are used to create the entire visible spectrum. These primaries, called the additive primaries, are red, green and blue (RGB) (Figure 26.6). When these colours are combined, the result is white.

Colour on paper is created from three primary *pigments* (inks) that are combined to create all other colours for printing. These primaries, called the subtractive primaries, are cyan, magenta and yellow (CMY) that when combined produce black. However, usually, black ink (symbolized K) is added separately to avoid a 'muddy' black from mixing the three other printing colours. Colour printing thus uses four colours of ink and is described as CMYK.

Therefore, because colours on a monitor and a printer are created differently, they will look different when printed. Mapping and illustration software have provision for converting RGB to CMYK colours in the printing process. Ideally, when converting a map from its display on a monitor to a printed product, the cartographer will make this conversion. However, since articles and maps are now available in both printed and electronic forms (particularly online), this conversion is often not made and the printed map usually suffers a loss of intensity and even clarity.

As with black and white, the number of shades of one colour is limited by perception, but many programs allow the user to create far more categories. Some maps printed from a GIS analysis have as many as 20 shades of a hue. It is impossible to distinguish between this many individual shades, and the human eye attempts to group them into a smaller number, so they serve no real purpose. If more than five categories are required, another hue should be added. (Cynthia Brewer's website ColorBrewer2.com and her book *Designing Better Maps* (2015) are both excellent resources.)

Legibility of type on colour maps, like black and white, is a major issue. This is true for maps viewed on a monitor as well, but because the user can zoom in on the words it may not be a problem. If the words must cross several areas represented in different colours, however, the text

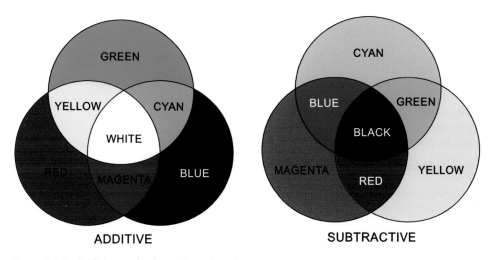

*Figure 26.6*  Additive and subtractive primaries

can be more difficult or even impossible to read. Black type is probably the easiest to work with, and coloured type on a coloured background should be avoided if possible.

## Conclusions

Despite the pervasive use of maps on electronic devices and computer monitors, the printed map still has a place in cartography, and when a printed map is used, the reader expects that map to be a useful tool. The cartographer must consider how the map will be viewed when beginning the design process, as the particular needs or expectations of the user may require printed or electronic media, or both to be met.

## Further reading

Dent, B.D., Torguson, J. and Hodler, T.W. (2009) *Cartography: Thematic Map Design* (6th ed.) New York: McGraw-Hill Higher Education.

Kraak, M-J. and Ormeling, F. (2010) *Cartography: Visualization of Spatial Data* (3rd ed.) New York: Guilford Press.

Krygier, J. and Wood, D. (2011) *Making Maps: A Visual Guide to Map Design for GIS* (2nd ed.) New York: Guilford Press.

MacEachren, A.M. (2004) *How Maps Work: Representation, Visualization, and Design* New York: Guilford Press.

Slocum, T.A., McMaster, R.B., Kessler, F.C. and Howard, H.H. (2005) *Thematic Cartography and Geographic Visualization* (2nd ed.) Upper Saddle River, NJ: Pearson Prentice Hall.

Tufte, E.R. (1990) *Envisioning Information* Cheshire, CT: Graphics Press.

Tyner, J.A. (2014) *Principles of Map Design* New York: The Guilford Press.

## Reference

Brewer, C.A. (2015) *Designing Better Maps: A Guide for GIS Users* (2nd ed.) Redlands, CA: ESRI Press.

# 27

# Internet mapping

*Ian Muehlenhaus*

## Introduction

Today an entire generation of GIS experts and map users is entering the workforce that has rarely, if ever, used paper maps. This is a watershed moment for cartographers and those participating in the billion-dollar online mapping business. Like landline telephones, paper maps are not dead, but they represent the past, not the future of the industry.

What exactly are Internet maps? What makes them different from paper maps? This chapter does two things. First, it argues that talking about Internet maps as separate from printed maps prevents us from better categorizing and analyzing the different roles that maps using the Internet play in society. Internet maps are not so much a separate category of maps as they are an offspring of print cartography. In that vein, the chapter segues into its second argument: that the development of maps on the Internet has been going through a growing process. With the ascent and widespread adoption of HTML5 technologies, we have entered an extremely stable period of map development. However, as will be argued, this period will be short-lived, as virtual reality and wearable Internet maps are just over the horizon.

## The fool's errand of defining *the* Internet map

Let's begin with a simple enough question. *What is an Internet map?*

> *A digital map dependent upon Internet technologies that uses a screen as its medium of communication.*

And then the obvious comparative question. *What was a paper map?*

> *An analog map created using print technologies with paper as its medium of communication.*

Both definitions are overly broad and vacuous. It is not the answers that are poor, it is the questions. Too often when speaking about Internet maps, cartographers emphasize the medium instead of the purpose.

Traditionally, maps have been characterized not by their medium but by their communicative objectives. Paper maps have generally been divided into purpose-oriented categories such as 'thematic', 'reference', 'marketing', and even 'spiritual' – for example, medieval T-and-O maps with Jerusalem at the centre. Internet maps are better defined and reviewed in this fashion as well. Just as there is no single type of paper map, talking about Internet maps as though they were homogenous is overly reductionist and leads to simplistic thinking.

Categorizing Internet maps by their purpose is crucial if we truly want to begin analyzing them in any useful fashion. This is particularly true now that screens are becoming ubiquitous, miniaturized, and immersive (i.e., 'wearables' and virtual reality goggles), as screen-type increasingly impacts what can be communicated on a map. The one common denominator among printed and Internet maps is the map itself. A good place to begin, therefore, is to categorize Internet maps using the established taxonomy of printed maps – expanding and branching the taxonomy as necessary.

There are two reasons for doing this. First, Internet maps did not develop in a vacuum; cartographers with experience originally created them in the medium of print. As will be shown, initially Internet maps were merely digital reproductions of printed maps. Thus, though the medium is different, the genealogy of Internet maps lies in print cartography.

Second, starting with the traditional categories, we can begin to compare the benefits and disadvantages of using print- and Internet-based mediums. We can also see what traits from print cartography linger throughout the evolution of Internet maps both up to the present and into the future as Internet maps eventually evolve to look nothing like their paper predecessors.

Though by no means exhaustive, the lists below act as a primer on how to better think about what role different types of Internet maps play in society. You will note that some online maps have morphed into their own unique categories that were not necessarily capable of existing in the paper medium – at least not as robustly.

## Reference maps

Google Maps – and subsequent competing map services – have become the ultimate reference map for people living in the digital realm. Constantly updated through crowd-sourced data and typically incorporating tracking technologies, these maps not only provide you with

*Figure 27.1* A screen shot of OpenStreetMap (*www.openstreetmap.org*) providing driving directions from the author's house to his office

directions but also allow you to explore an area in exquisite detail. With the added features of aerial photography, interactive street views, and real-time traffic reports, these maps have become some of the most powerful tools available to a daily commuter (Figure 27.1). They have become so powerful in fact that people may literally be losing their ability to think spatially without them (Wu *et al.*, 2008).

## Location-based service maps

Though often coupled with web-based reference maps, location-based service maps are focused on target-marketing. As people voluntarily allow themselves to be tracked, marketers can target audiences not only based on their demographic tastes but also their real-time location (Figure 27.2). The future of advertising is real-time and spatial. Users may think they are

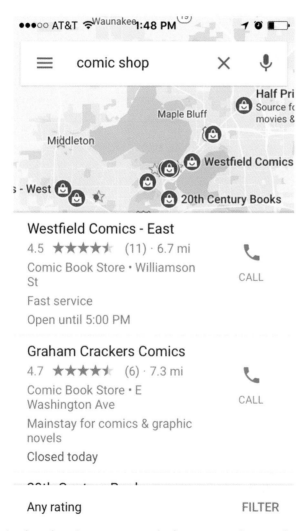

*Figure 27.2* Location-based services maps extend reference maps by providing information – typically consumer-oriented – to a map user based on location and needs (© Google Maps. Taken from iOS Google Maps App.)

searching for and choosing services, but increasingly, the options presented to them are based on algorithms founded on past spatial and buying practices.

## Thematic visualizations

Thematic visualization of spatial data is a natural fit for the Internet. The Internet allows access to millions of real-time databases and the web medium facilitates interactive tools for user-manipulated content. 'Interactivity' – the ability to transform information into different representations with the tap of a touchscreen – is also a benefit that the paper medium simply never had. No longer must cartographers aim for a single optimal representation using one type of thematic style. Instead, map users can select and change between multiple representations to garner a better understanding of the information. Though, whether the ability to switch between visualizations leads to cognitive overload and worse interpretation is something that is still to be determined (Oviatt, 1997).

## Exploratory geovisualizations

Though typically created and used in cubicles and behind closed doors – i.e., in the 'private realm' (MacEachren, 1995) – exploratory geovisualizations are increasingly web-based. This was not always the case, as early such visualizations largely depended upon discrete datasets and desktop applications. Exploratory geovisualizations are useful for analyzing data and turning it into useful information. They are often less useful for communicative purposes, however. Nonetheless, increasingly new Internet-based applications are being developed to not only help explore data but to then present it in an intelligible manner to end users (for example, Tableau and CARTO).

## Narrative maps

Narrative maps tell stories. Though maps have been used to tell stories for hundreds of years, since the dawn of Internet mapping there has been a corporate push to revitalize the genre (i.e., ESRI brands their Internet narrative maps 'StoryMaps'). Media outlets often use narrative maps to highlight adventures, expeditions, or contextualize current events. Journalism has increasingly turned to storytelling, through a combination of maps, text, and images, to better engage audiences and subsequently increase advertising revenue (Fisher, 2015). Narrative maps found on the Internet increasingly allow users to interact with rich multimedia storytelling components. Moreover, the interactivity of web-based narrative maps means that cartographers are increasingly able to tell stories in a non–longitudinal, hyper-textual, and graphic order.

## WikiMaps

Often overlooked as a category, wikimaps are powerful tools for collating community-based knowledge. In a sense, they act as a depository of mental maps and shared history – though, because the Internet proves ephemeral, the life of such depositories may be short. Wikimaps allow people to add their own personalized data, history, knowledge, and visual artifacts to the landscape. Thus far their use has largely been confined to helping indigenous populations, documenting memories before a landscape is reinvented, to help disenfranchised communities stake a claim on landscapes during contested political processes. They are increasingly being used to map emotions and memories. For example, a new start up called LifeMapping is marketing the ability to map loved ones' lives and memories before they die or lose their memory to Alzheimer's (see Figure 27.3).

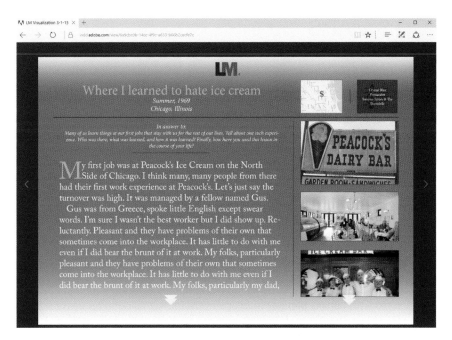

*Figure 27.3* An example LifeMapping memory embedded within a user's online map

*Source*: Dean Olsen (dean@deanwolsen.com). Used with permission.

## Persuasive and copy maps

Perhaps no industry in the world of Internet mapping has more potential for growth than persuasive maps (Muehlenhaus, 2014). Maps have always been used as visual copy, i.e., advertisements, news supplements, and emotional tools. However, as the tools for creating slick-looking interactive Internet maps become more accessible, any nefarious or maladjusted individual can make a map and distribute it for free to the world. Some of these persuasive maps inevitably go viral and spread their message much further and wider than any print map would go. In the United States where open records laws make it possible to collect information about people that in many countries would be considered private, a disturbing trend is to not only create propaganda maps but to target those you disagree with. Increasingly maps are distributed which mark individuals who hold opposing political views. Maps that target the home addresses of gun owners, anti-gay marriage proponents, and abortion doctors are all easily accessible on the Internet (see Figure 27.4).

As should now be evident, the idea that Internet maps can be discussed by their medium alone is trite. Some argue that because users can interact with Internet maps, cartographers should be able to design maps that achieve many communication goals at once. This is a backwards argument, however. It places map medium in front of purpose. Seasoned cartographers choose their medium based on what they are trying to communicative and on audience expectations. They do not create communicative goals and add them just because a medium allows them to do this.

Categorizing Internet maps is useful, but it should not overshadow the fact that the medium is not static. (Neither was the print medium, for that matter.) Technology keeps changing, allowing for different forms of communication, interactivity, and representation.

*Figure 27.4*    An online map mapping the location of every Proposition 8 donor (a proposal to constitutionally ban gay marriage in California). Clicking on the upside-down teardrops brought up the donor's name, place of work, and donation amount. The map has since been removed. © Google Maps, 2012

The Internet is central to this time-space compression, and it is evolving much more quickly than any communicative technology that ever came before it (Harvey, 1991). In the next section we review just how quickly Internet mapping has changed and will continue to change with its namesake medium.

## Don't blink! Internet maps are growing up fast

If the science of Internet mapping were a person, it would be a naïve, idealistic 20-something-year-old. Though Internet maps have been around since the beginning of the World Wide Web (WWW) – and even a bit before – one does not have to go back far to celebrate their birthday, after all, the WWW did not go public until 1994. Just as a parent can observe a child go through phases of life, cartographers can look back on the evolution of Internet maps from their infancy, into their adolescent phase, through their rebel teen years, and into their idealistic 20s. The good news? We survived the teen years!

### The infant years (mid to late 1990s)

Early on, Internet maps were simply static images. These were often scans of printed maps or image exports from GIS and graphic design applications. It is easy to forget, but even static images during the public Internet's infancy presented numerous obstacles to viewing. Using 56-kilobite per second or slower modem technology, a scanned map image could take several minutes to download into the browser for viewing. Looking at maps online was a time-consuming process, and it often made more sense to look up what you needed in a paper atlas.

Aside from JPEG, GIF, and PDF map images (the PNG file extension did not exist yet), the era was dominated largely by the innovations of a single company called MapQuest. MapQuest was the 'Google Maps' of the 1990s and early 2000s. As of 2015, it was the second largest web mapping platform in the world, only trailing Google Maps (Harlan, 2015). MapQuest was one of the first companies to promote interactive and scalable reference maps to web users.

MapQuest's innovation was that it designed an interface allowing people to explore detailed reference maps even with slow Internet speeds. When visiting mapquest.com, users were presented with a map interface and a search input for addresses. When one typed in an address, the interface loaded a simple image of the address at a particular scale. If a map required more or less detail, the user could click on a scale bar to get a larger or smaller scaled image of the same area. The real innovation came with how MapQuest facilitated in-map navigation. MapQuest placed directional arrows along all four sides of a map image. By clicking on a directional arrow, map users could load adjacent map images to the north, east, south, or west of any currently viewed image.

Though primitive by today's standards, it was a revolutionary step at the time. It facilitated multi-scale interaction with large areas. It also opened the world up to new mapping possibilities when Internet technology became faster and even more interactive.

The design of maps on the web at this time left much to be desired. Developers were largely stuck with 'web safe' colours – a palette including very few avenues for aesthetic nuance. MapQuest maps, for example, had jaundice-yellow backgrounds. Screen resolution was stuck at 72 pixels per inch – maximum. Essentially, there was no way an Internet map could come close to the design quality of a print map.

## Toddling along (late 1990s–2004)

In the late 1990s two things happened. First, broadband Internet started expanding into homes and businesses. (Though broadband speeds in the 1990s were anything but speedy by today's standards.) Second, personal computers were becoming increasingly powerful, allowing companies to begin proliferating proprietary systems for viewing online content.

Of course, the Internet had always been subject to brand battles. For example, America Online – the largest service provider of the Internet in the United States – attempted to convince users to log onto the Internet using its own desktop software instead of a simple web browser. Concurrently, the mid to late 1990s represented the first, multi-million dollar browser wars. Different web browsers – including Netscape Navigator, Mozilla, and Microsoft Explorer – had different HTML (HyperText Markup Language), CSS (Cascading Style Sheets), and scripting-reading standards, making it necessary to put disclaimers on websites that they were viewed better on such-and-such a browser than others.

The stratification of browsers and confusion over their different methods for interpreting scripts presented an opportunity for companies to develop and promote browser plug-ins. (Browser plug-ins are external programs that run within a browser.) In addition to allowing almost any browser running on any type of operating system to interpret one's website the same way, plug-ins also allowed for more interactive design and programming than was possible using HTML and a fledgling interactive scripting language named JavaScript. Quickly, two plug-ins became synonymous with interactive Internet maps – Flash and Java (currently, but not always, owned by Adobe and Oracle, respectively).

Both of these plug-ins revolutionized what was able to be shown, distributed, and done with maps on the Internet. Though now largely reviled as 'resource hogs' and 'virus magnets', plug-ins in the late 1990s and early 2000s pushed the envelope for interactivity and user interface design on the web. Flash in particular, originally created for professional vector animation (many children's cartoons are still created with Flash software today), allowed cartographers to begin emphasizing map design.

Perhaps nothing defines this era of map design more than animation and sound. Flash allowed cartographers to animate their data, map user interactivity, and to add sound effects to their maps. With the gift of hindsight, it was probably too much of a good thing. Many years were spent studying the benefits of animating map data – turns out there were not many (Harrower, 2007; Battersby and Goldsberry, 2010). People also began studying the use of sound on maps; although, in most cases, sound was just used to make irritating button click noises whenever one selected something on a map or to play dramatic music in the background. (The move to mobile inevitably eliminated the use of superfluous sound effects in maps, as people on smart-phones listening to music are irritated by clicks and clacks interrupting a favorite guitar solo.)

Though many of the maps remaining from this era were extremely gaudy, plug-ins did provide an avenue for Internet map experimentation. They also ushered in a series of thematic maps and exploratory visualizations that had largely been missing in the first iterations of Internet maps. The design was, however, still largely limited by monitor resolution and cross-browser incompatibility using HTML standards.

## The joys of adolescence (2005–2008)

Google Maps and Google Earth went public in 2005 and quickly revolutionized how maps were used on and made for the Internet. Google's acquisition and development of these technologies (originally developed by a company called Keyhole, Inc.) set Internet mapping on a trajectory that it is just now starting to evolve away from.

Within the context of the Web 2.0 revolution, Google (at the time known merely as a search engine company) refashioned the MapQuest model with AJAX (asynchronous JavaScript and XML). Its online site, Google Maps, allowed map users to interact directly with the map on the screen by clicking and dragging it instead of using directional arrows. Loading times were very fast compared to most map services and one could easily switch between aerial photography and generalized maps. The map design was more aesthetically pleasing than many of the competing online reference map companies as well. Finally, Google's user interface design was state-of-the-art for the time.

In addition to Google Maps, Google also introduced its stand-alone Internet-dependent, desktop program, Google Earth. Google was not the first company to attempt to map the world through aerial photography. Microsoft had actually beaten them to it by almost eight years with TerraServer – an online, black-and-white, aerial photography site. However, accompanying Google Earth was a new XML-based file format called Keyhole Markup Language (KML). KML was a geospatial data format specifically designed for web distribution. Beyond Google Earth it was the development of the KML data type that began revolutionizing online data visualization. Now people, with a bit of XML magic, could simply load spatial data files stored on the Internet. Google sped up the inevitable combining of accessible GIS data and online mapping.

In sum, this time-period was crucial for pushing Internet maps beyond the limitations of both print cartography and early web technologies. It also represented the final death-knell for the traditional, paper-based highway map/road atlas industry. Google not only gave directions, but you could early on print and email them to yourself. Later, you could send them to your mobile phone. All of this for free.

Due to its scale and global reach, Google set the benchmark for what Internet maps could be. Until recently, numerous other companies were playing catch-up and largely emulating Google's model. Google not only introduced the concept of tile maps in a real-time interactive environment, they opened developers' eyes to new types of spatial data formats, and they made cloud-based mapping easier. With benefits come setbacks, however. Google is also responsible for ushering in a ten-year period where the Mercator projection became the default representation of the world on the web. This following 20-plus years of research by academic cartographers critiquing the overuse of a projection originally devised for sea navigation (Monmonier, 2004).

## Terrible teens (2009–2014)

Following the rise of Google Maps the Internet changed very quickly. It went from computer screens and television sets to handheld devices. Of course, it had been possible to access the Internet in some form or another via mobile phones for many years. But with the release of the iPhone in 2007 and the subsequent development of the Android operating system by Google, Internet maps became far more mobile and far more useful than ever before.

The terrible teen phase of Internet maps was marked by insecurity over which technologies to use. Just as Flash map application programming interfaces (APIs) began to flourish after the release of ActionScript 3 in 2008, the door on Internet plug-ins came crashing down. In early 2010, Steve Jobs of Apple published an infamous letter where he outlined why plug-ins, and Flash in particular, were never going to be allowed on Apple's mobile devices (Jobs, 2010). It was the death-knell for plug-in-based maps. (It goes without saying that Java-based Internet maps do not work on most mobile devices either, unless written as native apps for the Android operating system.) Up to this point, plug-ins had been used to create richly dynamic web games and a majority of animated media on the Internet. The failure of plug-ins to adapt to gain widespread traction on mobile devices resulted in the rapid advancement of HTML5 technologies.

Beyond debates over plug-ins and APIs, the human-computer interface also underwent a major transition from mouse and keyboard to multi-touch. This had a major impact not only on how maps could be designed but with device compatibility. For example, many Internet maps possessed a rollover feature where when one rolled over an object on a map with a pointer device (e.g., the mouse) information would be displayed. However, with touch interfaces this no longer worked. Beyond touch, Google, Apple, and Microsoft all implemented powerful voice recognition features on their mobile devices, allowing users to input search information without the need for a keyboard.

Internet map design also underwent major changes during this period. Up until this point, many maps had been using skeuomorphism – design that mimics real-world features. However, beginning with Microsoft products (i.e., the Zune music player), then Google online apps, and finally Apple mobile devices, flat design became increasingly predominant. Flat design uses a modern, technical look, avoiding drop shadows, textures, and layering. It did not take long for Internet map design to resemble increasingly the mobile operating system interfaces they were embedded within.

In conjunction with flat design, this time-period also saw the rise of responsive web design. Responsive design automatically resizes and adapts a web page, mobile app, or Internet map to fit a given screen size. It allows a cartographer to design an Internet map once and not worry about how it will look on an iPad running iOS versus a Nokia handset running Windows. Responsive design is typically created using CSS or external JavaScript APIs. These days many map APIs, including Leaflet, incorporate responsive CSS files automatically.

The trend of Internet maps homogenizing around flat, responsive design and the Web Mercator had remarkably little to do with cartographers themselves. Instead, with the demise of established plug-ins, map-makers became increasingly dependent on JavaScript-based APIs to produce their Internet maps. Regardless of their origin, open or proprietary, APIs almost universally mimicked Google Maps tile-based system, including the use of the Web Mercator projection. APIs also limited a map-maker's control of map element styling. Thus, the graphical user interfaces of maps typically began adopting the flat, minimalist designs promoted by the web community broadly.

One final trend that largely began during this era was the obsession with mapping 'big data' – particularly data stemming from a variety of new social media sources. The ability to mash-up different APIs from both data providers and mapping services allowed map-makers to begin experimenting with big data visualizations (e.g., to tweet locations during large events using Twitter). The early results were often visually cluttered and less than perfect.

Traditionally cartography has been concerned with synthesizing data into useful information. After analysis and synthesis, maps are used to visualize useful information, which in turn leads to knowledge acquisition (DiBiase, 1990). However, with increasing access to gargantuan datasets on the Internet, it became increasingly acceptable for map-makers to dump large amounts of data into a map API and call their work done. Though mapping large datasets may prove useful for data exploration, in general many of these big data maps only communicated noise or population densities (Bollier and Firestone, 2010).

## Naïve twenties (2015–present)

Things have standardized quickly in the last few years. HTML5 technologies (i.e., HTML, CSS, and JavaScript) are open source and (for the most part) all major browsers are finally coalescing around a single set of standards. Many new online tools and services have been created for visualizing big data sets in manners other than upside-down tear drops and heat maps. And, although open source tools exist, Internet mapping remains a largely capitalist venture, with numerous angel-funded mapping start-ups clamoring for subscription-based memberships and ESRI steadily moving components of its desktop GIS platform online.

If any one word defines the current age of Internet mapping, it is 'open'. There is a large push among many in the cartography community to use open standards (e.g., GeoJSON) and open data (e.g., OpenStreetMap). Numerous companies have made it an initiative to incorporate open data into their services. (Notably, MapQuest was one of the first to incorporate wholly OpenStreetMap data into their business model.) Companies like Mapbox, CARTO, ESRI, and others now make use of crowd-sourced data and play well with open source JavaScript APIs such as Leaflet.

The design of Internet maps is changing perceptibly. Increasingly one finds Internet maps that do not simply use default Web Mercator projections and map APIs that allow for easy transitions between projections (see Figure 27.5). HTML5 Canvas can be used to create animated projections that shift on-the-fly as one zooms and pans a map (Jenny et al., 2016). Due to drastic increases in screen resolution and processing power, the aesthetic detail of Internet maps is increasingly on a par with, or even better than, those of print.

Also, map service providers are beginning to develop tools that allow for previously forgotten and oft-neglected methods of visualization. For example, CARTO and Mapbox now make designing hexbin maps (hexagonal binning aggregates data into coarser representation for display) extremely easy.

Finally, maps are no longer confined to what we think of as traditional screens. Cartographers are designing maps for micro screens found on wearables. Wearables include everything from wristwatches to glasses. From dashboard maps to virtual reality goggles, how we consume Internet maps is becoming more complex and embedded within our everyday experience. The next frontier in wearables is likely to be virtual reality goggles. Though bulky and wired as of this writing, it does not take much extrapolation to see that they will soon become micro, wireless, and wearable as well.

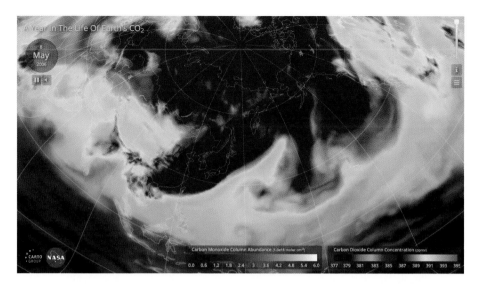

*Figure 27.5* A map created by the Oregon State Cartography Group and NASA using Canvas that not only animates carbon emissions over the course of a year but also allows the map user to re-project the map on-the-fly. Map available at: *https://co2.digitalcartography.org/*

## What will the next ten years bring?

Given how much has changed in the past twenty years, it is unwise to make predictions about the next five years, much less the next decade. However, there are several developmental avenues that will surely be pursued by Internet cartographers. Several of these have more predictable outcomes than others.

### Map-making has moved to the cloud

Cloud computing is here to stay. As I write, these words are being auto-saved to the cloud. Not only have maps as end products moved online; the tools used to make maps are there as well. Whether you create a map via HTML/CSS/JavaScript and an API or if you use someone's online application (e.g., CARTO, ArcGIS Online), the tools are cloud-based.

### Internet maps will become more refined

Internet maps have evolved. They have quickly become just as artistic and aesthetically pleasing as the best paper maps ever were – albeit the aesthetic is extremely flat. Interface designs will continue to evolve. Representations and symbols will be continuously tested, with the poor ones being weeded out and the best ones being replicated. Finally, purposeful generalization will become much more common. Online map services will be designed to communicate many things (directions, weather, traffic, crime rates), but will change aesthetically depending on what is being shown.

### Rise of equal-area projections

The need to use Web Mercator projections at small scales is decreasing. This is fortunate, as there is increasing evidence that looking at the Mercator projection so frequently is dementing an entire generation's mental map of the planet (Battersby *et al.*, 2014). It has long been established that equal-area projections are generally more appropriate for thematic maps than others (Slocum *et al.*, 2008); though, debate still continues about which types of equal-area projections are best in different situations. Different projections will become increasingly easier to implement in Internet map APIs.

Beyond projections, many new Internet APIs are advancing the usefulness of interactive cartograms. When mapping thematic data, cartograms allow map-makers to avoid projections altogether. Instead of representing data using intermediary symbols, cartograms resize enumeration units based on their values. Since all projections must grossly mutate either the shape or area of the Earth, there is a strong argument that cartograms do a better job than other types of data representation by mutating enumeration units in an immediately comprehensible manner (Vujakovic, 1989). New APIs such as D3 allow cartographers to experiment with different types of cartograms like never before.

## *Map use will be experiential*

There is one inevitable change that has the potential to disrupt all of the above. (This is because when it takes place, the idea of viewing the world in two-dimensions on a flat screen may become downright ludicrous.) Virtual reality maps are coming. These will not only allow map users to interact with maps but move around inside of them – likely as groups of networked friends.

Though adoption will probably proceed in pulses and be put to use for particular types of maps over others (e.g., data exploration before WikiMaps), it will undoubtedly influence all types of Internet maps eventually. Looking back at how drastically Google Maps, an innocuous invention just over ten years ago, reshaped all types of Internet maps, it is hard to believe that the virtual reality headsets currently invading the market will not quickly upend the mapping industry. Once virtual reality headsets shrink to the size of a regular pair of sunglasses, two-dimensional Web Mercator maps will naturally fade in importance.

## Conclusion

We have come to a point where talking about Internet maps as though they are something novel is trite. In the contemporary setting, Internet maps are just as ubiquitous, if not more so, than paper maps. They can be designed to achieve myriad communication goals and in most cases, have evolved beyond the capabilities of their print ancestors.

Though the technologies and style of maps produced for the Internet have gone through numerous fits and starts, stabilization is well under way. Standards have emerged in data model design, map aesthetics, and user interface techniques. Map-makers no longer depend upon proprietary plug-ins to design robust, interactive maps. Companies that develop their own proprietary APIs do not discourage interacting with open APIs and databases. Maps on the Internet have entered a stage of maturity.

Maturity does not mean stagnation, however. As the Internet evolves, so too do web and mobile maps. The rise of virtual reality headsets and immersive wearables will provide many opportunities for Internet maps to develop in directions we cannot even currently fathom. Unfathomable though they may be, we can all breathe a collective sigh of relief that new developments will inevitably spell the end of the Web Mercator projection for most small-scale maps.

# References

Battersby, S.E., Finn, M.P., Usery, E.L. and Yamamoto, K.H. (2014) "Implications of Web Mercator and Its Use in Online Mapping" *Cartographica* 49 (2) pp.85–101.

Battersby, S. and Goldsberry, K. (2010) "Considerations in Design of Transition Behaviors for Dynamic Thematic Maps" *Cartographic Perspectives* 65 pp.16–32, 67–69.

Bollier, D. and Firestone, C.M. (2010) *The Promise and Peril of Big Data* Washington, DC. Aspen Institute, Communications and Society Program.

DiBiase, D. (1990) "Visualization in the Earth Sciences" *Earth and Mineral Sciences Bulletin* 59 (2) pp.13–18.

Fisher, T. (2015) "Do Visual Stories Make People Care?" Available at: *http://blog.apps.npr.org/2015/11/19/sequential-visual-stories.html* (Accessed 1 May 2016).

Harlan, C. (2015) "'Does MapQuest Still Exist?' Yes, It Does, and It Is a Profitable Business" *The Washington Post* (22 May). Available at: *www.washingtonpost.com/business/economy/does-mapquest-still-exist-as-a-matter-of-fact-it-does/2015/05/22/995d2532-fa5d-11e4-a13c-193b1241d51a_story.html* (Accessed 1 May 2016).

Harrower, M. (2007) "The Cognitive Limits of Animated Maps" *Cartographica* 42 (4) pp.349–357.

Harvey, D. (1991) *The Condition of Postmodernity: An Enquiry into the Origins of Cultural Change* Oxford, UK: Wiley-Blackwell.

Jenny, B., Liem, J., Šavrič, B. and Putman, W.M. (2016) "Interactive Video Maps: A Year in the Life of Earth's $CO_2$" *Journal of Maps* 1–7. doi:10.1080/17445647.2016.1157323

Jobs, S. (2010) "Thoughts on Flash" *www.apple.com* Available at: *www.apple.com/hotnews/thoughts-on-flash/* (Accessed 15 April 2014).

MacEachren, A.M. (1995) *How Maps Work: Representation, Visualization, and Design* New York: Guilford Press.

Monmonier, M.S. (2004) *Rhumb Lines and Map Wars: A Social History of the Mercator Projection* Chicago, IL: University of Chicago Press.

Muehlenhaus, I. (2014) "Going Viral: The Look of Online Persuasive Maps" *Cartographica* 49 (1) pp.18–34.

Oviatt, S. (1997) "Multimodal Interactive Maps: Designing For Human Performance" *Human-Computer Interaction* 12 (1) pp.93–129.

Slocum, T.A., McMaster, R.B., Kessler, F.C. and Howard, H.H. (2008) *Thematic Cartography and Geographic Visualization* (3rd ed.) Upper Saddle River, NJ: Prentice Hall.

Vujakovic, P. (1989). "Mapping for World Development" *Geography* 74 (2) pp.97–105.

Wu, A., Zhang, X. and Zhang, W. (2008) "GPS Secure Against Getting Lost, or More Danger?" *Proceedings of the 2008 International Conference on Cyberworlds* pp.501–505.

# 28

# Maps and atlases for schools

*Stephen Scoffham*

By the time they start primary school, children (age 5) are already aware of the wider world and will have started to form their ideas about places beyond their direct experience. They will know that there are creatures such as tigers, whales and polar bears that live in different global habitats. They will have seen pictures of distant places on TV, films and electronic media and heard of other countries, perhaps through major news events or sporting connections. Some will have travelled abroad for holidays or to visit relatives. Others will have learnt about journeys to unusual places in picture books and stories. As children grow older and embark on their first years of schooling, their ideas about the world continue to develop. The information which they receive comes from many sources, but maps and atlases will undoubtedly be one of the key influences that will shape their images. This chapter explores some of the issues surrounding the production and use of school atlases in the UK, with a particular emphasis on the needs of younger pupils and adolescents. It focusses on print rather than electronic productions and is based on the author's own experience of writing and devising atlases for children and young adults over the last 25 years. It is argued that devising a school atlas involves making many compromises and recognising contradictions. Articulating the principles which guide content selection is fundamental in developing a deeper understanding of atlas production.

## What is a school atlas?

School atlases serve contradictory purposes. Traditionally they have been seen as providing spatial information about the location of places and different environments around the world. This function positions them as authoritative and comprehensive reference sources along with encyclopaedias and dictionaries. However, school atlases also have a role in supporting investigative teaching and learning, particularly in geography. As such they need to appeal to pupils imaginatively and provide creative spaces that will stimulate their thinking. The tension between these different approaches poses a challenge for atlas authors. They have to strike a balance between presenting a single, uncontested account of the world which is closed and precise with a much more open approach which acknowledges multiple and contested realities.

The tensions surrounding the purpose of an atlas are played out at both a structural and an individual page level. Atlases which consist almost entirely of physical and political maps of different parts of the world do not generally raise questions and pose dilemmas for pupils. A more varied layout which includes text, photographs, diagrams, charts, tables and other data can draw attention to issues and suggest a variety of interpretations. What then are the boundaries between an atlas, textbook and teaching pack? Publishers adopt a simple rule of thumb. Atlases have 50 per cent or more of their page space given over to maps. If the overall balance drops below this level, the publication falls into a different category.

## Pupils' world map knowledge

It is important for atlas authors to understand how children read and interpret atlas maps. The way that pupils progress from their initial confused and imprecise images of the world in infancy to a more systematic understanding of abstract cartographic conventions by early adolescence is a fascinating story. One of the most significant areas of study concerns children's growing under- standing of the world map. Research by Wiegand (1995) into English primary school children's (ages 5–11) free-recall map of the world suggests that there are a number of distinct stages. The most basic representations may simply consist of isolated and disconnected shapes representing places as varied as local towns, countries, the North Pole or places featured in stories. In the second stage territories begin to be differentiated by size and in the third sub-divided into areas. In the final stages the territories become increasingly recognisable and their scale and spatial dis- tribution increasingly accurate. Lowes (2008) has reported a broadly similar sequence from more recent research in the United States but notes that in the interim stages children either tend to draw land masses floating in the sea (island maps) or clinging to the frame (edge maps). The most sophisticated and recognisable maps also exhibit a tendency towards symmetry and alignment of land masses – something which has also been noted in adult responses (see Figure 28.1).

It is important to note that research which is based on children's free-recall maps of the world may well reveal more about their drawing ability than the extent of their geographical knowledge. Some children, too, are liable to be constrained by their desire for accuracy and feel inadequate when they recognise the gaps in their knowledge. Despite these limitations when interpreted judiciously free-recall maps are a rich source of information. As well as revealing gaps, mistakes and distortions, they also often provide clues to pupils' personal or private geog- raphies. Typically children include places where they have close family or personal connections. Furthermore, what is left out of a map is sometimes just as significant as what is included, as it can alert pupils to the limitations of their knowledge and prompt them to learn more. Engaging children as map makers, as Chave (2011) observes, helps them to recognise the subjectivity in their own maps and, by inference, the subjectivity in the maps of others. This opens the door to a discussion about the assumptions which underpin atlas maps and the principles on which they are based.

## Challenges for children

Atlas maps present many challenges for young children. Language is one of them. Insecure readers who are used to decoding text arranged in neat lines from left to right may find the labels which are scattered across the page in various type sizes hard to decode. Also, as they scan different maps they will come across unfamiliar place names derived from foreign languages which are difficult to read. Abbreviations present further difficulties as do homonyms (words

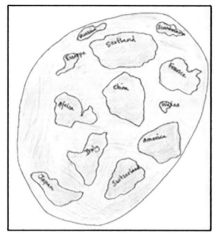

In this 'island map' the size and scale of the landmasses are almost entirely random (girl aged 8).

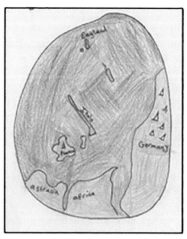

An 'edge map' is where the land clings to the border (girl aged 9).

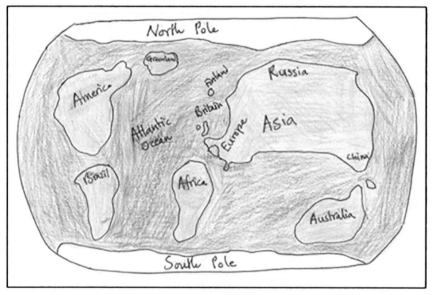

Land masses are distinguished by size, shape and distribution in this 'formal map'. Note how the polar regions frame the composition and the emphasis given to the northern hemisphere (boy aged 11).

*Figure 28.1* Children's free-recall maps of the world. In this 'island map' the size and scale of the land masses are almost entirely random (girl aged 8). An 'edge map' is where the land clings to the border (girl aged 9). Land masses are distinguished by size, shape and distribution in this 'formal map'. Note how the polar regions frame the composition and the emphasis given to the northern hemisphere (boy aged 11). Courtesy of pupils at Elham Church of England Primary School, Kent, UK

which sound the same but have different spellings and meanings e.g. 'more' and 'moor'). Linguistic 'false friends' such as the 'Alice Springs' in Australia or 'Salt Lake City' in the USA are a further complication as they suggest physical features but are actually settlement names.

At a more fundamental level when they use thematic maps children need to understand broad concepts such as 'climate' rather than more specific terms such as 'sun', 'wind' and 'rain'.

Scale and symbols are equally problematic. In order to show continents or large land areas, atlas maps are necessarily drawn to very small scales and involve considerable generalisations. Children who can make sense of maps of their local area which they have explored on foot and visited repeatedly have no such support when they are presented with a map of the world. Indeed, in order to interpret what they see they have to bring additional knowledge and under-standing to bear if they are to interpret cartographic conventions. To help young children understand ideas which are beyond their direct experience, atlas publishers sometimes use draw-ings on maps instead of abstract symbols. However, it appears these are not always readily understood. Wiegand and Steil (1996), for example, found in their research with 86 primary school pupils that in one picture atlas a car symbol, which was intended to represent car manu-facture, was variously interpreted more prosaically as a 'car park', traffic jam', a 'breakdown' or simply a 'place where people like cars'.

The notion of a country raises further complications. Not only are countries socio-political constructs but they are also part of a hierarchy in which smaller units such as towns and regions are nested within larger units such as continents. It was research by Piaget (1928) that first alerted educationalists to the complexities of nested hierarchies. Subsequent researchers such as Jahoda (1963), Harwood and McShane (1996) and Lowes (2008) confirm that pupils as old as ten or eleven commonly experience difficulties with the concept of a country and see the components of their address as isolated units rather than as components within a unified structure.

The difficulty that children experience in going beyond their direct observations is also illus-trated by their ideas about the Earth in space. Although infants may readily say that the Earth is a sphere, more detailed questioning reveals they often retain common sense notions about the sky above their head and the ground beneath their feet. When questioned, for example, about how people can stand on the Earth in the southern hemisphere when those in the north are the 'right' way up, Nussbaum (1985) found young pupils were typically unable to offer a satisfactory explanation. They advanced all manner of ingenious ideas but continued to believe that 'up' and 'down' were absolute rather than relative terms and appeared unable to entertain the possibilities of viewpoints other than their own.

The challenges which children experience in interpreting the world map can be seen as part of a much larger developmental process. Ideas about progress and development are intri-cately entwined with social, cultural and psychological factors and cannot be reduced to a linear sequence. Teachers know from their experience that learning often happens erratically and in unexpected ways. Discontinuities and misconceptions are the norm rather the exception. Despite these caveats children appear to pass through different stages, and it is possible that the middle years of childhood may be a particularly favourable time to teach children about the world map. At this age pupils have progressed beyond the ego-centric thinking of their infant years but have not yet entered the self-centred modes that characterise adolescence. As they begin to develop interests beyond their home and family and assert their independence, finding out about other parts of the world chimes with their needs (Kellert, 2002).

Age is just one of the factors which need to be taken into account when exploring children's ideas about countries and national groups. Barrett (2007) presents a comprehensive review of research into children's ideas conducted over many decades and draws attention to other key influences such as gender, culture, ethnicity, social class, overseas travel, personal experience and individual character traits. He also highlights how children's thinking is often associated with strong emotions which may develop before they acquire any factual knowledge or understand-ings. Such findings are important for educationalists as they suggest a complicated interaction between cognition and affect. Simply learning about another country is not necessarily going to change how pupils feel about it.

## Planning a school atlas

One of the challenges in devising a school atlas concerns the selection and sifting of content. This is not an open-ended process as decisions about the market and age range will have usually have been made in advance, along with costings for a specific number of pages and illustrations. It is also likely that there will be an anticipated publication date designed to catch a particular sales opportunity or to fit in with an established publishing programme. Sales and marketing are absolutely fundamental but contribute only marginally to the vision behind the publication itself. At a fundamental level what matters most is for the author and publisher to agree on its purpose. To what extent is it intended as a teaching resource, and what philosophy of learning underpins it? Why is the knowledge and information that it presents worth knowing? Does it present reasonably balanced images, and is there scope for creative engagement and a range of interpretations? Such questions are best addressed with reference to a deep understanding of teaching and learning and clearly acknowledged educational principles and values.

Surprisingly, one of the aspects of atlas planning which is often overlooked is what children themselves actually want. There is little direct evidence, but tangential research from the UK confirms that pupils associate geography with place and recognise the importance of maps (Kitchen, 2013). When it comes to the questions which children spontaneously ask about geography, a survey of 260 children aged nine and ten years old identified that the Earth in space, natural hazards, people and countries were key areas of interest (Scoffham, 2013). Interestingly, many of the children's questions involved interdisciplinary links (especially with science and environmental education) and revealed an interest in processes and personal identity as well as factual information. Earlier unpublished research into children's atlas preferences, also by Scoffham, confirms the focus on place knowledge and the notion of an atlas as a reference source. At an individual level pupils requested information on a very wide range of topics including the origin of foods, world languages, religions and global conflicts. School atlases and other educational resources are unusual in that they are often not bought by the users (either pupils or class teachers); purchasing decisions tend to be made by the subject co-ordinator in conjunction with the head teacher. This has the effect of breaking the direct link between the product and the consumer. However, it is also important to remember that survey results need to be interpreted with care as learners do not always know what they need in order to further their own learning.

The sequencing of material is another issue that needs to be considered. If an atlas is viewed as a reference source, then the way maps and topics are arranged makes little difference as long as information can be easily accessed. However, when an atlas is regarded as a learning tool, themes and topics are best arranged so that they lead into each other in a logical order. Traditionally, the opening pages of an altas deal with basic skills such as latitude, longitude, scale and map reading. This is followed by sections on the home country and surrounding areas, concluding with studies of more distant places and global themes. Could there be advantages in a sequence which acknowledges how children develop their knowledge of the world? Wiegand (2006) draws attention to research findings which suggest that map memory is encoded hierarchically with larger regions such as countries encoded before smaller ones such as cities. Lowes (2008) too draws on her findings to recommend that the highest class of objects such as continents and oceans are best taught first to provide a framework which children can then populate with more detailed information as their learning develops.

One of the main functions of an atlas as far as schools are concerned is to support the curriculum. In the UK, the National Curriculum has always made specific mention of atlases and locational knowledge. Despite these requirements there is good evidence that pupils' knowledge remains poor and that teachers are failing to contextualise studies. For example, Ofsted (2011) found that secondary school pupils were 'spatially naïve' (para. 36) and that primary pupils had 'exceptionally weak' knowledge of places (para. 10). Such failings are not confined to the UK. In a study involving 10-year-olds from different European countries, Schmeinck (2005) discovered that only around half of Swiss children and a third of Germans were able to locate their own country correctly on an outline map.

Concerns about poor place knowledge amongst both pupils and their teachers needs to be viewed alongside more general issues relating to curriculum planning. At junior school level in particular (ages 7–11), there has been a long standing discussion about whether it is best to focus on single subjects, integrated areas of learning or cross-curricular topics. This raises the question about which areas of the curriculum an atlas should support and whether to include perspectives from a range of disciplines. Understanding how teachers actually use atlases and assessing future trends are additional aspects which atlas authors have to take into account.

## Future trends

As well as keeping abreast of curriculum developments, atlas authors need to be informed by the latest academic thinking. Traditional approaches to geography which tended to favour clear cut descriptions and linear analysis have been replaced by multiple perspectives which recognise that knowledge is both contingent and contested. There has also been increasing interest in humanistic approaches and emotional responses to place. Bonnett (2008) makes the point that modern geography contributes to our understanding of many of the big issues that beset the world today, particularly in relation to the environment and international understanding. Meanwhile Lambert (2011), coming from an educational perspective, contends that geography needs to adopt a 'capabilities' approach which will help pupils to become competent and self-fulfilled individuals.

The dramatic advances in technology have also had a considerable impact on geographical enquiry. Pupils now have access to electronic maps and on-line school atlases which they can modify and manipulate to show different variables. One of the consequences is that the opportunities for pupils to identify patterns and interconnections relating to a particular line of enquiry have been greatly enhanced. At the same time, the range of cartographic material which they can access has multiplied. Flexibility and the ability to respond to change are key features of digitised data. Electronic maps can be regularly updated while print materials are bound to date.

These trends look set to continue, and they are having a huge impact on both the way that maps are compiled and the information that they portray. The view of the world which we carry in our heads – our geographical imagination – is evolving as a result. For a social geographer such as Danny Dorling, the potential for maps to show human rather physical data is enormous. As he puts it:

> When I look at a 'normal' map today I see a strange map. I see a map about places where people almost always don't live. I see almost all human life squeezed into a tiny part of the paper or screen … On our screens, on our phones, in our textbooks and magazines, our images of the world are changing faster than the world itself. This is because we are rapidly evolving … to think collectively, differently, to visualise better and to accept less meekly.
>
> *(Dorling, 2012: 98)*

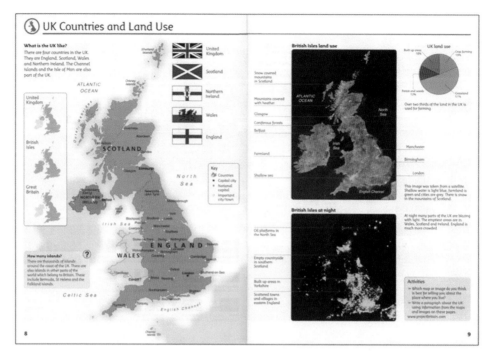

*Figure 28.2* **Setting maps and supporting information alongside each generates creative space for the reader**

*Source: UK in Maps © Harper Collins reproduced by kind permission*

Despite these developments, print atlases seem set to continue to have a meaningful educational role. They provide a clear global and regional framework for new knowledge which is invaluable for new learners and which supports more mature understanding. They also bring together a range of carefully graded and selected material in a single trusted publication on curriculum-related themes and topics. As Widdowson and Lambert (2006) argue, textbooks and print materials in general are part of the 'resource ecology' of the classroom and have an important role to play in contributing to the networks that support learning. There is, however, a real need for atlases to present information in ways that are open to interrogation. The sample double page shown in Figure 28.2 indicates how this might be achieved. Here three maps of the UK are set alongside each other. The maps focus on settlement, land use and light pollution. Information about political boundaries and physical geography is also included. The supporting text invites pupils to investigate the data and to use it as a resource in their own studies of the UK. There is plenty of scope for different interpretations as well as key information for initial learners.

## Design considerations

Much of the success of an atlas depends on the care and attention which is paid to its design. The way that information is presented on the page plays an important part in communicating it. Field and Demaj (2012) in their review of map design reaffirm that even in modern times, cartographers need to draw not only on science and technology but also on aesthetics if they are to create pleasing designs. This is perhaps especially relevant in relation to atlases for young children where visual communication takes precedence over text. If the pages are sufficiently attractive, they will capture the children's imagination and, by drawing them in, will encourage them to find out more.

Tyner (2010) identifies six main elements which combine to create a successful result – clarity, order, balance, contrast, unity and harmony. He compares good maps to good writing: both are 'enjoyable to view and satisfying to use' (2010: 41). When it comes to children's atlases, clarity and simplicity are particularly highly prized. However, the cartographer has to balance the need to provide information and sufficient detail with the confusion which comes from overload. How is it possible, for example, to show a world political map with over 200 countries with clarity and simplicity on a single double page spread? Consistency and logic in the use of type and symbols can help but does not necessarily solve problems. For example, using purple shaded lines for national borders on relief maps may work well in lowland areas but can obscure significant details in mountainous regions. Type conventions which are successful at one scale may not work so well at another. Technical factors also come into play. For example, the quality of the paper and printing can also have a significant impact on the clarity of a map and its readability.

Recognising the way that children respond to colour is another dimension of good design. Young children tend to see bright colours positively and dark colours negatively. There is, however, considerable variation between individuals and their responses are often highly subjective. For example, red may convey ideas of warmth and love, but can also be associated with danger and anger. Furthermore, there are significant gender differences. In their research with children aged seven to eight, Pope *et al.* (2012) found that girls tended to rate pink and purple as happy colours while boys viewed them negatively. Being aware of pupil responses is especially significant when it comes to political maps where countries are distinguished by colours which may unwittingly trigger prejudices.

One way of adding to the visual content of an atlas is to include photographs to supplement the cartographic content. Photographs are a powerful way of bringing a page to life and a valuable source of information in their own right. However, research shows that children often see details rather than the whole image and tend to interpret what they see differently to adults, partly due to their more limited life experience (Mackintosh, 2010). There is also a real danger of presenting stereotypical images, especially if there is only space for a single photograph to illustrate an entire country or theme. A useful question which stimulates debate amongst editors and map designers is how they would choose to represent their own country or area photographically. The overall balance of images across the atlas also needs to be considered. Is there a reasonable balance between men and women and are they portrayed in active or passive roles? Do the photographs give a negative impression of the developing world? Are the overall aims of the atlas supported by the images which have been selected?

Finally, when it comes to world map projections there has been extensive discussion amongst cartographers about the most appropriate approach with children. The consensus currently favours projections which seek to reconcile distortions of area, shape and direction. Eckert IV is widely used as the base for physical and political maps of the world, though equal area projections are essential for showing habitats and biomes. The question which remains is whether a school atlas should only represent the world in one particular way? Different projections and ways of representing continental and global information can extend pupils' thinking. Cartograms and satellite images have a similar role. There is much to be said for striking a balance. Using one projection for the majority of world maps establishes a basic outline in a pupil's mind. Variations challenge this perception and stimulate new thinking, as Vujakovic (2004) argues. The same principles apply to debates about whether to adopt a Eurocentric focus, orientated towards the north. Consistency helps children to consolidate their images but diversity is enriching.

## Gateways to knowledge

Maps are powerful resources which mediate our images of the world and are key gateways to geographical knowledge. They carry messages about what we value and what we think matters. As they study an atlas, children can be transported imaginatively to places around the world. This engagement can be a significant life experience. Research by Catling and others (2010) into the motivations of primary teacher educators has found that 50 per cent of respondents identified the love of maps which they developed in childhood as a key influence in their careers – 10 per cent specifically mentioned atlases.

Maps are also contentious. Questions to do with colonial relations, global inequalities and the future of the environment will inevitably emerge from any meaningful reading of the 'texts' which atlas maps provide. What is left out can also be highly significant – gaps and silences reflect ideological perspectives. Furthermore, the neatness and order which is shown in an atlas is often not reflected in what actually happens on the ground. For example, national boundaries are sometimes hotly disputed, especially in some parts of Africa and the Middle East, but are portrayed in most maps and atlases as static and defined. The population of cities and statistics about many aspects of human welfare are not precise and are open to question. As pupils develop a framework for new knowledge they need to be alerted to such issues. The narrative which an atlas presents is necessarily bounded by culture and time. It is a particular view from a particular place which, as Kent and Vujakovic (2012) remind us, will also be underpinned on an ethical level by a set of values and beliefs. It seems only honest to acknowledge these foundations and to recognise the hidden influence of ideology, power and politics. The alternative is to opt for what Adiche (2009) calls a 'single story' which as well as being misleading and disingenuous is educationally arid.

## Conclusion

Devising a school atlas involves making compromises. Once the basic principles and purpose have been established, the most obvious compromises concern decisions on regional and continental coverage, the balance between locational and thematic maps, and the amount of detail that they show. There are hard choices relating to page design, the use of photographs, charts and diagrams, and the way they can be used to complement each other. The budget which has been allocated and the availability of information provide inevitable limitations. The requirements of the curriculum and the practicalities of classroom use act as further constraints. On a more fundamental level, the tension between the dual functions of promoting flexible learning and definitive locational knowledge provides a further challenge for atlas authors. Some of the main factors involved are shown in the visual summary in Figure 28.3 and are grouped under four main headings (a) aims and values, (b) pedagogy, (c) curriculum, and (d) planning and design.

Atlas authors work as part of a team. The publisher, editor, cartographer, designer, artist, printer and sales team all have a part to play in the creation of a new publication. The way that this informal and extended group relate to each other and share a common purpose can make a significant difference to the final outcome. But perhaps the most important factor is the extent to which an atlas addresses the needs of schools and ultimately the children who use them.

Wiegand (2006) notes that school atlases are almost always out of date by the time they reach the classroom and that they tend to be used chiefly for locating places. He also expresses concerns that atlases are a form of establishment publishing which is necessarily conservative. While these are legitimate observations there is plenty of scope for innovation. For example, the *Usborne*

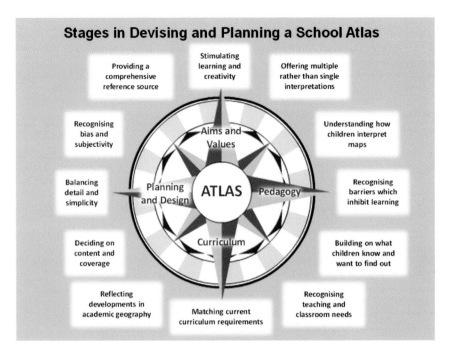

*Figure 28.3* A summary of the stages involved in devising and producing a world atlas

*Internet-linked First Atlas* (2004) broke new ground at the time by being one of the first blended atlases to make full use of electronic media. More recently, the *UK in Maps* and the *World in Maps* (published by Collins in 2013) has shown how a lively combination of cartography and graphics can be used to present pupils with data on a wide range of environmental issues.

Atlas authors can choose to confirm the status quo or they can decide to challenge current orthodoxies in the topics they select and the way that they cover them. They can engage pupils creatively by using text, data and visual images alongside maps to provide a rich resource that they interrogate and interpret. Furthermore, supporting teaching materials and website links can provide up to date information which will supplement the material which is found in the atlas itself. There are plenty of opportunities for printed publications and electronic data to complement rather than compete with each other. And there are many imaginative ways that schools and teachers can use an atlas to develop and extend their own locational knowledge, however weak it might be. As an entry point to learning about the world and as a resource which pupils revisit in different ways during their studies, school atlases deserve to remain highly prized resources which are not only valued but treasured.

## References

Adiche, C. (2009) "The Danger of a Single Story" Available at: *www.ted.com/talks/chimamanda_adichie_the_danger_of_a_single_story?language=en* (Accessed: 12 February 2015).

Barrett, M. (2007) *Children's Knowledge, Beliefs and Feelings about Nations and National Groups* Hove, UK: Psychology Press.

Bonnett, A. (2008) *What Is Geography?* London: Sage.

Catling. S., Greenwood, R., Martin, F. and Owens. P. (2010) "Formative Experiences of Primary Geography Educators" *International Research in Geography and Environmental Education* 19 (4) pp.341–350.

Chave, O. (2011) "Mapping the British Isles with Heart and Head" *Primary Geography* 75 pp. 14–15.

Dorling, D. (2012) "Mapping Change and Changing Mapping" *Teaching Geography* 37 (3) pp.94–98.

Field, K. and Demaj, D. (2012) "Reasserting Design Relevance in Cartography: Some Concepts" *The Cartographic Journal* 49 (1) pp.70–76.

Harwood, D. and McShane, J. (1996) "Young Children's Understanding of Nested Hierarchy and Place Relationships" *International Research in Geographical and Environmental Education* 8 pp.3–29.

Jahoda, G. (1963) "The Development of Children's Ideas About Nationality and Country" *British Journal of Educational Psychology* 33 pp.143–153.

Kellert, S. (2002) "Experiencing Nature: Affective, Cognitive, and Evaluative Development in Children" in Kahn, P. and Kellert, S. (Eds) *Children and Nature* Cambridge, MA: MIT Press.

Kent, A. and Vujakovic, P. (2012) "Maps for Growing Minds: Devising Atlases for Children" *Maplines* 18 (3) pp.4–5.

Kitchen, R. (2013) "Student Perceptions of Geographical Knowledge and the Role of the Teacher" *Geography* 98 (3) pp.112–122.

Lambert, D. (2011) "Reframing School Geography: A Capability Approach" in Butt, G. (Ed.) *Geography Education and the Future* London: Continuum, pp.127–140.

Lowes, S. (2008) "Mapping the World: Freehand Mapping and Children's Understanding of Geographical Concepts" *Research in Geographic Education* 10 (2) pp.1–37.

Mackintosh, M. (2010) "Images in Geography: Using Photographs, Sketches and Diagrams" in Scoffham, S. (Ed.) *Primary Geography Handbook* Sheffield, UK: Geographical Association, pp.120–133.

Nussbaum, J. (1985) "The Earth as a Cosmic Body" in Driver, R., Guesne, E. and Tiberghien, A. (Eds) *Children's Ideas in Science* Milton Keynes, UK: Open University Press, pp.171–192.

Ofsted (2011) *Geography: Learning to Make a World of Difference* London: Ofsted.

Piaget, J. (1928) *Judgement and Reasoning in the Child* London: Routledge and Kegan Paul.

Pope, D., Butler, H. and Qualter, P. (2012) "Emotional Understanding and Color-Emotion Associations in Children Aged 7–8" *Child Development Research* (2012) doi:10.1155/2012/975670.

Schmeinck, D. (2005) "Europe in Geographical Education: An International Comparison of Factors Influencing the Perceptions of Primary School Pupils" in Donert, K. and Charzynski, P. (Eds) *Changing Horizons in Geography Education* Torun, Poland: Herodot Network, pp.206–211.

Scoffham, S. (2013) "A Question of Research" *Primary Geography* 80 pp.16–17.

Tyner, J. (2010) *Principles of Map Design* New York: The Guilford Press.

Vujakovic, P. (2004) "World Maps: A Plea for Diversity" *Teaching Geography* 29 (2) pp.77–79.

Widdowson, J. and Lambert, D. (2006) "Using Geography Textbooks" in Balderstone, D. (Ed.) *Secondary Geography Handbook* Sheffield, UK: Geographical Association, pp.146–159.

Wiegand, P. (1995) "Young Children's Free Recall Sketch Maps of the World" *International Research in Geographical and Environmental Education* 11 pp.138–158.

Wiegand, P. (2006) *Learning and Teaching with Maps* London: Routledge.

Wiegand, P. and Steil, B. (1996) "Communication in Children's Picture Atlases" *The Cartographic Journal* 33 (1) pp.17–25.

# Part V
# Maps and society
## Use, uses, and users

# 29

# Mapping place

*Denis Wood*

Mapping place? Hard not to, when even the most ordinary topographic survey sheet, created with the maximum dispassion, can turn into what Brian Harley called 'a subjective symbol of place' when scanned by a human eye (Harley, 1987). At the time he was specifically referring to 'Ordnance Survey Map, Six-inch Sheet Devonshire, CIX, SE, Newton Abbot', a map of a piece of earth he had come to know intimately. He'd lived on it for seventeen years, both his children had attended school there, and that's where he buried his wife and son. The map had 'become a graphic autobiography', it restored 'time to memory', and it recreated for his 'inner eye the fabric and seasons of a former life', all this, a 'transcription' of himself, accomplished by what he called 'a very ordinary map'.

Twenty years later Florent Chavovet, who, thanks to his wife's internship was stranded in Tokyo for six months with nothing to do, sharpened his pencils and began drawing what he saw. As he did he discovered a visual style of his own, which led to a book, *Tokyo on Foot*, organized around the neighbourhoods in which he'd worked (Chavovet, 2009). 'Hand-drawn maps that are admittedly quite personal in their details introduce the neighborhoods', he writes, introducing the book itself with a map of the city on which he's located them. The maps are personal indeed, annotated with remarks like, 'This is where I was grilled over the matter of the not stolen bike'. They also situate the illustrations that flesh out the chapters, which are thus revealed as little more than map annotations.

'Devonshire, CIX, SE, Newton Abbot' and this collection of Chavovet's maps sketch a spectrum along which the mapping of place has fallen, from the most doctrinairely standard-ized to the most idiosyncratically individualized. 'Place', after all, generally makes reference to a location, but it makes reference to it with a purpose, with a function in mind; and the word is often used as a synonym for 'apartment', for 'home', for 'neighbourhood', for 'city', though only when preceded by 'my' or 'yours' or 'our'. That is, *my* apartment is my place, *your* neigh-bourhood is your place, *our* city is our place; and as Harley makes plain, it doesn't really matter if the map of that place is as devoid of 'my'-markers as can be. It's your place in any case, and a standardized map may be as capable – maybe even more capable – of evoking it as any other.

But because the *personal* character of place – Harley's 'subjective', his 'autobiography', his 'memory' and 'inner eye' – is implicit, the *mapping of place*, as opposed to the mapping of a loca-tion which merely *happens* to be someone's place, often implies some sort of personalization of the map*making* itself, and this is what 'mapping place' is usually taken to mean: the mapping

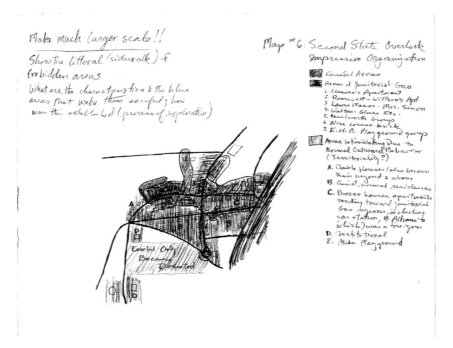

*Figure 29.1* **A map Denis Woods made in 1967 of a few blocks in Cleveland Heights, Ohio. Reproduced courtesy of Denis Wood**

of a location to reveal its significance to a *me*, to a *you*, to an *us*. Here, for example, is a map I made in 1967 of a few blocks in Cleveland Heights, Ohio (Figure 29.1). It was one of several I made as a graduate student at Clark University after having discovered J.K. Wright's thoughts about what he called geosophy, 'the study of geographical knowledge from any and all points of view' (Wright, 1947). I was eager to see what putting his ideas to work might feel like and I immediately started thinking about the geographical knowledge *I'd* evolved of Cleveland Heights since we'd moved there in 1956. I've written about this before (Wood, 2012, 2013) and published many of the maps I made at the time, but not this one, which strikes me as a particularly apt example of place mapping. It shows, in blue, the places where I felt comfortable ('Easeful Areas'), those where I felt less comfortable due to a variety of social constraints ('Areas Intimidating Due to. . .'), and those where the behaviour of janitors and custodians made me explicitly unwelcome, that is, places I'd been warned away from, places I'd been asked to leave ('Areas of Janitorial Gas'). The images flitting through my head were of sitting on those low walls that edge lawns, me and others, and being asked to leave . . . or not; of being allowed to yell and scream . . . or told to shut up; of being permitted on the premises – my apartment, those of my friends, my paper routes –or being warned off. This, of course, was no more than a sketch for a map I had intended to, to what? In any case I never did it, but this is *so* explicitly a mapping of place, of *my* place, of my place *in the world*, that it's a good place to begin.

## Feelings about places

Note that this isn't what landscape architects talk about when they talk about place: benches, gathering spots, water fountains. And it's not about character either, or about what Kevin Lynch called imageability (Lynch, 1960). It's not really *about* the world and so it can't be designed. It's about a relationship between a person or persons *and* the world. In the end it's about feelings. These don't have to be positive feelings (Kent, 2012). David Crouch and David Matless describe Conrad Atkinson's map of Cleator Moor, where he was born and raised, as 'a document of angry attachment, a lament rather than a celebration' (Crouch and Matless, 1996). Atkinson's map consists of gobs of colour daubed over an Ordnance Survey sheet of the town, with 'strontium', 'leukemia', 'ruthenium', 'invisible presence', 'cancer causal relation' and related phrases scrawled through them in a kind of graphic dirge (Figure 29.2). A coal mining town from the nineteenth century that employed generations of Atkinsons, most Cleator Moor employment these days is at the Sellafield complex, a nuclear fuel reprocessing and decommissioning site that's nine miles down the road. During the second half of the twentieth century there were twenty-one serious radiological releases from the site, and who knows how many lesser ones. Atkinson says, 'For me I would not have got to college if it hadn't been for Sellafield. My dad worked there in the 1950s and I worked there in 1957 between school and college' (McClounie, 2013). Atkinson loved Cleator Moor (though he lives in California). He has three huge sculptures in its market square (they're memorials to the town's mining past) but he's furious about the town's condition, and his fury is patent in his map.

Atkinson long taught at the University of California-Davis but it doesn't take *highly educated fury* to make a map like this; though it's true that the kids at Turners Hill Church of England Primary School exhibit something more like *resigned despair* than fury in their sweet map of Turners Hill with, if nothing else, its views over four counties. But there is *so* much else (Figure 29.3). Distressingly much of this much else are the better than 20,000 cars a day that stream through the town (only five miles from Gatwick), 52 of which run around the kids' map as a frieze, cars, vans, lorries (Figure 29.4). 'It is with some feeling that [the kids]

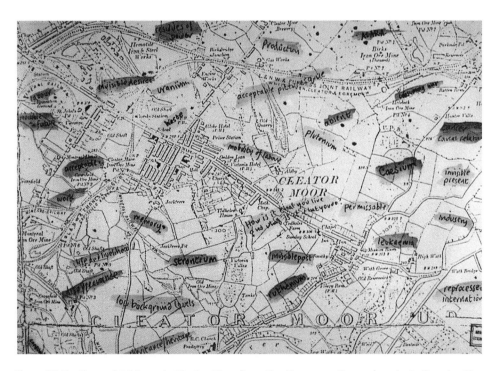

*Figure 29.2* Conrad Atkinson's *Cleator Moor* from the Common Ground project, *Knowing Your Place* (1987–1988). Reproduced courtesy of Common Ground

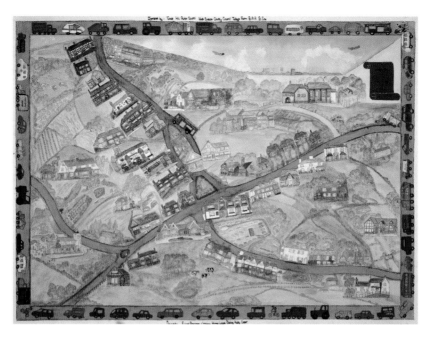

*Figure 29.3*   Map of Turner's Hill by children from Turner's Hill Church of England Primary
School (from Leslie, 2006). Reproduced with permission

*Figure 29.4*   Detail from the map of Turner's Hill (from Leslie, 2006)

show more wheels than buildings', Kim Leslie remarked in his *A Sense of Place: West Sussex Parish Maps* (Leslie, 2006), a collection of seventy-five of the parish maps he inspired for a project that was part of West Sussex's celebration of the millennium.

## Common Ground's parish maps project

Both Leslie's project, and Atkinson's map, grew out of Common Ground's parish maps project, which Common Ground had begun promoting in the mid-1980s. Sue Clifford and Angela King had created Common Ground in 1983 as a non-membership charity and lobby for what they thought about as *local distinctiveness*. Clifford has written:

> In forging the idea of *Local Distinctiveness* Common Ground has been working on libera-
> tion from preoccupation with the beautiful, the rare, the spectacular to help people explore
> what makes the commonplace particular and to build ways of demonstratively expressing
> what they value in their everyday lives. We contend this should be an inclusive process,
> encouraging local people to debate what is important to them as well as luring the experts
> to appreciate a broader view.
>
> *(England in Particular, n.d.)*

It was obvious to Clifford and King that these were things that could never be known, or even described, from the outside, and so it might be 'better to ensure that local culture has sufficient self-knowledge and self-esteem to be confident in welcoming new people and new ideas'. To this end they floated a slew of proposals and campaigns, among which was parish maps. By 'parish' they hoped merely to convey a useful sense of the local:

> [t]he smallest arena in which life is played out. The territory to which you feel loyalty,
> which has meaning to you, about which you share some knowledge, for which indig-
> nance and protectiveness is easily roused, the neighbourhood of which you have the
> measure, which in some way helps to shape you … It is in this sense of a self-defined
> small territory that Common Ground has offered the word parish, implying people and
> place together.
>
> *(Common Ground, 1991)*

Because they needed examples to show people what they were talking about, in 1986 they com-missioned eighteen artists – among them some big names (Anthony Gormley, Helen Chadwick and Conrad Atkinson) – to map places towards which they felt a particular attachment.

These maps travelled around the country in a 1987–88 show called *Knowing Your Place*, accompanied by a leaflet (Common Ground, 1991); the maps illustrated articles; and the maps appeared in Common Ground literature. A detail from David Nash's *A Personal Parish (Blaenau Ffestiniog)*, for example, decorated the cover of Common Ground's 1991 *Parish Maps* brochure; Ian Macdonald's *Echoes of Change (Cleveland)* took up most of the brochure's centrefold; and a detail from Simon Lewty's *Parish Map (Old Milverton)* concluded it. A larger detail from Lewty's map, in full colour, was wrapped around the cover of Common Ground's *From Place to PLACE: Maps and Parish Maps* (Clifford and King, 1996), where two of the artists, Lewty and Balraj Khanna, wrote about their maps. Six of the maps were turned into postcards, including Conrad Atkinson's *Cleator Moor*.

At the same time, a few parishes began making maps, and among these was one of Charlbury in Oxfordshire that Kim Leslie describes as 'a very modern and richly decorated parish map':

Steeped in detail through delicate pictures and text, it vividly brought to life this little Cotswold town and its surrounding countryside. And it wasn't made by professional map-makers, but local and very talented people who clearly had great affection for where they lived. Maps like this stir the imagination, they urge visits.

*(Leslie, 2006)*

It was only by chance that Leslie had come across a copy of this map as he was dipping into the map collection of the University of Sussex, but he was so taken with it that he made a point of visiting Charlbury and meeting its makers, who told him about Common Ground and the Parish Maps Project. Fired by the idea, Leslie proposed a Parish Maps project to West Sussex County Council after it began casting about for a way to celebrate the then forthcoming millennium. As inspired by the Charlbury map as Leslie had been, the council approved and authorized start-up money that let Leslie give talks all over the county, produce a fact sheet, organize a conference, and launch a newsletter. Elizabeth and Miles Hardy, who had led the Charlbury team, came down from Oxfordshire to share their experience, and of course Common Ground contributed.

The resulting maps were drawn, painted, stitched, embroidered, quilted, photo-collaged, even cast in bronze (Figure 29.5). Parish after parish participated: Aldwich, Apuldram, Arundel, Balcombe . . . Haywards Heath, Henfield, Highbrook, Hunston . . . Pulborough, Rogate, Selsey, Shipley . . . West Hoathly, Woolbeding and Linch, Yapton and Ford. By the time Leslie put an exhibition together in 2001, eighty-seven parishes had made maps of which the Worthing Museum was able to hang sixty-six, most of them originals. Over 2,000 volunteers had contributed to the making of the maps and, whether artists, calligraphers, gatherers of information, organizers, or fund raisers, all had given freely of their time. The money, from a variety of sources including local business sponsorships, treasure hunts, plants sales, and grants of various

*Figure 29.5*  Parishioners hold up part of the Parish Map tapestry of Thirsk, North Yorkshire, c.1989. Reproduced courtesy of Common Ground

kinds, largely went to the production of prints and postcards of the maps and the maps' professional mounting to costly conservation standards. The sale of these has raised surprisingly large sums of money for a range of parish projects. The Worthing exhibition was accompanied by a smart, full-colour catalogue that has helped to spread the word (Leslie, 2001).

The word has spread all over the place. There are way more than 2,500 maps of English parishes by now, and the idea has spread to Italy, where they're being promoted – as *mappa de comunità* – through the ecomuseum movement. Donatella Murtas, of the Instituto di Ricerche Economico Sociali del Piermonte (in Turin), who had come to see the Worthing Museum exhibition, later held exhibitions of a selection of the Sussex maps in Turin and Pietraporzio. Leslie in turn made presentations about the Sussex project in Turin, Biella, Genoa, and Argenta – Common Ground was also involved – and this has led to an expanding network of exchanges. It's a kind of marriage made in heaven because ecomuseums are explicitly about place and place identity, they're all about local participation, and they're committed to enhancing the life of their local communities (Davis, 2011). Through the rapidly expanding ecomuseum network, the Parish Maps idea is spreading around the world, with ecomuseums in Italy, France, Norway, Sweden, Portugal, Poland, Turkey, Iran, Japan, China, Vietnam, Mexico, and elsewhere. Even when they haven't embraced a parish maps practice, usually the maps the museums produce have a profound place-based character.

## Neighbourhood maps

So do most 'neighbourhood maps'. These are maps made of neighbourhoods, often by kids, often, but far from always, as part of a school curriculum. The map making promoted by Youth Voices of KCET – the US's largest independent public television station – provides a great example. Youth Voices is 'A digital literacy and civic engagement program that invites youth on an exploration of their neighborhood, where they investigate the social, cultural, and political history and take a critical look at the issues facing their community' (Youth Voices, 2016); and for the past few years maps have been an important part of this process. Typically asked to address a series of questions first ('What is a place that has deep personal meaning for or in relation to you? What is a place with great colours?'), the kids are then asked to make maps of the neighbourhood. They often follow this with a clear overlay on which they're encouraged to elaborate, and sometimes even more layers. The maps are then used in varieties of further neighbourhood-focused activities. In India a campaign called Humara Bachpan has organized 35,000 slum kids into more than 300 clubs in better than a dozen cities, all across India, for whom mapping is a central activity:

> Teams of young mappers and adult facilitators spend roughly 45 days traversing their slums. They learn the shape of their neighborhood, how streets interconnect (or don't), and the density of homes there. This information becomes the map's skeleton. Then, they fill in the specifics. They stake out what's needed through the eyes of children – where underserved public areas could become play spaces, where trash bins could be added in an area they regularly see littered with filth. Their ideal neighborhood is drawn and detailed onto the map. Then, after it's complete, leaders from the child clubs present their work to local officials.
>
> *(Sturgis, 2015)*

The thing about the maps is that beyond providing a platform for observing and thinking, they put the kids' conclusions in a form that the officials find hard to ignore (Figure 29.6). The kids' maps have actually begun to have an impact.

*Figure 29.6*　**Map of an ideal neighbourhood of Bhubaneswar, India, hand-drawn by slum children as part of child-led planning. Reproduced courtesy of Humara Bachpan**

Not all neighbourhood mapping is done by kids. *Tracing the Portola: A San Francisco Neighborhood Atlas* was put together by Kate Connell and Oscar Melara as an interpretive 'guide to finding the way through the Portola's present and its history, a way to understanding the natural and historical forces that have shaped our neighborhood' (Connell and Melara, 2010). They collected their neighbours' stories to develop their history of Portola as well as the neighbourhood's dynamics, and talked to others; and they've published their maps in poster form as well (Connell and Melara, 2011). One of their sources of inspiration was an atlas I'd made of Boylan Heights, the neighbourhood I lived in in Raleigh, North Carolina. Beginning in 1975 I used neighbourhood mapping as an exercise for the students I taught in landscape architecture studios, and the neighbourhood we worked in most frequently was mine. In the early 1980s we decided to make an atlas of the neighbourhood, for the neighbours, and eventually this was published, though only in 2010, as *Everything Sings: Maps for a Narrative Atlas*, 'narrative' because the intention was to tell the neighbourhood's story through maps:

> One way of thinking about Boylan Heights is as a place in Raleigh, North Carolina, bounded by a prison and an insane asylum and some railroad tracks and a little creek. But there are other ways of thinking about it too. You could think about it as a neighborhood; that is, as some sort of community, or as a marriage of community and place, or as those people in that place, their relationships, and their ways in the world; and thus, less a place than a process, a life process, a metabolic one. That would take an atlas to unravel: what a neighborhood is, what a neighborhood does, how a neighborhood works.
>
> *(Wood, 2010)*

The second edition of the atlas, which came out in 2013, contained sixty maps, maps of the streets, of course, and the sewer lines, but also maps of the colour of the leaves in the fall, of

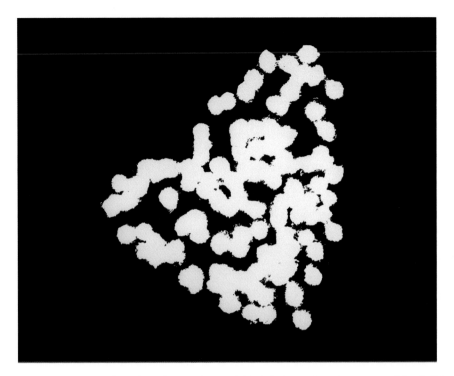

*Figure 29.7* From *Everything Sings* by Denis Wood (2010). Reproduced courtesy of Denis Wood

the lighted pumpkins the neighbours put on their porches on Halloween, of the wind chimes that sounded, of the ballet danced by buses that passed through the neighbourhood, of the pools of light cast by the street lights on the streets at night (Figure 29.7). If the map I made in 1967 of a few blocks in Cleveland Heights was a mapping of place, this atlas is a mapping of place times sixty, a straightforward transformation of feelings into maps.

## Indigenous place mapping

Though the number of place atlases has been growing, their development has not always been so straightforward. Most indigenous peoples – maybe all of them – have strong feelings about where they live, about their place in the world, about *their place*. But it's one thing to have a place, to fully inhabit it – and for generations upon generations – and another to keep it against the incursions of outsiders, another even to make a map of it capable of standing in court against the claims of those who may have no more interest in the place than strip mining it for coal or hydraulic mining it for gold, no feelings at all for the *place*. This was the situation of many indigenous peoples – maybe most of them – in the opening of the twentieth century, pushed out of their places or wholly deprived of any claims to them, allowed to live in them, if at all, only by sufferance.

Indigenous mapping changed this, appropriately enough, since it had been 'official' mapping – that is, government mapping, outsider mapping – that had created the situation in the first place. Essentially the mapping was about securing title, or some sort of security, over territory in which to live; but to the extent that such territory was in any sense traditional, it was necessarily about place as well.

409

This had been one of the problems: indigenous maps tended to be about place, about things that really mattered, while the courts could only understand . . . *maps of property*. Indigenous peoples had demanded their places from the instant they lost them – and they'd often mapped them as well – but it wasn't until legal frameworks shifted that non-indigenous authorities were able to see and hear them.

This was precisely the case with the mapping that led to the Canadian territory of Nunavut, home of most of the Canadian Inuit. This not only produced the three-volume *Inuit Land Use and Occupancy Project* (Freeman, 1976), with its *Land Use Atlas*, but a few years later the *Nunavut Atlas* (Riewe, 1992) which constituted the foundation for the territorial negotiations. These maps were all marked by novel mapping techniques as recalled by mapper, Peter Usher:

> We were no longer mapping the 'territories' of Aboriginal people based on the cumulative observations of others of where they were (as one would for mapping the ranges of wildlife species), but instead, mapping the Aboriginal peoples' own recollections of their own activities. The second innovation was to record peoples' own perceptions of the history and significance of their traditional lands. This was done through mapping geographical knowledge and oral history as exemplified by place names and ecological knowledge, all of which were used as supplementary indicators of use and occupancy.
>
> *(Usher, 2003)*

That is, though the goal was to be able to draw a line around Inuit territory capable of being understood by the courts, the only way to draw this line was . . . through place. Other Canadian Inuit live in somewhat similarly created spaces in Quebec and Labrador – Nunavik and Nunatsiavut – and still other areas are under negotiation for settlement.

Other Canadian indigenous groups had moved in related directions, some even earlier – it was the Nisga'a who had brought about the change in legal understanding that led to the Inuit efforts, and parallel efforts had been advanced by the Cree, the Dene, and others – but the publication of the *Inuit Land Use and Occupancy Project* in 1976, and the publication in 1981 of *Maps and Dreams* (Brody, 1981) by Hugh Brody, who'd worked on the project, gave indigenous mapping the exemplars and publicity it needed to sweep the world, and similar projects began to be common in Asia, Africa, Latin America, and Australia.

In fact it's Australia that makes it clear how far some courts have come in recognizing not only indigenous peoples, but the way they think about where they live. The aboriginal people of Fitzroy Crossing won their right to appear in court after presenting Australia's National Native Title Tribunal with a painting known as *Ngurrara II*:

> Frustrated by their inability to articulate their arguments in courtroom English, the people of Fitzroy Crossing decided to paint their 'evidence'. They would set down, on canvas, a document that would show how each person related to a particular area of the Great Sandy Desert – and to the long stories that had been passed down for generations.
>
> *(Brooks, 2003)*

The tribunal accepted the painting, one member commenting that the painting was 'the most eloquent and overwhelming evidence that had ever been presented' to them. In the end, of course, maps *were* made, though the court came close to expressing regret about the necessity:

Although the Court has to set boundaries in order to define the area of a native title determination, it is a fact that in the extremely arid region of the Western Desert boundaries between Aboriginal groups are rarely clear cut. They are very open to human movement across them. Desert people define their connection to the land much more in terms of groups of sites, thinking of them as points in space not as areas with borders.

Notwithstanding this concession, long lists of coordinates setting the boundaries concluded the decision.

## In the end

A painting such as *Ngurrara II* strikes some people as very much a map, others less so. So what? It's an image of the land imbued with all the stories, all the *life* these inhabitants had given it for generations; that is, it's an image of place and if their graphic conventions call for a painting instead of a map, what's the difference? Well, it's a big difference, one that underscores the importance of place *mapping*. Place mapping is about corroding or confounding or amplifying, in any event about *expanding* the reach of the map beyond its 'legal' role. As developed over the past 500 years by large, powerful, complicated societies, map-making has always been about propping up property and the nation-states that support, that guarantee it, that gain substance from its mapping. In the *News and Observer* this morning – my local newspaper – there were two articles about maps: one dealt with court-mandated revisions to voting maps that the court felt had been racially gerrymandered by the North Carolina legislature; the other with the state's imposition on property rights growing out of its 1987 Map Act. Those are the essential concerns of *maps*: state control (of voting rights in this case) and property. Froufrou about *feelings*, that should be left to poetry.

Place mapping challenges this whole perspective. Although, as Harley pointed out, the state's maps, when scanned by a human eye, are fully capable of transcribing anyone's relationship to the land. Maps created with this intention, that take it as their raison d'être, are enormously more capable, and so immensely richer in their ability to do so. The examples given here barely scratch the surface of a rapidly expanding world of mapping.

## References

Brody, H. (1981) *Maps and Dreams* Vancouver, BC: Douglas and McIntyre.

Brooks, G. (2003) "The Painted Desert: How Aborigines Turned Ancient Rituals into Chic Contemporary Art" *New Yorker* 28 July pp.63–67.

Chavovet, F. (2009) *Tokyo on Foot: Travels in the City's Most Colorful Neighborhoods* Rutland, VT: Tuttle.

Clifford, S. and King. A. (Eds) (1996) *from Place to PLACE: Maps and Parish Maps* London: Common Ground.

Common Ground (1991) *The Parish Maps Project* London: Common Ground.

Connell, K. and Melara, O. (2010) *Tracing the Portola: A San Francisco Neighborhood Atlas* San Francisco, CA: Book and Wheel Works.

Connell, K. and Melara, O. (2011) *Crossing the Street: Mapping Hidden Stories* San Francisco, CA: Book and Wheel Works.

Crouch, D. and Matless, D. (1996) "Refiguring Geography: Parish Maps of Common Ground" *Transactions of the Institute of British Geographers* 21 (1) pp. 236–255.

Davis, P. (2011) *Ecomuseums 2nd Edition: A Sense of Place* London: Continuum.

*England in Particular* (n.d.) Available at: *www.englandinparticular.info* (Accessed: 19 December 2009).

Freeman, M. (Ed.) (1976) *Inuit Land Use and Occupancy Project Report* Ottawa, QC: Supply and Services Canada.

Harley, B. (1987) "The Map as Biography: Thoughts on Ordnance Survey Map, CIX, SE, Newton Abbot" *The Map Collector* 41 pp.18–20.

Kent, A.J. (2012) "From a Dry Statement of Facts to a Thing of Beauty: Understanding Aesthetics in the Mapping and Counter-Mapping of Place" *Cartographic Perspectives* 73 pp.39–60.

Leslie, K. (Ed.) (2001) *Mapping the Millennium: The West Sussex Millennium Parish Maps Project* Chichester, UK: West Sussex County Council.

Leslie, K. (2006) *A Sense of Place: West Sussex Parish Maps* Chichester, UK: West Sussex County Council/Selsey Press.

Lynch, K. (1960) *Image of the City* Cambridge, MA: MIT Press.

McClounie, P. (2013) "Cumbrian Artist Conrad Atkinson's Nuclear Dump Vision" Available at: *www.news andstar.co.uk/news/cumbrian-artist-conrad-atkinson-s-nuclear-dump-vision-1.1030618* (Accessed: 21 February 2016).

Riewe, R. (Ed.) (1992) *Nunavut Atlas* Edmonton, AB: Canadian Circumpolar Institute and the Tungavik Federation of Nunavut.

Sturgis, S. (n.d.) "Kids in India Are Sparking Urban Planning Changes by Mapping Slums" Available at: *www.citylab.com/tech/2015/02/kids-are-sparking-urban-planning-changes-by-mapping-their-slums/385636/* (Accessed: 19 February 2015).

Usher, P. (2003) "Environment, Race, and Nation Reconsidered: Reflections on Aboriginal Land Claims in Canada" *Canadian Geographer* 47 (4) pp.365–382.

Wood, D. (2010) *Everything Sings: Maps for a Narrative Atlas* Los Angeles, CA: Siglio Press.

Wood, D. (2012) "Thinking about My Paper Routes" in Hall, B. and Bertron, C. (Eds) *The Known World, Pocket Guide* unpaginated.

Wood, D. (2013) *Everything Sings: Maps for a Narrative Atlas* (2nd ed.) Los Angeles, CA: Siglio Press.

Wood, D. with Bellamy, J. and Salling, M. (2013) "The Paper Route Empire" in *Where You Are: A Book of Maps That Will Leave You Completely Lost* London: Visual Editions, one of 16 booklets, boxed (22 pages with foldout map).

Wright, J.K. (1947) "Terrae Incognitae: The Place of Imagination in Geography" *Annals of the Association of American Geographers* 37 pp.1–15.

Youth Voices KCET TV (2016) Available at: *www.kcet.org/socal/departures/youthvoices/* (Accessed: 23 January 2016).

# 30

# Maps and identity

*Alexander J. Kent and Peter Vujakovic*

## Introduction: nested identities

Maps come in a range of scales, from small-scale world political maps that often use a patchwork of vibrant colours to clearly distinguish nations one from the other, to large-scale tourist maps that declare a distinctive sense of place in an attempt to attract visitors. Similarly, most people have multiple, layered or nested identities based on their home and their family, their neighbourhood, region, and nationality, as well as other affiliations and attachments which are not necessarily geographically determined. Germans, Austrians, and the Swiss, for example, have a sense of *Heimat*, a concept with no direct English equivalent that represents a shared space, generally a rural regional identity, associated with birth-place, family, and a community. The term, as a positive connection to place, can be traced back to the fifteenth century, although it now has some negative connotations with Germany's Nazi past (Blickle, 2002). In the UK in 1987, the environment and arts group Common Ground launched its first major public initiative – the Parish Maps Project (Crouch and Matless, 1996) – which used the term 'parish' to generate a similar sense of identity through discovering the 'local distinctiveness' of a place and to encourage local community groups to express this by producing maps of their 'parish', whether rural or urban (see Chapter 28).

Maps can play an obvious role in reinforcing or even constructing various forms of spatial identity, the most obvious being a sense of nation, but they also have a role in other forms of identity, from personal issues such as sexual orientation to 'virtual identities' as part of Internet communities. Yet a sense of national identity is the form most often associated with maps, and cartography certainly plays a part in the construction of this level of identification. This chapter examines and illustrates how maps play a role in forming or reaffirming identities and explains how this is achieved, from national atlases to parish maps.

## Mapping national identities

Most people are able to recognize their country's outline. The strength of the map as an 'icon' is its ability to work across both literate and semi-literate populations. Those with high 'legibility', i.e. those with distinctive outlines, work best; Cyprus is an obvious example, with the map being incorporated into the national flag.

The association of people with a specific 'national' space also lends itself to an elaboration of the basic map form, such as artistic embellishment in which the mapped territory is associated with an affirmative depiction. Classic examples are Nicolaes Visscher's (1650) *Leo Belgicus*, a map of United Provinces of the Netherlands drawn in the form of a courageous lion (Figure 30.1), or Aleph's (1868) descriptions of England as Britannia: 'Beautiful England, – on her island throne, – Grandly she rules, – with half the world her own'; and Italy as Garibaldi: 'Thou model chieftain'. Aleph's use of Britannia as representative of both England and Empire – but not of the whole of Britain – is interesting in its implication for national identity within Britain as a whole. The image is one of twelve 'humorous outlines' or maps of European nations by artist Lilian Lancaster and published by Hodder and Stoughton in 1868, with text by Aleph, the pseudonym of the well-known London medic and author William Harvey. The image portrayed of other British and European nations is much less flattering (see later discussion on satirical maps). Other positive associations generally involve mingling other icons of nationalism with the outline of the national space, for example, the state flag or currency (e.g. on banknotes, such as those of the US).

*Figure 30.1* 'England' by Lilian Lancaster, as appearing in *Geographical fun: being humorous outlines of various countries, with an introduction and descriptive lines* (Aleph, 1868). Reproduced courtesy of Library of Congress, Geography and Map Division

Maps are symbols, and, as such, they offer selective representations that preserve and promote some features while suppressing or obliterating others. There are few examples of cartography that can demonstrate effectively how maps build a narrative to construct a sense of national identity as the national atlas. A critical interpretation of national atlases (e.g. by Kent, 1986; Monmonier, 1994; Vujakovic, 1995; Jordan, 2004) suggests how nationality is communicated and expressed through maps in ways that are not necessarily subtle, by commenting on the cartographic treatment of certain entities and the range of themes that are included. As Monmonier (1994: 1) suggests, they 'may be viewed as the inevitable systematic publication of institutionally collected geographic knowledge', a position upheld in early surveys such as that by Yonge (1957). But they are also demonstrations of a nation's character and pride and portray a nation to its own people and the world (Kent, 1986: 122) by emphasizing symbols of national unity, scientific achievement, and political independence (Monmonier, ibid.). Moreover, according to Vujakovic (1995: 129), national atlases 'can be regarded as complex narrative structures, intricately weaving a story of national identity through a combination of words and images'.

In creating a visual representation of a historical-national space (Vujakovic, 1995: 131) and seeking to legitimize their nationhood, some atlases include a selection of reproductions of historical maps, which usually precede a main section devoted to a series of thematic maps covering a diverse range of themes, such as climate, vegetation, population density, and distribution of ethnic minorities. As national atlases are intimately associated with sovereign governments (Kent, 1986: 122), the expression of societal values in the range of topics included and the treatment of data (i.e. the relationship of the 'theme' with other data on maps) might be more pronounced here than in the design of other state mapping products, such as topographic maps, and therefore easier to identify. However, while they may provide the critical reader with more straightforward clues about the character of a society, national atlases lack the ubiquitous yet subtle 'objective' and 'natural' qualities of topographic maps which Wood (1992) unmasks.

If mapping is a cultural, political, and epistemological activity that is deeply imbricated in nations' narratives of their own formation (Rogoff, 2000: 74), political independence presents a significant opportunity for national self-determination and self-expression, and cartography can support the nation-building canon, particularly as a means of (re)classifying and (re)defining the landscape in the (re)construction of national identity. In his study of the maps of European countries provided by their respective embassies in the United States, Zeigler (2002) explicitly sought to detect a 'cartography of independence'. He concluded that post-communist countries seized the opportunity to formulate their own conceptions of a new European order in which they would enjoy greater centrality and higher status, being designed to inspire their own populations and to serve as iconographic nation-building tools. A comparative examination of maps produced in Latvia and Slovenia will serve to illustrate how maps construct a sense of national identity.

With populations of around two million each, Latvia and Slovenia both achieved political independence in 1991 from communist regimes where the ruling ethnic group was foreign, and both countries joined the European Union and NATO in 2004. Apart from these similarities, however, they exhibit some fundamental differences. They were at opposite ends of the political spectrum within communism: one was completely absorbed by the Soviet empire and the other was a republic within Yugoslavia; they have contrasting terrain and climate (Latvia has littoral plains and a Baltic, maritime climate, while Slovenia is mountainous and diverse with Alpine and Mediterranean climates); and Latvia has a large Russian minority whereas Slovenia is ethnically much more homogenous.

With the achievement of independence, Latvians gained unprecedented access to detailed topographic information. The impetus for its provision lies in the redistribution of power relations between citizen and state:

> While many atlases continued to show the Baltic States with other Soviet successor states (including Russia), their own maps showed them to be part of both the Baltic region and an undivided Europe. Their maps marginalized Russia and avoided the Russian language … The message of the new map is clear: Latvians want to resurrect their Baltic history and to change people's minds about where they are located in 'the new European order'.
>
> *(Zeigler, 2002: 676, 677)*

Zeigler's analysis of state-approved country maps, from which the above observation was made, found that Latvia's 'return to Europe' and its programme of re-orientation away from Russia and its Soviet past was manifested cartographically by placing an emphasis on the central location of the country within a wider European context. The latter is also clear in Figure 30.2 below, which is taken from a promotional presentation about the country produced by the state-funded Latvian Institute and aimed at an international audience via the organization's website. Produced since Zeigler's analysis was published, this map sustains his earlier observation that the country seeks to reposition itself within Europe. However, it could be argued that while centrality on a map suggests significance and power, it is unlikely that the principal subject of any small-scale thematic map such as this would be designed to occupy a peripheral space on the map surface.

National atlases are useful texts for analysing wider narratives of the state because they express the pride and independence of a country (Thrower, 1999: 192) through the selection and

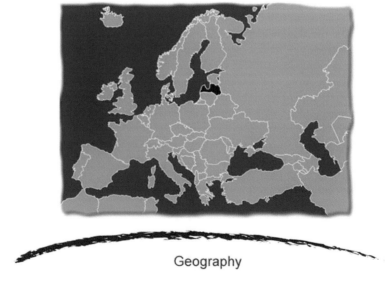

*Figure 30.2* Slide from the web-based promotional presentation 'Latvia' by The Latvian Institute (2003). Available at: *www.li.lv/old/prezent_en.htm* (Accessed: 20 November 2006)

presentation of thematic material. As part of the USSR, the Latvian Socialist Republic was portrayed in a series of atlases that included the whole of the Soviet Union, usually produced by GUGK (Main Department for Geodesy and Cartography) in Moscow. While these atlases embody a typical national atlas canon, for example through the inclusion of thematic material dealing with industry and productivity, they nevertheless aimed to direct the user's feelings of pride towards the Soviet Union. In 1988, a 'national atlas' of the Latvian Socialist Republic – a small paperback publication in landscape format called *Latvijas PSR Atlants* – was published in Latvian (the year in which Latvian also became the official state language) by GUGK and shortly after independence, a new version – *Latvijas Atlants* – was published by Latvijas Karte in Rīga in 1992. Given the small period between publication dates, there is very little difference in the content between versions and the design of the maps remains virtually consistent. However, in addition to the title of the atlas, there are some differences that would appear to indicate a move away from Soviet rule. First, an area to the east of Latvia, prominently shaded in yellow and white diagonal stripes, is shown on each map, which in the legend page provides the explanation (in Latvian): 'Territory which until 17th June 1940 belonged to the Latvian Republic and which the Latvian Republic Supreme Soviet, in accordance with the resolution on the 22nd January 1992, declared an illegal annex' (Klavins, 1992: 5). This refers to the Soviet invasion of Latvia during the Second World War and the subsequent retention of part of this territory by the Soviet Union after Latvia's independence in 1991. The boundaries between Latvia and Russia were not settled by treaty until October 1997 (Gladman, 2005: 364). Another prominent deviation from the 1988 atlas is the replacement of a map on p.35 of the atlas showing the distribution of organizational centres associated with the Soviet youth Pioneer movement in Latvia (Figure 30.3) with a map showing tourist destinations within the country (Figure 30.4).

In his survey of country maps, Zeigler (2002: 683) makes the following observation: 'Just as Poland, the Czech Republic, and other former satellite states have used maps to propel themselves out of a Moscow orbit, Croatia (and Slovenia) has applied similar techniques to escape from the gravitational pull of Belgrade'. Zeigler (2002: 682–683) also found that the Slovenian maps involved in his study affirmed the country's place in the middle of Europe and used pictograms to symbolize the folk cultural elements of the country's landscape, minus any symbols of the communist past. Indeed, if, as Gams (1991: 337) asserts, 'Slovenians have always looked upon themselves as Central Europeans, regardless of several regional geographies of the world and Europe', then such cartographic centralization would appear to reflect the campaign of re-orientation. Moreover, it is possible to find examples of this technique in state-approved maps published since Ziegler's analysis (see Figure 30.5). However, as previously argued, if the cartographer's interest of achieving 'balance' is as important as Chapter 22 of this Handbook suggests, this will oppose a marginalization of the main theme of a map, whether it is politically motivated or otherwise.

Slovenia's ethnic homogeneity and level of industrialization can be seen in the last 'national' atlas of Yugoslavia to be produced, the *Veliki Geografski Atlas Jugoslavije*, which used Serbo-Croat throughout and was published in Zagreb in 1987. According to Fridl (2004: 167), plans to publish a national atlas of Slovenia first arose in 1969, when Slovenia was still one of the republics of the former Yugoslavia, although the project was not realized until 1991. Following independence, some significant publications that represented the new country began to appear, such as a new edition of the *Krajevni Leksikon Slovenije* (Lexicon of Slovenian Places) in 1995, published in Ljubljana by Državna Založba Slovenije, the state publishing house. Written in Slovene and including photographs that help to describe the various features, it resembles an illustrated gazetteer whose primary aim is to encourage internal tourism, advance knowledge of the new republic and its cultural assets, and foster a sense of patriotism.

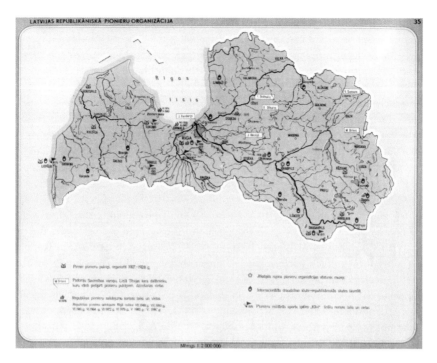

*Figure 30.3* Map of Pioneer organizations from *Latvijas PSR Atlants* (Pūriņš, 1988)

*Figure 30.4* Tourism map from *Latvijas Atlants* (Klavins, 1992)

*Figure 30.5*   Slovenia in Europe, published by the Slovenian Tourist Board (STO) in 2009. Map produced by the Geodetic Institute of Slovenia (reproduced courtesy of the Geodetic Institute of Slovenia)

Seven years after the project was initiated, the first national atlas of Slovenia was published in 1998, having been financed by the state publishing house and aimed chiefly at the domestic market. The *Geografski Atlas Slovenije* was written purely in Slovene and included 190 thematic maps. According to editor Jerneja Fridl (2004: 167), 'The *Atlas* is designed to present natural science, technical, social, economic, and humanistic-scientific knowledge about the natural, social, economic, spatial and environmental characteristics of Slovenia as comprehensively as possible'. Its initial print run of 35,000 copies rapidly sold out (ibid.: 173). Three years later, an English version of Slovenia's national atlas was produced, this time to commemorate the tenth anniversary of Slovenia's independence, and, according to Fridl (ibid.), 'to present our small but diverse and interesting country to the world'. Essentially a modification of the 1998 national atlas, it was somewhat condensed into 105 main thematic maps and a version in Slovene was also published. From the outset, the atlas openly presented a statement of national identity and Slovenia's European aspirations, as summarized in the introductory contribution by Borut Pahor, Speaker of the Slovenian Parliament:

> The National Atlas of Slovenia in many ways tells the story about renewal, about self-confidence, about a new behaviour. It tells of yesterday, of today, of places, and of people, who in their own way helped prepare Slovenia for its part in a shared European future.
>
> *(Pahor, 2001: 6)*

These aspirations are also expressed cartographically, particularly the desire to establish Slovenia as a natural accession state to the EU. For example, instead of a simple location map in which Slovenia might adopt a prominent, central position, there is a map of the EU in which members and associated members are contrasted rather sharply against the category of 'other' (Figure 30.6). The role of the location map has therefore been used as an opportunity for promoting Slovenia's natural progression towards EU membership. Unlike the map presented in Figure 30.5 above, it presents Slovenia not as a new, albeit isolated, entity with which to become acquainted, but as belonging to a wider community of European states. As Fridl (2004: 173) points out, the purpose of the atlas 'is to increase the recognition of the natural, cultural, and spiritual identity of Slovenia, especially now that Slovenia has joined the European Union'.

In addition to the usual maps dealing with population density, dialects (which, incidentally, extends beyond the northern border with Austria into Kärnten), and transportation, there are also maps that deal with rather distinctive topics, such as Alpine pastures, handicrafts, and linden (lime) tree blossoming. Like Triglav, the triple-peaked mountain that appears on the national coat-of-arms, the lime tree is a national symbol, appearing on the front cover of the *Geografski Atlas Slovenije*. According to Fridl (op. cit.: 167), 'Linden trees have always

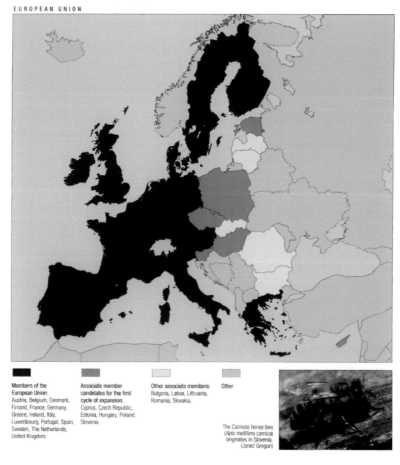

*Figure 30.6* Map of the European Union serving as a location map in the English version of the *National Atlas of Slovenia* (Fridl, 2001)

stood in the centre of Slovene villages or beside the roads leading to the wider world, and village elders assembled beneath them to make important decisions'. It would therefore appear that the choice of maps is also intended to suggest the uniqueness of Slovenia, to emphasize to the outside world what makes the country distinctive. Following his analysis of the content of the *National Atlas of Slovenia*, Jordan (2004: 157) suggests that maps concerned with the Alpine landscape stress the close emotional ties of Slovenes to their Alpine environment. Furthermore, Jordan (ibid.) concludes that the atlas conveys an image of a country which is composed of the following 'ingredients':

- It is an equal member of the international community.
- It is the heartland of a nation that extends beyond the current state borders.
- It is an old cultural region, which was also industrialized early.
- Although a small country, it is rich in natural and cultural landscapes.
- It has emotional ties with mountains and the Alps.

With independence, Slovenians took the opportunity to assert their identity cartographically and this process continues. In 2005, the Anton Melik Geographical Institute published reprints of the *Atlant* (the first world atlas written in Slovene, published between 1869–1877), which, according to Urbanc *et al.* (2006: 254, 266), helped to erase the boundaries between the Slovene community and those of the wider world and highlighted Slovenians as a national community well along the road to shaping its own identity.

Although the examples given above are less pronounced than the cartographic expressions of statehood originating from other post-communist countries and their associated struggle for independence, e.g. the National Atlas of Croatia (Vujakovic, 1995), they nevertheless display the urge for self-determination and distancing from their past identities. Moreover, the speed at which such expressions occurred is significant and indicates an awareness of the importance of maps as valid and persuasive texts in the nation-building process. This may be particularly important when national borders are more fluid. Many territorial boundaries are regarded as stable and unlikely to change radically, but some regions of the world suffer from greater levels of uncertainty. Batuman (2010), for example, examines both official and popular maps of the Turkish national space from the Cold War to the present.

## Mocking identities?

Maps within satirical graphics have a well-established history, with classic examples produced from eighteenth- and nineteenth-century caricaturists, epitomized by the work of James Gilray. However, it is with the advent of wide distribution print media, and more recently, electronic media, that the satirical graphics have become almost ubiquitous. Those containing maps have proliferated as audiences have become more geographically and geopolitically 'literate'. While occasionally used to celebrate national identity, the inclusion of maps is more usually used to parody it. Maps in satirical graphics are often used to subvert ideological positions rather than support them, and to denigrate individuals or groups rather than venerate. Various authors have addressed the significance of maps in political graphics (see, for example, Vujakovic, 1990, 2015; Holmes, 1991; Newman, 1991; Pickles, 1992; Herb, 1997; Jacobs, 2009).

Vujakovic (1990) developed a typology of map use in visual tropes within political cartoons, with a specific focus on geopolitical issues. He described several ways in which maps are used, most of which rely on national identification at some level. First, where the map represents the territorial arena in which some action is taking place. This scene-setting device has a classic example in cartoons which show a national territory being swept clean of some (supposedly) negative influence or group. Numerous examples exist, for instance a German cartoon from the 1930s showing a Nazi broom cleansing the German national space of Jews, communists, and intellectuals. This trope is invariably portrayed as a positive image, with the diminutive figures that are being swept away being represented as vermin-like. One of the earliest uses involves the map of Europe being swept clean of a diminutive Napoleon and army by two giant allied soldiers (Philippe, 1980). It is only with the recent inversion of the concept of 'ethnic cleansing' that this graphic device has itself been overturned; for example, a cartoon of former Yugoslav president Slobodan Milosevič sweeping men, women, and children away in order to create exclusively Serb areas (appearing in *Die Welt* in 1997). Interestingly, in this last example, the 'map' is implied rather than overt; perhaps the artist (Klaus Böhle) did not wish to dignify the act with a legitimate territorial authority.

A second use concerns the map as subject to some action, for instance, a map cut or ripped to represent a real, threatened, or desired territorial division. A classic example of the nation being ripped apart is the 1864 Currier and Ives lithograph of Abraham Lincoln and Jefferson Davis tearing apart a map of the United States (Philippe, 1980). A cartoon representing a more surgical dismemberment of national space showed the British Prime Minister Lloyd George, scissors in hand, implementing the 1920 Government of Ireland Act (Vujakovic, 1990). The Act was intended to establish 'home rule' for two subdivisions of Ireland; 'Northern Ireland' or Ulster, formed of the six north-eastern counties, while the remaining, and larger part would form 'Southern Ireland'. The territory as embodiment of the living nation makes the act of 'cutting' or 'amputation' an immensely powerful visual metaphor. A large number of maps in cartoons show the dismemberment of European states from the First World War to the collapse of the Soviet Union and Yugoslavia in the late twentieth century. Territorial fragmentation is illustrated in various ways (see Vujakovic, 1992); two archetypal examples from the UK and Germany picture Soviet leaders attempting to halt the disintegration of the Soviet Union; first Gorbachev trying to stick a torn map back together with sticky-tape (*The Economist*, 14 March 1991), followed by Yeltsin trying to stop a jigsaw style map of the region disintegrating (*Handlesblatt*, 30–31 August 1991).

Anthropomorphic and zoomorphic maps are a third category. This type of map has its roots in more positive images (as discussed earlier), for example, Visscher's *Leo Belgicus*, but the tendency has been for specific national types to be ridiculed or portrayed as a threat, e.g. the Russian bogeyman or ravenous bear. Maps with each state portrayed as an animal or a person became a popular format in which to satirize the political situation in Europe in the late nineteenth and early twentieth centuries. The *Serio-comic War Map* (for 1877 and 1900) produced by F.W. Rose are classics of this type (see Chapter 31), although similar images were produced in France and Germany. This trope is still frequently used today to ascribe negative characteristics to a nation which appears to threaten geopolitical stability (e.g. the bear or octopus to represent a threatening or acquisitive state continues to be used for Putin's Russia (Vujakovic, 2014)).

This device is also used to rework classic geopolitical motifs; for example, a cartoon satirizing German reunification by Peter Brookes (cover of *The Spectator*, 24 February 1990) which represented West Germany as the head of Chancellor Kohl literally 'consuming' East Germany. The image echoes theories of the 'state as an organism' that dominated geopolitical thinking in the early twentieth century and influenced German concepts of *Lebensraum*, and similar concerns prompted a drawing of a re-unified Germany as a 'vicious muzzled dog' (Holmes, 1991: 177).

Finally, the map as 'national icon' is often manipulated for satirical effect. An extremely clever series of images created by artist David Gentleman for his 1987 polemic book, *A Special Relationship*, uses a combination of flags, maps, and other icons to unpack the complex political relationship between the US and UK during the 1980s. Examples include, a spoof 22-cent postage stamp, which shows a map of the UK made up of the silhouettes of dozens of US aircraft, to commemorate the US bombing of Libya from UK airbases, and a picture of the US 'stars and stripes', of which one star is replaced by a minuscule map of the UK – the 51st state! The French cartoonist, Batellier, adopted a similar approach, with a cartoon based on the 'stars and stripes' (the blue field shaped like a map of mainland US); representatives of various developing nations are shown rolling up the red stripes (carpets!) as a sign of their disillusion with US global hegemony (Regan *et al.*, 1988).

## Mapping local and regional identities

Maps have always served as one form of local and regional identification. These are usually produced by institutions rather than local people. Parker (2004) has, however, identified a growing practice of community mapping, but notes that empirical research on this subject is limited, with most studies focused on 'counter-maps' (see Chapter 6) and indigenous maps, leaving locally produced maps unexplored. Parker examines the 2002 Portland Greenmap project (US) which aimed to increase community awareness of and connection to its urban ecology and social resources through volunteer-led mapping (other Greenmaps had been driven by existing organizations), with the deliberate intention of creating a 'grass-roots' community and sense of identity. Her study revealed a process that can yield positive outcomes, but can be 'fraught with tension and marked by unacknowledged privilege' (2004: 482), and her general conclusion is that the challenge facing community map-makers is the need to be truly inclusive, transparent, and empowering.

Parker's reservations can also be observed in maps produced through Common Ground's 'Parish Maps' project in the UK from the late 1980s onwards (see Chapter 29). The project sought to re-engage people with their locality or 'parish'. The organizers had gained experience in local protest groups and were convinced of the value of maps as part of the process of placing value on local environments. Parish mapping 'is presented as a process of self-alerting, putting people on their toes against unwanted change and producing an active sense of community' (Crouch and Matless, 1996: 236). Many communities engaged with the project and maps were produced in a variety of forms; sometimes the work of a single local artists or a group effort, sometimes in a traditional format or as textiles and ceramic maps (often displayed in village halls).

The desire to present an aesthetically conditioned view – with its inherent process of selectivity – is an intrinsic element of the 'authentic expression' of topographic cartography (Kent, 2012) and the extent to which the parish maps are alternative and challenging readings of the landscape can be disputed. Most of the maps were of rural sites and many fall into caricatured representations of the 'rural idyll'. While many people would wish to defend this aspect of many settlements, it provides a skewed view of the localities involved, celebrating the quaint and the historic. An interesting anecdote to this is to use Google Maps, another map form which has potential for community based identity-politics, and visit the site of such maps. The village of Marldon in Devon produced a map to celebrate the millennium and the map is typical of those that focus on the picturesque rural aspects of the locality (Figure 30.7). Only two modern buildings feature, and they are key community hubs (the primary school and the village hall), otherwise all are old buildings. Google Street View provides a very different image of the settlement, with a significant number of modern buildings surrounding the village core. The only form of transport shown on the map is horse-drawn, which belies the lines of cars parked along the village streets. Animal talismans hover around the map border and populate the map itself.

*Figure 30.7* Parish Map of Marldon, Devon (2000). Reproduced courtesy of Marldon Parish Map Group and Marldon Local History Group

Parish Maps such as these tended to bolster rather than contest the flight to the country of well-off commuters or second-home owners; none of the tensions that are likely to exist in many of these localities are expressed, rather a timeless sense of harmony. This contrasts with Denis Wood's Boylan Heights mapping project (Wood, 2010) in which his students mapped a range of topics related to the Bolan Heights neighbourhood in Raleigh, North Carolina. Their mapping exposed 'class' differences within the community; for example, the people who get mentioned in the local newsletter tend to live in the 'big houses' on the hill and are from the same households that carve and set out a pumpkin 'jack-o-lantern' at Halloween.

Other approaches have included local authorities or organizations developing maps to create a sense of community. In Edmonton (Canada), for instance, a 'walking map' as been developed as a valuable tool for not only encouraging physical activity but also motivating individuals to explore their communities and visit local community destinations (Nykiforuk *et al.*, 2012). More generally, tourist and other promotional maps also have an impact on local and regional identities. Attempts to rebrand a place, for example, sometimes draw on cartography as a means with which to establish, visualize, and communicate particular characteristics (Vujakovic *et al.*, 2014).

Another way in which maps can provide a positive role in identity formation is by bringing together scattered individuals or communities who might otherwise feel isolated. Google Maps, for example, has been used by the *Cirque de Soleil* to show where 'social circus' groups are located globally. The circus community claims that 'teaching children circus skills helps them to trust other people, improves their "physical literacy", and raises their self-esteem and sense of identity' (Pickles, 2015); being part of this helps children from disadvantaged areas and even war-torn regions like Afghanistan feel part of a caring worldwide community. Other groups have used the Internet to map places of safety, such as 'gay-friendly' bars and restaurants, where identities can be expressed without fear or prejudice. Social media and access to GIS will certainly expand the availability of mapping and production of group identities significantly.

## Conclusion

Maps are powerful tools for encapsulating, constructing, and communicating identity and their versatility allows them to address the national to the personal, connecting people with shared values with place. The emergence of counter-mapping initiatives and greater capabilities for disseminating maps in the last decades of the late twentieth and early twenty-first centuries have allowed identities to be defined and redefined using maps that serve or challenge the traditional narratives of the state. Whether the use of maps in this way will continue to grow and expand will be determined by whether the authenticity of their cartographic language is sufficient for expressing personal, local, regional, and national identities.

## References

'Aleph' (Harvey, W.) (1868) *Geographical Fun: Being Humourous Outlines of Various Countries, with an Introduction and Descriptive Lines* London: Hodder and Stoughton Available at: *www.loc.gov/resource/g5701am.gct00011/?st=gallery* (Accessed: 21 November 2016).

Batuman, B. (2010) "The Shape of the Nation: Visual Production of Nationalism through Maps in Turkey" *Political Geography* 29 (4) pp.220–234.

Blickle, P. (2002) *Heimat: A Critical Theory of the German Idea of Homeland* New York: Camden House.

Crouch, D. and Matless, D. (1996) "Refiguring Geography: Parish Maps of Common Ground" *Transactions of the Institute of British Geographers* 21 (1) pp.236–255.

Fridl, J. (Ed.) (2001) *National Atlas of Slovenia* Ljubljana, Slovenia: Rokus Publishing House.

Fridl, J. (2004) "National Atlas of Slovenia: Current Trends and Future Opportunities" *The Cartographic Journal* 41 (2) pp.167–176.

Gams, I. (1991) "The Republic of Slovenia: Geographical Constants of the New Central-European State" *GeoJournal* 24 (4) pp.331–340.

Gladman, I. (Ed.) (2005) *Central and South-Eastern Europe 2006* London: Taylor & Francis.

Herb, G.H. (1997) *Under the Map of Germany: Nationalism and Propaganda 1918–1945* London: Routledge.

Holmes, N. (1991) *Pictorial Maps* London: The Herbert Press.

Jacobs, F. (2009) *Strange Maps: An Atlas of Cartographic Curiosities* New York: Viking Studio.

Jordan, P. (2004) "National and Regional Atlases as an Expression of National/Regional Identities: New Examples from Post-Communist Europe" *The Cartographic Journal* 41 (2) pp.150–166.

Kent, A.J. (2012) "From a Dry Statement of Facts to a Thing of Beauty: Understanding Aesthetics in the Mapping and Counter-Mapping of Place" *Cartographic Perspectives* 73 pp.39–60.

Kent, R.B. (1986) "National Atlases: The Influence of Wealth and Political Orientation on Content" *Geography* 71 (2) pp.122–130.

Klavins, J. (Ed.) (1992) *Latvijas Atlants* Riga, Latvia: Latvijas Karte.

Monmonier, M. (1994) "The Rise of the National Atlas" *Cartographica* 31 (1) pp.1–15.

Newman, D. (1991) "Overcoming the Psychological Barrier: The Role of Images in War and Peace" in Kliot, N. and Waterman, S. *The Political Geography of Conflict and Peace* London: Belhaven Press, pp.192–207.

Nykiforuk, C., Nieuwendyk, L. Mitha, S. and Hosler, I. (2012) "Examining Aspects of the Built Environment: An Evaluation of a Community Walking Map Project" *Canadian Journal of Public Health* 103 (9 Suppl. 3) eS67–72.

Pahor, B. (2001) "Introduction" in Fridl, J. (Ed.) *National Atlas of Slovenia* Ljubljana, Slovenia: Rokus Publishing House, p.11.

Parker, B. (2004) "Constructing Community Through Maps? Power and Praxis in Community Mapping" *The Professional Geographer* 58 (4) pp.470–484.

Philippe, R. (1980) *Political Graphics: Art as a Weapon* Oxford, UK: Phaidon.

Pickles, J. (1992) "Text, Hermeneutics and Propaganda Maps" in Barnes, T.J. and Duncan, J.S. *Writing Worlds* London: Routledge, pp.193–230.

Pickles, M. (2015) "'Circademics/' Get Serious about Circus" *BBC News* Available at: *www.bbc.co.uk/news/business-33277907* (Accessed: 1 July 2015).

Pūriņš, V. (Ed.) (1988) *Latvijas PSR Atlants* Moscow: Glavnoe Upravlenie Geoezii i Kartographii.

Regan, C., Sinclair, S. and Turner, M. (1988) "*Thin Black Lines*" Birmingham, UK: Development Education Centre.

Rogoff, I. (2000) *Terra Infirma: Geography's Visual Culture* London: Routledge.

Thrower, N.J.W. (1999) *Maps & Civilization* (2nd ed.) Chicago, IL: The University of Chicago Press.

Urbanc, M., Fridl, J., Kladnik, D. and Perko, D. (2006) "*Atlant* and Slovene National Consciousness in the Second Half of the 19th Century" *Acta Geographica Slovenica* 46 (2) pp.251–283.

Vujakovic, P. (1990) "Comic Cartography" *Geographical Magazine* 62 (6) pp.22–26.

Vujakovic, P. (1992) "Mapping Europe's Myths" *Geographical Magazine* 64 (9) pp.15–17.

Vujakovic, P. (1995) "The Sleeping Beauty Complex: Maps as Text in the Construction of National Identity" in Hill, T. and Hughes, W. (Eds.) *Contemporary Writing and National Identity* Bath, UK: Sulis Press, pp.129–136.

Vujakovic, P. (2014) "Mapping the Octopus" *Maplines* Winter pp.10–11.

Vujakovic, P. (2015) "Maps as Political Cartoons" in Monmonier, M. (Ed.) *The History of Cartography (Volume 6)* Chicago, IL: The University of Chicago Press, pp.1162–1165.

Vujakovic, P., Hills, J. and Kent, A.J. (2014) "Turning the Tide? Visitor Maps and the Place Branding of Coastal Towns in Kent and East Sussex" *The Bulletin of the Society of Cartographers* 47 (1) pp.37–42.

Wood, D. (1992) *The Power of Maps* New York: The Guilford Press.

Wood, D. (2010) *Everything Sings: Maps for a Narrative Atlas* Los Angeles, CA: Siglio Press.

Yonge, E.L. (1957) "National Atlases: A Summary" *Geographical Review* 47 (4) pp.570–578.

Zeigler, D.J. (2002) "Post-Communist Eastern Europe and the Cartography of Independence" *Political Geography* 21 (5) pp.671–686.

# 31

# Maps, power, and politics

*Guntram H. Herb*

When J. Brian Harley exposed the rhetorical nature of all maps in the late 1980s (Harley, 1988, 1989, 1990), he destabilized the distinction between propaganda maps and scientific maps. This posed a significant problem for scientific cartographers who had always claimed to be the custodians of objectivity and accuracy in maps. Yet, the seemingly slippery slope from scientific maps to the work of clever propagandists who subtly distorted information for political ends still seemed to have a clear rupture at one point: German geopolitical maps. Surely, this case proved that some maps should rightfully be labelled as persuasive and even propaganda. Building on works by authors such as Quam (1943) and Speier (1941) who had already revealed their flaws in the 1940s, cartographers carefully traced the biased character of these Nazi era maps to the selective use of facts, falsified data, and misleading design tricks and juxtaposed them to more objective and accurate depictions (e.g., Tyner, 1982; Monmonier, 1991). Even critical geographers such as Raffestin (2000) and Boria (2008) – who exposed a carefully devised strategy behind German geopolitical maps and associated them with power politics of the political Right and masterful Nazi propaganda – helped elevate German geopolitical maps to the level of an exemplar of cartographic persuasion. Yet, as I will argue, such views are far too simplistic: persuasive political maps neither require a state apparatus to become the focus of orchestrated campaigns, nor are they necessarily right-wing, nor can they be easily dismissed as false or deceitful products. I will present my argument in three steps. The first is a case study of interwar Germany where a nationalist mapping campaign lent crucial support to Nazi expansionism. This lays the empirical groundwork for the next step, which critically engages with the theoretical literature in cartography and re-evaluates the dominant conceptual approach that focuses on the unique design and grammar of persuasive maps. Finally, I shift my focus to the organization and ideology of persuasive maps. Here I will apply social movement theory to better understand how non-state actors are able to develop orchestrated campaigns such as the one in interwar Germany and interrogate whether we need to reconsider the association of persuasive maps with old-style geopolitics and the political Right.[1]

## Nationalist mapping in interwar Germany

When Nazi Germany embarked on its course of expansionism in the late 1930s – the Anschluss of Austria, the annexation of the so-called Sudetenland, the occupation of the Memel district, the establishment of the protectorate over Czech lands, and finally the invasion of Poland – it justified its actions with the claim that it was merely repossessing German lands. Yet, the vast majority of these territories had been depicted in the green colour designating 'Slavic Peoples' in German school atlases since the turn of the century. So why could Hitler still count on the support and endorsement of this justification from the German people? Was this the achievement of the famous Nazi propaganda machine or were Germans simply ignorant about these facts? Neither of these two explanations suffices. Instead, we have to credit an alliance of German geographers and nationalists who had widely propagated the image of 'Greater Germany' during the Weimar Republic. It was based on new definitions of national territory and was aided by an intensive dissemination of maps. The trigger for this campaign was the Treaty of Versailles in 1919, which stipulated significant territorial concessions from Germany. When the conference that led to the Treaty commenced in 1918 in Paris, the German delegation did not expect larger territorial adjustments, but the German geographer Albrecht Penck was clearly worried. In late 1918, he started to map the ethnic distribution in the eastern parts of the German Empire and was thus ready to launch his attack on the new German boundaries when they became known in early May 1919 (Penck, 1921). Writing in the *Illustrierte Zeitung* on 22 May 1919, he insisted that the territories to be ceded were majority German and offered a dot map as evidence. He spoke of fraud and traced it to a map published by Jakob Spett in the French newspaper *Le Temps* on 19 March 1919, which he called a 'masterpiece of forgery' (Penck, 1919).

After Penck's volley, others took action. The geographer Wilhelm Volz and the *Deutschtumspolitiker* (pan-German activist) Karl C. von Lösch presented funding proposals to the German government for the establishment of an exchange centre (*Austauschstelle*) for scientists and politicians to aid the development of scientific 'weapons' – in particular maps – for the revision of the Versailles Treaty. They secured funding, organized 18 secret conferences between 1921 and 1931, and established the *Stiftung für Volks- und Kulturbodenforschung* in Leipzig. These efforts were crucial for coordinating the work of various research institutes on German ethnicity and territory, and for developing new definitions of German national territory. They also spearheaded goals for map design, such as prescribing the colour red for Germans. Yet, apart from giving financial support, the government did not become involved; it neither issued official directives, nor provided guidelines, nor oversaw the work (Herb, 1997: 65–75).

In their attempt to develop evidence for a revision of the Treaty of Versailles, German geographers and nationalists ran into problems right away. The most commonly used maps to depict national territory at Paris were (and still are today) ethnographic maps. These maps use language as an indicator because it is readily available data from censuses. When Germans consulted German ethnographic maps they noticed that even when they used the most up-to-date data they could not ensure a clear connection between the German mainland territory and East Prussia across a corridor along the Vistula River that had been awarded to the new Polish state. This was the most resented territorial loss because it cut Germany in two. Even newly developed 'accurate' methods of displaying language data, such as Albrecht Penck's dot map, did not help (Penck, 1919). Clearly, other maps were needed to support their cause and they turned their attention to the cultural landscape and the development of new territorial definitions of the German nation. The most influential concept was Albrecht Penck's notion of the German *Volks- und Kulturboden* (Penck, 1925). It postulated that the power of German culture was far superior to the culture of its Slavic neighbours so that a separate Czech cultural landscape could not develop on its own. As a result, the German cultural imprint extended far beyond the boundaries of the German Empire and included Bohemia and Moravia. The later establishment of the German Protectorate was thus already legitimized in 1925!

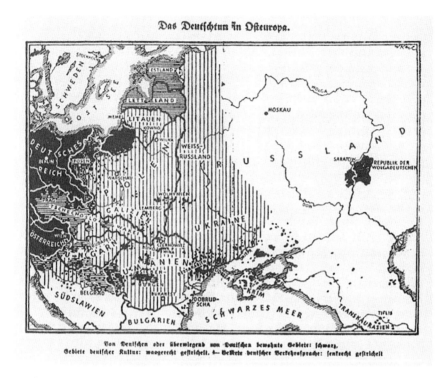

*Figure 31.1*   The German cultural imprint in Eastern Europe. A partial and reduced reproduction of the Penck-Fischer map (Spohr, 1930)

Penck made even further reaching claims when he defined the area of the use of German in commerce in a map he published with Hans Fischer (Penck and Fischer, 1925). Thus, he showed that German cultural influence extended from Brussels and Nancy in the west all the way to Lake Ladoga, Smolensk, and Kharkov, an eerie preview of German conquests in the Second World War (see Figure 31.1).

While Penck's maps of German cultural imprint outlined 'rightful' German territorial demands, other maps tried to show that a revision of the boundaries was a matter of life and death for the German nation. These maps presented German national territory as a wounded body, an idea that can be traced to the concept of the 'state as organism' developed by Rudolf Kjellen and Friedrich Ratzel (1897). There were two versions of this new definition of German national territory. One stressed the economy, the other military power.

Maps employing the economic version of the geo-organic definition showed the lost territories as integral parts of the economic organism. Road, water, and electricity networks were akin to a body's arteries that carried life-sustaining blood to essential organs. Thus, when the new borders cut these lines, they created 'bleeding borders'. However, this economic organism argument could not be used to claim Sudetenland or Austria, which had never been integrated with the German state. The military power version of the geo-organic definition focused on the vulnerability of the national body to attack. Boundaries were shown as threatening the survival of the German nation. Maps indicated that it was easy to cut Germany in two or to strike a deathblow to the heart of the nation because the capital Berlin was now very close to the borders (see Figure 31.2). Yet, this definition was also problematic since it was not possible to outline specific claims for a revised territory that was 'safe'.

Die Bedrohung des deutschen Ostens und Südens

*Figure 31.2*  **Threat to the German East and Southeast (Boehm, 1930)**

Maps of the German cultural imprint and of threats to the German geo-organism were widely disseminated in the middle 1920s to early 1930s in the Weimar Republic. They appeared in books, magazines, newspapers, and leaflets. The Penck-Fischer map was especially popular. It was published as a wall map and atlas supplement by the *Volksbund für das Deutschtum im Ausland (VDA)*; it was even included in school atlases, such as *F.W. Putzgers Historischer Schulatlas*, the most widely used historical atlas in Germany. The spread of these maps was unequivocally the result of a coordinated effort, an orchestrated map campaign designed to popularize an image of the Greater German nation. The creation of this new image even involved altering existing maps. The universal nature of changes again indicates that it was an organized effort. In edition after edition of ethnographic maps in school atlases and wall maps, Slavic areas in eastern parts of the German Empire became more and more Germanized. While the 'corridor area' that was ceded to Poland appeared as uniformly Slavic in school atlases published at the end of the First World War, it was depicted as mixed German-Slavic by the mid 1920s.

So how exactly was this campaign with maps coordinated? The main actors in the network behind the propagation of the image of Greater Germany were well-established professors of geography and well-connected pan-German activists. Central institutions were the *Stiftung für Volks- und Kulturbodenforschung* in Leipzig, which promoted the development of new research approaches and projects and monitored the publication of maps of German territory, and the *Deutscher Klub* in Berlin, which provided a place to meet and exchange ideas. The *VDA* and the *Deutscher Schutzbund* actively published and disseminated maps and – with the aid and scientific support of the *Stiftung* – also put strong pressure on publishers and editors who produced maps that did not conform to the new spatial conceptions of 'Greater Germany'. One cartographer was particularly influential and successful: Arnold Hillen Ziegfeld. He greatly advanced the development of geopolitical cartography and helped establish the publishing houses of Volk und Reich and Kurt Vowinckel as innovators and leaders in this map genre (Herb, 1997: 88–94).

Surprisingly, National Socialist organizations did not become involved in this map campaign and did not take over the network when they rose to power in 1933. They even prohibited maps of the German cultural imprint in Eastern Europe in 1942. Moreover, Ziegfeld, to his great dismay, was not recruited in wartime for his cartographic abilities. He complained to the Ministry of Propaganda and offered his services as a specialist in geopolitical cartography, but to no avail. His new position only employed his mapping expertise indirectly as a censor (Herb, 1997: 159–160). Of course, there were National Socialist propaganda maps, but these were usually more allegorical and illustrated threats to the German nation with a knife or a fist. They made arguments based on emotions rather than new definitions of national territory.

The carefully orchestrated map campaign thus contradicts the main arguments in the cartographic literature: first, that geopolitical maps are reprehensible because they use misleading design tricks to falsify the truth, and second, that map propaganda is the prerogative of totalitarian states (Prestwick, 1978; Tyner, 1982). Many of the maps that played a central role in the map campaign, such as the Penck-Fischer map in the *Putzger Atlas* and the Germanized ethnographic maps in school atlases, did not employ the emblematic style of geopolitical maps – for example, arrows – and they were based on scientific research. For example, the imprint of German culture was based on the spread of German house types or legal codes. Therefore, the maps of the campaign cannot simply be dismissed as lies. The second argument, that propaganda maps are associated with totalitarian states and hierarchical organizational structure, such as Portugal under Salazar (Cairo, 2006), Italy under Mussolini, or Germany under Hitler (Boria, 2008), must also be put into question. The National Socialist Party or the National Socialist State apparatus was not directly involved in the campaign. It was a more or less 'private' affair because influential individuals, such as professors of geography and members of pan-German associations, were in charge and not official institutions or regulations.

These empirical findings raise two important sets of questions, which will be addressed in the following two sections. First, what is the nature of persuasive or geopolitical maps? Are they inherently a means to justify expansionism, oppression, and war? Do we always have to consider them as aberrations, as lies? Second, what is the organizational structure and ideology of persuasive or geopolitical mappings? How can non-state actors become involved in effective campaigns and are they by default associated with the political Right?

## The nature of persuasive geopolitical maps

Authors of the most recent literature on geopolitical cartography, such as Claude Raffestin (2000) argue that geopolitical maps have two main characteristics:

(1) *Deception.* They have a unique style. They favour arrows and simple, block-print type representations in black and white. They are succinct, memorable, and convey their message without text; they 'speak for themselves'. These design elements are used to mislead and trick the map-reader.

(2) *Limited information*. They present simplistic correlations between political power and the geographic environment and are the preferred expression of German geopolitics under Karl Haushofer. Even authors who use a deconstructivist approach are eager to show how geopolitical cartography was enmeshed in the evils of German *Geopolitik* and Fascist ideas (Boria, 2008). As a result, the entire genre of geopolitical maps is ostracized and deemed malicious.

Anglo-American geographers first issued the charge of *deception* during the Second World War. These authors made a binary distinction between 'good' maps, i.e. objective, scientific maps, and 'false' maps, i.e. propaganda maps (Speier, 1941; Quam, 1943). They exposed the design tricks of German geopolitical maps and explained how the use of arrows or selective areas could be used to give a misleading impression; for example, a map that purported to show the encirclement of British troops in Dunkirk, but omitted to show the British Isles across the Strait of Dover (Quam, 1943: figure 9). This dichotomy fits well into the communication model in cartography, which was dominant into the 1980s and is still in use today. In this model, the map is conceived as a true representation of reality, and the cartographer (i.e. the sender of information) tries to depict the world or elements of world (e.g. the transportation infrastructure network in a road map) in such a way that the map user (receiver) can decode the information most effectively (Crampton, 2001). In other words, the goal of cartography is to make the communication of spatial information as effective as possible. Cartographic research used surveys and human subject tests to understand how people read maps and which designs work best. The scientific nature of this approach is revealed in the disciplinary terms of cartography: cartographic laboratory, cartographic technician, map construction, and design. By contrast, the authors of propaganda maps use – or rather misuse – the insights gained in this process to effectively communicate a *false* or biased image of reality to the map-reader.

A second and related theoretical concept is Jacques Bertin's cartographic semiology (Bertin, 1983). He compiled a grammar of signs based on the limits of human perception of graphic information. For example, according to Bertin, it is not intuitive for humans to distinguish different quantities of data by colour, but rather by the size of symbols. Some signs make it possible to differentiate elements, while other signs make it easy to associate elements and group them. As in the communication model, this approach stresses efficiency of design and strives to make maps that are the most accurate and effective representations of reality.

In the late 1980s, J. Brian Harley challenged the validity of these two concepts in a fundamental way. Inspired by the ideas of Foucault and Derrida, he considered maps as texts (Harley, 1989). So in contrast to Bertin, the issue was not what one could perceive or 'see' in a map, but what could be 'read' from them. Harley posited that maps reflect and reproduce the cultural and social context in which they are produced. He deconstructed maps and showed, for example, that topographic maps depicted castles and manor houses, but not slums, or that they showed golf courses, but not toxic waste dumps. Only that which was deemed important to a society, and especially to those with power, was mapped. The ideology thus inherent in all maps suddenly made it more difficult to distinguish between objective scientific maps, which showed a true image of the world, and propaganda maps.

More recently, Harley also became the object of critique. Despite his critical stance vis-à-vis the orthodox positivist approach in cartography, he did not put into question that maps are a representation of reality, that they communicate definite information about the world. Harley rejected the notion of an *objective* reality in positivist concepts, but still argued that maps represented a *social* and *cultural* reality. For Denis Wood (1992, 2010) and Jeremy Crampton (2001, 2002) this was an untenable stance. Arguing from a post-structuralist position, they viewed maps as social constructions that can be interpreted in multiple and even contradictory ways. But how could such an interpretation explain that people generally agree on what maps purport to show and that they accept maps as true, as reality? Judith Butler's theoretical work on performativity and reiteration can help us understand this quandary (Butler, 1993). The continued and repeated use of maps and the social context (dominant values, rules, and standards) in which this takes place, cements a shared meaning of maps. So, in addition to cognitive constraints on what we perceive in maps (what Bertin focused on), we also have social constraints on what we read from maps. Information in maps that conforms to our values, societal standards, and shared experiences will be uncritically accepted as true, as reality. My point here is that we have to view all maps as being embedded in performative and reiterative practices. Therefore, we cannot ostracize persuasive or geopolitical maps simply on account of their unique design.

I would also like to question the second main characteristic of geopolitical maps in the literature, *limited information*; that is the argument that they present simplistic correlations between political power and the geographic environment and that they are the preferred expression of German geopolitics under Karl Haushofer. Claude Raffestin (2000) claims that all geopolitical maps can be traced back to the German model and that they simplify the world to such an extent that they are 'uchronic and utopian'. What he means by this is that geopolitical maps do not show the rich character of places, but only locational information, and that they reduce complex historical processes to single dates. They homogenize time and space and reduce the world to 'political geometry'. Raffestin bases his argument on an analysis of design elements in German geopolitical maps. He develops a graphic grammar – derived largely from the theoretical works of Rupert von Schumacher (1934, 1935) – and then tries to show how certain design elements express and convey different dimensions of the ideology of German *Geopolitik*, such as territorial aggression, expansionism, and war. In other words, Raffestin confines his analysis to the cartographic objects themselves without fully considering the larger context of the development of geopolitical visualizations.

Geopolitical maps did not appear in a vacuum. They are part of the power tradition in political geography, not just in Germany, but also in Britain, the United States, and elsewhere (Herb, 2008). Authors in this tradition view the state as the most important actor and explain the world as binary opposites in terms of identity (e.g. us – them) and power (e.g. sea power – land power). Because the power tradition assumes that the relationship between peoples and states is defined by competition and conflict, its representatives are actively engaged in ensuring the global dominance of their own country. Of central importance in the power tradition are *geographs* or territorial codes, such as the heartland concept of Halford J. Mackinder or Alfred Thayer Mahan's writings on sea power (O'Tuathail, 1996). Geopolitical maps are simply visualizations of such territorial codes; they are *visualized geographs*.

But what exactly are territorial codes or *geographs*? They are best understood as structured explanations of how to interpret the world. They define which elements of the world are key to make sense of the complexity of global politics. So which elements of the world are kept and which are discarded in these territorial codes? There are three ways this selection happens:

(1) *Silencing*. That which is not important to us gets omitted, only what fits into our cultural context and value system is considered.
(2) *Othering*. The world is presented in binaries of identity and power.
(3) *Conflating*. Only one scale is generally considered: the level of states.

A good example of a *visualized geograph* is Mackinder's famous map of the pivot of the world (Mackinder, 1904), which he later called 'heartland' (see Figure 31.3). It shows that a state can

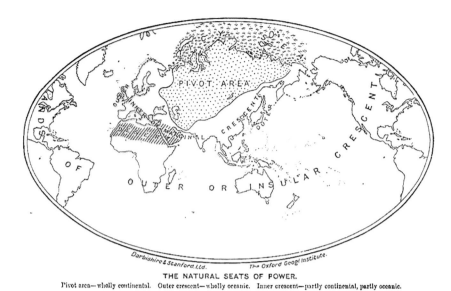

**THE NATURAL SEATS OF POWER.**
Pivot area—wholly continental.  Outer crescent—wholly oceanic.  Inner crescent—partly continental, partly oceanic.

*Figure 31.3*  Mackinder's geographic pivot of history (Mackinder, 1904)

only achieve global dominance when it controls this crucial area because of its strategic advantage in the contest between sea power and land power. In this map, global power equals the control of space. It is only important to take control of that area, not other aspects (*silencing*), conflict is reduced to a binary opposition between land and sea power (*othering*), and the world is divided into regions, such as the marginal crescent (*conflating*). Since Mackinder published his map long before geopolitical cartography came into existence in Germany, it makes it difficult to reduce this map genre to the model of German *Geopolitik*.

It is also important to take note that these territorial codes and their visualization in maps took place in a specific historical context. They were situated in the age of imperialism when European states competed for global dominance and when geo-determinism was a dominant philosophy. Visualized geographs, such as Mackinder's map of the heartland or the suggestive maps of the Weimar map campaign, were used and consumed in an uncritical manner because they represented shared values and fitted into the cultural context. Thus, geopolitical maps are best understood in light of Gramsci's (see Hoare and Nowell-Smith, 1971) concept of hegemony. German geopolitical maps represented ideas that already resonated with many Germans, but then reinforced, legitimized, and consolidated these ideas through the wide dissemination of maps via media, such as newspapers (see also Chapter 34).

Territorial codes and geopolitical maps are not a thing of the past. We still consume them uncritically today because they fit into our cultural context. Striking examples are ethnographic maps, which can be found in most school atlases, and maps of religious factions, which can be found in news reports on conflict zones, such as Syria, Iraq, or Northern Ireland. They can be considered 'geopolitical maps' because all three elements of geographs apply: these maps only show majority areas and thus omit minorities (silencing), they differentiate ethnic groups or religious factions as opposites (othering), and they imply that members of a given ethnic or religious group are homogeneous (conflating). Thus, in the context of the universally accepted concept of self-determination, such maps explicitly convey territorial demands and legitimize them. It can also be argued that they aid in the oppression of the minorities that do not even get represented on such maps. My point here is that geopolitical maps should not be ostracized because they provide limited information, and the entire genre should not be reduced to an evil aberration because there is a prominent case where they were misused.

## Organization and ideology of geopolitical maps

The National Socialist Party and state apparatus did not directly participate in the map campaign of the late 1920s and early 1930s that tried to create an image of the Greater German nation. It was essentially a 'private' affair that was carried by influential and respected individuals, such as professors of geography and pan-German politicians. There is neither archival evidence that the Nazis supported these activities nor that they took over the existing network even after they came to power in 1933. Why did the Nazis choose not take action and become involved? One probable explanation is that the members of the network were eager to ingratiate themselves to the new masters and their ideology. A representative example is Karl C. von Loesch, who rejected the notion of a racially based community of Germans in 1925, but after 1933 started to work intensively on the 'Jewish question' and the cartographic representation of Jews (Herb, 1997: 137). Another likely reason is that maps clearly define, and thus limit, territorial demands. National Socialists did not want to be constrained in their expansionism to the area of German cultural imprint. They would have had a difficult time explaining why territory beyond that area was rightfully German when cartographic evidence said otherwise (Herb, 1997: 177).

An even more compelling explanation is offered when the map campaign is analysed as a protest movement against the Treaty of Versailles. I want to stress at the outset that I am not trying to convey any legitimacy to the campaign by taking this approach; I am merely trying to take advantage of the insights gained from social movement theory to better understand the structure and effectiveness of the campaign. The most pertinent theories are resource mobilization, political opportunity structure, and frame alignment. These theories are founded on the premise that grievances are always present in societies, which means that grievances by themselves are insufficient to explain social mobilizations. Rather, protest mobilizations depend on available resources and opportunities and a consensus among individual and collective values, ideas, and ideologies (Klandermans, 1991; Hourigan, 2003). More specifically, there are four elements that determine protest movements and their actions:

(1) *Advantages and disadvantages of participation.* Individuals will take action when they believe it will benefit themselves or their collective group.
(2) *Organization.* Good organization is essential because it can remove potential obstacles early on and because it facilitates the recruitment of new members.
(3) *Potential Success.* Movements are more likely to succeed when there are favourable political conditions and opportunities, such as potential alliances with other powerful groups or promising new tactics.
(4) *Consolidation of goals.* Protest movements grow when individual grievances and aspirations receive collective support.

The main thesis of these social movement theories – that grievances are always present – and the four constitutive elements – prospects for participation, organization, potential success, and consolidation of goals – all apply to the geopolitical map campaign in Germany. Even though the territorial losses as a result of the Versailles Treaty were decisive, German nationalism has always been driven by the idea that German unification was incomplete. In other words, what Germans call 'the German question', that is, how the German state territory and the territory of the German nation could be brought into agreement existed long before and after the Versailles Treaty. The four elements apply as well:

(1) The main individuals in the network derived significant benefits from their activism. For example, Wilhelm Volz and other geographers received substantial financial support for their research and von Loesch's organization Deutscher Schutzbund profited from greater legitimacy as a result of its association with highly respected scientists.
(2) Existing institutions, such as the Deutscher Klub in Berlin greatly facilitated cooperation among members of the network, and the secret conferences organized by Volz, Penck, and von Loesch helped in the recruitment of new members and the development of new tactics.
(3) The success of the campaign was boosted because its views overlapped with those of other groups, such as advocates for a re-militarization of Germany, and especially because it adopted maps as a new tactic. Maps have always been associated with authority and science, which gave the demands of the network a cloak of scientific respectability and objectivity.
(4) The members of the network were convinced that the Versailles Treaty provisions were unjust and that Germans were hated by their neighbours who sought to destroy them. The wide dissemination of geopolitical maps in newspapers, school books, posters, post cards, and pamphlets made it appear that their beliefs were not on the fringe, but widely accepted by society.

Based on this discussion, I posit that the geopolitical mapping campaign in Germany is best understood as a protest movement and that, like all social movements, it was not hierarchically organized but rather *rhizomorphically* structured (see Deleuze *et al.*, 1987). A variety of groups and institutions produced these maps and while these actors had shared beliefs and the same goals, they did not directly coordinate their actions. Thus, following De Certeau, the creation and dissemination of these maps should be viewed as a tactic and form of resistance instead of a coordinated strategy (De Certeau, 2002). This also explains why the Nazis did not directly become involved: not only was there no concrete organizational structure that could be taken over, but the tactical approach of resistance is incompatible with the strategic approach of a state or a hierarchical organization, such as the National Socialist Party.

Viewing geopolitical cartography in the Weimar Republic as a protest movement helps free the genre from its customary banishment to totalitarian and fascist systems. Geopolitical maps can be effective tools for social movements and the dissemination of critical ideas. The French journal *Hérodote* is a good example. It unabashedly uses the term 'geopolitics' and employs maps whose design is reminiscent of German geopolitical maps of the Weimar era.[2] However, the approach of *Hérodote* is not old-style geopolitics. It differs from the power tradition of political geography because it stresses the need to examine different scales (what its founder, Yves Lacoste, calls spatial ensembles) and addresses questions beyond the state, power, and hegemony. Special issues in the journal have addressed women's rights, the environment, religion, local politics, and regional identities, and its editors were (at least in its early years) influenced by Marxist thought. Perhaps the best case is the cartographic work of *Le Monde Diplomatique*, which produces popular geopolitical atlases that take a clear critical stance and are decidedly Left (Gresh *et al.*, 2009). Their atlases are even published in translation into different languages and disseminated in many countries in the world, though not in English, which means they are still striving to find an audience in Britain and the US. Could it be that the old stigma of German geopolitical cartography still has too much sway in these countries? How unfortunate, because geopolitical maps are a most effective means to show the spatial dimensions of politics; they are didactically ingenious!

## Notes

1 I presented an earlier version of this argument in German at the conference "Kampf der Karten" at the Herder-Institut in Marburg in May 2009 (Herb, 2012). I am grateful to the Herder-Institut for their invitation and generous support.
2 See the map comparison in Herb (2008: 31–32).

## References

Bertin, J. (1983) *Semiology of Graphics: Diagrams, Networks, Maps* Madison, WI: University of Wisconsin Press.

Boehm, M.H. (1930) *Die Deutschen Grenzlande* Berlin: Reimar Hobbing.

Boria, E. (2008) "Geopolitical Maps: A Sketch History of a Neglected Trend in Cartography" *Geopolitics* 13 pp.278–308.

Butler, J. (1993) *Bodies That Matter: On the Discursive Limits of Sex* New York: Routledge.

Cairo, H. (2006) "Portugal Is Not a Small Country: Maps and Propaganda in the Salazar Regime" *Geopolitics* 11 pp.367–395.

Crampton, J. (2001) "Maps as Social Constructions: Power, Communication, and Visualization" *Progress in Human Geography* 25 (2) pp.235–253.

Crampton, J. (2002) "Thinking Philosophically in Cartography: Toward a Critical Politics of Mapping" *Cartographic Perspectives* 41 (Winter) pp.4–23.

De Certeau, M. (2002) *The Practice of Everyday Life* Berkeley, CA: University of California Press.

Deleuze, G., Guattari, F. and Massumi, B. (1987) *A Thousand Plateaus: Capitalism and Schizophrenia* Minneapolis, MN: University of Minnesota Press.

Gresh, A., Radvanyi, J. and Rekacewicz, P. (Eds) (2009) *L'atlas 2010: Monde Diplomatique* Paris: Armand Colin.

Harley, J.B. (1988) "Maps, Knowledge, and Power" in Cosgrove, D. and Daniels, S. (Eds) *The Iconography of Landscape* Cambridge, UK: Cambridge University Press, pp.277–312.

Harley, J.B. (1989) "Deconstructing the Map" *Cartographica* 26 (2) pp.1–20.

Harley, J.B. (1990) "Cartography, Ethics and Social Theory" *Cartographica* 27 (2) pp.1–23.

Herb, G.H. (1997) *Under the Map of Germany: Nationalism and Propaganda 1918–1945* New York: Routledge.

Herb, G.H. (2008) "The Politics of Political Geography" in Cox, K., Low, M. and Robinson, J. (Eds) *The Sage Handbook of Political Geography* London: Sage, pp.21–40.

Herb, G. H. (2012) "Das größte Deutschland soll es sein! Suggestive Karten in der Weimarer Republik" In Haslinger, P. and Oswalt, V. (Eds) *Kampf der Karten. Propaganda- und Geschichtskarten als politische Instrumente und Identitätstexte* (Tagungen zur Ostmitteleuropaforschung 30) Marburg, Germany: Herder-Institut, pp.140–151.

Hoare, Q. and Nowell-Smith, G. (Eds) (1971) *Selections from the Prison Notebooks of Antonio Gramsci* New York: International Publishers.

Hourigan, N. (2003) *Escaping the Global Village. Media, Language, and Protest* Lanham, MD: Lexington Books.

Klandermans, B. (1991) "The Peace Movement and Social Movement Theory" in Klandermans, B. (Ed.) *Peace Movements in Western Europe and the United States. International Social Movement Research (3)* Greenwich, CT: Jai Press, pp.1–39.

Mackinder, H. T. (1904) "The Geographical Pivot of History" *Geographical Journal* 23 (4) pp.421–444.

Monmonier, M. (1991) *How to Lie with Maps* Chicago, IL: University of Chicago Press.

O'Tuathail, G. (1996) *Critical Geopolitics: The Politics of Writing Global Space* Minneapolis, MN: University of Minnesota Press.

Penck, A. (1919) "Die Polengrenze" *Illustrierte Zeitung* 152 (3960) pp.536–537.

Penck, A. (1921) "Die Deutschen im polnischen Korridor" *Zeitschrift der Gesellschaft für Erdkunde zu Berlin* pp.169–185.

Penck, A. (1925) "Deutscher Volks- und Kulturboden" in Von Loesch, K.C. (Ed.) *Volk unter Völkern* (Bücher des Deutschtums 2, Breslau), pp.62–73.

Penck, A. and Fischer, H. (1925) "Der deutsche Volks- und Kulturboden in Europa" Berlin: Verein für das Deutschtum im Ausland.

Prestwick, R. (1978) "Maps and the Perception of Space" in Lanegran, D. and Palms, R. (Eds) *An Invitation to Geography* New York: McGraw-Hill, pp.13–37.

Quam, L. O. (1943) "The Use of Maps in Propaganda" *Journal of Geography* 42 pp.21–32.

Raffestin, C. (2000) "From Text to Image" *Geopolitics* 5 (2) pp.7–34.

Schumacher, R. von (1934) "Zur Theorie der Raumdarstellung" *Zeitschrift für Geopolitik* 11 pp.635–652.

Schumacher, R. von (1935) "Zur Theorie der geopolitischen Signatur" *Zeitschrift für Geopolitik* 12 pp.247–265.

Speier, H. (1941) "Magic Geography" *Social Research* 8 pp.310–330.

Spohr, W. (1930) *Deutsche Brüder Im Osten*. Auslanddeutsche Volkshefte 4 Berlin: Verlagsanstalt H. A. Braun.

Tyner, J. (1982) "Persuasive Cartography" *Journal of Geography* 81 (4) pp.140–144.

Wood, D. (1992) *The Power of Maps* New York: The Guilford Press.

Wood, D. (2010) *Rethinking the Power of Maps* New York: The Guilford Press.

# 32

# Persuasive map design

*Judith Tyner*

In 1942 the Hamburg schools produced a world atlas for the use of children, the *Atlas für Hamburger Schulen*. Among the maps was a dot map that showed German population in the United States. It appears to be a straightforward map. But on closer examination, some features are puzzling. There are no maps in the atlas of Irish, Italian, Spanish or any other national groups, only German. Some cities and towns are shown on the map, but only those with German names or with large German populations. Those in Ohio, Cincinnati and Cleveland are shown as well as a tiny German community of Van Wert, but not Columbus or nearby Dayton. The cover of the atlas has Nazi swastika symbols and we realize that the map and the atlas, as a whole, had an unstated purpose: it was designed to persuade the reader that the United States had a large German population, which, presumably, would be sympathetic to the Nazi cause.

Persuasive maps are a type of map whose main object or effect is to change or in some way influence the reader's opinion (Tyner, 1974). Such maps have existed for hundreds of years and they are used in advertising, in political campaigns, as wartime propaganda, in position papers and to illustrate books and articles. They are found as paper maps, on television and online. They are produced by governments, non-profit organizations, and advertising firms, among others. Persuasive maps are arguments that say 'this is true', 'vote for me', 'buy me', 'don't go here' and 'this is bad'. Persuasive maps are ubiquitous, but the ways in which they accomplish their goals have not been studied by mainstream cartographers until comparatively recently, largely because they do not fit the mould of 'scientific' cartography.

Persuasive maps can be used to promote 'good' causes as well as 'bad'. They are rhetorical devices. Because the public has great faith in maps but, for the most part, little training in their use and evaluation, it is easy to persuade with maps. To many, the map is the epitome of truth and accuracy. 'The map shows it', therefore it must be true – especially to those who commonly use maps for wayfinding. Beryl Markham said it well:

> A map in the hands of a pilot is a testimony of a man's faith in other men; it is a symbol of confidence and trust. It is not like a printed page that bears mere words, ambiguous and artful, and whose most believing reader – even whose author, perhaps – must allow in his mind a recess for doubt. A map says to you, 'read me carefully, follow me closely, doubt me not'.
>
> *(Markham, 1942: 245)*

Persuasive maps are like the printed page that bears symbols, 'sometimes ambiguous and usually artful' that allow in the reader's mind 'a recess for doubt'. This chapter attempts to explain how persuasive maps work, what makes a map persuasive and how are they designed.

## Rhetorical styles

Persuasive maps can be grouped by their function, e.g. advertising, politics, propaganda and the like, but Ian Muehlenhaus, in a recent study of persuasive maps, developed a categorization of rhetorical styles for persuasive maps based on their appearance and their use of map elements (Muehlenhaus, 2012). These styles are: *Authoritative, Understated, Propagandist* and *Sensationalist*. These categories are more useful for analysis than grouping maps by where or how they are used. Muehlenhaus' rhetorical categories are less ambiguous; an advertising map may be created in any of the rhetorical styles as can political maps, but the rhetorical categories are mutually exclusive.

*Authoritative* persuasive maps look scientific and official. They follow the guidelines for good map design used in academic or scientific cartography. This style is intended to engender trust in the map user. Rather than obvious persuasive design, the cartographers of these maps have modified the data through selection, generalization and classification (Figure 32.1).

*Understated maps*, like authoritative maps, look scientific. There is little or no extraneous information. The symbols used are typically geometric, not pictorial, and the data may be simplified. Colours are used symbolically such as red and yellow to denote danger (Figure 32.2).

The *Propaganda* and *Sensationalist* styles are what usually come to mind when thinking of persuasive maps. These two rhetorical styles most clearly have an idea to sell. Muehlenhaus describes propagandist maps as 'almost exclusively created to quickly and succinctly communicate certain

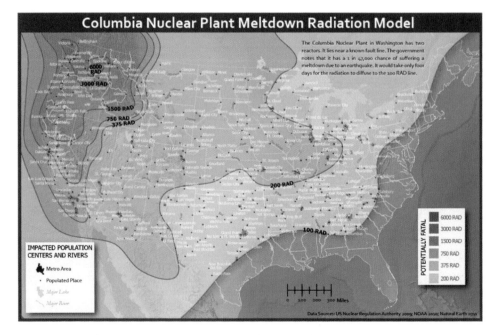

*Figure 32.1*   Authoritative persuasive map style, from Muehlenhaus (2012). © The British Cartographic Society

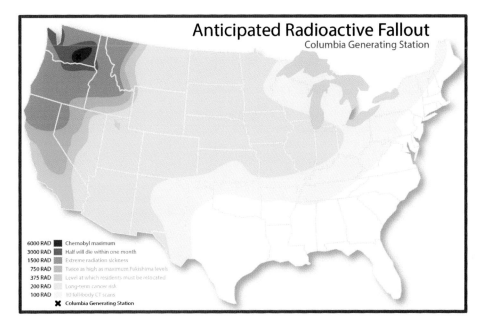

*Figure 32.2*   Understated persuasive map style, from Muehlenhaus (2012). © The British Cartographic Society

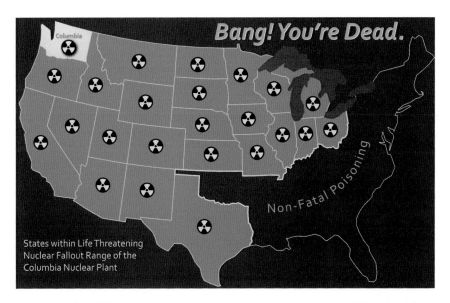

*Figure 32.3*   Propagandist persuasive map style, from Muehlenhaus (2012) © The British Cartographic Society

policies, agendas, ideology or jingoist messages' (Muehlenhaus, 2012: 364). The symbols used are dynamic or mimetic, there is high contrast in the colours and the colours chosen are ones with emotional impact. Data that do not support the message are omitted, while titles tend to be inflammatory and memorable (Figure 32.3).

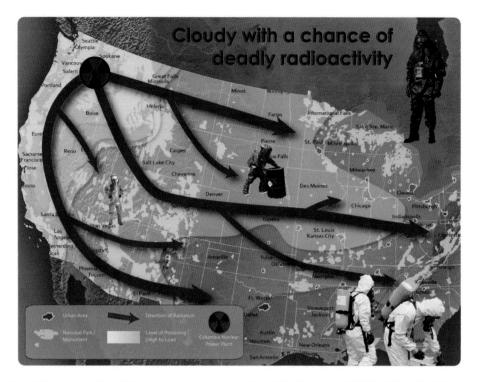

*Figure 32.4*   Sensationalist persuasive map style, from Muehlenhaus (2012) © The British
Cartographic Society

Sensationalist maps are eye-catching; they 'attempt to overwhelm one's senses with a barrage
of, often irrelevant, data and visualizations' (Muehlenhaus, 2012: 365). They frequently include
illustrative material and unusual orientations. The symbols are emotive and dynamic and much
irrelevant detail is included. Titles and text are 'catchy', memorable, and often also inflammatory.
Sensationalist style maps overwhelm the reader with information (Figure 32.4).

## Map elements

In order to achieve these rhetorical styles, the elements of the map are manipulated. The
elements that constitute the persuasive cartographer's toolbox are design and layout, generalization and scale, symbols, colour, typeface, text, and projection. A persuasive map may not
include all of the items described below and there is variation based on the rhetorical styles, but
a knowledge of these elements helps to identify persuasive maps.

### Design and layout

Probably the first thing one notices about a map is its overall appearance. Does it attract attention? Is
there much contrast? Do the elements lead the eye from one feature to another? Is there an unusual
orientation? Propagandistic and sensationalist styles scream for attention and often have a definite eye
path. Unlike written communications, which are read sequentially left to right and top to bottom,
maps can be read in any order. Eye path is a technique in design that attempts to direct the reader's
attention through a map or other composition by means of colour, lines, orientation and size.

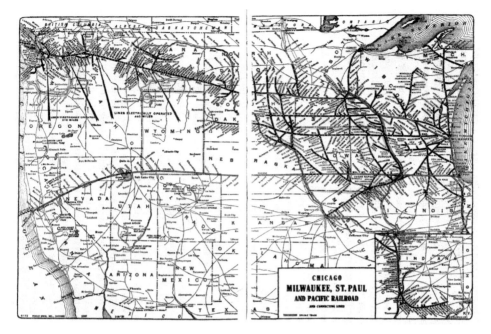

*Figure 32.5   Chicago, Milwaukee, St. Paul and Pacific Railroad* from Woodward, D. (1977), *The All-American Map* Chicago, IL: University of Chicago Press, p.34

## Generalization and scale

Generalization is, of course, required on all maps. Most maps are drawn to scale and cannot show everything. Persuasion generalizes with rhetorical purpose. Generalization involves selection, classification, simplification, smoothing, grouping, exaggeration and sometimes displacement. Selection involves choosing the types of feature to be represented, thus a conventional road map might show roads but not railways. It also involves choosing the amount of information shown within these categories; showing only major roads or only paved roads. Persuasive maps are highly selective and omit data or other information that does not support the cartographer's argument. The maps may be over-generalized, or, in some cases, under-generalized, overwhelming the senses with a wealth of extraneous information. There may be distortions introduced of size, location or shape. In the nineteenth century, US railroad companies would generalize to show their own routes as straight and direct while competitor's routes would be shown longer and more circuitous. Distorting the size of states and straightening the advertised routes accomplished this (Figure 32.5). Displacement has also been used to falsely locate features of strategic interest.

The scale of a map is the relationship of the size of a feature on a map to its actual size on the Earth. While most map-users realize that maps are drawn to scale, they may not realize that for maps of large areas, such as the world or a continent, the scale is not uniform throughout the map. On the Mercator projection, for example, the equatorial length appears the same as the 60th parallel, but on Earth the 60th parallel is one-half the length of the equator. Thus, a feature at the 60th parallel will appear twice as large as a similar feature at the equator. This is used to the persuasive cartographer's advantage and an expression of scale is frequently omitted on persuasive maps.

## *Distortion and projection*

Map projections convert the 3D surface of the Earth onto a 2D plane. All map projections have distortions (this is most obvious on maps of continents and the world) and the persuasive cartographer makes use of this fact. A common method is using a projection incorrectly. For example, when showing data that are areal dependent, such as population density, e.g. the number of people per square mile, the projection chosen should treat area equally, i.e. equal-area. But by showing the data on a non-equal-area projection such as a Mercator, an incorrect impression is given of density or sparseness according to the relative distortion of areas around the globe. Persuasive cartographers have used this method frequently. Another technique is the constant use of a single projection. For many years the Mercator projection was a stand-ard projection in schoolbooks. Generations of children grew up thinking that Greenland was about the size of South America. Persuasive cartographers have used this dulling of perception effectively. Omitting parallels and meridians has also been used so that the reader is not sure of the locations and possible distortions (see Dahlberg, 1961).

## *Symbols*

Maps are, of course, symbolic. Symbols represent towns, elevations, boundaries, data, in short, everything; symbols tell the story. Symbols can be classified in several ways. Most commonly, symbols are described as being points, lines or areas, for showing the loca-tion and type of each feature. These symbols are then divided according to the data they represent into nominal, ordinal, interval scaling (with nominal simply indicating that some-thing exists at that place, and ordinal and interval indicating respectively ranking and the quantitative difference between ranks). Conventionally, the appearance of symbols is often described as being either pictorial or abstract, but this distinction is too simplistic. They may be static, dynamic, abstract, pictorial (mimetic) or suggestive. A simple dot or line on a map is static and abstract. But a point symbol that looks like an explosion or fire is dynamic; a line with arrowheads is dynamic. Suggestive symbols are the most purely persuasive type of symbol. Suggestive symbols have pictorial elements with a high emotional impact – the meaning is suggested at a glance. The suggestive symbol may be a picture of an object, but it does not represent any quantities of the object. Instead, it illustrates a concept suggested by the object. Robert Chapin of *Time* magazine utilized a c-clamp symbol to represent the enemy being squeezed and sharks were shown cutting supply lines (*Time* Magazine, 20 April 1942). A suggestive symbol that has been found on maps for over 150 years is the octopus (Figure 32.6), which is used to imply that the tentacles of a country or company are taking over the world. All of these symbol types are used on persuasive maps and which type depends on the rhetorical style used. Propagandistic and sensationalist styles are most likely to use suggestive symbols.

## *Colour*

The psychology of colours is important in persuasion. Colours have cultural meanings that were studied in the early nineteenth century by Goethe (1810). Colour psychology is a much exploited tool in persuasion but is barely noted in most standard cartography texts, which tend to focus on colour perception and colour harmony. Colour printing used to be much more expensive, and colour symbolism was not a major issue. But now magazines, journals, newspapers and books commonly include colour maps, and, of course, colour is ubiquitous on Internet maps; thus,

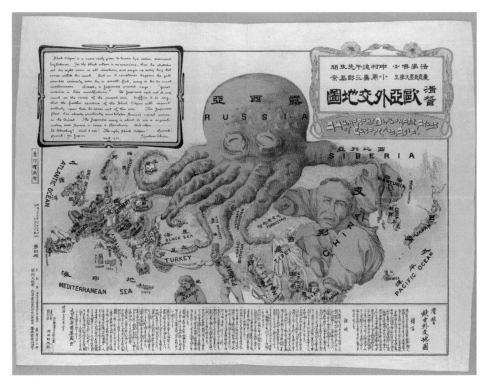

*Figure 32.6*   The octopus has been used as a suggestive symbol for over 150 years and the octopus head has been Germany, Russia, the I.G. Farben company, among others. Figure reproduced by permission of the P.J. Mode Collection of Persuasive Maps, Cornell University Library

colour symbolism has become more important as its use has become more widespread. Arthur Robinson, stated three reasons for the use of colour in cartography (Robinson, 1967):

1   It acts as a clarifying and simplifying element.
2   The use of colour seems to have remarkable effects on the subjective reaction of the map reader.
3   It has a marked effect on the perceptibility of the map.

Not listed by Robinson, but important are: colour attracts attention and it leads the reader's eye.

All of these are important in persuasive map design, but the subjective reaction and attraction are especially of interest. Colour associations and connotations are frequently used in this manner. Colours have been associated with a variety of characteristics, such as temperature, but also emotions. The most common associations are temperature, with reds, oranges and yellows perceived as warm; blues, purples and greens as cool. This association is so strong that colouring the North Pole red would invite confusion and probably outrage. There are also emotional connotations, such as red for danger and yellow for caution; we also feel blue or see red. We might be green with envy. Yellow is sometimes associated with cowardice, so to colour a country yellow on a political map can send a message about the people of the country. Using stereotypical skin colours for races on maps is generally considered offensive, but is used on some persuasive maps. Colour associations also vary by country or culture. For example, in the Western world, white is a happy colour that also symbolizes purity and is used for weddings; but in China, white is the colour of mourning and red is the colour of weddings.

## Typeface

Typefaces are said to have personalities. This is a subject much studied by infographics professionals. Some typefaces are considered to be 'honest' and believable; some are dainty and feminine; others do not invite trust. 'Scientific'-appearing type can lend an air of authority to a map that may actually be of dubious honesty. These associations have been utilized by persuasive cartographers. The size of the type also is a factor: large type shouts (especially when set in capitals); small type whispers. Dark type catches the eye and the orientation leads the eye – vertical type acts in much the same way as a vertical line and horizontal type leads the eye in a horizontal direction.

## Text

It is important to note that it is not just the appearance of the type, but also the choice of words used which is important. Text can be persuasive. Muehlenhaus (2012) has noted that propagandist and sensationalist maps have text styles that range from 'eye-grabbing' to 'inflammatory'. The text of the map, especially on the propagandist and sensationalist style maps, follows the rules of propaganda: use of stereotypes, substitution of names, selection, lying, repetition, assertion, pinpointing the enemy and appeal to authority (bandwagon technique) (Brown, 1963). Even on authoritative and understated maps, some of these techniques are used in the text, especially selection, assertion and lying. These maps by their very appearance send forth an announcement of truthfulness even if that announcement is false. The amount of text is also variable. While authoritative style maps may have very little text, sensationalist maps may bombard the eye with extraneous words. Colour can be combined with text and face to enhance the message. Thus, on a sensationalist map one might find a large, bright red title, with inflammatory words.

## Cartoon and humorous maps

A subset of the propagandistic and sensationalist map styles is cartoon maps and 'seriocomic' maps, such as those designed by Fred Rose in the late nineteenth century. Cartoon maps are often found as political cartoons on the op-ed pages of newspapers. They are blatantly sensationalist and make no attempt to appear anything but persuasive (Figure 32.7). Rose's maps are satirical and not usually discussed as 'real' maps, but are included in studies of cartographic curiosities; this is unfortunate since the persuasive impact is overshadowed by the sense that these are 'merely' humorous. One of the earliest uses of the octopus symbol is by Rose and shows Russia as the head of the octopus. Others turn countries into caricatures of the country represented, poking fun at the supposed characteristics of the inhabitants.

## Persuasive maps on the Internet

As might be expected, persuasive maps are now also found online. Muehlenhaus (2014) examines these and proposes three categories: *mashups*, *tailored dynamic* and *static*. *Mashups* utilize pre-fabricated tools that are used for Web map design, such as Google Maps or ESRI's ArcGIS

Figure 32.7    Serio-comic map by Fred W. Rose *Angling in Troubled Waters* (1899) London: G.W. Bacon. Comic maps are often not taken seriously, but they can present powerful messages. (Reproduced by permission of the P.J. Mode Collection of Persuasive Maps, Cornell University Library.)

Online (ibid.: 22), and they may allow panning and zooming. The *tailored dynamic* maps are designed 'from scratch' and have some degree of interactivity with the most elaborate allowing layers to be turned on and off, 'pop-up' windows, as well as panning and zooming. These maps are more time-consuming to make and utilize software requiring an HTML (hypertext mark-up language) browser plug-in (ibid.: 26). Dynamic maps may also be animated and use sound and even touch (instead of a pictorial symbol illustrating a bombing location, an animated symbol with sounds of explosion can be used, for example), but there has as yet been no study of how these techniques are used for persuasion. *Static maps* are produced by simple drawing techniques, GIS or graphic design software as much as paper maps. While these are simple, they can be distributed easily through e-mail attachments, social networking and photo-sharing sites and can also be printed easily (ibid.: 30). Online maps can generally be placed in the rhetorical categories discussed earlier, but the subject warrants more investigation.

## Accidental persuasion

While this chapter has focused on the deliberate creation of persuasive maps, one cannot always know the cartographer's intention. Persuasive maps can be, and are, created accidentally. Often this occurs when an inexperienced or untrained map-maker chooses a colour or typeface without recognizing the symbolism, or one who chooses an inappropriate projection for the map because it was available online. Inadvertent persuasion may be more prevalent with online maps (although this has not yet been tested) with the so-called 'democratization' of cartography. Neocartographers, who are generally described as non-professional cartographers using open source software for creating maps on the Web, may not be familiar with the principles of map design. They may use an inappropriate projection; they may use the default options for colour and symbols in the software or app just because it is easiest. Some maps are created by crowd-sourcing, which generates the data that are then mapped. Yet, these data are often biased in the sense of who generates them and the subject matter, and an inherent bias because the contributors come from a specific group who have access to the Web and know how to use it. Web use and knowledge are not yet universal.

## Conclusions

Persuasive maps have been made for centuries and with the proliferation of online maps will likely spread. Just as persuasion itself is not good or evil, persuasive maps may be used in either way. Tyner (1974) used the term 'persuasive map' rather than the more common term (at the time) 'propaganda map', because propaganda had a connotation of evil, of lies. This was later developed by Tyner (1982) into a continuum, i.e. from the unattainable ideal, scientific, objective map, to the propagandistic, untruthful map at the other end of the spectrum. All maps persuade to some extent and map-makers and map-users need to be more aware of this.

In the past 40 years, there has been increased study of persuasive maps and interest in them. In 2014, P.J. Mode, a collector of persuasive maps, donated his collection to the Cornell University Library. This collection is now online (see *http://persuasivemaps.library.cornell.edu*) and allows one to view persuasive maps from the 1600s through to the twenty-first century. The maps include the gamut of styles described above, including Rose's seriocomic maps.

## Further reading

Aberley, D. (Ed.) (1993) *Boundaries of Home, Mapping for Local Empowerment* Gabriola Island, BC: New Society Publishers.

Baynton-Williams, A. (2015) *The Curious Map Book* Chicago, IL: University of Chicago Press.

Boggs, S.W. (1947) "Cartohypnosis" *Scientific Monthly* 64 (June) pp.469–476.

Cosgrove, D. and della Dora, V. (2005) "Mapping Global War: Los Angeles, the Pacific and Charles Owens's Pictorial Cartography" *Annals of the Association of American Geographers* 95 (2) pp.373–390.

Francaviglia, R.V. (1995) *The Shape of Texas: Maps as Metaphors* College Station, TX: Texas A&M University Press.

Herb, G.H. (1989) "Persuasive Cartography in *Geopolitik* and National Socialism" *Political Geography Quarterly* 8 (3) pp.289–303.

Monmonier, M. (1977) *Maps, Distortion, and Meaning* Washington, DC: Association of American Geographers.

Monmonier, M. (1993) *Mapping It Out: Expository Cartography for the Humanities and Social Sciences* Chicago, IL: University of Chicago Press.

Monmonier, M. (1996) *How to Lie with Maps* (2nd ed.) Chicago, IL: The University of Chicago Press.

Muehlenhaus, I. (2013) "The Design and Composition of Persuasive Maps" *Cartography and Geographic Information Sciences* 40 (5) pp. 401–414.

Pickles, J. (1992) "Texts, Hermeneutics and Propaganda Maps" *Writing Worlds* New York: Routledge.

Ristow, W. (1957) "Journalistic Cartography" *Surveying and Mapping* 17 pp.369–390.

Tyner, J.A (2015) "Persuasive Cartography" in Monmonier, M. (Ed.) *History of Cartography Volume 6: Cartography in the Twentieth Century* (Part 2) pp.1087–1095.

Wright, J.K. (1942) "Map Makers Are Human" *Geographical Review* 32 (4) pp.527–544.

# References

Brown, J.A.C. (1963) *Techniques of Persuasion from Propaganda to Brainwashing* Harmondsworth, UK: Penguin Books.

Dahlberg, R.E. (1961) "Maps without Projections" *Journal of Geography* 60 (5) pp.213–218.

Goethe, J.W. von (1810) *Theory of Colours* (reprinted 1970) Cambridge, MA: MIT Press.

Markham, B.C. (1942) *West with the Night* Boston, MA: Houghton Mifflin Company.

Muehlenhaus, I. (2012) "If Looks Could Kill: The Impact of Different Rhetorical Styles on Persuasive Geocommunication" *The Cartographic Journal* 49 (4) pp.361–375.

Muehlenhaus, I. (2014) "Going Viral: The Look of Online Persuasive Maps" *Cartographica: The International Journal for Geographic Information and Geovisualization* 49 (1) pp.18–34.

Robinson, A.H. (1967) "Psychological Aspects of Color in Cartography" *International Yearbook of Cartography* 7 pp.50–59.

Tyner, J.A. (1974) *Persuasive Cartography: An Examination of the Map as a Subjective Tool of Communication* unpublished Ph.D. dissertation, University of California, Los Angeles

Tyner, J.A. (1982) "Persuasive Cartography" *Journal of Geography* 81 (4) pp.140–144.

# 33

# Schematic maps and the practice of regional geography

*Peter Thomas*

## Chorèmes, chorematic diagrams and schematic maps

With the application of scientific principles to map making since the eighteenth century, published maps have provided an increasingly precise representation of the Earth's surface. However the familiar map of the London underground system is just one reminder that a map does not necessarily need to be based on Euclidean principles in order to communicate effectively. Indeed, by eliminating extraneous information and focusing on a clear message, schematic maps can sometimes be more effective. One of the best-known examples is the French geographer Roger Brunet's representation of the core of the western European economy as a 'Blue Banana' extending from southern England to north Italy – a simple but provocative image and the focus of much public debate (Deneux, 2006).

The work of Brunet and his colleagues at the Groupement d'intérêt publique: Réseau d'étude des changements dans les localisations et les unités spatiales (hereafter RECLUS) research centre based at the Maison de la Géographie in Montpellier, France is discussed in detail by Andreas Reimer (2010) as part of a comprehensive review of what Reiner terms 'chorematic diagrams'. RECLUS was established under Brunet's direction in 1984 as a major research centre, employing up to 50 in-house researchers. RECLUS received French government funding until 1995, undertaking contract work commissioned by DATAR (the national spatial planning agency) and other similar bodies, while Brunet himself served as an expert adviser at government level. One of the central aims of RECLUS was to increase public awareness of geography, particularly through the use of innovative visual methods of presentation, and the journal *Mappemonde*, launched by RECLUS in 1986, became an important outlet in pursuing this objective (Clout, 1992). In the present context, the method which is of particular interest is the system of standard symbols or *chorèmes* devised by Brunet to represent the various elements of the spatial structure of a region. Individual *chorèmes*, representing features such as nodes, networks and hierarchies, could then be combined to construct a schematic map. A particularly striking example, which predates the establishment of RECLUS by several years, is a map representing France in the form of a hexagon (Brunet, 1973). The use of schematic maps appealed especially to regional planners as future development scenarios could be indicated in broad terms, without the need to specify detail. Typical examples include the maps prepared by planners to show development options in successive versions of the *Schéma Directeur* for the Paris region and maps prepared by planning agencies in Germany and the Netherlands, to which Reimer refers.

In proposing a taxonomy, Reimer suggests that, strictly speaking, the label 'schematic map' should only be applied to spatially distorted maps such as the map of the London underground system. However, he acknowledges that the term has also been employed more broadly, and in the present context it is used to embrace a wider range of maps, all of which rely on visual symbols to present a simplified representation of the world and to convey a clear message.

## Schematic maps and the changing nature of geography

Before referring in detail to specific examples of schematic maps, it is necessary to set Brunet's work in the wider context of methodological debates within the field of human geography. By the 1960s, the quantitative revolution in Anglo-American geography had led to a radical change of emphasis. Instead of the study of particular regions in isolation, innovative work now incorporated the use of spatial models and the formal testing of hypotheses (Haggett and Chorley, 1967). This change of paradigm embodied a law-seeking or nomothetic approach, at the expense of the descriptive and idiographic approach of traditional regional geography (Clout, 2003; Castree, 2009). Brunet himself was well aware of this methodological shift and his work reflected the influence of leading exponents of the 'new geography' such as Peter Haggett, whose seminal text *Locational analysis in human geography* (1965) had appeared in a French edition in 1973 (Haggett, 1973). Indeed, Brunet deliberately set out to transform French geography and he was a key figure in establishing the journal *L'Espace Géographique* in 1972 as an outlet for research reflecting the new paradigm. Brunet believed strongly in the existence of immutable spatial 'laws' which governed patterns of human interaction over space. But he also recognized that individual spaces differ from one another due to the particular mix of local factors in each case, and he argued that nomothetic and idiographic approaches could be integrated most effectively in specific geographical settings. Thus, for Brunet, the role of the *chorèmes* was to link general spatial theories with particular spatial contexts, thereby retaining a regional focus, albeit one which was expressed cartographically in a more contemporary format (Deneux, 2006; Reimer, 2010). Moreover, although the term *chorème* was coined by Brunet, it derives from chorography, a word of ancient Greek origin which refers to the art of regional description (Clout, 2003).

It is possible that Brunet's balanced approach may in part at least reflect the strong identification with regions in French society and the fundamental role of regional study in the evolution of French academic geography. The regional paradigm was firmly established in the late nineteenth century by Vidal de la Blache and his associates (Deneux, 2006; Castree, 2009) and many French geographers have retained a regional interest in the study of particular places, including areas beyond the borders of France (Clout, 2003). This continuing global perspective is exemplified by the publication of the new *Géographie Universelle* which appeared in ten volumes between 1990 and 1996 under Brunet's direction (Deneux, 2006). There is also an established French tradition involving the publication of popular atlases, exploring global themes and current issues in particular regional settings and in an accessible form, a typical example being Michel Barnier's geopolitical atlas of Europe (2008). Since 2005, the publishers of *Le Monde* have also produced a series of inexpensive atlases in A4 paperback format, each exploring a contemporary global theme such as migration or globalisation through the medium of 200 maps supported by a substantial written text. The *Le Monde* series also includes a wide-ranging *Atlas de la France et des français* (Denis and Giret, 2014). The publication of such an atlas can be seen as an affirmation of the continuing public interest in the geography of France and its component regions – a level of regional interest which has no real counterpart in the United Kingdom.

Meanwhile, Anglo-American human geography has adopted a non-regional paradigm in which highly specialized thematic studies have been favoured, largely to the exclusion of work which sets out to integrate a range of material in a specific regional context. Among British geographers, personal immersion in the geography and culture of a chosen country or major region had defined the academic identities of regional specialists such as W.R. Mead (Finland) and F.W. Carter (Eastern Europe). However, as Clout (2003) points out, by the 1980s area studies specialists had become an endangered species in British geography departments and, with the retirement of leading practitioners, there were no obvious successors with the inclination or necessary linguistic skills to sustain this level of expertise. Instead, a thematic Anglo-American human geography became dominant, and critics have claimed that the hegemonic status of this paradigm within the discipline has been maintained by a closed circle of leading academics and the commercial publishers of the most prestigious academic journals (Vandermotten and Kesteloot, 2012). Francophone geographers have been particularly critical of the dominant role of English-language journals within human geography, claiming that as a result, significant work written in other languages has scarcely been recognized. Moreover, this omission has consolidated the status of what is seen as a one-dimensional brand of human geography which fails to explore local interactions with global processes and offers limited scope for local or regional studies (Vandermotten, 2012; Lemarchand and Le Blanc, 2014). The recent relaunch of the Belgian journal *Belgeo*, in open access form, can be seen as a tangible response to such concerns (Vandermotten and Kesteloot, 2012).

Although the retreat from regional study in Anglo-American human geography was initiated by the 'new geography' of the 1960s, more recent trends have also reinforced this effect. In particular, the so-called 'cultural turn' in human geography has been accompanied by a blurring of disciplinary boundaries and this has led to increasing fragmentation within the discipline (Martin, 2001). This contrasts markedly with the regional phase in geography's development, when the focus on the region as a primary object of study ensured a greater degree of academic cohesion and integration. Globalization has also encouraged some observers to proclaim 'the end of geography', thereby prompting academic debate over the continued relevance of regional differences in an increasingly interconnected world (Graham, 1998; Castree, 2009).

There can be little doubt that the traditional regional geography of the 1950s was in need of revitalisation. In a comprehensive review, Clout (2003) refers to the 'ossification of a paradigm' and he is especially critical of encyclopaedic regional text books, in some cases extending to 800 pages, whose sheer bulk and overwhelming detail would deter even the most committed undergraduate student. But, at the same time, Clout insists that the academic pendulum has swung too far away from the study of specific places and that, as a consequence, British geography has 'rejected part of its birthright' (Clout, 2003: 267). Martin is also extremely critical of much recent work in social and economic geography which, in his view, largely ignores relevant issues in the real world while embracing instead 'the use of abstract, linguistically impenetrable constructs borrowed, usually uncritically, from the pantheon of postmodern French philosophy' (Martin, 2001: 196). A further consequence of this, to which Martin has drawn attention, has been the marked decline in the use of maps to accompany articles published in major geographical research journals. After surveying the content of several leading English-language journals, Martin reached the astonishing conclusion that 'most geography articles are completely devoid of maps' (Martin, 2000: 4). Kain and Delano-Smith (2003) also link the decline in the role of maps within geography to the parallel decline in the practice of regional geography. While, for Roger Brunet, a human geography without maps would have been utterly inconceivable.

## Schematic maps and the teaching of regional geography

Although fewer maps now appear in Anglo-American geography journals than was previously the case, maps have retained an important place in secondary education, especially in France, where students are asked to draw maps in responding to a set task in the History-Geography *baccalauréat* examination. The maps in question fall into two distinct categories, *croquis* and *schémas*, though both involve the use of visual symbols based on the *chorèmes* devised by Roger Brunet (Reimer, 2010; Jalta, 2012). The benefits of a similar approach, as an adjunct to the teaching and learning of regional geography at undergraduate level, will now be explored through a case study of Belgium, using a number of schematic maps. Figures 33.1 and 33.4 are highly diagrammatic *schémas* in which Belgium is represented as a simple polygon while, in contrast, Figures 33.2 and 33.3 are closer in style to the *croquis*, where symbols and annotations are added to a conventional pre-existing base map.

Despite its small size, Belgium provides a classic illustration of the geographies associated with successive phases of economic development, from the growth of trading cities such as Bruges and Antwerp in late mediaeval times through the nineteenth-century industrial revolution based on coal to the post-industrial age of the late twentieth century (Denis, 1992). Moreover, in the current phase, globalization and European integration have interacted in particular ways with a complex set of inherited geographies at the regional scale, thereby making Belgium an ideal setting in which to formulate a reinvigorated regional geography which gives due weight to the ways in which wider processes are responsive to local circumstances (Vandermotten, 2012).

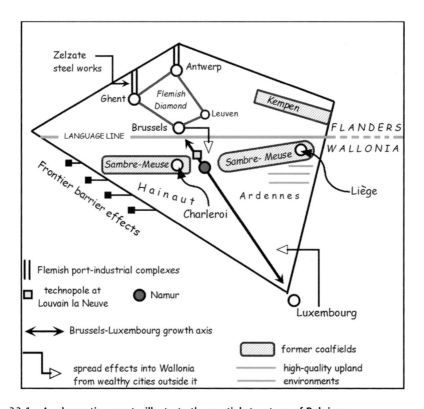

*Figure 33.1* A schematic map to illustrate the spatial structure of Belgium

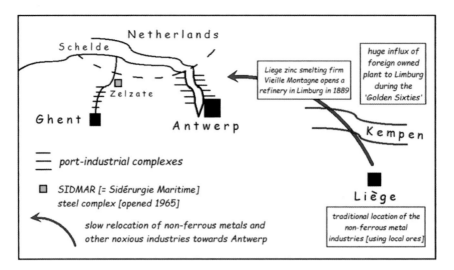

*Figure 33.2*  **A map to illustrate the northwards shift of Belgian industrial activity**

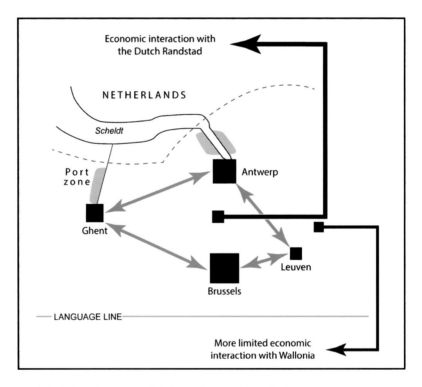

*Figure 33.3*  **The 'Flemish Diamond' in its wider spatial context**

But the study of Belgium also presents particular challenges. Belgium has frequently been mis-represented and even ridiculed in the popular media (see, for example, Le Sueur, 2014) and it remains a *terra incognita* to many non-Belgians. Moreover, Belgium's complex political structure and linguistic geography are further sources of possible confusion (Humes, 2014). For all these

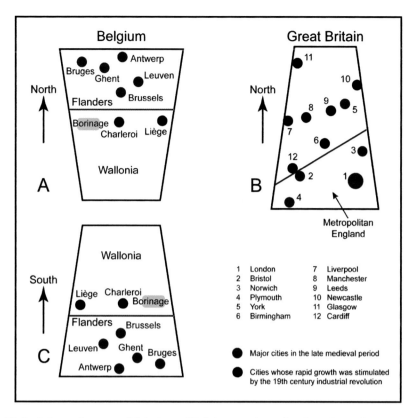

Figure 33.4 Schematic maps of Britain and Belgium to show the changing urban system

reasons, undergraduate students may be a little hesitant as they embark on the study of Belgium and they will certainly need support. It is in addressing this need that schematic maps such as those illustrated in this chapter have been devised and successfully used as an aid to learning.

For many proponents of traditional regional geography, regional studies represented the ultimate geographical synthesis. At its best, regional writing could show sensitivity and insight and, among British geographers, the work of W.R. Mead on Scandinavia (1958, 1981) was especially notable. Mead also appreciated the potential role of visual methods as an integral element of regional writing, and his use of diagrams to explore time-space relationships was imaginative and original (see, for example, Mead and Brown, 1962). But more typically there was a tendency for regional studies to be overwhelmed in a morass of local detail, often losing sight of the larger picture. In the case of Belgium, an extreme example is provided by F.J. Monkhouse's *A Regional Geography of Western Europe* (1959). The Belgian sections of this substantial text provide immense detail on the physical landscape and on regional economic development, but the text ignores the most fundamental regional division in Belgium – the language line which marks the boundary between the French speakers of Wallonia and the Dutch speakers of Flanders. However, the language line is clearly shown in the schematic map (Figure 33.1), and this also provides a concise summary of Belgium's spatial structure and major themes in its changing economic geography. Thus, the map is essentially a medium through which the essence of Belgium's economic geography has been summarized: indeed, the design of a successful schematic map requires a thorough understanding of the spatial patterns and processes which lie behind it, even though the end result may appear deceptively simple. For students, the use of a schematic map in which the outline of Belgium is represented as a diamond is extremely beneficial. While extraneous detail (such as sinuous political boundaries) has been eliminated, Belgium's essential spatial structure has been retained in a form which students can easily visualize, thereby providing a secure locational framework in which key issues (such as the political rhetoric of Belgium's north-south economic divide) can be explored more effectively (Thomas, 1990).

A further criticism of traditional regional geography is that regional studies often presented a rather static picture with an emphasis on the description of places in isolation from one another, rather than on the dynamic flows and connections between places as they interacted over time. In contrast, Figure 33.1 specifically highlights the importance of spatial relationships between places as dynamic factors in Belgium's changing economic geography. In particular, the spread effects generated by the cities of Brussels and, to a lesser degree, Luxembourg have contributed to the emergence of an economic growth axis extending in a south-easterly direction from the Belgian capital. Important elements which have boosted growth within the axis include the research activities based at the *technopôle* linked to the university of Louvain-la-Neuve and the important role of Namur as the capital of the semi-autonomous Walloon Region – both of which reflect the increasing rigidity of Belgium's linguistic division and the gradual adoption of a federal structure since the 1970s (Humes, 2014). The university town of Louvain-la-Neuve was developed in the 1970s as a new town, following Flemish objections to the use of French as a medium of instruction at the ancient university of Leuven, located in the Dutch-speaking territory, north of the language line. As well as positive spread effects, Figure 33.1 also highlights the negative impact of the Franco-Belgian border, along which positive flows into Belgium have been quite limited due to the depressed economy of many parts of the adjoining French region of Nord-Pas-de-Calais (Thomas, 2006).

Figures 33.2 and 33.3 also provide further examples of maps designed to highlight significant spatial and functional relationships within Belgium. Whereas the Sambre-Meuse coalfield had dominated the Belgian economy during the industrial revolution, from the late nineteenth century onwards, the economic impetus moved steadily northwards from Wallonia to Flanders (Thomas, 1990). This shift began as a response to the depletion of non-ferrous metal ores in Wallonia and was given a further impetus by the exploitation of the Kempen coalfield in the province of Limburg in the early twentieth century (Figure 33.2). As access to imported raw materials became increasingly important, a critical factor was the development of the port-industrial complexes at Ghent and Antwerp. As the seat of the European Commission, Brussels has also experienced rapid growth as a centre of tertiary sector activity and, together with the vibrant university city of Leuven, Brussels, Ghent and Antwerp define the limits of an economic core which now dominates the Belgian economy. The term 'Flemish Diamond' has been used by economic geographers as a convenient label to describe this zone (Figure 33.3), though Brussels itself is predominantly a French-speaking city, despite its location north of the language line, and administratively it is not part of the Flemish Region. Recent studies have highlighted the close functional connections between the cities of the Flemish Diamond, whereas interaction between the Flemish Diamond and Belgian cities south of the language line is much more limited (Aujean *et al.*, 2007). Figure 33.3 also illustrates the economic links between the Flemish Diamond and the Dutch Randstad (Dieleman and Faludi, 1998).

The ease with which schematic maps can be manipulated is a further advantage and this can be especially valuable in tracing successive phases of economic development and in developing comparative studies. In the late mediaeval period, the cities of Flanders ranked among the most prosperous in north-west Europe. However, during the industrial revolution, the Flemish economy stagnated while Wallonia was in the ascendant, before Flanders again established a dominant position in the twentieth century. In effect, the urban geography of late mediaeval Belgium has now re-emerged, leading to a stark economic divide between the prosperous cities of Flanders at one extreme and depressed former mining communities such as the Borinage on the former Sambre-Meuse coalfield at the other (Thomas, 1990; Boulanger and Lambert, 2001). This reversal of fortunes also has its parallel in Great Britain, where Halford Mackinder's 'Metropolitan England' (the dominant region in mediaeval times) is once again in the ascendant, after a relatively brief interlude during the nineteenth-century industrial revolution when the centre of gravity of the economy shifted northwards (Mackinder, 1902; Martin, 2004). Figure 33.4 identifies the cities typically associated with each phase of growth in Britain and in Belgium. Maps A and B adopt a familiar orientation, with north at the top, while Map C has been inverted and reflected, with south now towards the top of the Belgian map, thereby highlighting more vividly the close spatial similarities between the British and Belgian cases.

In addition to the use of schematic maps as a basis for simple comparative studies, the method can readily be extended to develop models which illustrate recurring spatial patterns and relationships (Deneux, 2006). This theme is elaborated by Brunet (1990) in the introductory volume to the *Géographie Universelle*, where the concept of the *chorotype* is introduced. The *chorotype* is essentially a spatial model made up a cluster of *chorèmes* which recur in a similar form in more than one location. A typical example is the estuary model, incorporating a major port located upstream (perhaps at the lowest bridging point on a river) with one or more out-ports located closer to the open sea. Brunet also proposes a tropical island *chorotype*, incorporating common elements shared by the Indian Ocean island of Réunion and French islands in the Caribbean (Figure 33.5). The *chorotype* illustrates similarities in economic development and shows how exposure to the prevailing winds has influenced internal patterns of economic activity within the islands to which the model relates.

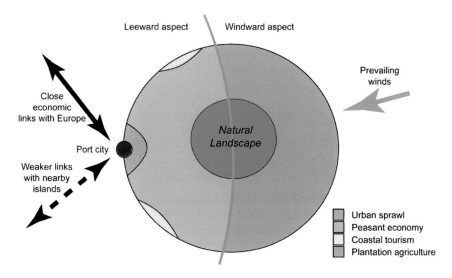

*Figure 33.5* A schematic model of a tropical island (adapted from Deneux 2006 after Brunet, 1990)

## Schematic maps and the world beyond the classroom

The previous examples have been used to illustrate a number of ways in which schematic maps can help to enhance students' understanding of the geography of particular places, primarily in the context of a regional geography course. This in itself is a worthwhile endeavour, though perhaps somewhat esoteric in its aims. But schematic maps also have the potential to fulfil a larger role by promoting a greater degree of engagement and interaction between the cartographer and the wider world to which the map relates, beyond the confines of the laboratory or classroom.

The distinction between the classroom and the wider world is of course a false one and schematic maps can play a useful role in linking the two. This is especially so where schematic maps are constructed by students, rather than by their tutor, thereby transforming the role of the student from passive consumer to active participant in the learning process. As Slack (1997) has demonstrated, the mere process of transcription of written text into sketch map form can in itself enhance students' understanding and increase their engagement with the content of regional geography courses. Whereas Slack describes the use of hand-drawn sketch maps, the same principle is equally applicable to maps drawn by students using standard computer packages, as illustrated in Figure 33.6. In this example, students were provided with an outline base map and then developed this by adding visual symbols and annotations as a synthesis of their understanding of the geography of Italy, reflecting to varying degrees their personal engagement with the published academic literature. Although more polished in this form, Figure 33.6 embodies elements of the *croquis* – maps which French students produce by adding symbols to a pre-existing base map. Alternatively, spatial patterns and dynamic processes within a chosen territory can be illustrated by means of a more abstract, diagrammatic representation or *schéma*. An extremely effective example is Dorel's (1991) schematic map of the western USA, which Deneux (2006: 241) warmly commends as an exemplar of a contemporary approach to regional study.

The connection between the classroom and the wider world is even more explicit where schematic maps are used to explore regional planning scenarios. For example, students might be asked to prepare and justify a schematic map of south-east England, incorporating the proposed location of a new London airport, major new housing developments or the likely impacts of rising sea levels due to climate change. An important factor in exercises of this kind is the ease with which schematic maps can be manipulated, thereby making it possible to 'play around with space' in a relatively uninhibited way, as professional planners are able to do when preparing similar maps. Moreover, such exercises demonstrate the potential role of the schematic map as an adjunct to a more relevant regional geography – a geography which focuses on real issues with which students can become actively engaged by exploring alternative futures in a specific regional setting. This approach is also consistent with Martin's recommendation (2001: 202) that human geographers 'need to develop a sense of intellectual cohesion around key social issues and problems' and with Brunet's focus on planning and regional development as central concerns of a socially relevant human geography (Deneux, 2006).

In contrast, a number of commentators have drawn attention to the relative failure of British academic geographers to influence public policy or to engage effectively in the public arena. Despite acknowledging the significant role played by certain individuals, Johnston (2003: 78) concludes that 'the discipline's overall profile has been relatively low', and Martin (2001: 191) asks why geographers are 'playing second fiddle to other academics and even journalists' in debates over public policy. Moreover, for Martin, this is a matter of real concern as many public policies have uneven spatial impacts at the local level which geographers are well-equipped to evaluate. Johnston (2003) identifies a number of possible reasons for geography's low public

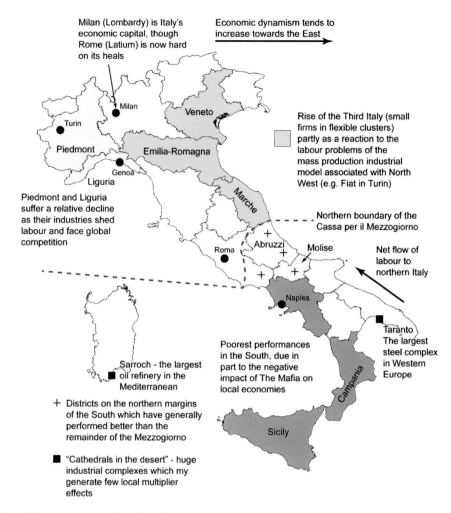

*Figure 33.6* **A map to show key themes in Italian economic geography**

profile, including its failure to establish a popular media presence to match that of other academic subjects, notably history. But it is possible that the retreat from the use of maps among Anglo-American human geographers, to which Martin (2000) has drawn attention, also provides part of the explanation. Maps represent the geographer's most effective tool as a means of communication with the public at large, and Lemarchand and Le Blanc (2014) suggest that maps could play an important role in the wider diffusion of the results of geographical research, while at the same time maintaining academic integrity. De Blij (2012) also highlights the need for a more adequate geographical perspective to inform public understanding of major global issues and emphasizes the fundamental role of the map.

## Conclusion

This chapter has attempted to show that schematic maps have an important educational role to play as an adjunct to a more relevant and academically engaging regional geography.

But, although the educational role of the *chorèmes* has been recognized, there has also been significant academic criticism of the use of schematic maps as an appropriate means through which to present geographical material in the wider public domain (Deneux, 2006; Reimer, 2010). Such maps of course embody a simplified representation of reality, and bold images such as Roger Brunet's representation of the Blue Banana were criticised by French academics for their alleged lack of scientific rigour (see, for example, Giblin-Delvallet, 1995). Critical reactions in France probably also reflected a degree of indignation as Brunet's Blue Banana model had relegated most of France to a semi-peripheral location relative to the core of the western European economy. The criticism of Brunet's maps also reflected a wider philosophical debate within French geography between the rival academic factions aligned either with Brunet or with the political geographer Yves Lacoste (Reimer, 2010). As Clout (1998) has shown, French academic geography has been characterised by marked ideological divisions, focusing in this instance on a fundamental difference of opinion between Brunet and Lacoste over the role of spatial laws within geography. Criticism of Brunet culminated in a 1995 issue of the journal *Hérodote* – the journal founded by Lacoste and expressing his personal conception of geography (Deneux, 2006).

In fact, Reimer (2010) suggests that the debate between Brunet and Lacoste was not primarily about cartography. Lacoste's concern was with the underlying analysis undertaken by Brunet as much as with the mode of presentation, and he was not fundamentally opposed to the use of maps. Moreover, despite the concerns raised over simplification, it can be argued that schematic maps are in fact especially well suited to the task of promoting geography in the public sphere by linking the geographer's traditional focus on the study of place to the wider context of policy debate. The ultimate proof of this is that, despite its many critics, Brunet's Blue Banana became perhaps the most talked about map in France and played a key role in increasing the visibility of geography as a discipline. While not overstating the role of schematic maps, it can certainly be argued that such maps have a number of potential applications, not only in sustaining a more engaging approach to regional geography but also in enabling geography to assert its identity more effectively as a place-focused discipline with much to contribute in the arena of public policy debate.

## References

Aujean, L., Castiau, E., Roelandts, M. and Vandermotten, C. (2007) "Le positionnement des villes belges dans le réseau global des services avancés" *Belgeo* 2007 (1) Available at: *http://belgeo.revues.org/11621* (Accessed: 26 February 2015).

Barnier, M. (2008) *L'Europe: cartes sur table – atlas* Paris: Acropole.

Boulanger, P.-M. and Lambert, A. (2001) "La dynamique d'un développement non durable: le Borinage de 1750 à 1990" *Espace, Populations, Sociétés* 2001 (3) pp.311–324.

Brunet, R. (1973) "Structure et dynamisme de l'espace français: schéma d'un système" *L'Espace Géographique* 2 (4) pp.249–254.

Brunet, R. (1990) "Le déchiffrement du monde" in Brunet, R. and Dollfus, O. (Eds) *Mondes nouveaux* (Géographie Universelle, Tome I) Paris and Montpellier: Hachette-Reclus, pp.10–273.

Castree, N. (2009) "Place: Connections and Boundaries in an Interdependent World" in Clifford, N.J., Holloway, S.L., Rice, S.P. and Valentine, G. (Eds) *Key Concepts in Geography* (2nd ed.) London: Sage, pp.153–172.

Clout, H. (1992) "Vive la géographie! Vive la géographie française!" *Progress in Human Geography* 16 (3) pp.423–428.

Clout, H. (1998) "L'état de la géographie en France" *Progress in Human Geography* 22 (2) pp.299–304.

Clout, H. (2003) "Place Description, Regional Geography and Area Studies: The Chorographic Inheritance" in Johnston, R.J. and Williams, M. (Eds) *A Century of British Geography* Oxford, UK: Oxford University Press, pp.247–274.

de Blij, H.J. (2012) *Why Geography Matters: More Than Ever* (2nd ed.) Oxford, UK: Oxford University Press.

Deneux, J.-F. (2006) *Histoire de la pensée géographique* Paris: Belin.

Denis, J. (Ed.) (1992) *Géographie de la Belgique* Brussels: Crédit Communal.

Denis, J.-P. and Giret, V. (2014) *L'atlas de la France et des français* Paris: Le Monde, hors-série.

Dieleman, D. and Faludi, A. (1998) "Randstad, Rhine–Ruhr and Flemish diamond as one polynucleated macro-region?" *Tijdschrift voor Economische en Sociale Geografie* 89 (3) pp.320–327.

Dorel, G. (1991) *Etats-Unis: la nouvelle donne régionale* Paris: La Documentation Française.

Giblin-Delvallet, B. (1995) " Les effets de discours du grand chorémateur et leurs conséquences politiques" *Hérodote* 76 pp.22–38.

Graham, S. (1998) "The End of Geography or the Explosion of Place? Conceptualizing Space, Place and Information Technology" *Progress in Human Geography* 22 (2) pp.165–185.

Haggett, P. (1965) *Locational Analysis in Human Geography* London: Arnold.

Haggett, P. (1973) *L'analyse spatiale en géographie humaine* Paris: Armand Colin.

Haggett, P. and Chorley, R.J. (1967) "Models, paradigms and the new geography" in Chorley, R.J. and Haggett, P. (Eds) *Models in geography* London: Methuen, pp.19–41.

Humes, S. (2014) *Belgium: Long United, Long Divided* London: Hurst.

Jalta, J. (2012) *Géographie Bac: croquis et schémas* Paris: Magnard.

Johnston, R. J. (2003) "The institutionalisation of geography as an academic discipline" in Johnston, R.J. and Williams, M. (Eds) *A Century of British Geography* Oxford, UK: Oxford University Press, pp.45–90.

Kain, R. and Delano-Smith, C. (2003) "Geography Displayed: Maps and Mapping" in Johnston, R.J. and Williams, M. (Eds) *A Century of British Geography* Oxford, UK: Oxford University Press, pp.371–427.

Le Sueur, A. (2014) *Bottoms up in Belgium: Seeking the High Points of the Low Lands* Chichester, UK: Summersdale.

Lemarchand, N. and Le Blanc, A. (2014) "Les langues de la diffusion scientifique: une question pour les géographes et les géographies" *EchoGéo* 29 Available at: *http://echogeo.revues.org/13941* (Accessed: 26 February 2015).

Mackinder, H.J. (1902) *Britain and the British Seas* New York: Appleton.

Martin, R.L. (2000) "Editorial: In Memory of Maps" *Transactions of the Institute of British Geographers* 25 (1) pp.3–5.

Martin, R.L. (2001) "Geography and Public Policy: The Case of the Missing Agenda" *Progress in Human Geography* 25 (2) pp.189–210.

Martin, R.L. (2004) "The Contemporary Debate Over the North-South Divide: Images and Realities of Regional Inequality in Late-Twentieth-Century Britain" in Baker, A.R.H. and Billinge, M. (Eds) *Geographies of England: The North-South Divide, Material and Imagined* Cambridge, UK: Cambridge University Press, pp.15–43.

Mead, W.R. (1958) *An Economic Geography of the Scandinavian States and Finland* London: University of London Press.

Mead, W.R. (1981) *An Historical Geography of Scandinavia* London: Academic Press.

Mead, W.R. and Brown, E.H. (1962) *The United States and Canada: A Regional Geography* London: Hutchinson.

Monkhouse, F.J. (1959) *A Regional Geography of Western Europe* London: Longmans.

Reimer, A.W. (2010) "Understanding Chorematic Diagrams: Towards a Taxonomy" *The Cartographic Journal* 47 (4) pp.330–350.

Slack, J. A. (1997) "Sketch Map Transcription of Standard Texts as a Means to Teaching Regional Geography" *Geography* 82 (2) pp.127–138.

Thomas, P. (1990) "Belgium's North-South Divide and the Walloon Regional Problem" *Geography* 75 (1) pp.36–50.

Thomas, P. (2006) "Images and Economic Development in the Cross-Channel Euroregion" *Geography* 91 (1) pp.13–22.

Vandermotten, C. (2012) "Cotation des revues de géographie, impérialisme scientifique anglo-saxon et culture de l'excellence marchandisée" *Belgeo* 2012 (1–2) Available at: *http://belgeo.revues.org/7131* (Accessed: 26 February 2015).

Vandermotten, C. and Kesteloot, C. (2012) "Editorial: *Belgeo* and the Four Crises of Geography" *Belgeo* 2012 (1-2) Available at: *http://belgeo.revues.org/6277* (Accessed: 26 February 2015).

# 34

# Cartography and the news

*Peter Vujakovic*

## Introduction: news from the frontline

'On 29 July 1693, during the Nine Years War (1688–1697), a particularly bloody battle was fought in the field of Flanders' (Luijk, 2008: 211). The broadsheet maps of the battle at Neerwinden that were swiftly produced by the victorious French as well as the defeated allies, the Dutch and English, are regarded by Luijk as both battle and propaganda maps, but also as 'news maps'. He notes, with regard to one of the broadsheets, 'Without a doubt, the map by Uytwert [representing the allies' viewpoint] was a news map, designed to cater for the public's demand for cartographical journalism' (2008: 213). Luijk provides a detailed analysis of the striking use of maps by both sides; first as a means to inform the public and to put their 'spin' on the result of the battle through the power of the image, and second, as a set of complex technical, economic and 'political' relationships between the publishers, engravers and printers, the state, and informants on the ground. In this he sums up much of the complexity that remains part of any understanding of news mapping.

Another example of this genre, representing an event over a century before Neerwinden, has recently been celebrated in an exhibition in the Maltese capital Valetta. The maps (four 'states' of a single map) provided information on the unfolding 'Great Siege' of Malta in 1565 by the Ottomans and its defence by the Knights of St John. Bernadine Scicluna (2016), the exhibition's curator, boldly states 'these maps were the most effective means of reporting the latest episodes of the siege on Malta – they were in effect equivalent to today's flash or breaking news coverage'.

In essence, very little has changed. In the complex and global political geography that emerged from the early modern period, with the advent of the modern state and overseas empires, and the concomitant and substantial innovations in weaponry and warfare, maps were to become and to remain a significant format through which publics are informed about conflict and inter-state competition. Numerous studies of contemporary news mapping have established this, both in terms of numbers of maps and of their complexity. Even print, as a medium, remains important, although maps in various broadcast and electronic formats are now also important. This chapter explores journalistic mapping and its significance in our current period of geopolitical and environmental uncertainty.

## News maps and the geography of news

Denis Wood (1992) coined the phrase, 'to live map-immersed in the world'.[1] By 'immersed', Wood means that individuals, particularly in Western industrialized societies, are so surrounded by and so frequently use maps that these become indistinct from other taken-for-granted consumer products. Maps are no longer special, but are regarded as 'apparently repro-duced . . . *without effort*' (Wood, 1992: 34). During a period characterized by mass production and consumption, they have become tools which are encountered, used or produced in every walk of life (Cosgrove, 2005). Mark Monmonier (1989), in *Maps with the News*, goes so far as to state that 'the news media are society's most significant cartographic gatekeeper and its most influential geographic educator'. This point is apt, given most individuals' reliance on news providers for their understanding of international issues. It is a reminder that our *personal geographies* are shaped by what we glean from the news; although this must be understood as a complex intertextual process in which maps, images and text continuously re-build our mental maps. The role of news maps must always be understood within the wider context of *news geography* (Wilke et al., 2012), but also within the technical and production constraints of the media in which maps are produced and disseminated. Maps are particularly important when other forms of visual information are unattainable or are suppressed, for example during wars, or social or environmental catastrophes. The part that maps play in news media has been long recognized, for example with Walter Ristow coining the term 'journalistic cartography' in 1939 (also see Ristow, 1957).

Monmonier (1986; 1989) details the historical growth in map use by the 'press' in the UK and North America. As well as their use in the press, maps have become embedded within a widening array of methods of reproduction, transmission and consumption. As communication technologies have evolved to allow more effective use of graphic images, maps have increasingly been used to provide spatial information, promote place knowledge, and to explain geographi-cal and environmental processes. Advances in technology have also led to the globalization of the news media, the compression of time and space through almost instantaneous transfers of information, effectively bringing places much closer together in what Marshall McLuhan, in the 1960s, famously dubbed the 'global village' (Robins, 1995).

The transition to web-based delivery has provided opportunities for interactive graphics, in which the reader is encouraged to actively engage with the news story. This may be little more than providing further detail; a good example is a map of the proposed high-speed rail link between London and Birmingham published by *The Telegraph* (UK)[2] where a series of 'click and reveal' icons along the length of the proposed route allow the reader to reveal detail on the reasons for proposed tunnels or diversions (e.g. to protect heritage sites) or photographs of key sites. More complex examples include *The Economist*'s interactive mapping of territorial disputes between India, China and Pakistan.[3]

Comprehensive studies of news media cartography have been relatively limited in number and the issue of the socio-political role of maps (e.g. in geopolitical discourse) has been largely neglected until recently. Nevertheless, several key strands of inquiry can be traced. First, a con-cern with technical, organizational and production issues; these are not explored in depth in this chapter as they have been examined in detail for the era of print and television by others (see, for example, Monmonier, 1986, 1989; Ferris, 1993; Perkins and Parry, 1996). Recent develop-ments in internet mapping and related formats are largely generic and explored elsewhere (see Chapter 27). Second, empirical surveys of media map design 'quality', frequency of use and thematic content; this includes some critical studies of the fitness for purpose of cartographic products. And, finally, the more recent cultural turn in the examination of news maps, in terms of their impact on public understanding and of discourse relating to important themes, for instance, geopolitics and the environment.

The general lack of research into the last of these areas, the socio-political role of media maps, may be due to several related factors. First, a lingering 'scientism' within cartography itself, in which map-making is still regarded by many of its practitioners as an objective, scientific enterprise disassociated from ideological concerns (Krygier, 1995). As Rundstrom (1993) pointed out, despite a growth of interest in cultural and social cartography, studies of the human element in cartography during the twentieth century were largely dominated by issues in experimental psychology related to the construction of 'better' maps. Indeed, critical discussion of journalistic cartography has often been aimed at deflecting criticism from cartography by noting that the designers of news maps are often 'artists untrained in cartographic principles' (Monmonier, 1989: 14). Examples of poor quality news maps, supposedly by graphic artists, led cartographic professionals to be dismissive of the quality and objectivity of maps drawn by people from outside of the profession. They saw such work as either unworthy of serious study, or caricatured it as 'propagandist' in comparison to maps produced by the 'impartial' cartographer (see, for example, Ager, 1977). Balchin (1985) epitomizes this view in the conclusion to his Media Map Watch initiative; he states: 'One suspects that only too often it is a graphic artist rather than a trained cartographer who is responsible' (1985: 343). Yet these criticisms ignore the innovations that artists and graphic designers such as Richard Edes Harrison and Charles H. Owens brought to media mapping during the mid twentieth century (Schulten, 1998; Cosgrove and della Dora, 2005; Barney, 2015).

The myth of the map as an objective representation of an external reality has been eroded by the critical turn in cartography (see Chapter 6). The debate concerning the 'cultural rules' by which cartography and cartographers operate and through which maps influence or reinforce particular 'world-views', has gathered momentum, led in particular by Harley's (1989a, 1990, 1992) adoption of a 'deconstructionist' approach to the history of cartography. He did much to increase awareness of the socio-cultural impact of maps, despite some justifiable criticisms (see, for example, Belyea, 1992; Rundstrom, 1991), but Harley's concern with the socio-political dimensions of maps provided a useful starting point for the examination of journalistic mapping.

Perhaps Harley's most important contributions to uncovering the socio-political implications of cartographic practices as they relate to the news are his concerns with the knowledge as power and with 'intertextuality'. Notwithstanding Belyea's (1992) criticism of Harley's understanding of Foucault concerning 'power/knowledge', Harley's discussion of the 'hidden power' of maps is valuable.[4] Harley refers to two forms of power. First, the 'external power' exerted *on* and *through* cartography according to the agenda of the map-maker or patron (monarch, state institution, news editor). Second, the 'internal power' of the map to convey an image of order through selection, abstraction and generalization, and hence normalize the 'world-view' of the authoring agency; see, for example, Vujakovic's (2002a) study of news maps as 'metaphor' for 'precision and accuracy' in support of NATO's peace-making action in Yugoslavia in 1999. Linked to this concern with power in and through 'texts' is a recognition that as 'the objects of enquiry . . . maps . . . must be approached intertexually; texts from other conceptual realms cross-cut, transform, and, in turn, are transformed by the texts in question' (Barnes and Duncan, 1992: 13). As Harley (1989b) noted, 'A textual approach alerts us to the shadows of other texts in the one we are reading' (1989b: 85). An intertextual approach is fundamental to any understanding of the ways in which maps operate as *part of* the knowledge circulated by news media and their relationship with other institutions (e.g. governments and independent 'think-tanks').

Rundstrom (1991), while agreeing with much that Harley had to say concerning the cultural role of maps, is critical of the tendency of so-called 'postmodernist' approaches to place too much emphasis on the discussion of maps as 'texts', while ignoring the wider technical *and* social processes within which they are embedded. This led Rundstrom to argue for a 'process cartography', consisting of two concentric ideas; the first, which situates the 'map artefact' within the realm of technical production, and a second which 'places the entire map-making process within the context of intracultural and intercultural dialogues over a much longer span of time' (1991: 6). Process cartography is posited on 'the idea that maps-as-artefacts are inseparable from mapping-as-process, and that the mapping process in turn, is made necessary and meaningful only by the broader context of the cultural processes within which it is located' (Rundstrom, 1993: 21). This approach acknowledges the importance of understanding the technological, organizational and design issues involved in production, but stresses also the need to locate the maps and their makers within their cultural *and* historical contexts. The link with process is especially important for news mapping, with its tight deadlines, reliance on a range of sources, complex organizational structures and editorial hierarchies, and fast changing technologies of production and publication/broadcasting. In media constrained by broadcast time or space, the struggle for resource between 'word people' and 'picture people' will be acute (Monmonier, 1989) and the 'opportunity costs' of one form of information over another will be critical (Ferris, 1993).

## News media maps: production and design issues

Some US authors (e.g. Gilmartin, 1985; Monmonier, 1986, 1989) have made a significant contribution to reviving interest in journalistic mapping, while in the UK, Balchin's UK 'Media Map Watch' was an early but flawed survey of news map design (Balchin, 1985, 1988). Undertaken in November 1984, it was essentially concerned with issues of technical accuracy, 'prompted by serious defects of map representation in certain TV programmes' (Balchin, 1985: 339). Unfortunately, the maps were generally taken out of context, ignoring the possibility that factors ascribed to inaccuracies (e.g. poor technique, or lack of cartographic understanding) may in fact have been attempts to manipulate 'reality' for a particular purpose. Harley (1992) went so far as to label those involved in the survey as 'cartographic vigilantes' defending an outmoded 'ethic of accuracy'. Perkins and Parry (1996) were also critical, believing that it 'epitomized [an] orthodox and prescriptive approach' (1996: 330), which disregarded the technical constraints of the media in which the maps were published.

In the UK, a significant contribution to the discussion of production and design issues (from a practitioner's point of view) was provided by Ferris (1993), based on her experience at a UK elite newspaper, *The Independent*. Ferris was particularly concerned with the constraints of producing maps for print; issues such as deadlines, print (image) quality and the opportunity cost of image space versus word space. Similarly, Daulby's (1988) comparison of constraints on information graphics in print and television, based on his BBC experience, provided useful insight into the practicalities of broadcast media map production at that time.

Perkins and Parry (1996) provided an excellent overview of news media mapping in *Mapping the UK: Maps and Spatial Data for the 21st Century*. They focused on two major issues, the constraints imposed by 'the press' and television, and design criteria. Their discussion of map production and design drew heavily on their own one-month-long study of UK news providers. Perkins and Parry's approach to the evaluation of map design concentrated on a systematic survey of its 'functional role'. They classified maps, for example, into 'locator', 'route', 'distribution' and 'explanatory maps'. Their findings (based on a survey of UK newspapers and television news during June 1995) showed that by far the most common form of map was the simple locator (67 per cent), with the more complex maps accounting for less than a third with no great variation across the range of elite and popular 'press'. Their typology was adapted by

Vujakovic (1999) for his three major studies of the elite press in the UK (the first six months of 1999, 2009, 2014). His findings showed a similar bias towards simple locator maps, but also noted the importance of more complex maps to supporting and developing complex geopolitical narratives (Vujakovic, 2002a, 2014).

Discussion of news maps has also taken place within the growing field of 'information graphics'. Examples include Holmes (1991), Sullivan (1996) and Case (1996). Holmes, then executive art director at *Time* magazine, soundly defended the use of what he called 'pictorial maps': 'This is not trivialization, it is teaching through demonstration, through metaphor, through symbolism' (Holmes, 1991: 130), an echo of Neurath's 'picture education'. More recent examples include Rendgen and Wiedermann (2012), although some of these are often uncritical of the cartographic design elements. Rogers (2012) provides a useful discussion of the recent growth of 'data journalism' in relation to news infographics. He makes a cogent case for the importance of graphics in a world in which evidence-based reporting using large numerical datasets has become ever more important, but points out that reporters who are not afraid of the numbers underlying important issues from finance to pandemics also need to ensure that they are communicated effectively in graphic form. Several essays in Antoniou *et al.* (2015) provide insight into map graphics in several news providers.

Several authors have extended the discussion of production and design to examine the impact on the audience in more detail. Monmonier's (1989) authoritative *Maps with the News*, is largely concerned with the impact of technical and design innovations on news mapping, but also explores the potential impact of effective or poor design on the public understanding of complex issues, making the point that 'words alone are both more awkward and less dramatic than maps for describing spatial relationships' (1989: 4). It is perhaps not surprising that he chose a world projection as his example in this context – a news map showing Canada's northern defence strategy. The effective use of map projections at world and world-regional scales is a classic example of the problems that can occur in presenting effective geographical information to the public. Several authors have focused on map projections as exemplars of the problems that occur where a sound cartographic understanding is lacking; see, for example, Gilmartin (1985) and Vujakovic (2002b). Gilmartin examined news maps accompanying stories of the shooting down by Soviet aircraft of Korean Air Lines (KAL) flight 007 in 1983. Of the twenty maps evaluated, many failed to appreciate the Great Circle routes taken by aircraft and simply showed the origin and destination linked by a straight line; as Gilmartin (1985) notes 'the plane's actual track is even more important to the story because it was its incursion into Soviet airspace which led to its destruction' (1985: 11). Mijksenaar (1996) notes similar problems with journalistic maps of the 'Gulf War' of 1990–1, believing this to be because graphic designers rarely have all the skills required to produce accurate representations in this specialist field.

As well as his general surveys of the use and abuse of world map projections in the news media, Vujakovic (1999, 2002b, 2010) has shown how specific issues, such as the mapping of missile threats from states such as China and North Korea, have been poorly communicated due to inappropriate use of world map projections. He noted that in general terms the UK news media (in 1999 and 2009) tended to adopt standard rectangular world map projections for most purposes. This is especially problematic when the story involves *area information* (e.g. the loss of habitat, extent of land degradation, climate change issues) and an inappropriate non-equal area map has been adopted; this can very often minimize the perceived 'threat' to low latitudes and exaggerate these in high latitudes. The use of standard rectangular maps to show circular missile ranges is particularly problematic; on these maps, it *appears* that the most direct route from North Korea to the US mainland, for instance, would track or parallel the line of latitude 40°N, while in fact, the direct route would follow a Great Circle across the

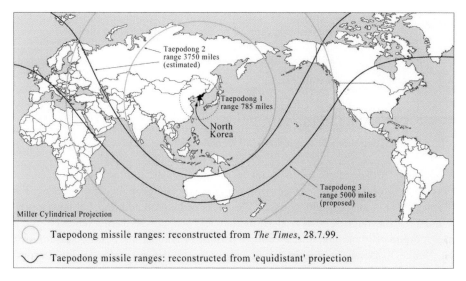

*Figure 34.1* **Taepodong missile ranges as plotted by** *The Times* **and reconstructed on a non-equidistant projection**

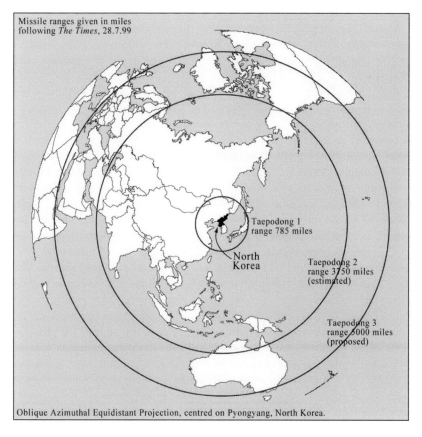

*Figure 34.2* **Taepodong missile ranges plotted on an equidistant projection**

Aleutian Basin (between 50° and 60°N). These missile ranges can *only* be shown as concentric circles if based on an *equidistant projection* centred on the appropriate launch site (see Figures 34.1 and 34.2). By 2009 there had been a slight improvement, with some news providers adopting equidistant projections, but many were still using poor maps despite appropriate models available from governmental and other sources.

## Thematic coverage: cartography and conflict

'Like bombers and submarines, maps are indispensable instruments of war' (Wright, 1942: 527).

Comprehensive studies of the use of maps in the news media have shown some consistent patterns in terms of thematic coverage, with conflict, war and geopolitical issues being highly represented. Monmonier's (1986) study of *The Times* of London and four US newspapers for 1940 and 1980 showed that news maps of military and geopolitical themes tended 'to be high both proportionately and numerically during periods of conflict and less significant at other times' (1986: 53). Despite this statement, his tallies for war, defence and geopolitical issues still remained the highest scoring categories in his study.

Perkins and Parry (1996) identified 'Military conflict, defence, geopolitics, threats and riots' as representing 27.2 per cent of all news maps (survey of UK press and broadcast news providers, June 1995). Vujakovic found similar tallies, with the highest scoring category for his 1999 six-month survey of the UK elite press being 'Military conflict/war, defence issues, territorial/ resource disputes', representing 28.4 per cent of all the published maps. This figure was inflated by the large number of maps devoted to the Kosovo crisis (Vujakovic, 1999, 2002b). Related surveys in 2009 and 2014 (Vujakovic, 2010, 2014) yielded figures of 29.6 per cent and 26.9 per cent respectively for the combined sub-themes 'Internal political conflicts' and 'Military conflict/war, defence issues, territorial/resource disputes', which was closer to Perkins and Parry's result, and may represent more 'normal' conditions. The results of both surveys clearly confirm the importance of maps in illustrating stories concerning geopolitics, resource/territorial disputes and conflict at both national and international scales. Table 34.1 provides a summary of all news themes surveyed in 1999, 2009 and 2014.

Studies by Gauthier (1997) for Canada and Tremols (1997) for Spain showed similarly high numbers of maps related to conflict, while Kowalski and Paslawski's (1997) study of the Polish press indicated high levels for some national newspapers, but not a uniform occurrence.

This becomes important when it is recognized that these issues are also associated with more complex forms of cartography. War, civil conflicts and geopolitical issues tend to involve both complex issues and dynamic circumstances. Holmes (1991) and Mijkenaar (1996) provide numerous examples of 'dynamic' and 'pictorial' maps associated with the Falklands War (1982) and the first 'Gulf War' (1990–1) respectively. Mijkenaar notes that the plethora of maps produced during such crises is often due to the lack of photographs and television images, which creates an immediate need for cartographic materials. He also notes that the Gulf War coincided with developments in computer and drawing software that allowed the press to develop 'infographics' teams to provide material in an affordable and timely manner.

Following the critical turn in cartography, a number of authors have undertaken more nuanced analyses of the role of news maps as a means of engaging audiences in what are often geographically distant and politically 'Byzantine' issues. As noted above, maps become increasingly important where other images are harder to obtain, during war, or tell only a partial story, for example in terms of complex geopolitical issues. Vujakovic (1999) has shown how the changing nature of the debate concerning European Union and NATO expansion

*Table 34.1* News map themes – UK elite press

| Themes | | Sub-themes | 1999a | 2009 | 2014 | 1999b |
|---|---|---|---|---|---|---|
| A Politics: internal | 1 | Government, legislation, electoral, parties, non-violent protest/strikes | **67** **6.9%** | **95** **7.6%** | **90** **8.4%** | 84 7.4% |
| | 2 | Riots, terrorism, civil conflict/war, secession movements, coups | **130** **13.4%** | **229** **18.3%** | **184** **17.2%** | 144 12.8% |
| B Politics: international | 3 | International relations, negotiations, agreements (non-trade) | **17** **1.8%** | **55** **4.4%** | **31** **2.9%** | 25 2.2% |
| | 4 | Military conflict/war, defence issues, territorial/resource disputes | **237** **24.5%** | **141** **11.3%** | **104** **9.7%** | 320 28.4% |
| C Disasters/ accidents | 5 | Large-scale disasters (earthquakes, floods, etc.), epidemics | **11** **1.1%** | **48** **3.8%** | **49** **4.6%** | 15 1.3% |
| | 6 | Accidents (transport, etc.), explosions & fires, industrial disasters, weather problems (e.g. avalanche) | **71** **7.3%** | **64** **5.1%** | **83** **7.8%** | 74 6.6% |
| D Environment and Science | 7 | General science, natural science, engineering, medical | **31** **3.2%** | **34** **2.7%** | **32** **3.0%** | 37 3.3% |
| | 8 | Environmental problems/impacts, pollution | **39** **4.0%** | **74** **5.9%** | **67** **6.3%** | 44 3.9% |
| | 9 | Transport systems, development and planning | **26** **2.7%** | **45** **3.6%** | **36** **3.4%** | 26 2.3% |
| | 10 | Land use/resource planning & conservation, public works, neighbourhoods | **38** **3.9%** | **40** **3.2%** | **25** **2.3%** | 41 3.6% |
| E Society | 11 | Demography/social trends, housing, employment, education | **29** **3.0%** | **53** **4.2%** | **55** **5.2%** | 30 2.7% |
| | 12 | Crime, courts/judicial, police, missing persons | **81** **8.4%** | **97** **7.8%** | **53** **5.0%** | 84 7.4% |
| | 13 | Social disasters (famine, refugees) | **8** **0.8%** | **17** **1.4%** | **21** **2.0%** | 8 0.7% |
| F Cultural affairs | 14 | History & archaeology, heritage, the arts and 'media' | **25** **2.6%** | **27** **2.2%** | **27** **2.5%** | 28 2.5% |
| | 15 | Travel, tourism, recreation and sport | **35** **3.6%** | **15** **1.2%** | **77** **7.2%** | 35 3.1% |
| | 16 | Human interest/'odd events', religion, VIPs/Royals, scandals (non-political), minor accidents (few people) | **53** **5.5%** | **72** **5.8%** | **40** **3.7%** | 56 5.0% |
| G Economics | 17 | Business & finance, industry | **41** **4.2%** | **92** **7.3%** | **58** **5.4%** | 43 3.8% |
| | 18 | Macro-economics, trade agreements, international monetary issues, aid and economic development | **30** **3.1%** | **50** **4.0%** | **35** **3.3%** | 34 3.0% |
| **TOTALS** | | | **969** **100%** | **1248** **100%** | **1067** **100%** | 1128 100% |

*Note*: 1999a covers January to June, 1999b includes July due to continued NATO activity in Yugoslavia.

during the 1990s was supported by UK news maps. The early euphoria following the fall of the Soviet Union saw a proliferation of maps that deployed myths of a new European political architecture based on Western core values of 'democracy' and 'pluralism', contrasting this with concerns about ethnic nationalism and 'tribal war' in the tidal lands of eastern Europe.

These maps metaphorically rolled out a 'red carpet' to eastward expansion based on 'idealist' conceptions of international relations to restore European unity. This was, however, replaced in the mid to late 1990s as a growing 'realist' view of European integration, with maps entitled a 'new Iron Curtain' or 'NATO's new frontier'.

This 'realist' turn was also examined in Vujakovic's (2002a) more detailed study of the role played by maps in the UK's news coverage of NATO intervention in Yugoslavia. In 1999, NATO prosecuted a peace-making action against the Yugoslav government with respect to the latter's treatment of Kosovo. The intervention was couched in terms of a 'just war' using 'precision warfare' and 'surgical strikes' to ensure that the capability of Yugoslavia to use military force in Kosovo was denied. Maps in the UK elite press became a guarantor of this position; the maps, as well as illustrating successful aerial bombardment of Yugoslav military and support infrastructure, provided a potent metaphor for *accuracy* and *precision*, reflecting Harley's concept of the 'internal power' of cartography. 'Objective' maps were often paired with ground-level photographs showing that the civilian population of Yugoslavia was unharmed and carrying on their daily lives while specific infrastructure (bridges, communications centres) and military equipment was successfully degraded. In April, at the height of NATO bombardment, 57 per cent of all the maps in the UK elite press were dedicated to the conflict. The role that these maps played was strongly underlined by the fact that once mistakes were made, such as the accidental targeting of a refugee convoy, the metaphorical power of these maps evaporated, and they were subsequently drastically reduced in number. In a later study, Vujakovic (2014) examined the role of maps in UK news discourses on the projection of power by China and Russia. Here maps played an important part in constructing a Western discourse of the geopolitics, relating to issues such as China's Indian Ocean strategy and Russia's intentions with regard to the Arctic and its 'near abroad'. Other authors have focused on news maps in the popular geopolitical discourse of the Middle East and the Israel-Palestine question in particular (Leuenberger and Schnell, 2010; Wallach, 2011; Leuenberger, 2012; Riopelle and Muniandy, 2013).

A number of other authors have since adopted this critical approach to news mapping, ranging from issues such as national identity to cross-border tensions; for example, Novaes' (2015) examination of popular geopolitics related to the Colombia/Venezuela border and the construction of what he refers to as the imagined 'geo-body' of the state. His paper is particularly interesting as it compares counter-mappings (see Chapters 6 and 7) through a collaborative arts project (see Chapter 38) with maps in the mainstream press that continue to reinforce the border as a line of separation. Novaes (2014) also adopts Harley's concept of the 'external power' of cartography in his examination of news mapping. He investigated the role of 'silences' in journalistic cartography to create a 'distorted geography' of Brazil's favelas as excluded or divided spaces.

Culcasi (2006) has argued that US cartographic constructions of 'Kurdistan', as an ambiguous 'space', has marginalized the Kurds and constructed them as either 'victims' or as 'violent rebels' depending on the US agenda at the time. Culcasi's analysis covered the period 1945 to 2002, incorporating Cold War as well as more recent discourses. Kosonen (2008) also takes a long-term historic approach in her study of Finnish press cartography and nation-building from 1899 to 1942, concluding that press maps have been important in stressing territorial integrity and in emphasizing (especially through caricature maps) differences from Russia as the significant 'other'.

## Future mapping?

While the discussion above has focused on war, conflict and geopolitics, it is important to stress the fact that news maps cover a very wide range of issues, from the environment and conservation, transport and economics, to crime and other social issues. While many of these maps are simple in form, there are ways in which they may still create a profound effect on their audiences. Even simple locator maps, through repetition, may create place-stereotyping in terms of crime or economic decline and deprivation. Like verbal stereotyping, news maps can contribute to 'bounded categories' (Jalbert, 1983) through which a particular people and/or territory is associated with certain characteristics, e.g. the Columbia border zone as a zone of conflict and violence (Novaes, 2015).

More important, as a form of *future mapping*, is the concept of 'cartographic precedent', and a fitting theme to draw this chapter to a conclusion. The term refers to a situation whereby maps convey a particular image of 'reality' in advance of that reality, potentially precluding or prejudicing alternative outcomes (Hall, 1981); examples can include simple route or locator maps, for example those showing proposed transport routes or locations for new airports. Precedent, however, becomes more important where it involves major geopolitical issues. Vujakovic (2000a) discusses the impact of the UK's maps of Yugoslavia in the 1990s. Whether the result of error or a deliberate attempt to graphically dismember the 'rump state' of Yugoslavia (Serbia and Montenegro), many news mappers in the late 1990s failed to use appropriate line symbols and typography to show the *de jure* political entities within the Balkan region. The *Daily Telegraph*, for example, failed in several instances to identity 'Yugoslavia' as the sovereign state entity, and 'symbolically' elevated Serbia to the same status as surrounding sovereign states, while treating Montenegro and Kosovo as if they had the same status *within* Serbia (Serbia and Montenegro were then constituent republics within the state of Yugoslavia). The cumulative effect was to create a confused and fragmented political situation or an image of Kosovo as a separate political space.

At the time of writing, perhaps the most interesting example of cartographic precedent with major international ramifications is the dissemination in the news media of China's so-called 'U-shaped Line' (also commonly known as the 'Nine-Dash Line', or derogatively, as the 'Cow's Tongue'). The U-shaped Line delineates the extent of China's putative maritime claims in the South China Sea (SCS). The line was first published by the Guomindang government in 1947, and has since been adopted by both the People's Republic of China and Taiwan. The problem is that China is equivocal about what exactly the line represents, creating a sense of ambiguity and insecurity within the region. This is the strength of the mapped line – China has created a sense of uncertainty which plays well with its approach to geopolitics (Perlez, 2014; Curtis, 2015), and, it could be argued, the recycling of the line by the Western and other news media simply reinforces China's hand.

Approaches to mapping the U-shaped Line in the news are variable. A survey by the author is illustrative. The influential *Financial Times* consistently adopted the use of the U-shaped Line during the period surveyed – January to July 2014. It marked the line boldly in red in four instances in May 2014 alone, and seven times in all. In no case were the other claims by Vietnam and others shown, despite the heightened tensions with Vietnam during that period. By contrast *The Observer* published only one relevant map and showed none of the claims, while *The Times* did publish both the Vietnamese and Chinese claims (in blue and red respectively), but no others. An interesting variant was *The Guardian*, which contrasted the U-shaped Line (in blue) with Vietnam's 200-nautical-mile Exclusive Economic Zone (UNCLOS) and the location of a Chinese rig in the disputed area. This range is not uncharacteristic of news coverage globally. The key question is 'What does this representation of China's chosen cartographic symbolism mean for the geopolitics of the SCS?' Does it create a sense that the international community needs to provide a diplomatic solution to the issue, or, if China is unwilling to concede its claims, will its circulation have worked in China's favour? Only time will tell.

In conclusion, journalistic cartography certainly remains one of the most important forms of geographic education, and the proliferation of image-based electronic providers will not dull this. But beyond the 'pedagogic' role assigned to news maps by Monmonier (1989), their real impact lies in their ability to reinforce, often subliminally, prejudices and precedent. The impact that journalistic cartography has on the individual's geographic knowledge and world-view (Vujakovic, 2009), and the consequences for developments from the local to the global, remains an under-researched and under-theorized aspect of news mapping that awaits further examination.

## Notes

1 Hans Speier (1941) had discussed the ubiquity of maps within the mass media and their impact on society decades before, including newspapers and magazines, in his paper 'Magic Geography'.
2 "More than half high speed rail line to be hidden" *The Telegraph* (UK) by David Millward, Transport Editor, available at: *www.telegraph.co.uk/news/uknews/road-and-rail-transport/9005599/More-than-half-high-speed-rail-line-to-be-hidden.html* (Accessed: 10 January 2012).
3 Fantasy frontiers: Indian, Pakistani and Chinese border disputes, *The Economist* online, available at: *www.economist.com/blogs/dailychart/2011/05/indian_pakistani_and_chinese_border_disputes* (Accessed: 8 February 2012).
4 Barbara Belyea (1992) provides a detailed examination of J.B. Harley's application of 'postmodernism', in particular the work of Michel Foucault and Jacques Derrida, to his new conceptual approaches to the study of cartography. Belyea is particularly critical of Harley's tendency to use secondary sources and commentaries rather than the original works of these two French writers. However, Harley makes it clear, especially with regard to his discussion of 'power/knowledge', that he owes his own adaptation of this concept to a reworking of Foucault's ideas by Joseph Rouse (Rouse, 1987).

## References

Ager, J. (1977) "Maps and propaganda" *The Bulletin of the Society of University Cartographers* 11 (1) pp.1–15.
Antoniou, A., Kiaten, R. and Ehmann, S. (Eds) (2015) *Mind the Map: Illustrated Maps and Cartography*, Berlin: Gestalten.
Balchin, W.G.V. (1985) "Media Map Watch: A Report" *Geography* 70 (4) pp.339–345.
Balchin, W.G.V. (1988) "The Media Map Watch in the UK" in Gauthier, M.J. (Ed.) *Cartographie dans les médias* Sillery, QC: Presses de l'Universite du Quebec pp.33–48.
Barnes, T.J. and Duncan, J.S. (Eds) (1992) *Writing Worlds: Discourse, Text and Metaphor in the Representation of Landscape* London: Routledge.
Barney, T. (2015) *Mapping the Cold War: Cartography and the Framing of America's International Power* Chapel Hill, NC: University of North Carolina Press.
Belyea, B. (1992) "Images of Power: Derrida/Foucault/Harley" *Cartographica* 29 (2) 1–9.
Case, D. (1996) "How to Design with Numbers" in Houkes, R. (Ed.) *Information Design and Infographics* Rotterdam, The Netherlands: European Institute for Research and Development of Graphic Communication, pp.23–34.
Cosgrove, D. (2005) "Maps, Mapping, Modernity: Art and Cartography in the Twentieth Century" *Imago Mundi* 57 (1) pp.35–54.
Cosgrove, D. and della Dora, V. (2005) "Mapping Global War: Los Angeles, the Pacific, and Charles Owen's Pictorial Cartography" *Annals of the Association of American Geographers* 95 (2) pp.373–390.
Culcasi, K. (2006) "Cartographically Constructing Kurdistan Within Geopolitical and Orientalist Discourses" *Political Geography* 25 pp.680–706.
Curtis, H. (2015) "Constructing Cooperation: Chinese Ontological Security Seeking in the South China Sea Dispute" *Journal of Borderland Studies* Available at: *http://dx.doi.org./10.1080/08865655.2015.106 6698* (Accessed: 21 March 2016).
Daulby, G. (1988) "The Similarities and Differences Between the Information Graphics of Print and Television" in Wildbur, P. (Ed.) *Information Graphics: A Survey of Typography, Diagrammatic and Cartographic Communication* London: Trefoil pp.107–110.

Ferris, K. (1993) "Black and White and Read All Over: The Constraints and Opportunities of Monochrome Cartography in Newspapers" *The Cartographic Journal* 30 (2) pp.123–128.

Gauthier, M-J. (1997) "Cartography in the Media: An Over-View of Canadian Developments" in Schaarfe, W (Ed.) *Proceedings of the International Conference on Mass Media Maps* Berlin: Freie Univesität Berlin pp.39–52.

Gilmartin, P. (1985) "The Design of Journalistic Maps: Purposes, Parameters and Prospects" *Cartographica* 22 (4) pp.1–18.

Hall, D.R. (1981) "A Geographical Approach to Propaganda" in Burnett, D. and Taylor, P.J. (Eds) *Political Studies from Spatial Perspectives* London: Wiley pp.313–330.

Harley, J.B. (1989a) "Deconstructing the Map" *Cartographica* 26 (2) pp.1–20.

Harley, J.B. (1989b) "Historical Geography and the Cartographic Illusion" *Journal of Historical Geography* 15 (1) pp.80–91.

Harley, J.B. (1990) "Cartography, Ethics and Social Theory" *Cartographica* 27 (2) pp.1–23.

Harley, J.B. (1992) "Deconstructing the Map" in Barnes, T.J. and Duncan, J.S. (Eds) *Writing Worlds: Discourse, Text and Metaphor in the Representation of Landscape* London: Routledge pp.231–247.

Holmes, N. (1991) *Pictorial Maps* London: Herbert Press.

Jalbert, P.L. (1983) "Some Constructs for Analyzing News" in Davis, H. and Walton, P. (Eds) *Language, Image and Media* Oxford, UK: Blackwell.

Kosonen, K. (2008) "Making Maps and Mental Images: Finnish Press Cartography in Nation-Building, 1899–1942" *National Identities* 10 (1) pp.21–47.

Kowalski, P. and Paslawski, J. (1997) "Polish Journalistic Cartography in 1995, A Preliminary Report" in Schaarfe, W (Ed.) *Proceedings of the International Conference on Mass Media Maps* Berlin: Freie Univesität Berlin pp.219–228.

Krygier, J.B. (1995) "Cartography as an Art and a Science?" *The Cartographic Journal* 32 (1) pp.3–10.

Leuenberger, C. and Schnell, I. (2010) "The Politics of Maps: Constructing National Territory in Israel" *Social Studies of Science* 40 (6) pp.803–842.

Leuenberger, C. (2012) "Mapping Israel/Palestine: Constructing National Territories across Different Online International Newspapers" *Bulletin du Centre de recherche français à Jérusalem* 23 pp.2–21.

Luijk, R.B. van (2008) "Maps of Battles, Battle of Maps: News Cartography of the Battle at Neerwinden, Flanders, 1693" *Imago Mundi* 60 (2) pp.211–220.

Mijkenaar, P. (1996) "Infographics at the Gulf War" in Houkes, R. (Ed.) *Information Design and Infographics* The Hague: European Institute for Research & Development of Graphic Communications, pp.55–62.

Monmonier, M. (1986) "The Rise of Map Use by Elite Newspapers in England, Canada, and the United States" *Imago Mundi* 38 (1) pp.46–60.

Monmonier, M. (1989) *Maps with the News* Chicago, IL: Chicago University Press.

Novaes, A.R. (2014) "Favelas and the Divided City: Mapping Silences and Calculations in Rio de Janeiro's Journalistic Cartography" *Social and Cultural Geography* 15 (2) pp.201–225.

Novaes, A.R. (2015) "Map Art and Popular Geopolitics: Mapping Borders between Colombia and Venezuela" *Geopolitics* 20 (1) pp.121–141.

Perkins, C.R., and Parry, R.B. (1996) *Mapping the UK: Maps and Spatial Data for the 21st Century* London: Bowker Saur.

Perlez, J. (2014) "China's 'New Type' of Ties Fails to Sway Obama" *The New York Times* 10 November 2014 p.A8.

Rendgen, S. and Wiedemann, J. (2012) *Information Graphics* Cologne, Germany: Taschen.

Riopelle, C. and Muniandy, P. (2013) "Drones, Maps and Crescents: CBS News' Visual Construction of the Middle East" *Media, War and Conflict* 6 (2) pp.153–172.

Ristow, W.W. (1939) "Geographical Information Please!" *Journal of Geography* 38 (8) pp.314–318.

Ristow, W.W. (1957) "Journalistic Cartography" *Surveying and Mapping* 17 pp.369–390.

Robins, K. (1995) "The New Spaces of Global Media" in Johnston, R.J., Taylor, P.J. and Watts, M.J. (Eds) *Geographies of Global Change* Oxford, UK: Blackwell, pp.248–262.

Rogers, S. (2012) "How Data Changed Journalism" in Rendegen, S. and Wiedemann, J. *Information Graphics* Cologne, Germany: Taschen pp.59–64.

Rouse, J. (1987) *Knowledge and Power: Toward a Political Philosophy of Science* Ithaca, NY: Cornell University Press.

Rundstrom, R.A. (1991) "Mapping, Postmodernism, Indigenous People and the Changing Direction of North American Cartography" *Cartographica* 28 (2) pp.1–12.

Rundstrom, R.A. (1993) "Introduction" (to Monograph 44, 'Introducing Cultural and Social Cartography') *Cartographica* 30 (1) pp.vii–xii.

Scicluna, B., (2016) *Siege Maps: Keeping Memory Safe* (exhibition brochure) Valletta, Malta: National Museum of Archaeology, in association with the Charles University, Prague.

Schulten, S. (1998) "Richard Edes Harrison and the Challenge to American Cartography" *Imago Mundi* 50 pp.174–188.

Speier, H. (1941) "Magic Geography" *Social Research* 8 (3) pp.310–330.

Sullivan, P. (1996) "The History of Newspaper Graphics" in Houkes, R. (Ed.) *Information Design and infographics* Rotterdam, The Netherlands: European Institute for Research and Development of Graphic Communication pp.9–22.

Tremols, M. (1997) "The Social Importance of Geocartography in the Spanish Media" in Schaarfe, W. (Ed.) *Proceedings of the International Conference on Mass Media Maps* Berlin: Freie Univesität Berlin, pp.219–228.

Vujakovic, P. (1999) "Views of the World: Maps in the British Prestige 'Press'" *The Bulletin of the Society of Cartographers* 33 (1) pp.1–14.

Vujakovic, P. (2000) "Balkan Borders: News Maps and 'Geographic Education' During the Kosovo Crisis of January to July 1999" *The Bulletin of the Society of Cartographers* 34 (1) pp.21–32.

Vujakovic, P. (2002a) Mapping the War Zone: Cartography, Geopolitics and Security Discourse in the UK Press *Journalism Studies* 3 (2) pp.187–202.

Vujakovic, P. (2002b) "What Ever Happened to the 'New Cartography'?" *Journal of Geography in Higher Education* 26 (3) pp.369–380.

Vujakovic, P. (2009) "Geopolitical Awareness: Geographic Knowledge and Maps in the News" *The Bulletin of the Society of Cartographers* 42 (1,2) pp.39–43.

Vujakovic, P. (2010) "New Views of the World: Maps in the United Kingdom 'Quality' Press in 1999 and 2009" *The Bulletin of the Society of Cartographers* 43 (1,2) pp.31–40.

Vujakovic, P. (2014) "The State as a 'Power Container': The Role of News Media Cartography in Contemporary Geopolitical Discourse" *The Cartographic Journal* 51 (1) pp.11–24.

Wallach, Y. (2011) "Trapped in Mirror-Images: The Rhetoric of Maps in Israel/Palestine" *Political Geography* 30 pp.358–369.

Wilke, J., Heimprecht, C. and Cohen, A. (2012) "The Geography of Foreign News on Television: A Comparative Study of 17 Countries" *International Communication Gazette* 74 (4) pp.301–322.

Wood, D. (1992) *The Power of Maps* London: Routledge.

Wright, J.K. (1942) "Map Makers are Human: Comments on the Subjective in Maps" *The Geographical Review* 32 (4) pp.527–544.

# 35

# VGI and beyond

## From data to mapping

*Vyron Antoniou, Cristina Capineri and*
*Muki (Mordechai) Haklay*

This chapter will introduce the concept of Volunteered Geographic Information (VGI) within the practices of mapping and cartography. Our aim is to provide an accessible overview of the area, which has grown rapidly since the mid 2000s; but first we need to define what we mean by VGI.

## Defining VGI

In a seminal paper published in 2007, Mike Goodchild coined the term Volunteered Geographic Information in an effort to describe 'the widespread engagement of large numbers of private citizens, often with little in the way of formal qualifications, in the creation of geographic information' (Goodchild, 2007: 217). At that point, rudimentary crowdsourced Geographic Information (GI) was created and disseminated freely with the help of innovative desktop applications (e.g. Google Earth) or Web-based platforms (e.g. Wikimapia, OpenStreetMap). By *crowdsourcing* we refer to the action of multiple participants (sometimes thousands or even millions) in the generation of geographical information, when these participants are external to the organization that manages the information and are not formally employed by it. Since then a lot has changed and VGI now has a deep and broad agenda that ranges from implicitly contributed GI through social networks to rigorously monitored citizen science projects. However, before we continue the discussion on this subject, it is necessary to shed light onto the key factors that have helped to create this phenomenon.

## Background technology and societal aspects

While it may seem that VGI was created suddenly with the birth of OpenStreetMap in 2004, the crowdsourced and collaborative creation of spatial content is not new. The current form of the VGI phenomenon emerged as several enabling factors matured, particularly regarding how people collaborate, with the key paradigms paving the way for similar endeavours coming from the field of software development. Open source software is a prime example of collaboration among otherwise unrelated programmers who aim to create software that is freely accessible by anyone. More importantly, the code used to create the software is also shared so that anyone can see, examine, reuse or improve it. This collaborative approach also started to grow in the geomatics domain, but, whereas in software development the purpose was to work together to create an application, here, the mentality was mostly oriented towards collaboration in order to solve a common problem.

VGI and citizen-driven data collection efforts are, however, not entirely new (Elwood *et al.*, 2012). For example, Stamp (1931) describes a group of voluntary teachers and students who participated in land use surveys in Britain and the United States, and Bunge (1971) mentions that city residents carried out local mapping efforts. More recently, examples of such collaboration can be found in what is known as Public Participation GIS (PPGIS) (Obermeyer, 1998) or in the broader area of citizen science (Haklay, 2013). In the former, PPGIS emerged as a means to help researchers and institutions work with local communities to investigate either controversial issues or to represent local knowledge. The collaboration takes place over a backdrop map where stakeholders or community members provide input in an effort to map and to better understand a phenomenon or a problem and then to find the best solution for all parties involved. In the latter, citizens are involved in scientific research activities where, often with the help of scientists, they collect, manage and analyse observations and data, and disseminate their results. Another enabling factor comes from the technology front. The removal of Selective Availability from the Global Positioning System (GPS) in 2000 (Clinton, 2000) signalled the proliferation of accurate and low-cost GPS-enabled devices. In turn, this enabled individuals to easily capture geographic data from their everyday activities (i.e. commuting, leisure activities, etc.) turning them into 'neo-geographers' (Turner, 2006) and citizen-sensors (Goodchild, 2007). It did not take long for this kind of crowdsourced GI to find its way to the World Wide Web. The turn towards a bi-directional Web, where the lines between content users and content producers were constantly blurring leading to what is known 'produsers' or 'prosumers' (Bruns, 2006: 2; Coleman *et al.*, 2009); the novel Web software techniques (e.g. AJAX and APIs) that gave to the Web a programmable façade; and the investment of technology giants (e.g. Google, Microsoft, Yahoo!) in spatial applications with the creation of global wide satellite imagery maps, provided a fertile ground for VGI to flourish. These factors enabled the crowd to instantly upload and consequently have access to spatial information on the Web. At the other end of the spectrum, National Mapping Agencies (NMAs) responsible for the creation and maintenance of Spatial Data Infrastructures (SDI) have continued to treat their geospatial data as valuable sources of income, with the effect of depriving access to these data by the general public. All these factors contributed to the appearance of the VGI phenomenon that was incarnated through the development of numerous Web-based and mobile applications.

## Redefining VGI

Against this backdrop of the technological and social factors that played a role in the birth and development of VGI, it is necessary to re-define it in a broader context. First and foremost, we need to understand that this area is highly interdisciplinary in that it intertwines the advances of many domains. VGI is at the centre of a wide scientific community that focuses on the harnessing of new sources of GI, and on satisfying the spatial turn fuelled by the neogeography revolution, a revolution which has put mapping within the grasp of almost anybody (Turner, 2006). VGI is the grafting of the underlying social, economic and technological situation with the geospatial domain. Especially from a cartographer's point of view, VGI can be viewed as a precious, yet so far elusive, map element: the user's perspective. Now, for the first time, the user's perception of space is tangible through the volunteered recording of spatial features or phenomena they consider important to have on a map. This presents a major change in cartography and has been made possible due to the factors explained earlier. Timely crowdsourced spatial content can fuel maps with constantly changing thematic layers which were impossible to have in the previous generation of paper or even online maps. The advantages of VGI, and what this phenomenon can bring to the cartographic domain, will be discussed further in this chapter.

## Sources of VGI and user participation

Today, VGI comes in many flavours and from various sources: toponyms, GPS tracks, geo-tagged photos, synchronous micro-blogging, social networking sites such as Facebook, blogs, gaming spaces, sensor measurements, complete topographic maps, and so on.

The creation of VGI is a process which involves (more or less) the voluntary participation of *produsers*. Indeed, VGI can be either voluntary/explicit or involuntary/implicit. The first type consists in all digital georeferenced footprints that are created by users who consciously interact with Web technologies of different kinds to locate events or to record elements (from animal species to artifacts) or experiences and perceptions. The second type of VGI is produced by users who interact with technologies but who are not aware of the fact that their interaction/activity will leave a digital footprint in cyberspace, such as most users of Twitter, Flickr, Foursquare (and similar social networking websites), who perform regular activities on the Web (e.g. talking to friends, booking a room) (Capineri and Rondinone, 2011: 561).

However, the voluntary element, in both cases, is a subject of debate in VGI research since it can affect the quality and the fitness for use of VGI data, since they are often collected publicly, without strict standardization, and every user inserts data according to his/her personal background and point of view (Coleman *et al.*, 2009; Haklay, 2010) (Figure 35.1).

*Figure 35.1*   Adding data to OSM after mapping Brighton Pier. Alexander Kachkaev: *https:// secure.flickr.com/photos/kachkaev/6448160479 – CC BY 2.0*, Wikipedia

In particular, end users who carry out *geographic volunteer work* (Priedhorsky and Terveen, 2008: 267–268) perform an active role when the volunteered information production is regulated by shared rules concerning geocoding (i.e. the process of transforming a description of a location to real-world coordinates), tagging (assigning a key attribute), and annotating the data. In this sense, VGI becomes part of citizen science, which has emerged from ecology, biology and nature conservation, whose projects are based on volunteering and contributing information for the benefit of human knowledge and science (Haklay, 2013). Citizen science reconsiders the separation between scientist and the public, and scientists need to adjust to this new character of the scientist as both mediator of knowledge and citizen and not as the sole repository of scientific truth: 'This might end up being the most important outcome of citizen science as a whole as it might eventually catalyse the education of scientists to engage more fully with society' (Haklay, 2013: 14). In this context, VGI may represent either the implicit or the explicit relationship of the individual with the whole; this relationship has long been a critical part both of the individual's motivation to act in some larger interest and of the group's ability to exhort the individual to take action and to participate (Curry 1997: 685–686).

## Advantages and disadvantages of VGI

From the early days of VGI, there was skepticism about its value. First, it was the validity of the term 'Volunteered'. Scholars challenged this term as misleading regarding the true intentions and motivation of the contributors (Elwood, 2008). Then, at a more substantial level, the criticism focused on the long-term sustainability of VGI and the trust, credibility and quality of the GI produced (Flanagin and Metzer, 2008). For the former, there were concerns that VGI was overhyped and it would share the fate of many new trends that disappear after a viral, yet short, life. For the latter, it was obvious that as VGI has deep social roots, the profile, motivation and the digital divide as well as the educational background of the contributors could severely affect the overall data quality. There were many issues that fuelled this discussion, ranging from the lack of any cartographic background or skills that professional geographers and surveyors have, to the appearance of malicious contributions – like those that are very common and largely anticipated – to the rest of the Web today.

However, most of the initial concerns have been dispersed by the evolution of VGI. Today the term is well defined and the evolution and possibilities of VGI in the geospatial domain have attracted the interest of academics and professionals alike, with thousands of scientific papers published on the topic and a growing number of governments and corporations leveraging this kind of GI knowledge. Notwithstanding the acceptance of VGI, the truth is that VGI comes with a number of caveats, mainly concerning the quality of the data collected. First, there are social and spatial biases. As VGI is technology driven, the digital divide directly affects the datasets compiled and thus we need to be very careful about the coverage and representativeness of the data that are being collected (Graham *et al.*, 2014). Second, is the GI itself. Lack of metadata, heterogeneity, patch work and fragmented contributions should be expected when handling such data. Moreover, in academic and professional discourse, spatial data quality is a mixture of tangible factors like thematic, positional and temporal accuracy, completeness, logical consistency and usability, but not all VGI sources can stand such scrutiny. However, existing spatial data quality measures can be adopted and new quality indicators have been devised to better describe the quality of VGI (Antoniou and Skopeliti, 2015).

A relevant aspect of VGI is the acquisition of local knowledge. People's contributions to VGI tend to be more accurate in places the contributor knows best and is closest to, according to

Tobler's Law which states that 'everything is related to everything else but near things are more related than distant things' (Tobler, 1970: 234). As such, some of the scientific literature associated with VGI hypothesizes that (a) contributors write about nearby places more often than distant ones, and (b) this likelihood follows an exponential distance decay function (Hardy *et al.*, 2012). However, according to recent research, about 50 per cent of Flickr users contribute local information on average, and over 45 per cent of Flickr photos are local to the photographer (Hecht and Gergle, 2010). Remarkably, local knowledge derived from VGI has been used in vernacular geography which 'encapsulates the spatial knowledge that we use to conceptualize and communicate about space on a day-to-day basis. Importantly, it deals with regions which are typically not represented in formal administrative gazetteers and which are often considered to be vague' (Purves and Hollenstein, 2010: 22). Also, it is worth mentioning that longstanding authoritative spatial products like gazetteers were traditionally based on knowledge that originated from local people (Kerfoot, 2006). Moreover, in some cases, especially when VGI is derived from social networks such as Twitter or Facebook, data are mostly produced in urban areas due to better Internet connections and higher population concentration. A recent analysis (Hecht and Stephens, 2014) shows that in the US there are 3.5 times more Twitter users per capita in core urban counties than in rural counties; the same authors have revealed also that urban users tweet more than their rural counterparts.

Another major contribution of VGI is that it sparked the creation of a virtuous circle having at its centre the linkage between society and spatial information. Indeed, technological advances have facilitated spatial data collection and online diffusion, and this has made people familiar with spatial content, cartographic products and location-based services. This, in turn, created the need for more spatial content that is both high quality and freely available, and thus VGI contributors were better placed to address this need, resulting in more sources of VGI becoming available on the Web. This positive feedback loop was also fuelled by the intrinsic characteristics of VGI data. First, the open and free access to VGI data was a major factor. VGI can provide a reliable alternative for many spatial and cartographic projects without the need to bear the high costs and restrictions that come with proprietary data. We should also consider the extended field of scope that VGI can cover. While VGI became known mainly from a handful of champion projects such as OpenStreetMap, Wikimapia and Geonames, examples also include data gathering for air pollution, urban noise, traffic and congestion maps, cycle maps, gpx-trail maps and soil mapping. Most of these topics were usually under the radar of the NMAs as their focus was on a few well-defined mapping products.

Another comparative advantage of VGI over proprietary data is the fact that accessibility seems quite easy and fast, thanks to computing architectures which enable data collection in a timely manner and to geoweb applications that foster citizen access and participation (Crampton, 2009). This reveals another important VGI feature: real-time recording and the proactive approach of the user. Timeliness in capturing and communicating an event (e.g. floods, fires), a problem (e.g. a traffic jam), an opinion, a feeling, and so on, enables the user to contribute to his/her environment in a participatory and collaborative way; these contributions help decision-making and problem-solving processes to be undertaken in a faster way than in the past (Bertot *et al.*, 2012). Moreover, ubiquitous sensor devices shorten the time horizons of the updating of geographical data as the time gap between data capture and data consumption is minimal. For example, a study by Antoniou *et al.* (2010) shows that more than 60 per cent of the photographs submitted to Flickr were actually taken in under seven days, while fewer than 9 per cent have longer than a one year gap between time of capture and time of upload. This creates a new environment for all cartographers in the sense that there is now an increased responsibility for delivering up-to-date spatial/mapping products. This is even more so in time-critical situations. Thus, it is unsurprising that the real value of VGI has surfaced after severe natural disasters such as wildfires in Santa Barbara (Goodchild and Glennon 2010) and the Haiti earthquake (Kent, 2010; Zook *et al.*, 2010).

All these comparative advantages gained ground for VGI, a fact that was further amplified by studies regarding the quality of the datasets produced, such as Girres and Touya (2010), Haklay *et al.* (2010), Antoniou (2011) and Neis and Zipf (2012), usually using OSM as a proxy for VGI sources. Moreover, new Geographic Information Retrieval (GIR) techniques have been developed that focus on the use of implicitly generated GI in existing spatial processes such as the validation of land use/land cover (LU/LC) maps (Antoniou *et al.*, 2016). In this context, it is clear that while, at least in the near future, VGI cannot replace proprietary data, it can play a crucial role in correcting, enriching and updating existing datasets (Craglia, 2007) or provide the basic information layer for new products. This change is crucial in many ways, but for the first time in the geomatics domain, a grassroots process has such an impact on well-established processes. The impact of VGI on NMAs, on society and on new mapping products is further discussed in the next section.

## The impact of VGI

### VGI and NMAs

The open data policies adopted in many countries around the world inevitably led to the free usage of spatial data. While this can severely affect the economic resources of NMAs, their mission to provide up-to-date datasets nationwide has not changed. On the contrary, they have to step up their efforts and keep up with the technological advances and demands for more data. In a sense, today, NMAs need to do more with less. Moreover, as Goodchild (2007) notes, the arguments made by Estes and Mooneyhan (1994), about a mistaken popular notion of a well-mapped world, are still true and thus NMAs are facing some pressing challenges.

Cartographers in NMAs should be prepared to change their conceptual apparatuses used so far in the front of data management and organizational structure alike. In many NMAs today, the focus is on developing new and agile in-house processes and organizational structures that can foster the creation of VGI data and then can easily incorporate such input with spatial data derived from traditional sources (Antoniou, 2011). Indeed, a growing number of NMAs (including Ordnance Survey in the UK, the Institut national de l'information géographique et forestière in France, Dienst voor het Kadaster en de openbare registers in the Netherlands, and so on) have started to explore the potential use of VGI in map production and evaluation activities. The intertwining of NMAs and VGI can take place in many levels. Examples include the collaboration of the Natural Resources Canada with the OpenStreetMap community to facilitate data updates (Bégin, 2012) or the development of an efficient alert reporting system like the one developed by IGN France. Similarly, the National Land Survey of Finland provides its entire topographic database freely to the public and also provides a Web and mobile application to collect feedback from citizens. Moreover, the Netherlands cadastral agency has developed an in-house system to collect feedback from its users and also provides a change-detection mechanism for the 1:10,000 dataset (TOP10NL) based on local administration datasets and OSM (Olteanu-Raimond *et al.*, 2016). In another case, vernacular place names are mixed with authoritative toponymic datasets to provide a detailed gazetteer to support search-and-rescue missions from the UK Maritime and Coastguard Agency (Haklay *et al.*, 2014). However, the conflation of VGI datasets with authoritative data should utilize the best of both worlds: the rigorous quality and standardization processes applied in NMAs and the overwhelming flow of timely spatial content that is generated by volunteers.

## VGI and citizens

The spread of sensor networks, the growing availability of Internet connections and the creation of wireless connections have all enabled information and services to be reached virtually from any place. These technological innovations have also affected the mode of communication on the Internet by creating communities of interest who are potentially free from the bond of physical proximity and are bound together by mutual interests, practices and passions. The community-building process has also been described as 'communities of crickets' (Buchanan, 2002: 49) and 'members of a food web' (Buchanan, 2002: 17). This process has been reinforced by the emergence of the Geoweb, which is characterized by a high level of interaction between the Web and the users; their activity is the footprint of daily routines, movements, ideas and values which reveal new forms of sociability, social activism and engagement. Moreover, daily life is local by its nature: daily issues ranging from health to entertainment, to education, to the supply and security of goods and services, relate largely to the local scale. Traditional types of data and information (i.e. census data, on the spot surveys) are not good at managing and dealing with knowledge at this local scale or asking timely questions. Most of the data and information have a geospatial component and are converging with other digital tools to help institutions manage services such as transportation, water supply, sanitation, public safety, public health and energy to improve the quality of life. This proactive ability has attracted the interest not only of citizens but also of institutions and enterprises. In particular, public institutions are being called upon to reinvent the governance of public affairs and to update the means for interacting with their communities; at the same time they become protagonists in a complex scenario that requires new professional skills (e.g. Internet capability, willingness to collaborate) and new abilities to accommodate and 'exploit' the 'collective intelligence' of Web 2.0.

## VGI as point of departure for new cartographic activities

From the early days of VGI, it was understood that if this phenomenon managed to last and evolve it could spark shift changes in the geomatics domain, and thus there was considerable research interest from the beginning. However, Goodchild (2012) explains that initially research focused on the phenomenon itself: the type and origins of VGI and contributors, data quality, social context and so on. As VGI evolved, the research focused on using VGI in areas such as social and environmental sciences or citizen science. In other words, VGI served as a research apparatus that provided a novel way for approaching existing and new domains. In this context, VGI played a key role in expanding the horizons of cartographic activities, introducing new thematic areas and data sources or redefining existing ones.

An indicative example can be found in the field of urban sensing and smart cities. Today, with ubiquitous sensor networks, our living environments are being transformed into smart cities where the flow of VGI in terms of volume and currency opens the opportunity to monitor and understand, in an unprecedented way, what exactly takes place in every corner of the urban fabric. In a sense, this in turn redefines the scope of urban maps. While static street-level mapping products are still useful, today it is possible to map, analyse and monitor the heartbeat of most urban areas. Another field where VGI can offer a fresh view is emergency mapping. When disasters occur, authoritative geospatial support is not always available, and

*Figure 35.2*　Crowdsourced noise mapping around City Airport, London, UK (*http://mapping forchange.org.uk/*)

experience has shown that the most effective efforts come from volunteers in the field. Haklay *et al.* (2014) discuss a number of cases where crowdsourcing and VGI provided the needed mapping products. Of special interest are the efforts by the Humanitarian OpenStreetMap Team (HOT), a group of volunteers who provide base maps and cartographic products to relief organizations in emergency cases via the OSM platform. Their projects include emergency mapping for the Haiti earthquake in 2010, the Ebola outbreak in 2014 and the Nepal earthquake in 2015.

Another field that has been stimulated by VGI is the creation of applications for maps and mapping. Today, leveraging the opportunities that low-cost technology is providing, the citizen sensor can capture and thus map any phenomenon at a local or a global level. Thus, novel mapping products and applications have started to emerge. From noise maps around airports (Figure 35.2),[1] problem-reporting to local authorities[2] to global environmental monitoring,[3] the scope of mapping applications is getting broader with the help of crowdsourced GI and this puts geospatial engineers and cartographers into the driver's seat for the next generation of mapping applications.

## The art of mapping VGI

Mapping VGI requires art, just as it has always been with any cartographic activity. The maps of VGI fall into the category of thematic mapping, and the recent burgeoning literature offers many experimentations of VGI representation on maps where the map is either a tool supporting some other actions (e.g. emergency relief in disaster management) or highlighting a discovery (e.g. the boundaries of vague places). In extreme synthesis, placing VGI on a map implies a process which can be either very simple, such as a basic representation of data through the many intuitive and attractive Web mapping tools, or more sophisticated according to the aim of VGI's use (e.g. distribution areas of a phenomenon). Neogeography has certainly opened up the black box of maps for a larger public and for much more diversified uses than in the past; moreover, traditional paper maps are being substituted by a new publishing genre which raises the question of the maps' maintenance and storage. But when VGI is employed as raw information to be transferred onto a map, either on a piece of paper or on screen, then cartographic principles and methodologies are required just as much as before. Nevertheless, among the many mapping practices, at least four main types of representation seem to emerge, with each type referring to a different interpretation of the data.

The first type is the neutral or locational representation. If the map is simply carrying positional information (i.e. an address) of the VGI data by using the georeference attribute (e.g. coordinates, geo-name), this kind of representation satisfies the need to know where contributors or the recorded items are located (Figure 35.3).

The second is the representation of spatial hot spots, which aims to identify areas with unusual high data densities by means of clustering techniques; this allows the discovery of areas of interest/activity independently of given boundaries (Figure 35.4).

The third type is a hybrid representation where VGI data are grouped according to existing boundaries. Figure 35.5 shows the concentration of tweets in Florence's enumeration districts weighted according to the resident population. In this case, VGI data may augment existing information if they are integrated with other authoritative sources.

The fourth type (Figure 35.6) is the creative representation which derives from the interaction of the citizen/producer with a base cartographic map upon which information is added according to the aim of the investigation or field work. This is the type of map generally used in problem-solving applications, such as PPGIS or citizen science.

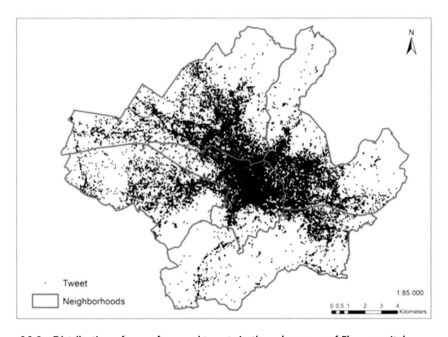

*Figure 35.3*  Distribution of georeferenced tweets in the urban area of Florence, Italy

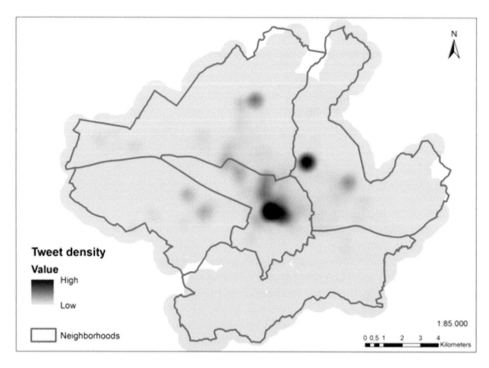

*Figure 35.4*   Kernel density distribution of tweets in the urban area of Florence, Italy

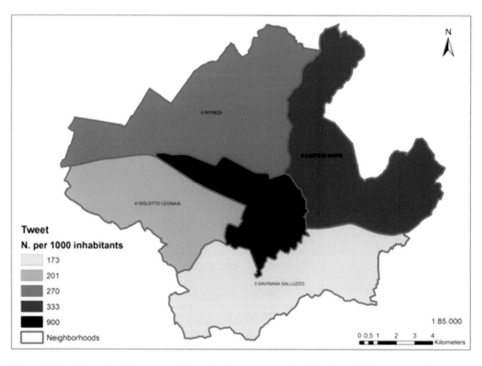

*Figure 35.5*   Tweet/resident population in the enumeration districts of Florence, Italy

*Figure 35.6*   Community mapping for non-technical users (from Ellul *et al.*, 2009)

## Conclusions and future of VGI

It is always a challenge to predict the future. Nevertheless, here we make an effort to envisage the future of VGI by taking into account the technological and the societal trends that can affect this phenomenon. These two intertwined factors generated the VGI phenomenon and nothing suggests that they will not be its driving forces in the future.

First, on the technological front, while it is expected that communications bandwidths will keep increasing, the cost of hardware will keep dropping and the number of people online will keep growing, the true revolution of VGI is expected to come from the development of devices for capturing spatial data that will proliferate or be introduced in the future. On the one hand, we have the ubiquity of sensors that passively collect spatial data, mostly in an urban context. The transformation of our living environment into smart cities inevitably gives rise to a better understanding and a more detailed recording of space and human activity. Thus, future smart cities should be spatially enabled and develop new spatial skills (Roche, 2014). This development is based on the idea that location and spatial information are common goods and promotes their availability in order to stimulate innovation (Roche *et al.*, 2012). Thus the flow of spatial data generated in a smart city will present new challenges to urban planners and cartographers.

On the other hand, we have the individually controlled devices. For example, we have recently seen an increase in the use of drones, and we are still exploring their capabilities to contribute to systematic data gathering despite surveillance and privacy issues which we do not discuss here (see, for example, *http://opendronemap.github.io/odm/*). The transition of the ability for immediate image capturing from authorities and private companies to citizens is expected to fuel the creation of more user-generated spatial data, similarly to what happened with the availability of Yahoo! Imagery and its role in the evolution of OSM. Moreover, the evolution of wearable technology, while still in its early days, is expected to contribute to the evolution of

VGI. The omnipresence of wearable sensors will multiply the availability of spatial data on the Web. A similar impact is expected from the development of indoor positioning and mapping (see, for example, Google's Tango project) which will extend VGI into new fields.

It is obvious that, technology-wise, there are already various means for collecting VGI and there will be even more in the future. However, what transforms the device-capturing capabilities into a phenomenon that is reshaping the geospatial domain is how people are using these means. Thus, of high importance are the social factors that drive the creation of VGI. The future of volunteerism, of crowdsourced efforts and of the mentality of collaboration and the proactive role of citizens who are called to participate in the management and improvement of the quality of life (e.g. environmental quality, security, crisis mitigation and so on), will prove equally important factors in the evolution of VGI as the technology advances.

To an extent, this is a time of very rapid change. Most of our prior efforts in map making have been for the relative long term, but we now have a much more immediate sense of how space is changing in terms of physical features, value and importance. What we need to do, and VGI is a key enabling factor in this, is to replace the notion of static geospatial data, which served the idea of authoritative maps rather well in the past, with the comprehension of the spatiotemporal environment that we live in. Once this is achieved, then it is easy to imagine a new breed of online maps that escape from the restricted notion of base or topographic mapping towards more agile and flexible forms that better suit contemporary societal requirements.

## Notes

1 *http://mappingforchange.org.uk/projects/royal-docks-noise-mapping/.*
2 *www.fixmystreet.com/.*
3 *www.geo-wiki.org/.*

## References

Antoniou, V. (2011) *User Generated Spatial Content: An Analysis of the Phenomenon and its Challenges for mapping Agencies* unpublished Ph.D. thesis, University College London.
Antoniou, V. and Skopeliti, A. (2015) "Measures and Indicators of VGI Quality: An Overview" *ISPRS Annals of the Photogrammetry, Remote Sensing and Spatial Information Sciences* II-3/W5 pp.345–351.
Antoniou, V., Morley, J. and Haklay, M. (2010) "Web 2.0 Geotagged Photos: Assessing the Spatial Dimension of the Phenomenon" *Geomatica* 64 (1) pp.99–110.
Antoniou V., Fonte, C.C., See, L., Estima, J., Arsanjani, J.J., Lupia, F., Minghini, M., Foody, G. and Fritz S. (2016) "Investigating the Feasibility of Geo-tagged Photographs as Sources of Land Cover Input Data" *ISPRS International Journal of Geo-information* 5 (64) pp.1– 20.
Bégin, D. (2012) "Towards Integrating VGI and National Mapping Agency Operations: A Canadian Case Study" paper presented at *The Role of Volunteer Geographic Information in Advancing Science: Quality and Credibility Workshop, GIScience Conference* Columbus, OH, 18 September.
Bertot, J. C., Jaeger, P. T. and Hansen, D. (2012) "The Impact of Polices on Government Social Media Usage: Issues, Challenges, and Recommendations" *Government Information Quarterly* 29 (1) pp.30–40.
Bruns, A. (2006) *Towards Produsage: Futures for User-Led Content Production QUT ePrints* Available at: *http://eprints.qut.edu.au/4863/* (Accessed: 18 March 2016).
Buchanan, M. (2002) *Nexus: Small Worlds and the Groundbreaking Science of Networks* Cambridge, UK: Perseus Publishing.
Bunge, W. (1971). *Fitzgerald: Geography of a Revolution* Cambridge, MA: Schenkman.
Capineri, C. and Rondinone, A. (2011) "Geografie (in) volontarie" *Rivista geografica italiana* 118 (3) pp.555–573.
Clinton, W. (2000) *Improving the Civilian Global Positioning System (GPS)* Office of Science and Technology Policy, Executive Office of the President.

Coleman D. J., Georgiadou Y. and Labonte J. (2009) "Volunteered Geographic Information: The Nature and Motivation of Produsers" *International Journal of Spatial Data Infrastructures Research* 4 (1) pp.332–358.

Craglia, M. (2007) "Volunteered Geographic Information and Spatial Data Infrastructures: When Do Parallel Lines Converge?" *Position Paper for the VGI Specialist Meeting* Santa Barbara, CA, 13–14 December.

Crampton, J.W. (2009) "Cartography: Maps 2.0" *Progress in Human Geography* 33 (1) pp.91–100.

Curry, M. (1997) "The Digital Individual and the Private Realm" *Annals of the Association of American Geographers* 87 (4) pp.681–699.

Ellul, C., Haklay, M., Francis, L. and Rahemtulla, H. (2009) "A Mechanism to Create Community Maps For Non-Technical Users" *The International Conference on Advanced Geographic Information Systems & Web Services GEOWS 2009* pp.129–134 Cancun, Mexico, 1–9 February.

Elwood, S. (2008) "Volunteered Geographic Information: Future Research Directions Motivated by Critical, Participatory, and Feminist GIS *GeoJournal* 72 (3,4) pp.173–183.

Elwood, S., Goodchild, M. and Sui, D. (2012) "Researching Volunteered Geographic Information: Spatial Data, Geographic Research, and New Social Practice" *Annals of the Association of American Geographers* 102 (3) pp.571–590.

Estes, J.E. and Mooneyhan, W. (1994) "Of Maps and Myths" *Photogrammetric Engineering and Remote Sensing* 60 (5) pp.517–524.

Flanagin, A. and Metzer, M. (2008) "The Credibility of Volunteered Geographic Information" *GeoJournal* 72 (3,4) pp.137–148.

Girres, J. and Touya, G. (2010) "Quality Assessment of the French OpenStreetMap Dataset" *Transactions in GIS* 14 (4) pp.435–459.

Goodchild, M.F. (2007) "Citizens as Sensors: The World of Volunteered Geography" *GeoJournal* 69 (4) pp.211–221.

Goodchild, M.F. (2012) *VGI Research Overview* Available at: *http://web.ornl.gov/sci/gist/workshops/2012/vgi_documents/2012_VGIworkshop_Goodchild.pdf* (Accessed 30 October 2016).

Goodchild, M.F. and Glennon, J. A. (2010) "Crowdsourcing Geographic Information for Disaster Response: A Research Frontier" *International Journal of Digital Earth* 3 (3) pp.231–241.

Graham, M., Hogan, B., Straumann, R. and Medhat, A. (2014) "Uneven Geographies of User-Generated Information: Patterns of Increasing Informational Poverty" *Annals of the Association of American Geographers* 104 (4) pp.746–764.

Haklay, M. (2010) "How Good is Volunteered Geographical Information? A Comparative Study of OpenStreetMap and Ordnance Survey Datasets" *Environment and Planning B: Planning and Design* 37 (4) pp.682–703.

Haklay, M. (2013) "Citizen Science and Volunteered Geographic Information: Overview and Typology of Participation" in Goodchild, M.F. and Sui, D.Z. (Eds) *Crowdsourcing Geographic Knowledge* Dordrecht, The Netherlands: Springer, pp.105–122.

Haklay, M., Basiouka, S., Antoniou, V. and Ather, A. (2010) "How Many Volunteers Does it Take to Map an Area Well? The Validity of Linus' Law to Volunteered Geographic Information" *The Cartographic Journal* 47 (4) pp.315–322.

Haklay, M. E., Antoniou, V., Basiouka, S., Soden, R. and Mooney, P. (2014) "Crowdsourced Geographic Information Use in Government" *Report to GFDRR (World Bank), London.*

Hardy, D., Frew, J. and Goodchild, M. (2012) "Volunteered Geographic Information Production as a Spatial Process" *International Journal of Geographical Information Science* 26 (7) pp.1191–1212.

Hecht, B.J. and Gergle, D. (2010) "On the Localness of User-Generated Content" *Proceedings of the 2010 ACM Conference on Computer Supported Cooperative Work* pp. 229–232.

Hecht, B. and Stephens, M. (2014) "A Tale of Cities: Urban Biases in Volunteered Geographic Information" *Proceedings of ICWSM 2014* pp.197–205.

Kent, A.J. (2010) "Helping Haiti: Some Reflections on Contributing to a Global Disaster Relief Effort" *The Bulletin of the Society of Cartographers* 44 (1,2) pp.39–45.

Kerfoot, H. (2006) *Manual for the National Standardization of Geographical Names* New York: United Nations.

Neis, P. and Zipf, A. (2012) "Analyzing the Contributor Activity of a Volunteered Geographic Information Project: The Case of OpenStreetMap" *ISPRS International Journal of Geo-Information* 1 (3) pp.146–165.

Obermeyer, N. (1998) "The Evolution of Public Participation GIS" *Cartography and Geographic Information Science* 25 (2) pp.65–66.

Olteanu-Raimond, A., Hart, G., Foody, G., Touya, G., Kellenberger, T. and Demetriou, D. (2016) "The Scale of VGI in Map Production: A Perspective on European National Mapping Agencies" *Transactions in GIS* doi: 10.1111/tgis.12189.

Priedhorsky, R. and Terveen, L. (2008) "The Computational Geowiki: What, Why, and How" *Proceedings of the 2008 ACM Conference on Computer Supported Cooperative Work* pp.267–276.

Purves, R. and Hollenstein, L. (2010) "Exploring Place through User-Generated Content: Using Flickr to Describe City Cores" *Journal of Spatial Information Science* 1 pp.21–48.

Roche, S. (2014) "Geographic Information Science I: Why Does a Smart City Need to Be Spatially Enabled?" *Progress in Human Geography* 38 (5) pp.703–711.

Roche, S., Nabian, N., Kloeckl, K. and Ratti C. (2012) "Are 'Smart Cities' Smart Enough?" in Rajabifard, A. and Coleman, D. (Eds) *Spatially Enabling Government, Industry and Citizens: Research Development and Perspectives* Needham, MN: GSDI Association Press, pp.215–236.

Stamp, L.D. (1931) "The Land Utilization Survey of Britain" *The Geographical Journal* 78 pp.40–47.

Tobler, W. (1970) "A Computer Movie Simulating Urban Growth in the Detroit Region" *Economic Geography* 46 pp.234–240.

Turner, A. (2006) *Introduction to Neogeography* Sebastopol, CA: O'Reilly.

Zook, M., Graham, M., Shelton, T. and Gorman, S. (2010) "Volunteered Geographic Information and Crowdsourcing Disaster Relief: A Case Study of the Haitian Earthquake" *World Medical & Health Policy* 2 (2) pp.6–32.

# 36

# Maps and imagination

*Peter Vujakovic*

## Introduction: the view from Mr Clevvers Roaling Place

'To ask for a map is to say "Tell me a story"' Turchi (2004: 11).

From the crest of the chalk downs, on the precipitous edge of Mr Clevvers Roaling Place, I can look down on my house in the middle distance. Directly below me, I can see Widders Dump, and further out Bernt Arse. On a clear day, I can see all the way to Dunk Your Arse. You will not, however, find any of these places named on the Ordnance Survey's maps of East Kent, but they do, in a manner, exist. They are a corruption of contemporary place-names conjured in the imagination of the author Russell Hoban. They are very real locations in the landscape I inhabit. The places, in order, are the Devil's Kneading Trough (a deep chalk coombe), the hamlet of Withersdane, the town of Ashford, and Dungeness, a massive shingle spit. Other degraded names in Hoban's (1980) cult novel *Riddley Walker* and the accompanying map include 'Horney Boy' for Herne Bay, 'Monkeys Whore Town' for Monks Horton, and 'Hagmans Il' for Hinxhill – itself a corruption of Haenostesyle (c.1100) (the hill of the stallion, or a man called Hengest from the Old English *hengest + hyll* according to Mills, 1998) and part of the parish in which I live. Hoban tells the tale of a post-apocalyptic, post-nuclear world 'regressed to the level of the Iron Age' (Self, 2002). The narrator of the story, the eponymous Riddley Walker, is the supposed author of the map accompanying his narrative; he notes 'This here is jus places ive tol of in this writing. I dont have room for the woal of every thing there is in inland [England]' (precedes title page). The whole book is written in this degraded language – and is most easily understood if read slowly and aloud. Through his mapping, Walker stamps the authority of cartographic knowledge as power on his narrative – maps in 'inland' must be few and far between.

The map in Hoban's book involves more than a simple alteration of place-names. Hoban has created a world in which the configuration of the coast, as well as language and culture, has changed due to sea-level rise. In creating this partially drowned landscape, he acknowledged the help of scientist Paul Burnham (Wye College, University of London) in preparing the map. *Riddley Walker's* coastline is very similar to that which existed in late Iron Age Britain (see Young, 2004), when the Isle of Thanet was separated from the rest of Kent by the Wantsum Channel; *Riddley Walker* is a work of imagination, but it is also one of cartographic, toponymic and topographic logic. The map is an important reminder to its readers that the world is not static, but that over millennia the landscape will change and so will its appellations. There are, however, problems with mapping change where real world processes are poorly understood. Hoban's sea-level rise may map reasonably well to the contours of the Stour valley, but the world is not a plastic bucket and it is extremely unlikely that the shingle spit at Dungeness would retain its current shape with the anticipated sea-level rise. Imagination has its limits.

This chapter examines the various 'creative' connections between maps, cartography and literature. Maps and the imaginative realms of fiction have been almost constant bed-fellows since the advent of the modern novel. The link takes several forms: maps created as an adjunct to the literary work (often illustrating an itinerary, or providing a regional setting); maps cited and sometimes included graphically as part of the narrative, for example, in Stevenson's *Treasure Island* or Conan Doyle's *Lost World*; or the narrative itself as a 'map' – a textual 'deep geography' – Joyce's *Ulysses*, for instance. There have also been attempts by others than the author to 'map' the narrative, whether entirely fictional or based on a real geography; for example, Barbara Strachey's (1981) *Journeys of Frodo: An atlas of J.R.R. Tolkien's The Lord of the Rings*, or attempts to map the protagonists' wanderings in Joyce's Dublin (see, for example, Hogan, 1996).

Maps are no mere adjunct to a story in a novel; they are often the stimulus to the author's imaginings and the connection with an intense geographic 'authenticity' that captures and enhances the reader's imagination. Brogan (2004) notes this in his study of Arthur Ransome's children's books, as well as other authors, and states 'they are drawn to give concreteness . . . the map is a powerful tool of what Tolkien calls "sub-creation"; the invention of a secondary world' (2004: 151). Maps help construct a distinct 'sense of place' as well as a tangible *spatial framework*, and often act as a visual metaphor for accuracy, authenticity, or for surveillance and control. The latter is exemplified by many of the imperial narratives in which Western protagonists appropriate the territory of others, either directly or through their hegemonic 'gaze' (see later in this chapter).

Brian Harley's (1989) concepts of the 'external and internal power' of cartography and the role of silences remain valid for maps of fictional places, as well as maps of real places incorporated into fictional narratives. The author (or other authoring agent) exerts external control on and through the map, i.e. governs what information the map contains and what is suppressed. The map, through the power internal to cartography produces a sense of order and detail that in turn exerts its influence over the reader's imagination. As Hegglund (2003), writing about Joyce's *Ulysses*, notes that once information (fact or fiction) is mapped 'it instantaneously acquires the aura of fact and "reality"' (Hegglund, 2003: 167).

## Once upon a time. . .

Maps, geography and literature have a long and complex history. Miguel de Cervantes' *Don Quixote* is widely considered to be the first modern European novel, published in two volumes in 1605 and 1615; the third edition, for *La Real Academia Española*, included a detailed map that displayed the route taken across Spain by the eponymous hero (Shaw, 1996).[1] This is an important concept in modern literature; the map is not an incidental embellishment – it anchors the narrative in the real world (a series of numbered notes cross-reference with Don Quixote's adventures along the route). This use of the map is an example of the concept of 'gratuitous detail', in which speculative or fictional representations are rendered in *exceptional detail* (Gifford-Gonzalez, 1993). The wealth of detail on the map of the 'real world' sustains the plausibility of the narrative.

The *Don Quixote* map is but the first of many that have been used to describe an imagined itinerary on a plan of the real world. Other famous examples include 'A MAP of the WORLD on wch is delineated the Voyages of ROBINSON CRUSO', supposedly drawn by Daniel Defoe for his famous novel based loosely on the life of Alexander Selkirk (Bradbury, 1996; Harmon, 2004 cites the map as 'artist unknown'). The map is based on a projection that shows the old and new worlds on two separate hemispheres – the attention to cartographic minutiae in terms of the graticule and other map conventions reinforces the veracity of the narrative through the application of rigorous detail. Similarly, the very detailed map of the voyage of the brig *Covenant* and the 'wanderings' of David Balfour, the hero of Robert Louis Stevenson's' *Kidnapped* (1891 illustrated edition). The detail involved is exquisite, from the landing on Mull to the eventual arrival at Queensferry. These maps can also be seen as examples of the panoptical-power of the author. Gratuitous detail does not have to involve the 'real world' to work; fantasy worlds are often mapped in such high detail that they become convincing. Maps are the territory – even if the territory is imagined.

## Yoknapatawpha County: the map is the territory?

A number of authors have created detailed maps of the milieu in which their novels are set to help in plotting their narratives. Yoknapatawpha County is the setting for what William Faulkner called his 'little postage stamp of native soil' based on Oxford, Mississippi (Bradbury, 1996). Faulkner rechristens this 'Jefferson'. The map he drew framed his world using two natural features which retained their real names, the Tallahatchie and Yoknapatawpha rivers. As Bradbury notes, Faulkner's world was 'like Thomas Hardy's Wessex, it was an imaginary place, but laid over real facts . . . 2,400 rural square miles of crossed destinies and complex genealogies' (1996: 197). Faulkner's strong connection with place and its history is evident in his evocation of the American South. Harmon's (2004) comment on Faulkner's map – that 'his fictional world seem[s] *almost* real' (2004: 183, emphasis added) – underestimates the power of cartography. Faulkner, by mapping his 'place', assumes a hegemonic and panoptic position as its creator and controller and draws his readers into a 'reality' with which they can actively engage.

Others, in the 1920s, did for the mid-West what Faulkner did for the South, for example, Sinclair Lewis in *Main Street* and Sherwood Anderson in *Winesberg*. Anderson's *Winesberg*, a series of twenty-five linked stories, dissected the lives of the people trapped in a small town whose life is also focused on another Main Street and its institutions. The accompanying map of Winesberg has something of the specimen held under a magnifying glass. The 'map', drawn from an oblique viewpoint, is focused on just two streets with their back alleys, and a railway line; each building and plot (with their picket fences) is shown in detail, but no transient elements such as a car or person are included, giving the view a feeling of *breathless stillness*. Its intimacy and isolation (a large amount of white space surrounds the detailed map of the town centre) reinforces the feeling of stifling entrapment. The map not only frames the action, but creates a sense of ethnographic survey.

Hardy's 'Wessex' is perhaps the embodiment of this narrative link with a particular geography. Hardy is the best-known of a group of regional novelists whose work focused on south-central England. Hardy revived the regional term 'Wessex' (based on an Anglo-Saxon kingdom) in 1874 as part of his belief in the need to provide a strong link between his characters and their environment (Birch, 1981). Real places were given new names, and so their locations became an object of fascination for his readers. Hardy created a map of Wessex in 1895, and other versions have been included in various editions of his novels to satisfy the demands of his readership.

Hardy, acknowledging the power of his creation, noted that twenty-one years after adopting the regional title of Wessex, his 'dream-country' had 'by degrees solidified into a utilitarian region which people can go to, take a house in, and write to the papers from' (Thomas Hardy Society, n.d.). The map is the territory – cartographic representations of the region have solidified this 'sense of place' further, with contemporary government agencies now being named for this literary region, for example the 'Wessex Area Team' of the NHS and the Crown Prosecution Service Wessex.

What is more interesting, however, is the complex manner in which Hardy uses geography and cartography. As Kolb (2014) points out, the circles at work in Hardy's plots are also often represented by 'circling journeys' taken by his characters made 'possible partly because of the intense geographical specificity that Hardy affords his novels' (2014: 596). The map is more than a backdrop to Hardy's world. It provides the very basis for structuring many of his novels. Kolb makes a strong case for Hardy's representation of the binary choices that his characters often have to make in terms of 'games' of chance or based on imperfect information which effectively 'randomizes' decisions. She draws attention specifically to the 'snakes and ladders' genre, many of which had map-like structures (e.g. 'RIGHT ROADS and WRONG WAYS', 1848). These dice-based games are entirely based on chance and have the players 'tracing circle after circle before the game's close' (2014: 616). The roads in Hardy's novels are full of forks, both literal and metaphorical. Here the map is more than the territory; it supports the directorial power of the author to shape destinies.

Perhaps the most famous map to form the detailed basis of a fictional narrative is Robert Louis Stevenson's *Treasure Island* (1883) (Figure 36.1), largely because of his own reflections on the role of this and other maps in the creative process. In his later essay on the book, Stevenson not only acknowledges the part the specific map played in stimulating his novel but also waxes lyrical on the importance of maps in general as a means of stimulating the imagination:

> I made the map of an island; it was elaborately and (I thought) beautifully coloured; the shape of it took my fancy beyond expression; it contained harbours that pleased me like sonnets; and with the unconsciousness of the predestined, I ticketed my performance 'Treasure Island'. I am told there are people who do not care for maps, and find it hard to believe. The names, the shapes of the woodlands, the courses of the roads and rivers, the prehistoric footsteps of man still distinctly traceable up hill and down dale, the mills and the ruins, the ponds and the ferries, perhaps the *Standing Stone* or the *Druidic Circle* on the heath; here is an inexhaustible fund of interest for any man with eyes to see or twopence-worth of imagination to understand with!
>
> *(Stevenson, 1905: 111)*

This map is so much a part of the story that the artist and caricaturist Ralph Steadman, despite his iconoclastic style, does not deviate from the original detail of Stevenson's map for his illustrated version of the book. Turchi (2004) reminds us that the map reproduced in the original *Treasure Island* was itself a *re*-creation by Stevenson of his original – supposedly mislaid by the publisher or lost in the mail. Bushell (2015) provides a much more detailed analysis of the significance of Stevenson's narrative of a 'lost original' for the authorial myth-making associated with the map.

But yet again, the map is more than simply the territory, as Bushell (2015) makes clear in her detailed analysis of the role played by the map/chart in the novel. The map has three distinctive roles: its navigational value as a chart, its 'geocentric' role as topographic map and its 'egocentric' role as something equivalent to Flint's 'will', with a value to its holder 'that

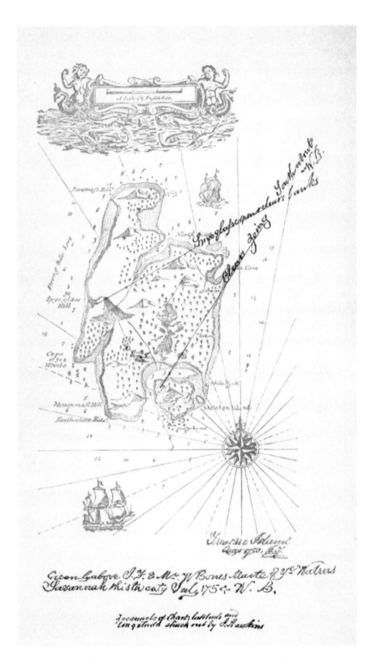

*Figure 36.1  Treasure Island –* Robert Louis Stevenson's inspirational map

exceeds its informational value' (2015: 617). She also makes an interesting point suggesting that the map is 'as duplicitous as those who fight over it' (2015: 619), because while the map proved 'accurate' in leading the protagonists to the island, it fails to lead them to the treasure, which has been moved. She is effectively pointing to the internal power of maps and the importance of 'accuracy' as a perceived matter of trust in cartography, despite her lapse into anthropomorphism with regard to the map! (Brogan, 2004).

## The 'Great Game'

The old adage that 'a picture paints a thousand words' works equally well for cartography, and the author's ability to summon a well-known map into the reader's imagination can encapsulate a world-view that would otherwise take pages to describe. In Joseph Conrad's *Heart of Darkness* (originally published in 1899), his hero, Charlie Marlow, gazes upon a political map of the world, a Victorian world-view that would have been extremely familiar to Conrad's readership, and makes the requisite connotations as Marlow prepares to sail to the Belgian Congo:

> There was a vast amount of red – good to see at any time, because one knows that some real work is done in there, a deuce of a lot of blue, a little green, smears of orange, and, on the East Coast, a purple patch, to show where the jolly pioneers of progress drink the jolly lager-beer. However, I wasn't going into any of these. *I was going into the yellow.* Dead in the centre. And the river was there – fascinating – deadly – like a snake.
>
> *(Conrad, 1983: 36, emphasis added)*

As Edward Said (1993) notes, despite Conrad's semi-outsider status, and his scrupulous record of the differences between Belgium's dreadful treatment of indigenous people and relatively more benign 'British colonial attitudes, he, as author, could only imagine the world carved up into one or other Western spheres of dominance' (1993: 27). This is perhaps unsurprising in a world in which maps of the empire had become a potent and dominant viral meme circulating from the classroom to the boardroom (Vujakovic, 2009, 2013). Marlow leaves us in no doubt that he is going into a *toxic space*. Yellow has habitually carried negative connotations, for example a German propaganda map of the 1940s skilfully subverts a standard map of the British Empire by substituting the traditional use of 'red' for yellow (against a black background, nature's warn-ing colours), evoking an image of Britain as a predatory power and a danger to world peace. The British Empire was traditionally displayed as rose-pink or red implying health and vigour (Vujakovic, 2014). In an earlier passage, Marlow's fascination with the colonial enterprise is aroused by another map, in this case by:

> [a] mighty river, that you could see on the map [in a shop window], resembling an immense snake uncoiled, with its head in the sea, its body at rest curving afar over a vast country, and its tail lost in the depths of the land … it fascinated me as a snake would a bird.
>
> *(2014: 33)*

This role of the map in affirming hegemony, generally Western over 'subaltern' populations and regions in the Americas, Asia and Africa – the 'other' – is not uncommon, neither is the concept of the 'map' as the playing-field to be contested by great powers. Brogan (2004) suggests that it is probably no coincidence that use of maps in British literature coincides with both the cheaper reproduction techniques and the fact that the British Empire was at its zenith; even *Treasure Island* 'looks back joyously to one moment in the rise of the empire . . . the age of pirates on the Spanish Main' (2004: 151) – the 'Age of the Predatory Powers' of which England was one.

Kipling's *Kim* (publ. 1901) involves the eponymous hero in learning the 'tricks of the trade' of surreptitious surveying methods as part of the 'Great Game', the contest between Britain and Russia for the north-west frontier region. In Erskine Childers' *The Riddle of the Sands* (publ. 1903) two maps and two charts of German East Friesland are critical to the narrative involving German plans for an invasion of Britain. The 'authenticity' of this fictional narrative is rein-forced by a preface (initialed E.C.) in which Childers claims he is merely the editor of material

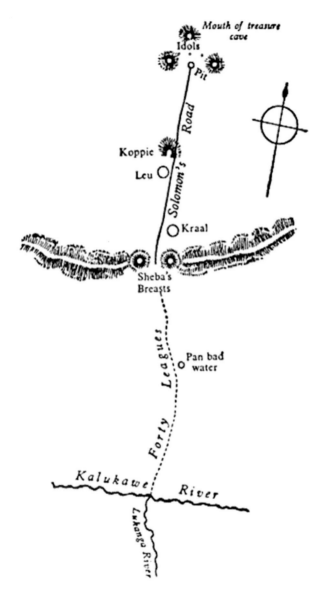

*Figure 36.2* The African landscape eroticized – map from H. Rider Haggard's *King Solomon's Mines*

provided to him by one of the protagonists, and to which he adds a note that the charts are based on British and German Admiralty charts, with superfluous details omitted.

In other novels of empire, sketch maps form an important element of Western 'penetration' of the space of the other. Conan Doyle's *The Lost World* (publ. 1912) includes two sketch maps of a remote region of South America. The maps serve as evidence of the 'discoveries' made by the European explorer-scientists. More fascinating, perhaps, is the sketch map in H. Rider Haggard's *King Solomon's Mines* (publ. 1885) (Figure 36.2). This is a treasure map like Stevenson's (another example of the map as 'will'), but the form of the map is outstanding in its iconography of the imperialist urge. Several authors note the fact that the map 'displays'

495

the topography of the setting as a female body that succumbs to the wish of male adventur-ers (Stott, 1989; Cheng, 2002). Anne McClintock (1995) intimately details the symbolism of bodily penetration associated with the map, in which the European males pass between Sheba's Breasts (two distinctive hills) before passing a 'Koppie' (a small hill corresponding in the bodily metaphor to the navel), before descending into the mouth of the mine surrounded by a v-shaped set of three hills covered in 'dark heather'. As McClintock states, 'the body is [shown] spread-eagled and truncated – the only parts drawn are those that denote female sexu-ality' and 'is literally mapped in male body fluids' (1995: 3), the blood of an earlier explorer. Africa becomes a passive body waiting to be possessed (Bushell, 2015). Haggart's map may be the most overt of the genre, but the metaphor of 'penetration' has been a common feature of nineteenth- and early twentieth-century geopolitical discourse.

## Dense geographies: the text as the map

As Matthew Edney notes, there is actually no such thing as 'the map' – maps do not have to take a graphic form, but may be gestural, spoken or performed (see Chapter 5). Some literature can, on this basis, be read as 'dense geography', in which the milieu is 'thickly mapped' within the narrative. An obvious example is Joyce's *Ulysses*. Kiberd (1992) notes 'from earliest child-hood, Joyce had been obsessed with making lists' (1992: xxi) and quotes his father as saying 'if that fellow was dropped in the middle of the Sahara, he'd sit, be God, and make a map of it' (1992: xxii). The intensely cartographic nature of his work is acknowledged by many authors (see Hegglund, 2003). Both Bulson (2001–2) and Hegglund (2003) provide interesting perspec-tives based on Joyce's use of maps based on the Ordnance Survey's mapping of Ireland as an imperial project.

'Precision, accuracy, authenticity, realism: these are all terms that have circulated in discus-sions of Joyce's Dublin to identify his almost heroic attention to navigational details' (Bulson, 2001–2: 96). This has led to a number of authors, including Joyce's friend Frank Budgen, to trace the geography of Joyce's Dublin. Budgen recorded Joyce's assertion that 'I want . . . to give a pic-ture of Dublin so complete that if the city one day suddenly disappeared from the earth it could be reconstructed out of my book' (Budgen, cited in Hegglund, 2003: 164). Studies that make extensive use of maps include Hart and Knuth (1975), Seidel (1976) and Delaney (1982). Several of these, according to Hegglund (2003), treat the cartographic element of the work in a 'factual' manner, whose prime purpose is to catalogue places. Hegglund takes the discussion further and examines Joyce's *rhetoric of cartography*. He uses recent ideas in critical cartography, specifically the idea of the 'creative map reader', to unpack Joyce's ability to destabilize the imperialist mapping of Dublin and Ireland more broadly, and, like pidgins and other dialects that adjust a hegemonic language, re-signify space. Hegglund (2003: 189) concludes: 'The incorporation of cartographic rhetoric in *Ulysses* need not be seen as an imperial complicity, nor can it be reduced to a revo-lutionary postcolonial mapping of resistance'. But as challenging and re-routing its readers in unpredictable ways, Joyce opens the geography of Dublin to multiple possibilities that challenge both imperial power and blinkered nationalism. Beautell (1997) provides a similar analysis of cartographic practice in Canadian literature; mappings which offer alternatives to the colonial writing of maps to construct alternative spaces of geography and identity in a contested 'nation'.

Like Joyce, the German author Günter Grass was another complier of lists, and this also extended to his 'dense geography' of the city and surrounds of Danzig (now Polish Gdansk). Grass was born in Danzig in 1927 and his autobiography (2007) acknowledges his debt as a writer to Joyce. Grass's works deal with the region around the city and the Vistula river basin. His so-called Danzig trilogy (*Tin Drum*, *Cat and Mouse*, and *Dog Years*) is particularly associated with this 'unique cultural setting' (Frank, 2006: 1). Frank regards these books as part of the 'movement' referred to as *Vergangenheitsbewältigung* ('a coming to terms with the past') which deals with the rise of Nazism, and the German people's experience of the war and the post-war era.

Siegfried Mews' (2008) comprehensive review of Grass' works and his critics addresses Grass' compunction to list as a part of his construction of a 'fantastic realism'. He notes that Grass has 'an almost pedantic penchant for rendering historical, geographical, political, and other details with a high degree of accuracy' (2008: 87), which is then overlain with the fantastic to 'explode' this hyper-realism. Grass' deep geography, with its gratuitous detail, serves to heighten the catastrophe of Nazism as it emerges from the meticulously described lived-experience of his various protagonists. Mews notes that *Heimat* (the German concept which has no clear English equivalent, but is a positive relationship between a people and a specific place, a 'home' or 'homeland') is an important element of Grass' work, but that he does not idealize or sentimentalize his own *Heimat*, Danzig. Rather, its subsequent loss, as Germans were expelled from territories to the east of the Oder-Neisse line, is seen as the inevitable consequence of middle-class German support for the Nazi programme. Grass is interested in this blurring of geographies; his Danzig/Gdansk is a quintessential cross-cultural border territory (Frank, 2006) and his writing plays with the concepts of continuity and change. The inclusion of Cassubians within the family histories (Slavic-Pomeranian peoples) adds extra detail to Grass' setting, 'introducing a completely different reality than that of the petit bourgeoisie in the city of Danzig' (ibid.: 12). This, together with Grass's continuous references to the historic changes in political power within the region, serves to disrupt the hegemonic mapping of racialized space.

Grass also invokes real maps as a means of understanding processes of geopolitical change:

> Circles described by compasses on military maps. Schlangenthin taken in counterattack. Antitank spearheads on the road to Damerau. Our troops resist heavy pressure northwest of Osterwick. Diversionary attacks of the twelfth division of airborne infantry thrown back south of Konitz. In line with the decision to straighten the front, the so-called territory of Koshnavia is evacuated. Last-ditch fighters take up positions south of Danzig.
>
> *(Grass, 1969: 140)*

This use of maps and geographic lists is an essential element of Grass' cosmos, which treats human history in the 'tidal lands' of Europe as a long-drawn process of invasion and occupation in which the local populations show a stoic forbearance and maintain local loyalties to space and place. Cartographic performances even extend to the dog Harras (in *Dog Years*) and his sense of territory:

> All over Langfuhr, in Schellmühl, in the Schichau housing development, from Saspe to Brösen, up Jäschkenter Weg, down Heiligenbrunn, all around the Heinrich Ehlers Athletic Field, behind the crematory, outside Sternfeld's department store, along the shoes of Aktien Pond, in the trenches of the municipal police, on certain trees of Uphagen Park, on certain lindens of the Hindenburgallee, on the bases of advertising pillars, on the flagpoles of the demonstration-hungry athletic field, on the still unblacked-out lampposts of the suburb of Langfuhr, Harras left his scent marks.
>
> *(Grass, 1969: 139)*

Grass, like Joyce, challenges the rhetoric of cartography through his varied cast of misfits that subvert the Nazi appropriation of territory for the racially and bodily pure. These mappings offer alternative perspectives on colonial inscriptions of space.

## Conclusion

While this discussion has mainly focused on what may be deemed imperialist and counter-imperialist, or at least alternative, mappings of empires, it is clear that most forms of mapping have a hegemonic aspect, including, for example, Hardy's appropriation of south-central England. Whole genres, such as fantasy with its proliferation of maps, or children's literature have largely been ignored, especially those for younger children. These also offer fertile ground for examination of their role within narrative structures and beyond (for example, in children's environmental learning). Pavlik's (2010) study of maps in children's literature recognizes that mapping is an act of creation of territory, but generally regards the experiential relationship between map reader and the map as an imaginative activity. Sundmark's (2014) in-depth examination of the maps in Tove Jansson's *Moomintroll* series also suggest that maps provide a playful and creative space for child readers. Yet, the Western enlightenment urge to catalogue space as a form of control remains evident when Moominmamma muses that she might create a map of their island (in *Moominpappa at Sea*, publ. 1974, trans. of *Pappan och havet*, 1965) 'showing all the rocks and shallows and perhaps the depth of the water as well'. Jeremy Black (1997), in *Maps and Politics*, similarly notes that A.A. Milne's Winnie-the-Pooh appropriates territory through exploration and cartographic record, while the maps accompanying recent editions of Kenneth Graham's *The Wind in the Willows* use cartographic 'silence' to remove the 'other' – the weasels and stoats – from Rat, Mole and Toad's rural idyll (much in the manner that Black African townships were removed from some maps of Apartheid-era South Africa, as noted by Stickler, 1990).

To ask for a map may be to ask for a story, but as the preceding discussion suggests, the story may be complex, reinforce dominant world-views and involve hidden forms of power and control, or may be open to highly contested readings. As Said (1987) observes, an Indian reader would generally focus on very different aspects of Kipling's portrayal of people and landscape in *Kim* from that of English and American readers at the time of publication. The same will be true for the cultural filter that each reader applies to their relationship with the cartographic rhetoric in works of literature, graphic or textual.

## Note

1 A high resolution of the map can be viewed at the Newberry Digital Collection for the Classroom, available at: *http://dcc.newberry.org/items/map-of-don-quixotes-route-through-spain* (Accessed: 14 March 2016).

## References

Beautell, E.D. (1997) "Spatial/Identity Deconstructions: Nation and Maps in Canadian Fiction" *Atlantis* 19 (2) pp.45–50.

Birch, B.P. (1981) "Wessex, Hardy and the Nature Novelists" *Transactions of the Institute of British Geographers*, New Series 6 pp.348–358.

Black, J. (1997) *Maps and Politics* London: Reaktion Books.

Bradbury, M. (Ed.) (1996) *The Atlas of Literature* London: De Agostini.

Brogan, H. (2004) "The Lure of Maps in Arthur Ransome" in Harmon, K. (Ed.) *You are Here: Personal Geographies and other Maps of the Imagination* New York: Princeton Architectural Press, pp.150–153.

Bulson, E. (2001-2) "Joyce's Geodesy" *Journal of Modern Literature* 25 (2) pp.80–96.

Bushell, S. (2015) "Mapping Victorian Adventure Fiction: Silences, Doublings, and the Ur-Map in Treasure Island and King Solomon's Mines" *Victorian Studies* 57 (4) pp.611–637.

Cheng, C.-C. (2002) "Imperial Cartography and Victorian Literature: Charting the Wishes and Anguish of an Island-Empire" *Culture, Theory and Critique* 43 (1) pp.1–16.

Conrad, J. (1983) *Heart of Darkness* Harmondsworth, UK: Penguin Books (originally published 1902).

Delaney (1982) *James Joyce's Odyssey: A Guide to the Dublin of* Ulysses New York: Holt.

Frank, S. (2006) "Territories and Histories: Transgressive 'Space Travels' and 'Time Travels' in Grass's *Dog Years*" *Quest* 2 (20pp.) Available at: *www.qub.ac.uk/sites/QUEST/FileStore/2010JanEdit/Fileto upload,178939,en.pdf* (Accessed: 2 April 2016).

Gifford-Gonzalez, D. (1993) "You Can Hide, But You Can't Run: Representation of Women's Work in Illustrations of Palaeolithic Life" *Visual Anthropology Review* 9 (1) pp.22–41.

Grass, G. (1969) *Dog Years* Harmondsworth: Penguin Books (trans. by R. Manheim and first published in Germany in 1963, as *Hundejahre*)

Grass, G. (2007) *Peeling the Onion* Orlando, FL: Houghton Mifflin Harcourt.

Harley, B. (1989) "Deconstructing the Map" *Cartographica* 26 (2) pp.1–20.

Harmon, K. (2004) *You are Here: Personal Geographies and other Maps of the Imagination* New York: Princeton Architectural Press.

Hart, C. and Knuth, L. (1975) *A Topographical Guide to Ulysses* Colchester, UK: A Wake Newslitter Press.

Hegglund, J. (2003) "'Ulysses' and the Rhetoric of Cartography" *Twentieth Century Literature* 49 (2) pp.164–192.

Hoban, R. (1980) *Riddley Walker* London: Jonathan Cape.

Hogan, D. (1996) "James Joyce's Dublin" in Bradbury, M. (Ed.) *The Atlas of Literature* London: De Agostini, pp.166–169.

Kiberd, D. (1992) "Introduction" to Joyce, J. (1960) *Ulysses* (The Bodley Head reset edition) Harmondsworth, UK: Penguin.

Kolb, M. (2014) "Plot Circles: Hardy's Drunkards and Their Walks" *Victorian Studies* 56 (4) pp.595–623.

McClintock, A. (1995) *Imperial Leather: Race, Gender and Sexuality in the Colonial Contest* New York and London: Routledge.

Mews, S. (2008) *Günter Grass and His Critics: From* The Tin Drum *to* Crabwalk Rochester, NY: Camden House.

Mills, A.D. (1998) *Oxford Dictionary of English Place-Names* (2nd ed.) Oxford, UK: Oxford University Press.

Pavlik, A. (2010) "'A Special Kind of Reading Game': Maps in Children's Literature" *International Research in Children's Literature* 3 (1) pp.28–43.

Said, E. (1987) "Introduction and Notes" to Kipling, R. (1987) *Kim* (Penguin Classics edition, first published in 1901) Harmondsworth, UK: Penguin.

Said, E. W. (1993) *Culture and Imperialism* London: Chatto & Windus.

Seidel, M. (1976) *Epic Geography: James Joyce's* Ulysses Princeton, NJ: Princeton University Press.

Self, W. (2002) *Introduction to* Riddley Walker London: Bloomsbury.

Shaw, P. (1996) "Cervantes' Spain" in Bradbury, M. (Ed.) *The Atlas of Literature* London: De Agostini, pp.28–31.

Stickler, P.J. (1990) "Invisible Towns: A Case Study in the Cartography of South Africa" *GeoJournal* 22 (3) pp.329–333.

Strachey, B. (1981) *Journeys of Frodo: An Atlas of J.R.R. Tolkien's* The Lord of the Rings London: Unwin Paperbacks.

Stevenson, R.L. (1905) *Essays in the Art of Writing* London: Chatto & Windus.

Sundmark, B. (2014) "'A Serious Game': Mapping Moominland" *The Lion and the Unicorn* 38 (2) pp.162–181.

Stott, R. (1989) "The Dark Continent: Africa as Female Body in Haggard's Adventure Fiction" *Feminist Review* 32 pp.69–89.

Thomas Hardy Society (n.d.) *Hardy Country* Available at: *www.hardysociety.org/about-hardy/hardy-country* (Accessed: 17 April 2016).

Turchi, P. (2004) *Maps of the Imagination: The Writer as Cartographer* San Antonio, TX: Trinity University Press.

Vujakovic, P. (2009) "World-Views: Art and Canonical Images in the Geographical and Life Sciences" in Cartwright, W., Gartner, G. and Lehn, A. (Eds) *Cartography and Art* Berlin: Springer, pp.135–144.

Vujakovic, P. (2013) "Warning! Viral Memes Can Seriously Alter Your Worldview" *Maplines* Spring 2013 pp.4–6.

Vujakovic, P. (2014) "The Power of Maps" in *The Times Comprehensive Atlas of the World* (14th ed.) Glasgow, UK: Times Books, pp.42–43.

Young, C. (2004) "The Physical Setting" in Lawson, T. and Killingray, D. (Eds) *An Historical Atlas of Kent* Chichester, UK: Phillimore, pp.1–6.

# 37

# Mapping the invisible and the ephemeral

*Kate McLean*

Within the scientific paradigm of cartography, urban space is represented as 'directly *analogous* to actual ground conditions', and as such, 'maps are taken to be "true" and "objective" measures of the world' (Corner, 2011: 90). And so urban space is portrayed in what Deleuze and Guattari (1987) refer to as a *tracing* that always involves an 'alleged competence'; recreated point-by-point onto a physical or a digital substrate and disseminated as an artefact. Thus in such representations the city comprises a series of structures, fixed points in both space and in time.

But as well as the topographical and tangible features of the cityscape, our understanding of urban space is also full of ephemeral experiences – the dust of building constructions rising and tumbling, the smells emanating from restaurants, cafés and street markets, traffic flows pulsing according to the rhythms of labour, the ringing out of bells. Add to the ephemeral datasets listed above some 'urban invisibles' (Amoroso, 2010) such as crime statistics, SMS traffic, population density and property values, and we come to appreciate that aspects of urban landscapes are neither static nor fixed, but rather repeated across a multitude of contexts.

Advocating new analytical models for how urban planners might conceive of cities, Batty (2002) proposes a model in which the city can be regarded as 'a series of spatial events' in place of the traditional fixed infrastructures. His analytical tools also serve as a framework to consider recent work in mapping ephemeral and invisible datasets. Similarly, in her attempts to understand a fundamental political question of how humans are going to live together, social scientist and geographer Doreen Massey moved away from the idea of space as a flat surface to a more active definition in which space becomes 'the dimension of things being, existing at the same time: of simultaneity. It's the dimension of multiplicity' (Massey, 2013). With this in mind, I consider how some current urban mapping practices of the non-material engage with the multiverse qualities of the spatial.

As a designer of smellscape maps (Porteous, 1985), the olfactory equivalent of a visual land-scape, my early maps reflected on a fundamental contradiction that smell is an urban presence that forms a significant component of the city, and yet smells are invisible and ephemeral enti-ties, many of which rarely linger long enough to be identified quantitatively. I pinned specific smells to a time and a place on a physical map. The work featured in this chapter, 'Smellmap: Pamplona', seeks to address the simultaneity of qualitative perception and the dynamic nature of this urban ephemeral.

URBAN SMELL COLLECTION & DISAPPEARANCE
*'We had a smell but we lost it' commented a participant during one of a series of five smellwalks in the city of Pamplona in October 2014.*

Losing a smell in such a way has been observed during an urban smellscape mapping smellwalk; on one occasion a smell was noticed, lost and rediscovered for just long enough to give it a descriptor, 'leather', prompting recollections of summers in the Greek sandals.

Could smell be mapped? Should smell be mapped? And if smell should be mapped, from whose standpoint should it be? Which characteristics of smell might be mapped that would draw attention to our understanding of smell in the city? The urban environment is astonishingly complex. Which elements of smell might be mapped to meet Wurman's (1989: 260) tenet that 'a map is a pattern made understandable', as he advocates for simplicity over complexity in a quest to understand our cities? In considering mappings of invisible and ephemeral information as 'urban spatial events', I suggest it may be useful to consider them as fluid and transitional patterns – as Sant and Johnson (2015) advocate, we can 'consider mapping the city through its use patterns, rather than illustrating it as an assembly of static landmarks'. In this text I draw attention to how some of the sets of ephemeral and invisible information flying through our cities have been mapped by contemporary designers and artists, and include a more detailed account of a three-month-long performative, participatory, phenomenological smellmapping project of my own in Pamplona from October 2014.

## Urban mappings of invisibles and ephemerals

The contemporary city abounds with invisible data and abstract forces that shape our lives. Amoroso discusses how quantitative datasets such as crime statistics and property values 'can be rendered artistically, spatially and informatively in the form of alternative "maps" which represent urban dynamics' (Amoroso, 2010: 117). Her 'Densityscape' series of maps of New York, London and Toronto (Amoroso, 2014) use digital imaging to transform the number of people per unit area plotted onto a topographic map of the city. The points are then joined to form three-dimensional surfaces, referred to as 'data landscapes'. In these depictions, New York's population is an island of thrusting, jagged peaks; London has a series of lower blocky rock-like forms around its periphery and steep mountains in its centre; whereas Toronto is altogether smoother and hill-like. Amoroso's 'Crimescape' series applies the methodology to plotting crime statistics for individual boroughs in New York and London and inverts the 'mountains' of the population density to draw our attention beneath the flat surface, highlighting the negative connotation of the subject matter. Amoroso compares 'Crimescape' (rudi.net, 1997) to the submerged elements of an iceberg and furthers the analogy with an icy grey colour palette. These maps can be viewed digitally from multiple directions and angles but also CNC rendered as physical sculptures, affording a tangible reality. Using an aesthetic device of rendering urban statistics as dimensional forms appeals to our familiarity with natural landscapes; we can envisage ourselves walking amongst the data, experiencing transitions as we pass through neighbourhoods, enabling us to make decisions about where we would prefer to be. When seen as an ensemble, the spatial dimension of 'multiplicity' is readily apparent.

Where Amoroso depicts the data of the immediate past, the Dutch urbanism and architectural practice MVRDV visualizes sustainability requirements for cities of the future. Dimensional visualizations of urban data, or 'datascaping', is central to MVRDV's methodology for improving the environment, 'Through these datascapes MVRDV is able to create unexpected forms that go beyond artistic intuition or known geometries, replacing them with completely data-driven designs' (MVRDV, 2015). These theoretical visuals of a city's energy requirements, drinking water usage and household waste depict imagined and dimensional worlds, where resource data-forecasts are piled onto a planet's surface in direct contrast with single red living units that represent the human presence. Drinking water, a vital urban invisible, is rendered as a solid mass above ground, contained inside a series of glass towers reaching into a white sky. A year of water supply, normally hidden below ground, is aggregated and frozen in time as we stand back in awe, gazing from below. Digital 3D imaging and a futuristic aesthetic are employed to create vast data-filled, imagined landscapes in a world that is both sublime and anonymous.

Where MVRDV deals with urban invisibles of the future, the MIT Media Lab concentrated on the present with a 'You Are Here' campaign (Kamvar, 2014b). The project's goal was to initiate social change through mapping the small details of cities, details that go beyond the urban fabric of buildings and roads. The repeat nature of each mapping, to create 100 maps of 100 cities, speaks to the simultaneity of space. Once an API source for data collection and data-driven programming solution is established for one city, the process is rolled out across other cities in the USA. The 33 forms of invisible data topics explored include bicycle crashes, births, coffee shops, income mapped to subway, fastest mode of transportation and noise complaints. 'Urban Pyramid' (You Are Here, 2014b) visualizes the estimated sum of the building volume in each city as stacked floors, enabling comparative studies of activity in buildings at different elevations. 'Awake' (You Are Here, 2014a) displays shop opening hours over a 24-hour period, revealing intensities of commercial activity in direct correlation to the diurnal nature of the city. Dynamic digital mapping affords the option to play the day at speed as 24 hours are condensed into less than a minute. All the data-driven maps of the project unveil the seemingly insignificant through beautiful patterns, luring the viewer to question and explore nuances of urban life. San Francisco has far fewer tall buildings than either Boston or Chicago; does this mean I will spend less time in elevators and have unimpaired views of the horizon when walking? Honolulu's opening hours extend later than those in Las Vegas, the city whose livelihood is dependent on being open 24 hours every day. Why is this so?

The urban invisibles discussed above derive from data and statistics, factual information, the quantitative invisibles of urban living. However, there are also invisible qualitative aspects of urban environments that play a significant role in our city experience – the subjective, ephemeral and evanescent elements that may manifest as personal memory or story, an emotion, a sensation or a perception.

There are points in the city of quantitative–qualitative intersections; Kamvar and his team draw attention to how mass transportation networks are 'powerful orientation features. Because they are so familiar, they form a nice backbone over which we can communicate data in an intuitive way' (Kamvar, 2014a). But to Ng-Chan, the mass transportation networks go beyond being a functional data communication tool. She suggests that personal memories and stories also deserve their place on the city's mass transit maps. Her collaborative project, '*Detours*: Poetics of the City' (Ng-Chan et al., 2012) includes video poetry, mobile media music and photography to recount urban ephemerals of riding the metro and the bus, site-specific sounds and paths created by walkers and cyclists in Montreal. The work explores personal formations of the 'image of the city' (Lynch, 1960) through artistic renderings of contemporary urban familiars, from cinematic narratives of the empty seats on mass transit in sunlight to photographs of the trodden paths cutting corners in curated park spaces. The authors suggest that the digital narrative mappings 'are optimally viewed or listened to in the places that inspired them, though they also beckon to the armchair tourist. *Detours* is meant to be a mobile media project, accessible through tablet computers and smart phones (headphones recommended)' (Ng-Chan et al., 2012). Through re-experiencing mappings of small moments of dissent and vernacular knowledge, the 'everydayness' (Lefebvre and Levich, 1987) of the city is revealed and re-revealed. At these moments, shared and collective minutiae of urban living unmask the extraordinary in the ordinary time after time.

Human emotion is a further layer of subjective urban mappings in Nold's referential work in the field of bio-mapping, which uses galvanic skin response sensors connected to GPS devices correlating physiological arousal with physical location, developed with a 'participatory methodology for people to talk about their immediate environment, locality and communal space' (Nold, 2006).

The data from volunteers walking in designated urban areas are mapped using Google Earth, using height as an indicator of arousal level. The resulting post-representational (Kitchin, 2010) map of Greenwich combines the visualized data with annotated notes from the volunteers (Nold, 2007). Comments such as 'Felt calm around here – quiet, peaceful', 'Crossing illegally', 'entered another factory' encourage personal exploration of emotions and empathy with those who were recorded, and a curiosity as to the lived perspective of the participant. In collecting a series of spatial events (as a combination of visual markers and succinct text descriptors) presented on the common ground of the base map, Nold achieves his primary objective: 'I'm trying to use 3D visualization as a way of talking about the space' (Nold, 2006). Nold's mapped space is active; the data speaks and occupies volume as it emerges from a surface.

Other ephemeral mappings include those drawn from sensations and perceptions derived from the 'active detecting systems' (Gibson, 1983) of our bodily senses. Practitioners in this field of mapping the city through the senses include the 'London Sound Survey' (The London Sound Survey, 2015) who have crowd-sourced a comprehensive collection of historical and contemporary sound recordings that can be digitally accessed via a series of basemaps, from a black and white planimetric view to a topological variation on Harry Beck's famous London Underground map. One criticism of mapping sensory information in which human experience forms the data source is that the perceived environment is the lived and vernacular, whereas the resulting map takes a panoptical top-down approach (Margioles, 2001). With more sophisticated interface design and a turn to the multisensory, the purity of geolocated positioning gives way to a more holistic generation of ambience. Babbar combines complex sets of visual information with sound recordings of Mumbai, and says 'I tried to capture the extraordinary in every ordinary' (Babbar, 2015). Using a static/dynamic contradiction of still photography, a single pushpin on a map and the recorded sound experiences enables him to portray subjective data that is primarily dependent on the state of individual persons perceiving it. The result is not a site, not a map, but a hybrid multisensory digital experience, and just like the city, it continues to play on an eternal loop.

## Smell as a mappable phenomenon

Celebration of the small, the undetected and the unexpected is also a focus for my own mapping practice of an ephemeral and invisible set of information that exists in the everyday. Humans breathe 24,000 times per day, affording 1,000 opportunities every hour for detecting a smell, scent and odour, and recent research indicates that humans have the capacity to discriminate up to a trillion olfactory stimuli (Bushdid *et al.*, 2014). But the majority of the urban olfactory dataset volatilizes, unnoticed, into the atmosphere.

Smell is the double-invisible sense: we are blind to smells in the landscape, unable to see their presence, and when we remark upon them they are frequently no longer in the space where we found them. At a second level, smell as a human sense was relegated in the sensory order in the Western world during the Enlightenment, at which point in time the rational thinking sought evidence in the empirical, and where the senses of sight, touch and sound could be retrieved and replicated, whereas smell and taste disappeared from common cultures of knowing (Classen *et al.*, 1994). In the contemporary city, odour monitoring is an accepted method by which we record and understand smell, as a pollutant being monitored real-time (Cheshire West and Chester Council, 2014). In Western cultures we have an increasing desire to distance ourselves from smells, either through elimination, masking or deliberate scenting of environments (Henshaw, 2013). My research investigates smell perceptions of the city environment, depicting the findings in a variety of artistic, cartographic forms, and in doing so seeks the means by which we can share our olfactory perceptions.

Smell is a relatively new area of study for designers. The idea of re-exploring a known environment with our noses has captivated me since 2010, when I set out on my first olfactory exploration, resulting in an installation artwork/mapping – a virtual olfactory dérive entitled Smellmap: Paris (McLean, 2012). This participatory installation asked people to take the tops off tiny bottles and spray scent diffusers in order to sniff the natural, homemade (in my kitchen) scents of everyday life in the French capital, and in doing so re-imagine themselves on a walk through urban space. Bottled urban smells are invisible, but captured inside glass containers they form one approach to urban olfactory mapping (Figure 37.1).

My evolving practice of mapping the olfactory dimension of cities is a collaborative exercise in which I collect the 'data' using a sensewalking methodology, similar to that of soundwalking (Westerkamp, 2001). I regard my own smell observations ('smellervations') as insignificant; Pamplona for example is not my city, neither am I familiar with its smells. So in looking to explore and communicate any city's smellscape, I engage the local population to detect the city's aromas and record their findings. I facilitate this process through leading participants on smellwalks (Henshaw and Cox, 2009), indicating sniffing techniques and providing smellnotes (McLean, 2017) as recording devices. Taking part in a smellwalk requires a re-orientation of the

*Figure 37.1*    'Smellmap: Paris' an installation of olfactory experiences in Paris, bottled and ready for sniffing by those undertaking a virtual dérive. © 2010 Kate McLean

*Table 37.1* Listing of artist-led smellwalks in Pamplona in October 2014, run by Kate McLean with translation between English and Spanish conducted by Marta Agote Agúndez

| Date/Time | Group | Figure photo | Route | No. of participants |
|---|---|---|---|---|
| Oct 26 1100h | Open access/general public | Figure 37.2 | Ensanche | 12 |
| Oct 27 1100h | IES Pedro de Ursua (High School) | Figure 37.3 | Mendillorri – Soto Lezkairo | 14 |
| Oct 28 1100h | Once (visually impaired citizens and their support group) | Figure 37.4 | Casco Viejo | 15 |
| Oct 29 1215h | Profesores Upna (university lecturers) | Figure 37.5 | Barrio Universidad | 9 |
| Oct 29 1900h | Enólogos (wine makers) | Figure 37.6 | Casco Viejo | 8 |
| **TOTAL** | | | | **58** |

senses on the part of the leader and the participants, temporarily emphasizing the information received from the sense of smell over all other senses. Five smellwalks (Figures 37.2–37.6), with a cross-section of the general public, took place in different sections of the city of Pamplona in October 2014, each walk lasting 90 minutes (see Table 37.1).

*Figure 37.2* Smellmap: Pamplona; experiencing a variety of urban smells in Ensanche.
© 2014 Kate McLean

*Figure 37.3* Smellmap: Pamplona; experiencing zero smell from metal until human contact is made with its surface in Soto-Lezkairo. © 2014 Kate McLean

*Figure 37.4* Smellmap: Pamplona; smell detecting on the open streets of the Casco Viejo. © 2014 Kate McLean

*Figure 37.5*   Smellmap: Pamplona; smell detecting on building plots of the Barrio Universidad.
© 2014 Kate McLean

*Figure 37.6*   Smellmap: Pamplona; smell detecting in the evening back at the Casco Viejo.
© 2014 Kate McLean

Smellnotes (Figure 37.7) translated into Spanish were the means by which smellwalkers manually recorded their 'smellervations'. I find the deliberate action of inscribing with a pen initiates careful thought and consideration, adding to the quality of the response, encouraging reflexivity on the part of each participant. Since 'The relationship with odours depends on the smeller's judgement (taste, quality, memory) and on the odour's duration (repeated stimuli lead to inurement) and concentration' (Barbara and Perliss, 2006: 28), individual means of recording are important. Batty proposes the following metrics for recording cities 'as being clusters of "spatial events", events that take place in time and space, where the event is characterized by its *duration, intensity, volatility,* and *location*' (Batty, 2002: 1). I designed the Pamplona smellnotes using semantic differential scales as a means of reducing acquiescence bias (Robson, 2011: 310), specifically asking the participants to record intensity on a scale of weak to strong, duration on a scale of stationary to dynamic, and hedonic response on a scale of disagreeable to pleasant. Participants recorded textual descriptors for the 'smellname' and any personal associations/thoughts. Whether the smell was expected or not was noted as a binary option.

In order to extend the range of neighbourhoods where citizen smell data was collected beyond that of the five planned smellwalk routes, two blank maps with instructions and

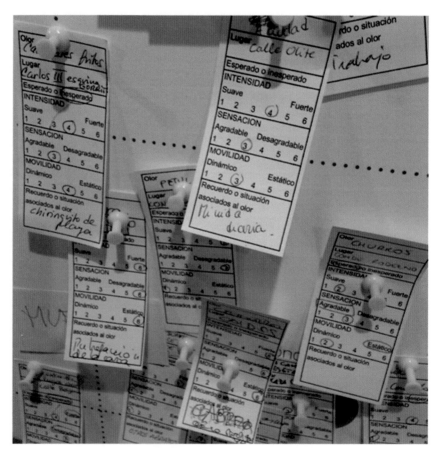

*Figure 37.7*  SmellNotes – used by human sensors to record their urban smell perceptions of Pamplona. © 2014 Kate McLean

*Figure 37.8* Smellmap: Pamplona; blank maps for the general public to record noted smells within the city. On exhibit at Sala Conde Rondezno. © 2014 Kate McLean

smellnotes were placed in public spaces; the Sala Conde Rodezno exhibition space and the Universidad Pública de Navarra foyer (Figure 37.8) to enable visitors and passers-by to add their perceptions. I divided the city into quadrants so as to geolocate the smell data.

To walk and sniff is to know, in an unexpectedly fine and detailed way. The experience of smellwalking comprises a series of infinitesimally small moments; it demands mindfulness and concentration to refocus and reorient sensory awareness. Through the process of actively smelling we are both sensitized to the cadence of the urban smellscape and understand how it coincides with our everyday lives. The associations connected with the smells form another stream of riches and granularity worthy of further independent study. By engaging in conversation about odour perception we discover both the rich complexity of smells that distinguish our cities and the individuality of perception and responses. Before detailing how I generated the mapping, here are some observations taken from the smellnotes and from the smellwalking experiences:

*WHEN THE ABSENCE OF SMELL = ABSURDITY*
Place: *Pastry shop.*
Smell: *Nothing.*
Comment/association: *Absurd.*

This single recorded smellnote from a general public smellwalk on Sunday October 26 around the secundo ensanche epitomizes the feeling of unease when two senses do not correlate. Which sense should we believe?

\*\*\*

*SPECULATIVE CONJURINGS OF IMAGINED WORLDS*
*'Cooking… frying…'*
*'Tomato!'*
*'And meat…'*
*'Bolognese?'*
*'No. Definitely not Italian…'*
*'A meat stew perhaps then… with tomatoes and onions…'*

Unseen smells generated a hotly debated, imagined recipe and cooking scenario amongst smellwalkers near to the Universidad Pública de Navarra. Each one of the group temporarily relocated themselves into an imagined kitchen somewhere in the block of flats above our heads. An 'un-visual' moment occurred, drawing from memory, conjectures about the dish being cooked were both varied and emphatic. Any contestation was minimized as we all wrote down 'cooking smells' on the smellnotes.

**\*\*\***

*SMELLY FOCAL POINTS*
A greater number of smells were perceived on street corners where human activity, architecture and wind pattern combine indicating a vernacular olfactory density in urban space. If smell indexes life, vernacular smell in the city indexes activity.

I recorded the date, wind data and quadrant of the city for each smellnote. The completed smellnotes were transcribed into a database and codified into a category based on Henshaw's (2013) urban smell classification system (see Table 37.2). This proved problematic as it quickly became apparent that two further urban smell categories needed to be added: *Complex* where two or more categories were combined, e.g. 'Wood/stone/moisture' and 'Fresh rubbish with vegetables' and 'Rusted iron/burnt tyre', and *Emotion/abstract* for instances such as 'Empty wine cellar' and 'Pre-university exam' and 'Summer day' where the description was not concrete.

I allocated each urban smell category a colour, based on my own interpretation drawing on inspiration from photographs of Pamplona's landscape, architecture and design (Figure 37.9).

*Table 37.2* Urban smell classification categories and Pamplona exemplar data

| Code | Category name | Exemplar data | Urban smell classification source |
|------|---------------|---------------|-----------------------------------|
| TR | Traffic emissions | Car, fumes, gasoline | (Henshaw, 2013: 53) |
| ID | Industry | Coal | |
| FD | Food/beverage | Candy floss, lemon, chocolate with churros | |
| SK | Tobacco smoke | Cigarette, smoke, marijuana | |
| CL | Cleaning | Bleach, cleaning, cleaner | |
| SY | Synthetic | Air freshener, scented candle, medicine | |
| WS | Waste | Bin, litter, garbage | |
| HU | People/animals | Body, human, BO | |
| NT | Nature | Tree, rain, fresh air | |
| CN | Building/construction | Drain, dust, plaster | |
| NF | Non-food | Cardboard, leather, new clothes | |
| EM | Emotion/abstract | Empty wine cellar, pre-university exam, disgusting | Added by McLean, |
| CX | Complex | Three unrelated words or more | 2014 |

*Figure 37.9* Smellmap: Pamplona; design elements for the mapping including smell colour reference, urban smell classification and smell icons. © 2014 Kate McLean

My selection of colours specific to each city ensures that the mapping directly references the local environment from which it emanates.

Taking a phenomenological approach to lived experience means my primary interest is in the smellscape as perceived, so 'analyses' are based on first-person accounts of the 'data'. And while it is possible to compare the numbers of smells in each category, my aim is not to quantify. My overall aim with the mapping of urban olfactory environments is to generate an experience of being inside the city's smellscape on the subjective part of the viewer, rather than to create a visual representation of a dataset, so more akin to the notion of an atmosphere, 'perceived through our emotional sensibility' (Zumthor, 2006: 13) than a scientific chart for analysis. Confusion as to the apparently scientific nature of my work may emanate in the cartographic conventions that I adopt and the resulting maps' resemblance to topographic contour lines. One particular goal of this project was to portray the smellscape as dynamic and contested, where one smell replaces another in space as it volatilizes. Diaconu suggests mapping of a smellscape can be further problematized as paradoxical in that it 'endeavours to objectify, to visualise, to order and to stabilise smells' (Diaconu, 2011: 229). While this argument may be applied to traditional mapping practices, I suggest that a creative engagement bringing individually perceived smells to ephemeral life through motion graphics might partly alleviate any stability.

The following is a synopsis of my 'olfactory interpretation' – a methodology for bringing individually perceived urban smells to temporary life. It uses graphic elements to create a moving and volatilizing smell presence/absence that combines external olfactory existence with personal olfactory experience.

## Olfactory translation: mapping smell through changing the angles of view

In 'Smellmap: Pamplona', individual smells morph through three stages of being, each designed from a different angle of view (Figure 37.10):

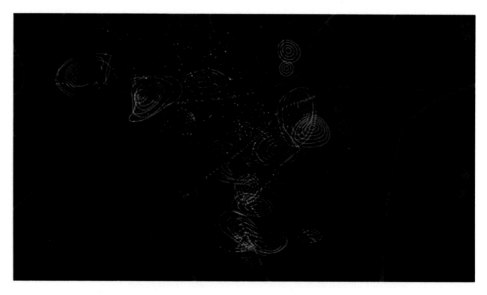

*Figure 37.10* Smellmap: Pamplona; still image from the motion graphic in which smells come into being, morph and move in the wind and volatilize. © 2014 Kate McLean

1 Smell appearance
ANGLE OF VIEW = BIRD'S EYE, PLANIMETRIC
The smell comes into being, emerging at a geographical source, appearing as a pinprick and growing to form a series of coloured, concentric rings. The number of rings (and thus overall size) correlates to 'smell intensity' data from the smellnotes

2 Smell perception
ANGLE OF VIEW = ISOMETRIC 45°, VIEWED FROM ABOVE
Each smell moves across the cityscape in the direction of the day's wind on the day the data is recorded, the distance is an estimate based on the wind speed as taken from local weather data and the movement lasts for the 'smell duration' data as indicated in the smellnotes

3 Smell volatilization
ANGLE OF VIEW = LIVED HORIZONTAL
Each smell instance then volatilizes, and the viewpoint of the smellwalker trying to clutch at, and identify, an evanescent aroma that simply disappears

## Contestation in the smellscape

'Smellmap: Pamplona' is designed to be as ephemeral as the data; a series of volatile components, representing moments in time, a contested smellscape – where single moments contain multiple smell perceptions in, and of, the same place. Smells are fleeting, lasting no more than a couple of seconds and individual to each one of us. As Hecht describes after a smellwalk in Brooklyn:

> Being paired with a buddy taught me that I will not always smell what you smell. Sometimes that's because the odor is moving, like the passing of cigarette smoke or recently made pad thai – the smell is intense when fresh, but you might miss it if you walk over even a minute or two later.
>
> *(Hecht, 2014)*

In order to map the dynamic nature of the space of the smellscape, I place an animated graphic rendering of smell on a timeline. Each smell comprises a series of three animated life stages: growth, movement, volatilization. The map addresses contestation through a series of sequential smells appearing in each quadrant of the city, as one smell volatilizes so another one takes its place until there is no more data.

The map stops becoming when the data collected comes to an end – the last smells to disappear are those of food, people and plants. Finally, a listing of the vernacular descriptions and associations of the smells completes the motion graphic. 'Smellmap: Pamplona' motion graphic can be viewed online at: *http://sensorymaps.com/portfolio/smellmap-pamplona*.

## Discussion

Mappings of urban invisibles cited early in this chapter are based on quantitative datasets – emanating either from discrete one-off recordings or a continuous dynamic feed. Urban invisibles can be mapped as datascapes – dimensional and seemingly tangible presentations similar to a physical geography – in which the flat surfaces of space across which we may seem to walk, are transformed into landscapes with suggested angles of view. These maps are dimensional and serial, and static. By contrast, the urban ephemeral mappings are qualitative, single-occurrence events that are layered onto a basemap using moving imagery and sound to arouse the curiosity of the perceivers, enabling the latter to delve more deeply into their own perceptions of the space. The ephemerals generally contain dynamic elements, but rest flat on the landscape as events and moments that exist despite the underlying formal cartographic structure. There are commonalities – both quantitative and qualitative mappings use a Cartesian grid basemap as a starting point, often subsequently eliminating other any geographic reference.

The hybrid digital space of Mumbai's soundscape (Babbar, 2015) and my smellmap of Pamplona may offer alternative ways to approach dimensionality in mapping ephemerals as simultaneous collections of angles of view with dynamic elements. The field is open to further analysis and discussion, artistic renderings and interpretations. 'Smellmap: Pamplona' starts to explore possible methodologies and cartographic languages for collecting and displaying olfactory datasets. This interpretative mapping of an olfactory world inhabits a theoretical dimensional space that can be understood as a set of simultaneous spatial events.

The invisible, the transient and the fleeting are integral elements of a lived city experience. This chapter starts to address some of the strands and approaches across art and design, and there is a need for further research and practical activity and creativity. The field of invisible and ephemeral mapping has an exciting potential for future work, and to this end August 2015 saw an 'Ephemeral Cartography' workshop (Ng-Chan *et al.*, 2012) held prior to the ICC conference. It focussed on current practices and methodologies followed by a practical collaborative project and the creation of an ephemeral map, asking another pertinent question:

'For what length of time should the mapping of ephemeral phenomena remain in existence?'

## References

Amoroso, N. (2010) *The Exposed City: Mapping the Urban Invisibles* New York, Routledge.
Amoroso, N. (2014) "Data + Mapping Viz" Available at: *www.nadiaamoroso.com/* (Accessed: 8 July 2015).
Babbar, T. (2015) "Sounds of Mumbai" Available at: *http://soundsofmumbai.in* (Accessed: 30 June 2015).
Barbara, A. and Perliss, A. (2006) *Invisible Architecture: Experiencing Places through the Sense of Smell* Milan: Skira Editore.

Batty, M. (2002) "Thinking About Cities as Spatial Events" *Environment and Planning B: Planning and Design* 29 (1) pp.1–2.

Bushdid, C., Magnasco, M.O., Vosshall, L.B. and Keller, A. (2014) "Humans Can Discriminate More than 1 Trillion Olfactory Stimuli" *Science* 343 (6177) pp.1370–1372.

Cheshire West and Chester Council (2014) "Real-Time Monitoring Graphs" Available at: *www.cheshirewestandchester.gov.uk/residents/pests_pollution_food_safety/pollution_and_air_quality/real-time_monitoring_graphs.aspx* (Accessed: 5 September 2014).

Classen, C., Howes, D. and Synnott, A. (1994) *Aroma: The Cultural History of Smell* London: Routledge.

Corner, J. (2011) "The Agency of Mapping: Speculation, Critique and Invention" in Dodge, M., Kitchin, R. and Perkins, C. (Eds) *The Map Reader* Chichester, UK: Wiley, pp.89–101.

Deleuze, G. and Guattari, F. (1987) *A Thousand Plateaus: Capitalism and Schizophrenia*. Minneapolis, MN: University of Minnesota Press.

Diaconu, M. (2011) "Mapping Urban Smellscapes" in Diaconu, M., Heuberger, E., Mateus-Berr, R. and Vosicky, L.M. (Eds) *Senses and the City: An Interdisciplinary Approach to Urban Sensescapes* Vienna, Austria: Lit Verlag.

Gibson, J. (1983) *The Senses Considered as Perceptual Systems* Westport, CT: Praeger.

Hecht, J. (2014) "Put Your Nose First: Smellwalks for You and Your Dog" Available at: *http://blogs.scientificamerican.com/dog-spies/put-your-nose-first-smellwalks-for-you-and-your-dog/* (Accessed: 8 July 2015).

Henshaw, V. (2013) *Urban Smellscapes: Understanding and Designing City Smell Environments* New York: Routledge.

Henshaw, V.A. & M Cox, T.J. (2009) "Researching Urban Olfactory Environments and Place Through Sensewalking" Available at: *www.manchester.ac.uk/escholar/uk-ac-man-scw:122854* (Accessed: 21 December 2014).

Kamvar, S. (2014a) "Subway Boston" Available at: *http://youarehere.cc/j/subway/boston.html* (Accessed: 28 June 2015).

Kamvar, S. (2014b) "You Are Here" Available at: *http://youarehere.cc/#/maps* (Accessed: 25 June 2015).

Kitchin, R. (2010) "Post-Representational Cartography" *Lo Squaderno* 15 pp.7–12.

Lefebvre, H. and Levich, C. (1987) The Everyday and Everydayness *Yale French Studies* pp.7–11.

Lynch, K. (1960) *The Image of the City* Cambridge, MA: MIT Press.

Margioles, E. (2001) "Vagueness Gridlocked" *Performance Research: A Journal of the Performing Arts* pp.88–97.

Massey, D. (2013) "Doreen Massey on Space" Available at: *www.socialsciencespace.com/2013/02/podcast doreen-massey-on-space/* (Accessed: 25 June 2015).

McLean, K. (2012) "Emotion, Location and the Senses: A Virtual Dérive Smell Map of Paris" in Brasset, J., McDonnell, J. and Malpass, M. (Eds) *Proceedings of the 8th International Design and Emotion Conference* London, 11–14 September.

McLean, K. (2017) "Smellmap: Amsterdam – Olfactory Art & Smell Visualisation" *Leonardo* 50 (1) pp.92–93.

MVRDV (2015) "MVRDV - Design Philosophy" Available at: *www.mvrdv.nl/en/about/Design_Philosophy* (Accessed: 26 June 2015).

Ng-Chan, T., Akrey, D., Blomgren, L., Allen, G.N., Gorea, A., O'Brien, E. and Thulin, S. (2012) "Poetics of the City" Available at: *http://agencetopo.qc.ca/detours/index.html* (Accessed: 28 June 2015).

Nold, C. (2006) "Bio Mapping" Available at: *www.biomapping.net/interview.htm* (Accessed: 28 June 2015).

Nold, C. (2007) "Bio Mapping: Christian Nold" Available at: *www.biomapping.net/new.htm* (Accessed: 28 June 2015).

Porteous, J.D. (1985) "Smellscape" *Progress in Physical Geography* 9 (3) pp.356–378.

Robson, C. (2011) *Real World Research: A Resource For Users of Social Research Methods in Applied Settings* Chichester, UK: Wiley.

rudi.net (1997) "The Exposed City: Inspiration for New Urban Form" Available at: *www.rudi.net/books/8940* (Accessed: 8 July 2015).

Sant, A. and Johnson, R. (2015) "The Studio for Urban Projects: Mapping the Ephemeral Landscape" Available at: *www.studioforurbanprojects.org/storefront/teaching/mapping-the-ephemeral-landscape/* (Accessed: 25 June 2015).

The London Sound Survey (2015) "The London Sound Survey Featuring London Maps, Sound Recordings, Sound Maps, Local History, London Wildlife" Available at: *www.soundsurvey.org.uk/* (Accessed: 30 June 2015).

Westerkamp, H. (2001) "Soundwalking" Available at: *www.sfu.ca/~westerka/writings%20page/articles%20 pages/soundwalking.html* (Accessed: 6 January 2015).

Wurman, R.S. (1989) *Information Anxiety* New York: Doubleday.

You Are Here (2014a) "Awake" Available at: *http://youarehere.cc/#/maps/by-topic/awake* (Accessed: 27 June 2015).

You Are Here (2014b) "Urban Pyramid" Available at: *http://youarehere.cc/#/maps/by-topic/urban_pyramid* (Accessed: 27 June 2015).

Zumthor, P. (2006) *Atmospheres: Architectural Environments – Surrounding Objects* Basel, Switzerland: Birkhäuser GmbH.

# 38

# Mapping in art

*Inge Panneels*

Since the 1960s contemporary visual artists have increasingly been using maps and mapping strategies in their work as part of a post-modern idiom. The 'expanded field' started to define both arts practice and theory with ideas, which are traditionally geographical terms; notions of space, place and site. Artists creating work informed by the locale have turned artists into producers of new knowledge, of makers of alternative forms of data, and this has been a key factor in the changing field of geography itself (Hawkins, 2014). This chapter will explore how and why artists use maps and mapping and where this trend in mapping in art is heading.

## Mapping in art: how we got to now

There has always been a close relationship between art and cartography; artists were traditionally employed to draw and embellish maps. The mapping of our world has long been a significant scientific undertaking which has influenced how we think about the world and our place within it. The history of geography from early global navigation and enlightenment exploration, to the institutional geographies of the nineteenth and twentieth centuries and the focus of recent history of spatial thinking in human geography, illustrates how maps have played a key role in the Modern Era. The *spatial turn* in philosophical thought spearheaded by French philosophers such as Michel Foucault, Henri Lefebvre and Michel de Certeau's discourse on spatial theory, brought about an understanding of the relationship between space and power in the latter half of the twentieth century. The understanding of how maps *work*, and the diverse ways in which *space* and *place* are conceptualized and analytically employed to make sense of the world, have been analysed more in the last few decades.

The use of maps and mapping techniques in the lexicon of contemporary visual art occurred at the point of arrival of post-modernism in the aftermath of the Second World War, and a reconsideration of values of the Modern Era.[1] Both cartographers[2] and art critics have witnessed a rise in cartographic language and visual idiom within contemporary arts practice. Journalist and writer Katharine Harmon defines it as symptomatic of the post-modern era, 'where all conventions and rules are circumspect' (Harmon, 2009: 9). Cartographer Denis Wood and geographer Denis Cosgrove have been particularly key in raising and examining the role of cartography within broader critical discourses about landscape, culture and art. Whilst the mapping instinct has been evidenced in archaeological maps since the rise of the agricultural settled lifestyle (Brotton, 2012), map *knowledge* does not come to us naturally but through complex cultural understanding. Each map speaks of its time and the value of its culture; the tacit knowledge of ocean faring embedded in the sticks and shell system of a Polynesian chart, the spiritual and geographical knowledge passed on through millenia-long Aboriginal oral tradition, the territorial claims of competing Portuguese and Spanish empires staked out on maps and globes in the Age of Discovery or the global dominance of internet giant Google in online digital mapping. The artist and academic Ruth Watson wrote a comprehensive paper *Mapping and Contemporary Art* for the British Cartographic Society in which she calls for a 'new history of the map in art to be written that upends the usual suspects from their comfortable nodes on a one-sided cultural map' (2009). Non-Western cultures emergent on the geopolitical landscape bring with them other mapping paradigms. However, Western thought and culture dominate the mapping discourse discussed in this chapter.

Understanding the historical provenance and the rationale of maps to claim territories, mark property and denote political boundaries, make maps potent tools of power. Wood (1992) dissected the map, not as an impartial neutral tool but as a biased, active and potent agent, and artists have been singularly astute in challenging the perceived neutrality of the map.

The French Situationists, with protagonist French writer and artist Guy Debord, had literally up-ended the map through the challenge of the dérive in the mid-1950s. This technique of transient passage, allowed the psycho-geographer to explore the terrain through an element of chance. This *drift* (Cosgrove, 1999: 231) subverted dominant authoritarian readings of the urban landscape, and instead honed in on the ephemeral, fugitive and sensory spatial experience. The 'tactics of consumption', to paraphrase Michel de Certeau (1980), thus places the agency of mapping back onto the consumer (map user). This embodies the essence of the Situationists: the power of the map claimed by the consumer rather than the producer (map maker).

Artists have explored the subversive use of the powerful semiotics of cartography more noticeably since the 1960s. That mapping became much more pronounced during the Land Art movement is no coincidence as it was part of a broader shift in artistic practice; by moving art out of the gallery and into the landscape itself. The *expanded field* (Krauss, 1979) had started to define both arts practice and theory with ideas, which are traditionally geographical terms; notions of *space, place* and *site*. The American sculptor Robert Smithson used the Earth itself to make monumental artworks – such as the iconic *Spiral Jetty* from 1970 – and is considered a pioneer in the tradition of *earthworks*; a phrase he coined to describe artworks which engage and are defined by its geographical location.

The British artists Richard Long and Hamish Fulton were among the pioneers to walk the map and map the walk. They used earth, found objects, photography and poetry to create artworks informed by the terrain through the act of walking: 'No Walk, No Work' (Morrison-Bell et al., 2013: 61). Long embodied the walk through the use of language in visual form in his textworks. *Language* in mapping is closely tied to the land it describes; colloquial words and place names give us access to a meaning of landscape, often forgotten. The etymology of place names enriches our understanding of landscape beyond mere description, but also as a way to *know* it and walking is the prime means of doing so. As the Situationists before him, Long used maps in a performative manner in which to experience the landscape by walking, a tradition which continues today. Tracking movement across a landscape through the availability of GPS (global positioning system) tracking has led to artists logging their daily movements and plotting them over a long range of time in different media and in both rural and urban landscapes. Jeremy Wood and artist duo Plan B (Sophia New and Daniel Balasco Rogers) used GPS to track their daily movements and made these tracks tangible on and in paper. Artists such as James

Hugonin and Rachel Clelow map their daily routines into colourful datasets, each block of colour representing a feeling or geographical area of their day's journey. The GPS technology has thus been deployed to map movement in space and the artists have made this movement visible.

American sculptor Maya Lin trained as an architect, and her large-scale installations and monuments are strongly rooted in site. Her sculptural works suggest natural organic landforms; rolling hillsides evoked by stacked wooden blocks *2x4 Landscape* (2006) or undulating metal tubing suspended in space evoking geography as in *Water Line* (2006), constructed using GIS (geographical information systems) technology. The geographical data is visualized into three dimensions using computer-aided design-driven (CAD) digital manufacturing technologies, such as laser cutting, rapid prototyping and others, a methodology used by an increasing number of artists.

These works chart the move away from the paper map towards digital mapping and the embracing of the technological developments of mapping techniques. Crampton and Krygier called this the 'undisciplining of cartography' (2005) led by two distinct cartographic developments. First, mapping as the consolidation of disparate fields of knowledge and second, the democratization of mapping, of anything, by anyone. This latest technological transition of cartography has not been enabled so much by new software but by a mixture of open source collaborative tools, mobile mapping applications and geotagging. The exponential growth of mapping is a by-product of a data rich era, where mapping has become the go-to methodology for making sense of all this information, and the world at large. The etymology of geography, geo-graphē, as the *scribing* or *drawing* of the world, hints that geography as such would be best placed to visualize this: 'I map, therefore I am' wrote Katharine Harmon (2004). It was a means of defining the *modus operandi* of the artists described in her book, who use maps and methodologies to make sense of the world. The cultural geographer Harriet Hawkins observed that these arts practices, however, also became to be valued for the alternative form of data they created and the epistemological model they offered, 'providing a model of interdisciplinarity that was not afraid to challenge existing geographical or artistic approaches' (Hawkins, 2014: 238). She argues that the trans-disciplinary field of the art-geography nexus is changing the field of geography itself.

## The materiality of maps

The ubiquitous paper map and historical maps in particular, have proved a fertile ground of inspiration for many artists. Australian artist Ruth Watson (Figure 38.1) has used the cordiform maps of the sixteenth century as a recurring theme for her installations, such as the *Lingua Geographica (South)* (1996) – a heart-shaped map constructed from thousands of cibachrome photographs of the surface of a tongue infected with *lingua geographica* pinned to the wall – or made out of pink plastic shopping bags in *Natural State* (2001), using the map projection to challenge the way the world is represented through careful choice of materials.

The map of Utopia drawn by sixteenth-century cartographer Abraham Ortelius[3] in 1596 in response to Thomas More's eponymous story has inspired Grayson Perry's *Map of Nowhere* (2008), Chinese artist Qiu Zhijie's *Map of Utopia* (2012) and Stephen Walter's *Nova Utopia* (2013) (Figure 38.2). All three artists painstakingly draw their maps by hand. They are using the visual language of the map – of a land, which is essentially unknowable – to visualize political and social commentary on the societies they each live in. Perry had adopted the map iconography and language for a number of years in his prints and etchings, as a means to subvert and flouting social taboos. Walter critiques More's story through a contemporary commentary on the commodities of the free market economy and freedom of travel and speech in his detailed

*Figure 38.1    Lingua Geographica (South)* **by Ruth Watson (1996)**
**Materials: collage of cibachrome photographs, pins**
**Image credit: Ruth Watson**
**Work shown at: The World Over exhibition at Stedelijk Museum, Amsterdam, 1996**
**Collection: Museum of New Zealand Te Papa Tongarewa, Wellington**

drawn map, which was exhibited for the first time alongside the original sixteenth-century map which inspired it at Museum aan de Stroom in Antwerp in 2015.[4]

Some artists have simply used the inherent beauty of the paper map as the basis of their works but untethered from place; the delicate cut bird forms by Claire Brewster (Figure 38.3), the layered cut maps of Emma Johnson or cut atlases of Etienne Chambaud. Others, such as American artist Nina Katchadourian amend the cartography of maps by drawing over them or Clare Money inters them for Nature to deface them and digs them up years later. Others use the specific geographies of the map to great effect: Georgia Russell's *Britain: March 2003 (Britain on Iraq)* (2003) is the outline of a UK map cut out from a paper map of Iraq, thus creating a projection of the UK map on the void behind the map. Layla Curtis cuts and re-assembles maps into familiar but disturbing collages, such as *The United Kingdom* (1999) which

519

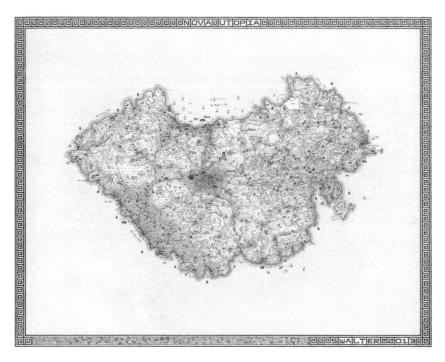

*Figure 38.2*   *Nova Utopia* by Stephen Walter (2013)
                Materials: paper
                Image credit: Images courtesy of Stephen Walter © & TAG Fine Arts, London

redraws the geography of Britain by transposing Scottish landscape features into the South, and dense urban English townscapes into the North. It was produced during the devolution of 1997–8 and seemed prescient of the political ramifications of the Scottish Independence Referendum of 2014.

The geographer Doreen Massey considered that *how* space and place are conceptualized, shapes the understanding of it. Those who define the space therefore have the power. As such an analysis of spatial relations between, for example, people, cities, jobs, is key to an understanding of politics and power. 'If time is the dimension of change, then space is the dimension of the social: the contemporaneous co-existence of others' (2005), so therefore the social and spatial need to be conceptualized together. Like maps, *space* was no longer considered neutral. Therefore the materials of which maps are made, and whom they are made by (the social), are an integral part of the map. The series of *Mappa* tapestries by Italian artist Alighiero e Boetti, were woven by craftswomen in Afghanistan between 1971 and 1994, and reflected on a rapidly changing geopolitical world. Artists such as Palestinian Mona Hatoum and Chinese Ai Wei Wei have used the map of the world and the globe as a recurring motif to confront issues of globalization and geopolitical tension. Wei Wei used ironwood reclaimed from dismantled temples from the Qing Dynasty and worked with carpenters to construct the jigsaw-like *Map of China* (2006). The configuration of its heterogeneous pieces could be read as a symbol of China's cultural and ethnic diversity, whilst asserting that whilst China remains distinctly singular, it is a fusion of countless individuals. Hatoum's *Present Tense* (1996) is a large floor-based installation of 2,400 blocks of square soap into which tiny little red glass beads are embedded, outlining the

*Figure 38.3* *The Green Green Grass* by Claire Brewster (2013)
Materials: vintage map of South East of England, pins
Image credit: Paul Minyo

disconnected territories, which were agreed to be returned to Palestinian control under the Oslo accords of 1993. The olive-oil soap is a typical Palestinian product, which she called 'a particular symbol of resistance' (Hessler, 2015: 143) yet is of course also a product prone to dissolve.

Mapping has also emerged as a strong theme among artists from South America as a visual language with which to tackle big themes such as social justice and inequality. The Argentinian artist Guillermo Kuitca has used maps extensively in his work. His trademark maps are painted onto mattresses, evoking a sense of dislocation. The works of Brazilian artist Vik Muniz used the map motif in several works including his *www (world map) Pictures of Junk* (2008) challenging the uneven distribution of wealth, of power or as in the above piece to challenge our consumer society. The political undertone of these works find resonance in the work of Susan Stockwell who deploys the map motif regularly in her work, such as *World* (2010) made of discarded electronic equipment, sorted by colour to represent traditional topographical maps (Figure 38.4). *Afghanistan A Sorry State* (2010) an embroidered map made up of dollar bills sewn together to make a commentary on consumer society. In addition to social justice, inequality and geopolitical commentary, climate change has become a recurring theme in mapping in art. But these are of course linked. In Hatoum's *Hot Spot* (2006), a flickering red neon outline of the world is mounted on a steel globe, as both a comment on zones of conflict and climate change. In *Venezia, Venezia* (2013) the Chilean artist Alfredo Jaar made a scaled model of Venice submerged into murky green water before rising it again, challenging both the inherent inequality of the political power structures of the arts establishments of the Venice Biennale and the city's on-going battles with the incoming water, which are set to increase with climate change. The German artist Mariele Neudecker makes evocative topographical maps which invite the viewer to reconsider the natural world as a source of spiritual transcendence and whose work has been included in the ArtCop exhibitions; cultural events which are timed to coincide with the UN Climate Change summits. Engagement with environmental discourse emerged in the Land Art movement. Pioneers such as the American artist couple Newton and Helen Mayer Harrison, are at the forefront of the eco-art movement and established a collaborative and interdisciplinary way of working with communities to make site-specific projects all over the world. The American artist Eva Mosher is part of the new generation of artists employing mapping methodologies to engage with climate change. Mosher's HighWaterLine projects in New York (2007) and Bristol (2014) used scientific data, modelling rising water levels due to climate change, and made this visible by literally drawing the high-water line on the ground of the city. Making change *visible* is important to engender discussion on a topic which the philosopher Timothy Morton (2013) called a 'hyper object'. He argues that part of the problem with climate change – and capitalism – is that we cannot

*Figure 38.4*   *World* by Susan Stockwell (2011)
　　　　　　　Materials: recycled computer components
　　　　　　　Image credit: Susan Stockwell
　　　　　　　Location: Bedfordshire University, Luton

perceive these directly. These are entities of such vast temporal and spatial dimensions that they defeat what a *thing*[5] is in the first place. New modes of philosophy are trying to grapple with new ways of looking at the world. Through the action of artists such the Harrisons and Mosher, we can perhaps begin to visualize a subject so vast.

## Globalization

Cartography has been the visual language of choice through which to visualize and make visible the complexities of a networked and connected world, defined by the enduring image of the globe as a uniting visual metaphor. The image of the globe has also come to signify a different kind of globalism, no longer defined by economic trading, supply and shipping routes or even telecommunication or travel, but instead as a symbol of environmentalism. The aerial view, the view of the Earth from above, has defined the way in which we view the world and by implication how we map it. The democratization of air travel in the early twentieth century had shifted the aerial view away from the god's eye view to a specific human position of what cultural geographer David Matless called a 'sky-situated knowledge' (Cosgrove, 1999: 212) where the ability to travel afforded a more personal understanding of our place on Earth. It is only with the advent of the Space Age that the first objective images as seen from above gave a true depiction of Earth. The *Apollonian View*, a phrase coined by Cosgrove defined how the iconic *Blue Marble* photograph, taken in 1972 at 45,000 kilometres above the Earth by Apollo 17, became a defining image of the twentieth century and changed our perspective and by implication our understanding of our planet and our place within it. Cosgrove wrote that:

> [e]arthbound humans are unable to embrace more than a tiny part of the planetary surface. But in their imagination they can grasp the whole of the earth, as a surface or a solid body, to locate it within infinities of space and to communicate and share images of it.
>
> *(Cosgrove, 2001)*

The film *The Power of Ten* (1968) by artist Ray Eames and designer Charles Eames, nine years after the first moon landing, but a year before the first human stood on the moon and preceding the *Blue Marble* image by four years, depicts the relative scale of the Universe according to an order of magnitude based on the factor of ten. The film starts with a picnic scene at human scale and zooms out into the infinity of space and then zooms back in until a single atom and its quarks are observed; the limits of human knowledge at the time. This film was remarkable for the scope of it in trying to encompass emerging new scientific knowledge into a coherent visual.

Space exploration also gave rise to the emergence of satellite communication. Surveillance technology has been at the heart of the work of the American artist Trevor Paglen, whose investigative work into the surveillance culture of the National Security Agency in the USA and the complicity of the international governments in this, including the UK, gained particular relevance after the Edward Snowden revelations of June 2013 and the release of the Oscar winning film *Citizen Four* in 2014 by artist filmmaker Laura Portas, in which Paglen featured as a contributor. The implied critique of the work by *mapping* the inherent power structures in the prevailing surveillance culture, reveals the transformative role of cartography in creating geographical meaning.

Historian Jerry Brotton wrote that 'the discourse of globalization has been able to chart its development across the surface of the terrestrial globe, yet this globe is an increasingly inadequate form in representing the rapid and complex transfer of information across its surface' (Cosgrove, 1999: 71). The globe's image as a cultural signifier of globalism within the global economy of late twentieth-century capitalism has started to wane. Instead alternative visions are being sought which can embody and signify the 'reduction of time-space'[6] where data and information travel instantaneously, silently and invisibly across the world, via 'extra-terrestrial' satellites and deep-sea data cables. The artist Lise Autogena (Figure 38.5) explored this in the *Black Shoals Stock Market Planetarium* (2004), a constellation of live stock market data to make an ever-changing map, where a terrestrial image was not adequate but a celestial map embodied the complex system of data packets. For the traders, *Black Shoals* is:

> [l]ess a spectacle than a more or less efficient visualization model for viewing the movement of a large number of individual stocks and shares. That movement is difficult to represent in a coherent visual field, though great prizes await those who can do so since people are generally much better at understanding what changes are taking place when presented with a picture than with an array of numbers.
>
> *(Stallabrass, 2015)*

The ability of the map to visualize incredibly complex and live data is one of the map's most accomplished achievements.

*Figure 38.5*   *Black Shoals Stock Market Planetarium* by Lise Autogena and Joshua Portway (2004)
Materials: live data spherical projection
Image credit: Nikolaj Copenhagen Contemporary Art Centre, 2003/4
Location: Nikolaj Copenhagen Contemporary Art Centre, 2003/4

The work of Autogena, Paglen, Eames and others signals a move away from the map towards mapping: away from the topographical towards the topological, where mathematical data points rather than geographical points construct the map (Mattern, 2015).

## Beyond the map

'This is the moment of the map', wrote artist Lize Mogel in 2008. 'The enormous amount of recent cultural production involving maps and mapping is reaching a critical mass' (Mogel, 2008: 107).[7] The survey by New Zealand art critic and curator Wystan Curnow on mapping in art (Cosgrove) in 1999, noted an increase in mapping exhibitions from the 1980s to the 1990s. Ruth Watson expands on his survey and brought it up to date in her 2009 article and includes the seminal exhibition *An Atlas of Radical Cartography* co-curated by artist Lize Mogel in early 2008, and the exhibition *Experimental Geography; Radical Approaches to Landscape, Cartography and Urbanism*, curated by Nato Thompson and Independent Curators International later that year, both in the USA. There have been a number of significant exhibitions in public galleries and museums in the since the mid-2000s in the UK alone, which have discussed the use of cartography by contemporary artists; from *Mapping the Imagination* at the Victoria and Albert Museum (October 2007–April 2008); *Magnificent Maps: Power, Propaganda and Art* at the British Library (April–September 2010);[8] *Uneven Geographies*[9] at Nottingham Contemporary (May–July 2010); *Mind the Map; Inspiring Art, Design and Cartography* at the London Museum of Transport (May–October 2012); and *Mapping It Out; An Alternative Atlas of Contemporary Cartographies* produced by the Serpentine Gallery to coincide with their *Map Marathon* (16–17 October 2010). This was preceded by nearly a decade (2001–2009) of exhibitions with the title *The Map Is Not the Territory*[10] curated by Jane England at the London-based private gallery England and Co, which featured many of the artists exhibited in the subsequent exhibitions, such as Grayson Perry and Stephen Walter. It included works by many eminent British artists such as Langlands and Bell, Cornelia Parker, Tracy Emin and Peter Greenaway; artists who have used the mapping language as part of their broader contemporary practice. It also included the work of artists such as Susan Stockwell, Kathy Prendergast, Layla Curtis, Georgia Russell and Satami Matoba; artists who have made maps and mapping a defining motif of their artistic oeuvre.

From the site-specific art of the 1960s and 1970s Land Art, and the personal, identity politics of the 1980s and 1990s, a more global, sometimes utopian, activism addressing environmental and political concerns has emerged. From cutting and reassembling, to walking, making and writing the map towards *mapping*: 'is a generational shift away from the map (and associated problems of the image and representation) towards mapping as a process, with a concomitant focus on action and activism' (Watson, 2009). The emergence of *critical cartography*, the linking of maps with power, and thus an active and political agent (Crampton and Krygier, 2005) as had been theorized by Wood, has emerged as a distinctive critical tool for artists as is evidenced in the burgeoning cartographic work produced by visual artists in the last four decades, and more significantly in the twenty-first century. Nato Thompson, the curator of *Experimental Geographies* exhibition (2008) observed that the new emerging mapping practices in art could be described as 'operating across an expansive grid with the poetic-didactic as one axis and the geologic-urban as another'. Visual cultural theorist Irit Rogoff wrote that:

[i]t is precisely because art no longer occupies a position of being transcendent to the world and its woes nor a mirror that reflect[s] back some external set of material conditions, that art has become such a useful interlocutor in engaging with the concept of geography, in

trying to unravel how geography is an epistemic structure and its signifying practices shape and structure not just national and economic relations but also identify constitution and identity fragmentation.

*(Rogoff, 2000: 10)*

Critical cartography in the hands of artists has become *radical cartography* (Mogel, quoted in Thompson, 2008), responsive to a political moment, temporal and anti-monumental as a form of resistance, whilst at the same time it has become an *experimental cartography* (Thompson, 2008), where mapping has been used to visualize radically new ideas of world making.

## Images

All images reproduced by kind permission of the artists.
All copyright reserved by the artists.

## Notes

1 The Modern Era is referred to here as the historical era from the Renaissance until the twentieth century. However, Modernity is the period from 1898 until the Second World War and defines a period of revolutionary movement in politics, philosophy, design and art defined by a questioning of the Romantic Era. Modernity is often referred to in common parlance as 'modern'. There is an ongoing debate as to when the Modern Era finished and the post-modern began. This refers to the era of Western civilization and philosophy in particular, rather than the era of Modernity in Art.
2 The *Art and Cartography Commission* of the International Cartographic Association was established in 2008 specifically to 'explore the art element in cartography'. Whilst its aspirations 'to facilitate interdisciplinary cross fertilization of ideas and concepts and to disseminate information about developed theory and ontologies related to the interaction of art with cartography and cartography with art' are a welcome acknowledgement of the intertwining of both disciplines, it is perhaps not as active as the Royal Geographical Society which has an active strand on cultural geography evident in its annual conference programme. The Royal Holloway University of London also established a new centre for the Geohumanities in June 2016 to reflect this.
3 Abraham Ortelius (1527–1598) was a Flemish cartographer in the Age of Discovery, conventionally recognized as the creator of the first modern atlas (*Theatrum Orbis Terrarum: Theatre of the World*) in 1570. The origin of the word 'atlas' is however attributed to another Flemish cartographer, Gerardhus Mercator (1512–1594) who used it in 1585 to denote a book of bound maps, as we know it today.
4 The exhibition *The World in a Mirror* is the story of the world in maps and was displayed from 24 April until 16 August 2015 in the MAS and coincided with the 'International Conference on the History of Cartography (ICHC)' in Antwerp in July 2015.
5 My emphasis.
6 Brotton, J. *Terrestrial Globalism: Mapping the Globe in Early Modern Europe* chapter 3, p.71; Cosgrove, D. (1999) *Mappings* London: Reaktion Books.
7 Lize Mogel co-edited the eponymous catalogue *An Atlas of Radical Cartography* (2008) together with Alexis Bhagat; see Bhagat, Alexis and Mogel, Lize *Journal of Aesthetics and Protest Press*, Sip edition (January 2008).
8 The exhibition at the British Library in 2010 featured the work of several international contemporary artists including Stephen Walter's and Grayson Perry's work, and was also featured in *The Map Is Not the Territory*.
9 'Uneven Geographies', Nottingham Contemporary, exhibition held 8 May–4 July 2010, curated by Farquharson, Alex and Demos, T.J. Online digital archive of eponymous exhibition, first accessed in October 2014. Alex Farquharson was subsequently appointed new director of Tate Britain in July 2015.
10 *This Is Not the Territory*: (i) July–September 2001, (ii) October–November 2002, (iii) (May–June 2003), followed by *The Map Is Not the Territory Revisited* (November 2009).

# References

Bhagat, A. and Mogel, L. (Eds) (2008) *An Atlas of Radical Cartography*, exhibition catalogue, touring July 2007–October 2009, *Journal of Aesthetics and Protest Press, Sp. edition* March 1, 2008.

Brotton, J. (2012) *A History of the World in Twelve Maps* London: Allen Lane.

Cosgrove, D. (Ed.) (1999) *Mappings* London: Reaktion Books.

Cosgrove, D. (2001) *Apollo's Eye: A Cartographic Genealogy of the Earth in the Western Imagination* Baltimore, MD: The Johns Hopkins University Press.

Crampton, J.W. and Krygier, J. (2005). "An Introduction to Critical Cartography" *ACME: An International E-Journal for Critical Geographies* 4 (1) 11–33.

De Certeau, M. (1980) *L'invention du quotidien, Vol I, Arts de Faire* (1980) Translated as *Practice of the Everyday* by Randall, S. (1984) Berkeley, CA: University of California Press.

Eames, C. and Eames, R. (1997) "The Power of Ten" (film, version 1977, original dated 1968), now in National Film Registry of Library of Congress) Available at: *www.youtube.com/watch?v=0fKBhvDjuy0* (Accessed: 1 September 2012)

Farquharson, A. and Demos, T.J. (n.d.) *Uneven Geographies* [online] Digital Archive Of Eponymous Exhibition, 8 May–4 July 2010, Nottingham Contemporary, Available at: *www.nottinghamcontemporary. org/art/uneven-geographies* (Accessed: 1 October 2014).

Harmon, K. (2004) *You Are Here: Personal Geographies and Other Maps of the Imagination* New York: Princeton Architectural Press.

Harmon, K. (2009) *The Map as Art: Contemporary Artists Explore Cartography* New York: Princeton Architectural Press.

Hawkins, Harriet (2014) *For Creative Geographies: Geography, Visual Arts and the Making of Worlds* Abingdon, UK: Routledge, 2014.

Hessler, J. (Ed.) (2015) *Map: Exploring the World* London: Phaidon.

Krauss, R. (1979) "Sculpture in the Expanded Field" *October* 8 (Spring, 1979) pp.30–44.

MAS (2015) *The World in a Mirror: World Maps from the Middle Ages to the Present Day* (Catalogue for Eponymous Exhibition, 24 April–16 August, MAS) Kontich: BAI.

Massey, D. (2005) *For Space* London: Sage.

Mattern, S.C. (2015) *"Gaps in the Map: Why We're Mapping Everything, and Why Not Everything Can, or Should, Be Mapped"*, wordsinpace web log, published 18 September 2015 Available at: *www.wordsinspace. net/wordpress/2015/09/18/gaps-in-the-map-why-were-mapping-everything-and-why-not-everything-can-or-should-be-mapped/* (Accessed: 25 September 2015).

Morrison-Bell, C., Collier, M., Ingold, T. and Robinson, A. (2013) *Walk On: From Richard Long to Janet Cardiff 40 Years of Art Walking* Sunderland, UK: Arts Editions North.

Morton, T. (2013) *Hyperobjects: Philosophy and Ecology after the End of the World* Minneapolis, MN: University of Minnesota Press.

Rogoff, I. (2000) *Terra Infirma: Geography's Visual Culture* London: Routledge.

Stallabrass, J. (2001) *A View from the Fishtank*, critique of Black Shoals. Available at: *www.blackshoals.net/writing/* (Accessed: 30 July 2015).

Thompson, N. (Ed.) (2008) *Experimental Geography: Radical Approaches to Landscape, Cartography and Urbanism* Brooklyn, NY: Melville House.

Watson, R. (2009) "Mapping and Contemporary Art" *The Cartographic Journal* 6 (4) pp.293–307.

Wood, D. (1992) *The Power of Maps* New York: The Guilford Press.

# 39

# Gaming maps and virtual worlds

*Alison Gazzard*

For centuries, cartographic maps have been used to store and represent geographic knowledge about the world and beyond. They form an integral part of how we understand and explain the world.

*(Dodge and Kitchin, 2001: 65)*

Alongside this more traditional interpretation of the cartographic map, maps now often form the basis of games and virtual worlds. Maps may act as simulated experiences, as sites for competition and reward, as spaces for exploration and as a means of way-finding through an unknown territory. The virtual world offers two distinct, yet often related forms of mapping: the map as a designed, constructed site to be exposed and traversed, and the annotated map as a tool to aid users in some of these traversals. One of the joys and purposes of interactive media is to allow the user a sense of agency (Murray, 1997), a sense of being able to change and see the results of what they are changing in those experiences. One of the simplest and most common ways of doing this is through translated movement. Up, Down, Left, Right, Click and Swipe have all become actions associated with how we engage with games, as has moving a mouse to control a cursor across a screen. The desired input device acts as a means for the player to navigate their avatar or character through the world and to traverse the virtual landscape presented to them.

Instead of being a landscape moulded in part by natural processes, the shifting of plates, weathering of the land and the changing seasons, the virtual landscape is totally anthropogenic. As virtual worlds, videogames involve a process of design and architecture (Schweizer, 2013) as they are modelled, animated and programmed, and the boundaries of exploration and negotiation are set through the rules of the designer and the 'algorithms' (Manovich, 2001) of action presented to the player. At the same time, the virtual space is bounded in such a way as to limit this exploration onscreen in a multitude of ways whilst still allowing for a negotiation of space.

The real world can now be augmented with virtual objects as applications can geo-locate the user of a smartphone and create a gameworld in their local vicinity. Here the natural landscape becomes a game map as virtual objects are placed around them, such as the *Augmented Reality Ghost Hunter* application (launched 2009). Yet in these instances we are still bounded by the physical objects and buildings that might get in our way. It is through an exploration of games, play and the properties of the virtual space that this chapter will uncover some of the ways that maps present themselves to the player, how the player can create their own maps, and how player and designer work together to expose and explore these generated worlds.

Much like the real world, the virtual world is one constructed of paths to be traversed, choices to be made and new places to be found, understood and explained. Writing about non-digital games in the 1950s, Roger Caillois distinguishes between 'ludus' or what is often seen as purely ruled play and the act of exploration, discovery and the pleasure of 'paidia', the notion of 'wild, free-from improvisational play' (Caillois, 2001). It is not to say that these categories remain as fixed, binary oppositions, but instead work as a sliding scale of playfulness across a range of game-playing experiences. It is these categories that allow us to understand the map as a 'site of exploration' and the map as a 'site for competition' whilst both exposing ideas of territorial gain. Whereas some virtual worlds offer maps as a form of simulated experience, such as the iOS game, *Plague Inc* (Ndemic Creations, 2012) in which the players spread deadly infections to end humanity as we know it, other game spaces focus on more intricate narrative elements, exploration and the potential for getting lost.

## Getting lost

'There is much more to be said about losing oneself in worldly space than can be referenced – or remedied – by recourse to the abstract objectivity of a map' (Sobchack, 2004: 15).

Not all games have representational maps built into them and many players had to, and often still have to, map the game space in order to remember not only where they have been, but where the enemies may lie and where secret or hidden paths may be located. The process of walking in unknown real-world spaces is one that involves a mental mapping on behalf of the walker, a way of organizing the various paths into distinct places that have been encountered along the way. Whereas in a virtual space the places have to be marked with constructed objects, in the real-world place they are often marked by nature itself. However, the process of way-finding and mental mapping remains the same between virtual world and real-world landscapes, even if the visual clues that enable us to do this may change across platforms and landscapes. This kind of mapping can occur on the fly, in real time, as a personal form of navigation. For human geographer, Yi-Fu Tuan mapping in real time is having 'spatial awareness and spatial skill' saying that, 'walking is a skill, but if I can "see" myself walking and if I can hold that picture in my mind so I can analyse how I move and what path I am following, then I also have knowledge' (Tuan, 1977).

This type of mental mapping becomes extremely important in the playing of early text adventure games, such as *Zork* (Infocom, 1977). Text adventure games operate through the typing of onscreen commands to gain a text response detailing the area and position the player's character is in within the game. In the case of Zork, the initial screen presented to the player informs them that they are 'West of House. This is an open field west of a white house, with a boarded front door. There is a small mailbox here'. Players have to type in short commands such as 'open mailbox' to interact with the gameworld presented to them and start to navigate their way through it. Instead of having graphical clues shown onscreen to depict what is happening, the player has to rely on clues within the text that start to reference various locations and describe the immediate surroundings. In analysing another text adventure game titled *Adventure* (by Crowther and Woods in 1976), Nick Montfort notes, 'the pirate's maze offers rooms that are all uniformly described as "a maze of twisty little passages all alike". To figure out which room is which, the player character must drop objects to mark the different rooms' (Montfort, 2003: 77). Here the player is using a similar technique to that of walking the real-world path in a confusing space (following Hansel and Gretel's breadcrumbs or Ariadne's ball of string). These physical, player created maps depict the paths between places, based on North, South, East and West coordinates and the relationship of one object/room to another within the text descriptions displayed. They rely on the placement of distinct objects to remember where they have been and where they need to go. The significance of not having an object placed in a room denotes a space yet to be explored or recorded by the player. By overcoming these obstacles, and by the player representing them in some sort of mapped form separate to the words appearing on the screen, the game can continue in the moment or be continued after turning off the game machine and restarting another day. The obstacles and successes of way-finding allow the player to feel a sense of achievement as they are rewarded for overcoming these mental or digital stumbling blocks in the game's landscape.

Chaim Gingold (2003) continues these suggestions of exploration in games stating how videogames are constructed of hierarchies of microworlds, existing as separate levels within the overall structure of the game. Although many gameworlds may appear to be vast and expansive, such as the more open-world games of *Grand Theft Auto: San Andreas* (Rockstar North, 2004) and *Red Dead Redemption* (Rockstar San Diego, 2010), the world is condensed into bytes of data that unfold as the player progresses through various stages of reward, retrieval and discovery. Open-world computer games seemingly allow for a greater sense of movement and exploration without the need to always progress through set levels all of the time. In this instance, our natural instinct to explore virtual environments continues to stem from our associations with real-world spaces. To explore the spaces around us is a way of interpreting them. Returning to Yi-Fu Tuan's writing about spatial navigation, he notes, 'when space feels thoroughly familiar to us, it has become place' (Tuan, 1977: 73). By exploring spaces we can start to place them and make associations between the spaces that once were and the places in between that are now formed from those associations. The familiarities of known spaces are apparent in examples such as the home places in *Grand Theft Auto: San Andreas* that allow players to save their progress. Here the term home is used not only as a 'safe place' that we associate with this type of place in the real world, but also as a place where the player knows they can create a safe place for their game data to be saved before moving on to further exploration or a new task in the game.

This act of space becoming place within a home-like environment can also be understood through the example of the first-person interactive narrative-based game *Gone Home* (The Fullbright Company, 2013). *Gone Home* is based around exploration. Players do not know what is in each room when they first enter the house and attempt to trace stories of the player character's family and what exactly has happened in her year away. Upon entering the unknown environment, the only way to start putting the pieces together is by exploring the spaces, finding clues in the form of written notes, boxes of books, Super Nintendo Entertainment System (SNES) games, cassette tapes, and so on, as well as turning on lights in each room to start placing where the character is amongst the seemingly similar door fronts (and gaining access to a mapped view to help them along the way). The combination of recognizable objects helps the player establish both where they have been in the house and also piece together the story that starts to unfold. Here the concept of the 'story map', 'the result of . . . reading the game space in combination with the directed evocative narrative elements encountered along the way' (Nitsche, 2008: 227) becomes another categorization of space as the game allows for exploration leading to a sense of slightly more ruled play as access to a mapped view of the house becomes available.

Access to a map within the game also shows that although videogames can be about exploration, they may also be concerned with conquering space. Spaces are there to be moved through and interpreted, but ultimately gained as a way of rewarding the player in terms of both their own personal and potentially socially situated goals (Gazzard, 2011). As Jenkins notes in his early discussions of 'games as narrative architecture', 'game designers don't simply tell stories, they design worlds and sculpt spaces' (Jenkins, 2004: 121). This concept of designer as architect can allow us to see virtual spaces to be constructed in such a way as to engage the player (via the avatar) in problem solving and way-finding through these designed worlds. In the same light, the player themselves can also be seen as an 'architect' of the space, albeit with a different role to that of the designer. The 'player as architect' rearranges the designed spaces of the gameworld in order to fit their own play styles and possibilities. For many players, the experience of moving through the map is an act of territorial gain and the unexplored areas of the newly played gameplay need to be conquered in order to progress and move forward.

We can see this in the mapped world of *PacMan* (Namco, 1980), where the mapped-out pills of the gameworld start to be conquered by the player. This causes a new player-generated map on each play that works with the remaining pills, where they are, and how the player will reach them whilst battling the ghosts. The underlying maze structure remains fixed as the walls of the space are not moveable, but conquering the map through pill collection changes the shape and direction of play each time the game is loaded. Here the player map is dynamic, changing with the player actions and marking out the various sections of the gameworld that need to still be captured. The multi-layered map acts as a reward, with the disappearing pills signifying the near-completion of the level. It is for this reason that the game designed map becomes the basis of any game experience, and is transformed by player action and the rules of the game into various uses. This especially happens in three-dimensional worlds where maps start to mimic real-life uses and beyond.

## Deliberate maps

As in the example of *Gone Home*, it is often the case that some type of visual map is included within the majority of videogames, so what does this mean in terms of player experience? King and Krzywinska (2005) comment on the different types of mapping on offer as they discuss the player's restrictions on exploration within the gamespace. They differentiate between maps that are seen as 'in-game' and are on the screen whilst the player is navigating the gameworld, and 'out-of-the-game-world maps' that are accessed by the player 'pausing' the game in order to view them. In terms of the maze-like paths of the first or third person videogame, the player is now in many ways able to appreciate the design of the paths as well as experience them through in-game mapping.

'Out-of-the-gameworld' maps can be accessed by 'pausing' the gameplay to allow the player to get their bearings, much like real-world navigation. However, the separation of map and experience does not stop the gameplay, but instead creates various windows of play through the switching between different game screens. This map is accessed deliberately through button presses, pausing and staring at what is displayed onscreen. The use of this type of map requires the player to memorize parts of it in order to remember another sequence of possible spatial events. This becomes as much a part of the problem solving of the game as the navigation itself. It is the 'out-of-the-gameworld map' that temporarily breaks the path of the player's avatar, whilst the player navigates a separate menu outside of the gameworld. These maps can exist either as overviews of the gameworld, or show the immediate area and some of the linking passages of areas yet to be explored or unlocked by the player. Maps accessed out of the gameworld may also be enhanced with other items found within a player character's itinerary. This can be seen in the notebook available in the crime investigation game *L.A. Noire* (Team Bondi, 2011) that is used to uncover various missions and allow the player to look back on to the aspects of each crime they are examining. *L.A. Noire* includes a standard pause command that allows the player to access a map of the streets of Los Angeles that are navigable within the game. This map allows for places to be pinpointed as a way of navigating to them, much like an in-car satellite navigation system. The map gives an overview of all the areas that it is possible to drive to. In this instance the map is 'dynamic' (Dodge and Kitchin, 2001) and can be updated through both player progression and player pin-pointing of locations they wish to go to. At the same time players can also access a notebook, again by pausing the game, that shows the places found and places that need to be visited within each mission dialogue. This list of places works in conjunction with the map, as by accessing each place, the player is able to move their character to that position. The player can either choose for the player's character to drive the car and be guided to the place via the in-game navigation system providing arrows showing the way to the place, or via the non-player character co-driver taking the player directly to the place. In the latter example, the places in between are not shown and the places are joined almost instantly without movement. However, when the player controls the character and moves them through the city, new landmarks are uncovered and placed onto the map. These become useful in later puzzles where the player has to uncover clues and visit the right place in order to reveal the next clue sequence. The player gradually starts to place the world through these associated landmarks, adding layers of familiarity and allowing for places to be accessed more quickly through this recognition. As Calleja (2011: 87) notes in his discussion of spatial involvement of player's in the gameworld:

> Cognitive maps of the area start being built on the basis of chains of landmarks and recognizable routes one moves through over and over again. As these are learned, a sense of comfort and belonging settles in, creating an attachment between player and game environment.

Calleja relates this recognition of landmarks in the virtual world to Lynch's (1960) concept of 'imageability' that references the visual qualities of structures present in the physical environment. For Calleja, the visual mapping and subsequent recognition is a key component of virtual world navigation where 'nonvisual cues like heat, smell, and variety of sounds are either absent or restricted' (Calleja, 2011: 88). Therefore, the visuals exposed by the game map and the player's cognitive map, alongside the actual player-character experience, evolve out of and enhance each other throughout the game.

Now that gameworlds have become larger, the use of onscreen maps and navigational devices (such as compasses) have also become commonplace within the user interface. Games, such as *L.A. Noire* also allow the player to see a constant map of their immediate area on the screen. When trying to complete missions, this map is complemented with onscreen advice in the form of yellow lines on the in-game map showing the paths to take to get from point A to point B. This is much like a satellite navigation device people can place in their real-world cars that give directions to the driver and show a colour coded route to be taken. In some respects, games such as *L.A. Noire* offer the player both design and experience, although the design is experienced on a small scale as the map is limited to the area where the player's avatar is currently situated. The onscreen mapping helps the player situate their avatars within the gameworld. Once again, the homes of characters start to become more recognizable along the directed paths through the connection between the building landmarks and the subsequent events that occur within the buildings. Parts of the game landscape start to become more familiar through the shaping of certain road paths, and this familiarity leads to the game space becoming 'placed' for the player; a process that is enhanced and cemented by using the map in this way.

In contrast, the overall map of *Grand Theft Auto IV* (Rockstar North, 2008), accessed by pausing the game, shows the player the world of the gameworld without having to wait for areas to be unlocked. The player becomes more aware of areas that can be explored through the adding of further landmark icons as new places are accessed and missions completed. As King and Krzywinska (2005: 79) state, the viewing of the map shows that 'maps can indicate areas that simply appear interesting to visit, at any given stage in a game that offers relatively large scope for movement, without the anticipation of any other immediate pay-off'. In the case of *Grand Theft Auto IV*, the map acts as a reminder of those areas that can be explored freely as a reward for completion of pre-defined missions and the learning curve associated with such a game. It is this concept of learning that can lead to other types of map creation by players as they continue to expose their mastery of the game.

## Walkthroughs and making maps

'Let's Plays' and video-captured walkthroughs of games also provide players with another type of mapping of the gameworld. Let's Plays are videos of player's capturing their own play-through of a game often with spoken audio commentary. They do not necessarily expose all hints and tips like a walkthrough would, but they start to show the player's engagement, learning and exploration through the space of the game. In these instances, the map is not only an outline of the world and how to overcome it but it also acts as a way of capturing those particular players' moments of navigation and negotiation. As Iain Simons and James Newman note:

> More than commentaries on the game, walkthroughs serve at least three purposes. First, they frequently offer maps detailing the full extent of the gameworld including 'secret areas'. Second, they offer narrativised, egocentric accounts of the ways in which the player may tackle the game that present a relational space much like the pirate's treasure map (take ten paces forward, you will come to a rock, take three paces left…) that indicates the ways in which, for example, secret areas may be uncovered. Thirdly, the production of walkthroughs, as well as FAQs (Frequently Asked Questions) and Glitch Lists, represent a significant and visible mastery of the game and must be seen as existing at least in part to signal the position of the author as expert user.
>
> *(Simons and Newman, 2003: 8)*

The video capture in this instance records the moments of avatars jumping over platforms, dying, respawning and moving in the right direction to forward the level. The memory of play is shared amongst communities of people through posting the clips to social sites such as YouTube, enabling the content to be 'spreadable' (Jenkins *et al.*, 2013) amongst others. These act as *maps of mastery* in their own right, with the player being able to expose their achievements for others to see. Each instance of the walkthrough further opens up the player's notion of the possibility space available within the game and the discovery of what lies on and beyond the screen.

Mastering the space of the game can also be shown through players creating their own maps, once again exposing the fluidity of fictional game spaces. *Lode Runner* is often referenced as being one of the first games to include level editing functionality. Released in 1983 by the software company Broderbund, *Lode Runner* is a platform style game that involves dodging enemies and collecting the rewards. The player character is able to climb ladders, go along wires and create holes in the brickwork to temporarily trap the enemies in order to succeed. The simplicity of the design can see the game being likened to other platform style arcade games such as *Donkey*

*Kong* (Nintendo, 1981) yet the options screen on loading up the game portrays a different story. Here the menu system not only presents the user with a 'Game Generator' option, allowing the user to edit, test and move the content. When editing the game, the user is then able to construct new walls, poles to climb on and re-position enemies and rewards. On completion, levels can be saved and shared on the many microcomputing platforms it was available for.

Beyond this early example, the game world of *Doom* (id Software, 1993) is often discussed as one of the milestone games that allowed for new levels or assets to be created for the games. These practices are defined by Sue Morris (2003) as being 'co-creative media' where 'neither developers nor player-creators can be solely responsible for production of the final assemblage regarded as "the game", it requires the input of both'. Examples such as these continue today, most notably through the game of *Little Big Planet* (Media Molecule, 2008), allowing users to 'Play, Create and Share' as per the game's tagline. Here, not only is the game played, but new levels can be created through the numerous 'tools' (Wirman, 2009) inherent in its design, seeing the player become both cartographer and designer as they shape the landscape of new spaces to be traversed by others. This is an increasing theme of other recently released games, such as the *Minecraft* (Mojang, 2009) phenomena. *Minecraft* allows players to craft and explore a world by using virtual building blocks to construct new areas as well as dig out, or mine, ones that may already exist. Much like the traditional physical toy of Lego blocks, *Minecraft* allows players to use their imaginations in the same way to create different spaces for their own use and for playing with others (so much so that Lego released a range of Minecraft-related products). Once again, in *Minecraft*, the space is explored in a multitude of ways through different acts of play as player's start to make their own sense of it.

It is this continual searching of the 'algorithm' (Manovich, 2001), the underlying features of the game, the sense of wonder that emerges through walking the virtual world, without an objective, without a map to guide us that occurs in many current game designs. The need for mental mapping becomes apparent once again, as new spaces are remembered and places are secured. In *Proteus* (Key and Kanaga, 2013), the player wanders the landscape and is allowed to become lost in the limited space of the virtual island landscape, as this allows for the sense of more freeform play. Whereas conquering the map gives players a sense of reward, a sense of social achievement in amongst their peers, in *Proteus*, the absence of the navigational map allows for the freedom of exploration once again, the freedom of play that we started this chapter with, except this time the exploration occurs purely through the paths created via moving along the predetermined island mapped landscape. It is these qualities that allow the game map and the virtual world to provide unique experiences, as not only does the map lie at the heart of these experience, it can continue to be explored in a multitude of ways.

# References

Caillois, R. (2001) *Man, Play and Games* Champaign, IL: University of Illinois Press.

Calleja, G. (2011) *In-Game: From Immersion to Incorporation* Cambridge, MA: MIT Press.

Dodge, M. and Kitchin, R. (2001) *Mapping Cyberspace* London: Routledge.

Gazzard, A. (2011) "Unlocking the Gameworld: The Rewards of Space and Time in Videogames" *Game Studies* 11 (1).

Gingold, C. (2003) "Miniature Gardens and Magic Crayon: Games, Spaces & Worlds" unpublished M.A. dissertation, Georgia Institute of Technology, GA.

Jenkins, H. (2004) "Game Design as Narrative Architecture" in Wardrip-Fruin, N. and Harrigan, P. (Eds) *First Person: New Media as Story, Performance and Game* Cambridge, MA: MIT Press, pp.118–120.

Jenkins, H., Ford, S. and Green, J. (2013) *Spreadable Media: Creating Value and Meaning in a Networked Culture* New York: New York University Press.

King, G. and Krzywinska, T. (2005) *Tomb Raiders and Space Invaders* London: I.B. Tauris.

Lynch, K. (1960) *The Image of the City* Cambridge, MA: MIT Press.

Manovich, L. (2001) *The Language of New Media* Cambridge, MA: MIT Press.

Montfort, N. (2003) *Twisty Little Passages* Cambridge, MA: MIT Press.

Morris, S. (2003) "WADS, Bots and MODs: Multiplayer Games as Co-Creative Media" *Level Up Conference Proceedings* DiGRA, University of Utrecht, The Netherlands.

Murray, J.H. (1997) *Hamlet on the Holodeck* Cambridge, MA: MIT Press.

Nitsche, M. (2008) *Video Game Spaces* Cambridge, MA: MIT Press.

Schweizer, B. (2013) "Moving Through Videogame Cities" *Mediascapes Journal* Available at: *www.tft.ucla.edu/mediascape/Fall2013_MovingThroughCities.html* (Accessed: 21 October 2014).

Simons, I. and Newman, J. (2003) "All Your Base Are Belong to Us: Videogame Culture and Textual Production Online" *Level Up Conference Proceedings* Utrecht, The Netherlands.

Sobchack, V. (2004) *Carnal Thoughts: Embodiment and Moving Image Culture* Berkeley, CA: University of California Press.

Tuan, Y.F. (1977) *Space and Place: The Perspective of Experience* Minnesota, MN: University of Minnesota Press.

Wirman, H. (2009) "On Productivity and Game Fandom" *Transformative Works and Culture* 3 Available at: *http://journal.transformativeworks.org/index.php/twc/article/view/145/115* (Accessed: 21 October 2014).

**Part VI**

# Reflections on the future of mapping and cartography

# 40

# Hunches and hopes

*Mark Monmonier*

Confident that nearly a half century of writing and thinking about maps would yield insights worth sharing, I had planned, early on, to title this essay 'Forecasts' and pepper it with bold predictions about how, say 30 years from now, maps that look markedly different will have radically altered the way we see the world. Insights, of course, are not forecasts, and bold predictions are often saturated with wishful thinking. What I can do is identify some emergent trends and plumb their plausible effects, beneficial or otherwise, on map use. Though the discussion occasionally verges on Orwellian science fiction, my hope is that an informed public, skeptical of geospatial technology, will both defend its privacy and preserve the cartographic record.

This reluctance to forecast reflects my failure, three decades ago in *Technological Transition in Cartography (TTC)*, to recognize the proliferation of personal computers and mobile devices as well as the far-reaching impacts of networked communication (Monmonier, 1985). Although I had mentioned the PC (forerunner of the laptop) and the ARPANET (forerunner of the Internet) and had predicted unprecedented customization by map users empowered to choose (or at least to strongly influence) what a map showed at a particular level of detail for a particular area, I didn't anticipate that around 1990 (five years after publication of *TTC*) Tim Berners-Lee would invent the World Wide Web and radically reframe the way we use information. The Internet provided the conduit, but it was the Web that truly revolutionized cartography (and lots of other things) both by giving website developers and content providers a highly integrated arena for exchanging information quickly, and by giving users an efficient way to search and explore. More efficient processors, wider bandwidth, and more commodious digital storage were also important, but the Web was the catalyst.

Enhanced connectivity is both a boon and a threat. Consider, for example, cloud storage, which lets us deposit our photos, maps, travel diaries, and other memorabilia in electronic vaults readily accessible wherever we go. The price of this enhanced access is diminished privacy insofar as information entrusted to so-called third parties is more readily vulnerable to government and corporate snooping than if we kept it in our homes and offices – technology made this vulnerability possible, and the courts made it legal, to promote national security as well as let employers manage their workers more closely. What's more, the cloud's custodians, in it for profit and not accountable to voters, can mine our data for clues useful in influencing purchases and opinions – it's hardly surprising that the technological revolution that brought us free customized maps and no-fee online searching also brought us highly targeted advertising. Information might want to be free, as some opponents of intellectual property rights maintain, but someone – usually advertisers in this case – must pay the bills. Hardly a new thought, of course.

Compromised privacy intersects cartography in several ways. For example, our history of Web browsing might suggest a perverse interest in unwholesome places, local or overseas; overhead imagery can reveal an unapproved, untaxed swimming pool in someone's backyard or marijuana growing in middle of a cornfield; and our cell phone's GPS chip, in revealing where we've been, might also suggest socially or politically suspicious actions or intentions. That none of these examples necessarily involves a graphic map in no way contradicts the cartographic connection. In the electronic era, cartography is no longer focused on printed assemblages of lines, words, and symbols: once mapping morphed into geospatial technology, the traditional visible map became just another form of geographic information.

In *TTC* I told the story of how GPS was conceived as a surveyor's tool, whereby a few satellites in conveniently configured orbits could triangulate the latitude, longitude, and elevation of a particular ground station. A useful tool for putting features on existing maps as well as for creating new maps, GPS found a role in missile guidance systems, otherwise dependent on a comparatively crude technology that kept cruise missiles on course by comparing digital elevation maps stored onboard with terrain profiles captured on the fly with a radar altimeter. If what the altimeter reported didn't match the terrain along the programmed route, the system calculated a course correction. Because reliable terrain correlation demanded reliable elevation maps, the Defense Mapping Agency (DMA) fostered development of the digital elevation models now widely used for environmental modeling and flyover animations – an example of the trickle-down from military to civilian applications.

Evolving technology is reflected in the renaming of America's military mapping organizations. In succession, the Army Map Service became the Army Topographic Command in 1968, the DMA in 1972, the National Imaging and Mapping Agency (NIMA) in 1996, and the National Geospatial-Intelligence Agency (NGA) in 2003. The transition from *Map* to *Mapping*, the addition of *Imaging*, and the eventual substitution of *Geospatial-Intelligence* are not mere public relations ploys. As the new labels imply, traditional map production became a back-office chore, high-resolution imagery collected with specialized sensors became tactically essential and strategically influential, and cartographic data sufficiently enhanced and vetted to merit the term *intelligence* reflect a profound shift from merely identifying and plotting targets to more holistic ways of knowing the enemy.

No less dramatic is the trickle-down of declassified military technology into the civilian realm. In May 2000, largely in response to lobbying by the consumer electronics industry, the government abolished the policy of 'Selective Availability', a mandatory blurring of GPS coordinates that would have made satellite navigation useless for vehicle navigation systems, which proliferated over the next decade and now cost less than a hundred dollars (Monmonier, 2002: 14–15). Another technological hand-me-down reflects the unblurring of satellite imagery, once reserved for spying on our Cold War enemies but now readily available online to anyone curious about a neighbor's fenced-in backyard. Even so, mapping websites cheated a bit when they called their most detailed pictures 'satellite' imagery – aerial imagery sufficiently detailed to identify yard plants was most likely captured from an airplane, not from space, and is typically sharper than a spy satellite's 10-centimetre resolution.

Another kind of trickle-down occurred when the private sector took over the state's role as the prime provider of geographic information. Inspired by Washington's infatuation with privatization and public-private partnerships, and helped along by bureaucratic inertia and venture capital, mapping corporations like Google appropriated the Geological Survey's longstanding role as the nation's principal maker and seller of topographic maps. But in a markedly different way: the contour lines and other brown-ink relief symbols are gone, and in their place are street names (for which fixed-scale USGS maps had no space in developed areas), as well as the ability to look up destinations by street address and craft customized routes that avoid tolls and sluggish traffic as well as minimize distance. What's more, a single click could swap in an overhead snapshot of land cover or a 'street view' showing front shrubbery, kids at play, and cars in driveways. Although the imagery was hardly real time, it was much more current than the average USGS topographic map, typically two or three decades out of date. And aside from what we pay our Internet and wireless providers, it's free.

Google's Street View has been blamed for lost privacy, especially by the Germans and the Swiss, who forced the company to blur facades or signage if the owner requests (Segall, 2009–10). Although merely capturing what's easily visible from the road is hardly akin to staring through walls and draperies, people (and perhaps entire nations) vary markedly in what they don't want others to see. Google routinely blurs faces and license plates – an algorithm does this automatically – but the process might not be foolproof. Carl Hiaasen (2013), in a recent tale of murder and mischief in south Florida, described how a fictional forensic pathologist, fascinated with Google Maps, discovered that her husband was having an affair when Street View revealed a strange Honda Accord in their driveway next to his Ram pickup. Able to read part of the license plate – Hiaasen has an impish mind – she asked a police acquaintance to run it through the state's registration database, which fingered a female paddleboard instructor from whom she had purchased three lessons as hubby's birthday present. Frankly, I'm more annoyed that Google shows the empty trashcans one or two of my neighbors had neglected to take in the previous evening.

More ominous a privacy threat than Street View is your cell phone's GPS chip, originally intended to promote reliable responses to emergency calls. Thanks to a convergence of display-screen, memory and tracking technologies, the smartphone became a commonplace navigation instrument, easily enhanced by social media apps that alert us when favorite friends are nearby or keep track of how many times we visit a particular location – 'check in' frequently at your local Starbucks and Foursquare might declare you the café's 'mayor'. Although voluntary check-ins might seem harmless because you appoint your friends and can turn off the app, the mobile phone's GPS can track your movements with remarkable precision, far better than with older instruments that relied on differences in signal strength to triangulate from the nearest cell towers. The growing number of LBS (location-based services) apps increases the likelihood that your comings and goings might become a commodity, available at the right price to stalkers as well as advertisers. Get used to it or demand laws to protect locational privacy.

Because location data are retained, ostensibly to troubleshoot complaints about poor coverage or dropped calls, anyone armed with a subpoena can track you retroactively, or even in real time, and mine your locational history for potentially incriminating details. And there's the National Security Agency (NSA), eager to keep tabs on whom we Americans call. If you don't want to be tracked, you can, of course, leave your phone at home – easy to do accidentally, when hurried or distracted – or put it inside a metal can that serves as a Faraday cage, to block radio-frequency signals. Shoplifters employ a similar concept when they stuff merchandise into a 'booster bag', lined with metal foil to thwart entryway sensors designed to pick up signals from RFID (radio-frequency identification) tags. Be wary that such ploys can invite suspicion, and don't get caught in a shopping mall with a booster bag.

Because college professors are not the only victims of absentmindedness, there's surely a market for a flexible electronic leash that doesn't let people get too far from their phones. Some clever entrepreneur is bound to devise a proximity-alert chip, embedded in the skin or attached as a body piercing and programmed to administer a tiny pinch or trigger an audible warning when you step out of the house with your phone still on its charger.

Embedded chips? Like the ones used to reunite errant dogs and cats with their people? Yes, but a lot more powerful once nanotechnology asserts its power as a tracking technology. In 2002, in *Spying with Maps* (subtitled *Surveillance Technologies and the Future of Privacy*), I reproduced an illustration from a patent awarded in May 1997 for a subdermal GPS implant powered by body heat that reported its location over the wireless network. At least one company was marketing a battery-operated wristwatch version as a strategy for tracking children and Alzheimer's patients – they named it Digital Angel – and conceptually similar systems were already in use for monitoring convicted sex offenders and criminal defendants free on bail.

What's easily available with current technology – though I know of no specific implementation – is a system that sends an alert when a tracked subject enters a no-go zone configured as an invisible circular fence around a school, a playground, or the home of an ex-spouse, former girlfriend, obsessed fan, or frequent burglar – sex offenders are not the only vicious stalkers. Once the system determines a potential predator is within the same county as his likely victim – statistics justify the masculine pronoun – it's easy to continually calculate straight-line distance between the moving subject and the fixed location. If this distance drops below the minimum allowable proximity, the system would alert the local police or the potential victim, or even the tracked subject, who could be offered an opportunity to retreat. What's more, a dynamic protection zone is possible if the protected person carries a GPS device.

A passive alert – *Bad guy nearby!* – is not the only response. In sharp contrast to travel restrictions on sex offenders or 'orders of protection' against harassment, a proactive surveillance system might administer an annoying tingle as a warning whenever the subject approaches a no-go area and follow up with a sharp pinch should he cross the boundary. And if the invasion is dangerously rapid or the intruder seems reluctant to retreat, the system might even inject a soporific into his bloodstream. *You've been warned, Asshole. Now take a nice little nap while we come to pick you up!* Whoever said geospatial tracking had to be nice?

No-go areas can have an irregular shape represented by a series of straight lines joined at points to form a polygon. The 'point-in-polygon' test, which (as its name implies) determines whether a point is inside a given polygon, is among the oldest and simplest GIS operation. It can be used dynamically to alert authorities whenever a registered sex offender or compulsive harasser (the point) is inside a forbidden zone around a school, park, playground, or victim's home (the polygon). Similarly, thin corridors defined along a prescribed route connecting the subject's home and various approved venues could be made available on certain days and times, to allow travel between home and work as well as trips for grocery shopping, church services, and support-group meetings. Frighteningly Orwellian, to be sure, but arguably more humane than incarceration: prisons are nasty places in which sex offenders and weaker inmates are often assaulted, and experienced criminals can share tricks with younger inmates.

*Geofencing*, as it's called, is a low-cost alternative to incarceration: taxpayers spend less on prisons and guards, and subjects can even contribute to the cost of monitoring. And it's not just for sex offenders: GPS could keep pickpockets away from crowds, discourage gang members and political dissidents from congregating, and strictly monitor the movements of assorted felons on parole – are thieves any less deviant than flashers and gropers? Clearly geospatial tracking's greatest threat is ease with which those in power can define deviancy upward. Civil libertarians need to be privacy advocates, and vice versa.

Whether tracking is a threat might depend on whether the electronics and the people in charge of monitoring do what they're supposed to, which doesn't always happen. In early 2013, for instance, David Renz, a 29-year-old Syracuse resident awaiting trial for possessing child pornography, disabled the ankle bracelet of his 'tamper resistant' GPS monitor, carjacked a school librarian in a shopping mall parking lot, killed the woman, and raped her 10-year-old daughter (Eisenstadt, 2013). Adept at electronics, Renz cut through and then quickly reassembled the bracelet, which he left at home for an evening of mayhem. His system sent a tamper alert to the monitoring service in Colorado, which notified the U.S. Probation Office in Syracuse, which ignored the warning after it received an all's-well signal five minutes later. Probation officials had apparently ignored 45 earlier alerts, too readily dismissed as false alarms while Renz practiced his escape. A report by embarrassed court officials included a photograph of Renz's reassembled bracelet, held together with duct tape and a screw.

Another implication is *geoslavery*, a term geographers Peter Fisher and Jerry Dobson introduced in 2003 to describe the oppressive geospatial monitoring of children, females, low-wage workers, and political dissidents. They warned of 'a practice in which one entity, the master, coercively or surreptitiously monitors and exerts control over the physical location of another individual, the slave' (Dobson and Fisher, 2003: 47–48). One of their hypothetical scenarios involved a Third World village in an unnamed country where a young girl was executed by family members who felt embarrassed because she attended a movie without permission. GPS might make this kind of so-called honor killing more common, they argue, insofar as a local entrepreneur who recruited a mere 100 subscribers need invest only $2,000 for the central unit and $100 for each tracking unit, 'amortized over ten years' (Dobson and Fisher, 2003: 50). A decade later the technology is a lot less expensive.

Let me push further into the future by imagining a personal identification system similar to the collision-avoidance and automatic-identification schemes devised for aircraft and ships. In much the same way that a freighter or warship approaching a nation's defense perimeter could be required to broadcast its position, registry, size, and contents, a person strolling down the street could be made to announce electronically his name, ID number, and other potentially pertinent information (height, weight, hair color, language, criminal history, etc.) to anyone equipped to listen. A network of strategically positioned receivers would let Big Brother track people's movements, and electronic screeners carried by passers-by could warn of the approach of dangerous individuals or old friends.

Another implementation, akin to what's called enhanced reality, might use a display embedded in an eyeglass lens to remind the wearer of a forgotten acquaintance's name, its pronunciation, and the circumstances of their prior interaction. And to antagonize anyone wary of Street View, a system linked to the local property assessment database could instantaneously tell the wearer looking at a house not only the name of the owner but also the purchase price, the number of bathrooms, and whether property taxes had been paid on time. A more fully integrated system might even reveal the locations and types of crimes reported on the block and what specific neighbors told a news reporter. Erroneous or malicious postings can have a long life on online.

Why not bypass the eyes altogether with an electronic implant that funnels map data directly to the brain? Research underway since the 1970s with support from the National Science Foundation and the Defense Department can be found in bibliographic databases using search terms like *brain–machine interface*, *brain implant*, and *neural implant* – think of this as biomedical engineering meets nanotechnology. The fundamental approach is hardly new: audio implants, in particular the well-known Cochlear Implant, have been a boon to people with hearing impairments for over half a century, and visual prosthesis seems to be catching up. Although neural implants for normally sighted people remain a topic for the military, science fiction authors, and conspiracy theorists leery of mind control, I'm sure a way can be found to integrate a self-contained bionic eye with a GPS map viewer and connect it to the wireless network.

The wearer's brain gets a dynamic mental map with a 'You are here' symbol, and anyone with system access and the right password can add the person to his own mental map, further enhanced by targeted searching and computer-generated directions.

I can understand why conspiracy theorists feed on these possibilities. Although wide dependence on neural implants would have a serious downside – an individual's transmitter could die at an inopportune moment, or an electromagnetic pulse from a low-altitude nuclear explosion could spread chaotic disruption across a wide area – automated identification and tracking of some sort seems inevitable. And because some blackguard will no doubt find a way to thwart his or her electronic annunciator, networked systems would be needed to identify and report any UWO (unidentified walking object), who might be a criminal, an escaped prisoner, a terrorist, an enemy invader, a robot, a citizen whose system unexpectedly shut down, or simply a large wild animal. Livestock and pets would, of course, be appropriately tracked.

In the near run a more plausible approach is spatial biometrics, notably the iris scan, which U.S. Customs and Border Protection adopted (along with fingerprints) for its Trusted Traveler program, whereby low-risk travelers enjoy expedited clearance at selected high-volume airports. Australia, Germany, and Japan have implemented similar systems, and many high-security government and private facilities also use iris identification. Because a person's eyes are highly unique, iris scanning might well become the standard way to deny unauthorized persons ready access to schools, apartment buildings, and other facilities with a single, carefully monitored entry point. In addition to trusted travelers, we'll have trusted students, trusted residents, and perhaps even trusted shoppers and trusted diners. But can we trust the people charged with certifying who is or is not to be trusted?

A broader but less precise identification technology suitable for pole-mounted street cameras is facial biometrics, which first measures the size, shape, and relative locations of the eyes, nose, mouth, chin, eye brows, and other facial features and then attempts to match these attributes with the corresponding measurements for known individuals. Unless the person is facing the camera, the reliability is only so-so but sufficient for a network of cameras perhaps designed to track a subject through a neighborhood or shopping mall. Not a new idea, though: in *Spying with Maps* I mentioned Britain's use of street cameras for law enforcement and suggested the possibility of a web of video cameras 'as dense as streetlights' if a convincing argument can be made for its effectiveness (Monmonier, 2002: 115). The argument is advanced every time street cameras play even a minor role in reconstructing heinous incidents like the Boston Marathon bombing of 2013. Better at catching criminals than preventing crime, these systems are too easily sold to a fearful public.

A wholly different approach looks down from above and relies on differences in thermal radiation to identify people and other mammals as well as the banks of light bulbs used to grow large amounts of cannabis indoors. Although a FLIR (forward looking infrared radar) instrument carried aboard a helicopter can spot an exceptional amount of 'waste heat' leaving a building, it's currently illegal for police to routinely fly around looking for enclosed marijuana plantations. Under the Fourth Amendment, a person's home is a private space, off limits to sophisticated scanning technologies not generally available to the public. The U.S. Supreme Court reaffirmed this protection in 2001, in its landmark 5–4 decision in *Kyllo v. United States*. Danny Lee Kyllo had more than 100 cannabis plants in his attached garage, but the high court overturned his conviction because the warrant police used to enter his home had been based on unauthorized FLIR surveillance. Writing for the majority, Justice Antonia Scalia declared that when FLIR or a similar technology is used 'to explore details of the home that would previously have been unknowable without physical intrusion, the surveillance is a "search" and is presumptively unreasonable without a warrant' (Kyllo v. United States, 2001: 40).

When international terrorism might be even remotely involved, getting a warrant is a piece of cake. At least that's what recent revelations about the super-secret FISA (Foreign Intelligence Surveillance Court) suggest (Gellman and Poitras, 2013; Greenwald, 2013). Authorized under the Foreign Intelligence Surveillance Act of 1978, the court's official name is the United States Foreign Intelligence Surveillance Court. Leaks to the media in 2013 reveal that the FISA court never rejects an intelligence agency's request for a subpoena. With the court's endorsement, the NSA has been collecting information on the origin and destinations of telephone calls as well as similar 'metadata' for text messages and email. It's unlikely that the FISA court would reject a federal agency's request, made 'in the interest of national security', to authorize other forms of surveillance presumably banned by the Fourth Amendment.

Although government stockpiling of the origin–destination records of our telephone calls and emails has little direct impact on most people's privacy, I understand why many of us feel violated. Personally, I am not as upset as I am curious about exactly how the NSA uses this information to protect national security, how effective this surveillance really is, and how they know it's effective, how carefully the data are protected from uses outside the FISA court's ruling, and how long the NSA intends to keep the data. If the retention period is longer than the 72 years that our Census questionnaires remain confidential, these so-called metadata could become a goldmine decades from now for social science researchers.

No less frightening is the possibility that the FISA court might allow routine high-resolution imaging of the insides of buildings. I don't know enough about three-dimensional imaging technologies like MRI scans to say whether this type of invasive surveillance is physically possible or how it might be done. This is idle speculation, of course, but imagine the political stink if the NSA or some other alphabet agency were caught looking through walls in search of assault rifles. I'm inclined to dismiss this threat because at the current rate of creeping acquiescence, in-the-home sensors akin to the two-way 'telescreen' in George Orwell's *1984* – mandated, of course, to promote health and safety – seem far more plausible than intrusive overhead imaging. If Orwell had not given 'Big Brother' a disgustingly negative connotation, it would make a great marketing slogan.

A more immediate concern is the imminent proliferation of unmanned aerial vehicles (UAVs), commonly known as drones. What concerns me more than the Defense Department's and the CIA's Predator and Global Hawk are the smaller drones akin to the toys demonstrated in shopping malls and sold online for a few hundred dollars. The more expensive mini-UAVs are useful mapping tools, the moderately priced ones provide amateur photographers with new vistas, and the cheap ones are no doubt useful for annoying one's neighbors with an unspoken threat. Overhead surveillance is no longer the prerogative of government agencies and big corporations.

Another element of the Orwellian surveillance state is HUMINT, for HUMan INTelligence, which in the cartographic context is an inherently benign endeavor called crowdsourcing or VGI, for Volunteered Geographic Information. Unlike the tattletale society in *1984*, today's informants might well be school children, motorists, gardeners, bored retirees, and other participants in what academics are calling neogeography. The activity is not entirely new insofar as map users have eagerly reported errors to map publishers for centuries, typically when they discovered a missing or misnamed street. As far as I can tell, this reporting was wholly unorganized until 1994, when the U.S. Geological Survey established the National Map Corps to help keep its integrated topographic database, known as The National Map, as up to date as possible (McCartney *et al.*, 2015). The movement attracted over 2,000 volunteers, but a decade and a half later Twitter, Facebook, and other social media were fostering clever and timely online maps depicting events as diverse as traffic accidents, earthquakes, plant fungus, and good buys on gasoline, all based on geotagged comments and photos posted by enthusiastic mobile users, often unaware that their posts included geographic coordinates calculated by their cellphone's GPS. No doubt useful, these maps typically lacked the systematic, methodical coverage of conventional cartographic surveys, and thereby challenged academic neogeographers to devise new concepts of accuracy, reliability, and uncertainty.

Threats to location privacy are no more troubling than the fragility of digital maps, which can be wiped out instantly by an electromagnetic pulse (EMP) from a low-altitude nuclear blast, and printed maps that can disintegrate as the paper slowly oxidizes (Carafano *et al.*, 2011). Electronic data are susceptible to diverse threats, some from nature and others from humans. Natural nemeses include disc rot, which contaminates magnetic storage media, and geomagnetic storms, which disrupt the Earth's magnetic field, the power grid, and telecommunications, including signals from GPS satellites. Anthropogenic threats are more varied and include ionizing radiation from nuclear explosions, massive cyber attacks, and computer hackers, whether malicious or merely mischievous.

A successful cyber attack could disable the power grid, crippling everything that depends on it, and even destroy geospatial databases not backed up on a secure server. By contrast, a more targeted attack might demonstrate the perpetrator's cleverness with a newsworthy stunt like switching Cleveland and Detroit, reshaping the Grand Canyon, or renaming streets after hacker's friends – visual evidence of having broken into Google Maps. This type of mischief has precedents in the pre-digital era in the minor tampering occasionally carried out by bored low-level employees at street mapping firms. Because even the best map is never error-free, adding a fictitious town or two probably did little to undermine a map's usefulness.

A wholly different concern is the protection of the cartographic record from the gradual, inevitable processes of physical decay. All paper oxidizes – a fancy word for burning – but this deterioration can be slowed markedly when the maps are printed on acid-free paper and kept away from direct sunlight. Older maps on vulnerable paper can be treated chemically or encapsulated, to choke off the supply of oxygen.

Scanning is only a temporary solution. Digital media are more fragile than cheap paper, and must be recopied every decade or so to avoid corruption. It's ironic that the most efficient way to preserve an electronic map for a thousand years or more is to print it – with ink, not toner – on high-quality paper. Or better yet, etch a smaller version onto a minimally corrosive metal plate. Nanotechnology might be the key to reliable preservation of the cartographic record.

Not every map merits long-term preservation. The vast number of ephemeral electronic maps, displaying optimized routes for specific journeys or exploratory combinations of land-use and topographic overlays, makes preservation a difficult goal, not than many people seriously believe the task, even if possible, is at all desirable. Moreover, preserving the data and related software would be pointless without also preserving the hardware and operating systems needed for interactive exploration and display. For instance, my antique Atlas Touring demos, still useful in showing my Map Design students what dynamic maps were like in the early 1990s, will probably be lost for good once the cranky Macintosh Powerbook 1400cs laptop on which they're stored fails to boot up (Monmonier, 2014: 58–61). I still have backup copies stored on obsolete 3.5-inch diskettes, but these too will become either terminally corrupt or functionally useless. Operating systems move on, leaving orphaned software with no stage on which to perform. Although my files could be moved to another aged but still functioning computer, each year the possibilities shrink.

The narcissist in me longs for a national museum of cartography, perhaps a Smithsonian affiliate, at which my primitive animations and others like them might be kept alive for anyone interested in digital maps as historical artifacts. Sad to say, map historians have had little respect for digital developments. The Smithsonian has a credible collection of surveying instruments and related historical artifacts but lacks a serious interest in dynamic maps and mapping software, and the Library of Congress and the National Archives are no better equipped. Electronic ephemera based on magnetic storage and micro circuitry lack both the physical durability and the visual appeal that inspires collectors of old maps and scientific instruments to rescue worthy artifacts from the dustbin and bequeath them to a museum or library, at which conscientious curators could orchestrate preservation and fill gaps in the collection.

On a more optimistic note, scholarly and technical journals, along with textbooks, technical manuals, government reports, theses, dissertations, and assorted environmental impact studies constitute a comparatively durable historical record, at least of static images. However useful for map historians and lay enthusiasts, this resource seems unlikely to intrigue many younger scholars, who in their avid pursuit of improved techniques and new knowledge, routinely ignore anything published before 2000 – anything *so last century* – however relevant. And a scholarly record preserved primarily on digital media is no more durable than the maps it describes.

I'd welcome a concerted effort to carefully preserve and critique the cartography of the first quarter of the twenty-first century. Best to do this now, when oral histories, ephemeral documents, and surveys of uses and usefulness can be collected systematically, with apparent gaps filled and contrary interpretations included. Although much but not all of this effort could be conducted online, a physical facility with exhibits that relate older developments to more recent ones would be an asset to educators, scholars, and map lovers of all types. Given a choice, I'd opt for a tourist-oriented approach, like the Baseball Hall of Fame, which combines a museum with an impressive archive and manuscript collection. Mapping might not have the hoards of fans that follow team sports, but it's no less pervasive and visually engaging.

## References

Carafano, J.J., Spring, B., and Weitz, R. (2011) "Before the Lights Go Out: A Survey of EMP Preparedness Reveals Significant Shortfalls" Available at: *www.heritage.org/research/reports/2011/08/before-the-lights-go-out-a-survey-of-emp-preparedness-reveals-significant-shortfalls* (Accessed: 20 March 2016).

Dobson, J.E. and Fisher, P.F. (2003) "Geoslavery" *IEEE Technology and Society Magazine* 22 (1) pp.47–52.

Eisenstadt, M. (2013) "One Minute Separated Suspect from Monitor" *Syracuse Post-Standard* (20 March 2013) p.A-1.

Gellman, B. and Poitras, L. (2013) "U.S. Intelligence Mines Data from Internet Companies in Broad Secret Program" *Washington Post* (6 June 2013).

Greenwald, G. (2013) "Collecting Phone Records of Millions of Verizon Customers Daily" *The Guardian* (6 June 2013).

Hiaasen, C. (2013) *Bad Monkey* New York: Alfred A. Knopf.

*Kyllo v. United States* (2001) 533 U.S. 27.

McCartney, E.A., Craun, K.J., Korris, E., Brostuen, D.A. and Moore, L.R. (2015) "Crowdsourcing The National Map" *Cartography and Geographic Information Science* 42 (Supp. 1) pp.54–57.

Monmonier, M. (1985) *Technological Transition in Cartography* Madison, WI: University of Wisconsin Press.

Monmonier, M. (2002) *Spying with Maps* Chicago, IL: University of Chicago Press.

Monmonier, M. (2014) *Adventures in Academic Cartography: A Memoir* Syracuse, NY: Bar Scale Press.

Segall, J.E. (2009–10) "Google Street View: Walking the Line of Privacy – Intrusion upon Seclusion and Publicity Given to Private Facts in the Digital Age" *Pittsburgh Journal of Technology Law and Policy* 10 (1) pp.1–32.

# 41

# Can a map change the world?

*Danny Dorling*

## Maps colour in our imaginations

Maps can change the world because it is through maps that the world is imagined in the minds of those who change it. Change the map and you change how the world is viewed. Change how the world is viewed and you change the prejudices of those who can change the world. Change their prejudices and they will then change the world differently to how they might otherwise have behaved.

All behaviour matters when it comes to the influence of maps. From the smallest aside in a conversation about how people in a particular place are – what you think of them – to a commander moving an army in one direction rather than another because of what he (it's almost always a he) perceives as opportunities or constraints displayed on a paper map. Maps, from ephemeral 'performance' to hard copy or digital images, colour in our imaginations.

Figure 41.1 shows a view of the world in which China is represented as a pig, Turkey as a tiger, Spain as a dog, Russia as a sleepy bear and Japan itself as a Samurai warrior. The boundaries of colonies being demarked in Africa by imperial powers are shown as being stitched together in ways that may not last. It is one of very many images from the late nineteenth and early twentieth centuries which presented states as if they were species of animal, and as if the boundaries between them were somehow natural and flexible. Dominant above them all was Russia, sometimes as lethargic, as in this image, sometimes as aggressive. Drawn in similar maps, Russia has been represented as an octopus about to infiltrate the rest of the continent of Eurasia with its tentacles. The United States is depicted as largely insignificant, placed on the lower right-hand corner, observing from the side-lines.

Figure 41.1 was published a few months after the outbreak of the First World War, at a time when nobody could have known what the eventual outcome of the conflict would be. These anthropomorphic maps, many produced before the war, would contribute to people's perceptions of the protagonists and influence their attitudes. The map focuses on various contemporary stereotypes: Russia as a lazy bear (with its hobnail boot hovering over eastern Europe), China as a fat pig, India as a sleepy elephant, all near or neighbouring Japan represented by a Samurai warrior, located close to Korea (a docile peasant). Few of these images are positive except Japan, and even Japan appears to be sporting a tiger's tail. All are being eyed up by Uncle Sam from

*Figure 41.1* **A humoristic/racist Japanese world map, created 3 September 1914** Available at: *https://upload.wikimedia.org/wikipedia/commons/3/3e/World_around_1900.jpg* (Accessed: 10 July 2016)

the distant Americas, largely off-stage. With his telescope and top hat, Uncle Sam is depicted as a small and insubstantial rodent-like figure.

In the event, it would be the United States that invaded Japan just thirty years after this map was drawn. The map was prescient in its depiction of Japan as a small warrior poised to energize its larger and lethargic neighbours into action; this may have been one of myriad tiny visual and verbal shoves in the direction of future war. The other tiger in the image represents what is now Turkey, but was then part of the Ottoman Empire; scowling beneath the underbelly of a complacent Europe in which Germany is a hog impaled with many arrows and the United Kingdom is a sea horse.

Images such as Figure 41.1 may appear bizarre to us today. Reifying countries as animals, with each nation-state being presented as if it were a single creature with a particular temperament, is less common a century on, but can be remobilized as circumstances dictate; for example, the visual conceits of the acquisitive octopus or the rampant bear continue to be used today to portray Putin's Russia and its penetration of power into eastern Europe and the Middle East (see Vujakovic, 2014). Perhaps in the future people will look back at the images we produce today and ask how we could have been so naïve, so stereotyping, so manipulative? And yet we still talk of countries as if each were an object controlled in a singular way by a single mind: Putin's Russia, Obama's America, and Cameron's (or Theresa May's) Britain.

Just over thirty years after the 'humourist' (racist) Japanese world map was published, the United States military would drop the first of two nuclear bombs on Japanese cities. Figure 41.2

shows an image of the firestorm that followed the nuclear detonation over Hiroshima on 6 August 1945. It is an image that needs very little embellishment to still have a menacing effect today.

The red area in Figure 41.2 shows the extent of the firestorm that killed so many people, those who did not die in the initial blast. Air was sucked in from the surrounding hills after the initial blast, as the heat pored upwards – the inrush of the oxygen turned the central city into vast crematoria. Over 50,000 people died in Hiroshima that day. More than 50,000 others would die within the next few months as a result of their burns, other injuries, radiation sickness, lack of medical care and starvation. Almost all of the dead were civilians. Nine out of ten doctors in the city were killed or injured that day and an even higher proportion of nurses. No significant help came. If this story is new to you, look again at Figure 41.2. The single most deadly bomb ever dropped in the history of humanity (Truman, 1946). To what extent stereotyping, cartographic or other, contributed to an 'othering' of the people targeted for destruction can only be conjecture, but when a person or group is constructed as less than human, their fate becomes easier to justify.

To me cartography is the means by which we make sense of our world. We have to make sense of our world to be able to change our world. The better a sense we have, the less likely we will do damage in what we try to change. Almost all of the earliest cartographic images were

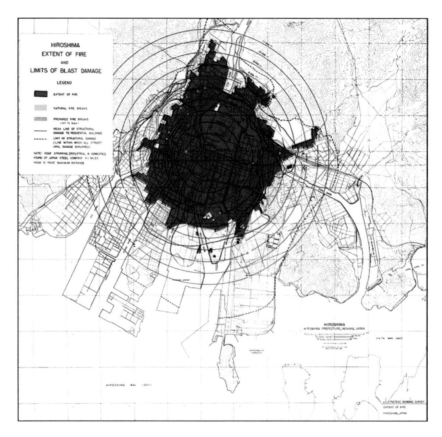

*Figure 41.2*   **The extent of the firestorm, Hiroshima, Japan, 6 August 1945 Available at:**
*http://fly.historicwings.com/wp-content/uploads/2012/08/HighFlight-LittleBoy3.*
*jpg* **(Accessed: 10 July 2016)**

not simply way-finding maps, but contained messages of how human society was ordered at the time and what the threats and opportunities to that society might be, as well as providing messages for how much better things could be; and how much worse – if disaster struck. There is every reason to believe our maps today serve similar purposes, but that these are less obvious to us as we tend not to consider our outlook subjective.

## New maps show when the world changes

One of the earliest maps of anywhere in the world is of a city, Çatalhöyük (Figure 41.3). The map depicts houses built check by jowl, with no streets or even alleyways between them and with a threatening volcano looming over this earliest of human concentrations. Recently archaeologists have suggested that what we see today as a volcano may, in fact, have been a rather lurid leopard skin dress, drawn above the city for some other reason. We may never know if that dot-filled-form is dress or volcano, but we know that this map – revealed on a 9,000-year-old plaster wall – served a purpose greater than simply being a remarkably accurate depiction of the buildings around it, and which for many thousands of years had been buried and out of sight. The map shows how this ancient people thought that their city and that part of the world was organized.

The original image is augmented in Figure 41.3 by two modern-day plans. These show how the city without streets might have looked had anyone then been able to fly over it when it first existed at its full extent and how it was laid out in plan form. The houses of Çatalhöyük were denser than those built in Hiroshima, but not much denser. Millennia would pass before the region in which the city existed (now called Turkey) would be depicted as a tiger ready to leap up towards Europe from Arabia. It would take that same amount of time for us to start spacing our cities out more, to be at lower risk from fire.

Today we presume that people got to their homes in Çatalhöyük by walking over the roofs of other people's property. Also almost certainly property will have had a different meaning then. There were no countries, as we know them now, and the idea of giving generic names to masses of water, the entire lengths of river networks, and maybe of towns and cities will have all been later inventions – spawned from thoughts that have come to human minds long since Çatalhöyük was first built, along with both the idea of streets and, in some cases a very long time later: sewers (Figure 41.4).

Modern maps of the world privilege coastlines as being important, almost universally they put north uppermost and leave much space blank. These illustrate how the 'bottleneck' between the great continental landmasses of Africa and Eurasia could have been the place where human innovation flourished, as interaction was maximized, but it might also mislead.

If, in the past, people found moving long distances over water far easier than moving long distances over land, then we would need to turn the map inside-out (in some way) to see the topography and topology that actually led to Çatalhöyük being, for a time, near the centre of things. Cartography can obscure as much as it reveals. Use a modern-day map to look at the past, or even look at the present without getting the metric right, and you might not see things as clearly as you could.

Cartography changes as our conception of how we are all connected to each other and how that is changing. We can easily think that how we draw maps today is how they should always have been drawn, if only people in the past had benefited from what we now know (we might think), but it is easy to disprove this as both a fanciful and narrow-minded viewpoint. In many ways, the modern mechanization of map making has held innovation back because computer software mimics what cartographers of a few decades ago regarded as 'good cartography'.

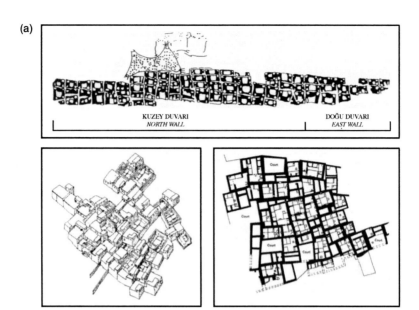

*Source*: Redrawn from many reconstructed images that now appear on the web.

Where three continents join

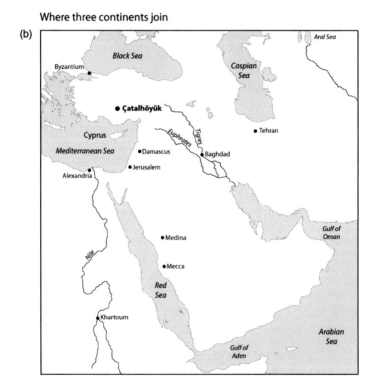

*Figure 41.3* Çatalhöyük – the world's first city (without streets)

*Source*: Re-drawn from author's map.

*Note*: Selected cities that were founded and seas and rivers that were named after Çatalhöyük in lighter type

*Figure 41.4* The Manchester sewers imagined beneath the shiny station platform

Figure 41.5 shows one depiction of the migration of humans out of Africa, through what is now Turkey and around the coasts towards Australia; back (much later) to Madagascar; across to the Americas and, at a similar time into cold, wet, western Asia – later to be christened 'Europe'. We had no idea that such a story of our past was possible just six generations ago. We did not know our origins as a species. Just three generations ago we had no computers and no reliable knowledge of DNA. Most people in the world did not come to accept modern ideas on biological evolution until long after the theory was propounded by Darwin; many still do not accept it.

The projection underlying Figure 41.5 is too complex to be created by a person working with paper and pen, no matter how long they might have been given to try to draw it. The world is changing because our understanding of it is changing and so our maps of the world

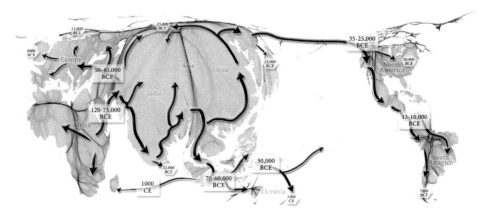

*Figure 41.5* The migration of people out of Africa on a map sized by the descendants. Drawn by the author with help from Benjamin D. Hennig, Carl Lee and Wikipedia

are beginning to change very rapidly in response. In maps thousands of years old we painted animals on cave and plaster walls, or leopard skins maybe. In maps just over a century old, animals depicted countries all printed on paper. Maps then became more 'scientific', more accurate but still they were about what we believed was around us in our 'hearts and minds'. Even as the world is being stretched to show where people live, using a scientific algorithm, a political choice is being made – that Africa, India and China need to be shown with more prominence. Not as a set of borders being stitched up, or as a sleepy elephant and a greedy pig, but as hope to the bulk of humanity.

## Travel time mapping in spacetime

Should you be a teacher, here is an exercise you can set your students. Ask them to draw a new map of the world. This is a map they can draw without needing a computer to draw it. They need to take a list of the world's largest modern cities. Many such lists exist; all are to a degree arbitrary. Begin with the largest city, in the case of the list given below I use Tokyo, and then calculate how long it takes to fly from there to the second largest city in the world. In 2012, when the data used to draw Figure 41.6 were collected, this was Guangzhou.

Here are the kinds of instructions you can give students for them to try to draw a new world map:

> There are many calculators on the web that will give you the shortest travel times between two cities. Use one of these to draw each city as a circle with area in proportion to its population and the distance between each pair of city centres made proportional to the quickest possible travel time between their most central airports. A third and fourth city can be added by taking a pair of compasses and drawing circles from each existing city centre the correct distances away. Where those compass lines intersect is where the next city down in the hierarchy should be placed. Keep adding cities until it becomes too hard to work out the appropriate location of the next by hand.

The map below is based on this set of instructions and is just one attempt among many possibilities to create a new world map of the major cities, with each city sized by population, where fastest travel time is the geometry, not traditional over-land and sea distance. What is most remarkable about this map is that the world is flat when it comes to its most populous cities, although it wraps around from top to bottom (if east is put uppermost again as it once was in most ancient maps).

Tokyo is at the top of Figure 41.6. Now no longer drawn as part of a Samurai, but as the largest circle on the map. In many ways, it is the world's most successful city; a place that is home to more people living at a higher standard of living than anywhere else on the planet. It is remarkable to think that in just the space of the last century a part of the world that was one of the later places to be colonized by humans (Figure 41.5), a proud, but politically marginal state just a century ago (Figure 41.1), and which suffered so much just over half a century ago (Figure 41.2) should now be at the centre of the most significant cluster of cities in the world today – which are all now in east Asia.

The second most significant cluster of world cities in Figure 41.6 lies just beneath the first, but in and around that supposedly sleepy elephant of India. The Greater Indian cluster includes Tehran on its western border. In contrast, Europe today can only boast three world-sized cities within its borders and another three on (or just outside) its borders, including one in Africa which now has four in total; a number unimaginable when that continent was under colonial control and being portrayed as a 'stitch up' (as in Figure 41.1). But travel times within the African-European cluster of nine world cities are spread-out, as are their economic fortunes of course. Those nine cities do at least benefit from being in similar time zones, and the same can be said for the thirteen world cities of the Americas which in turn connect more easily to the largest east Asian grouping than to greater India. East Asia is now the economic centre of the world.

Try to imagine how this map might have looked a century ago when people mostly did not fly. Back then places were connected most quickly by sea and the dominant ordering of cities was very different. Now try to imagine how Figure 41.6 might look in a century's time. World population is now predicted to rise from its current 7.2 billion, by 2, 3, 4 or even 5 billion people in that time – possibly more – or it may shrink to less. No one really knows! In many

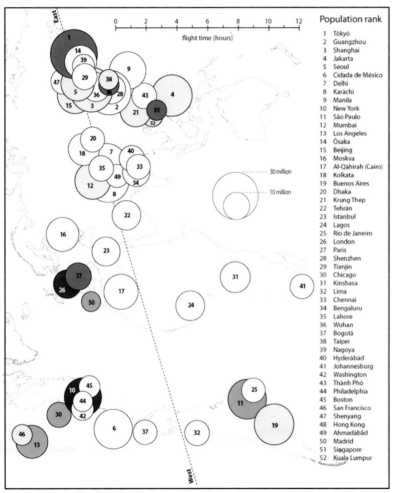

### The top 52 world cities by population
(Shading is rank in world city hierarchy)

Population rank

1 Tōkyō
2 Guangzhou
3 Shanghai
4 Jakarta
5 Seoul
6 Cidada de México
7 Delhi
8 Karāchi
9 Manila
10 New York
11 São Paulo
12 Mumbai
13 Los Angeles
14 Ōsaka
15 Beijing
16 Moskva
17 Al-Qāhirah (Cairo)
18 Kolkata
19 Buenos Aires
20 Dhaka
21 Krung Thep
22 Tehrān
23 Istanbul
24 Lagos
25 Rio de Janeiro
26 London
27 Paris
28 Shenzhen
29 Tianjin
30 Chicago
31 Kinshasa
32 Lima
33 Chennai
34 Bengaluru
35 Lahore
36 Wuhan
37 Bogotá
38 Taipei
39 Nagoya
40 Hyderābād
41 Johannesburg
42 Washington
43 Thành Phố
44 Philadelphia
45 Boston
46 San Francisco
47 Shenyang
48 Hong Kong
49 Ahmadābād
50 Madrid
51 Singapore
52 Kuala Lumpur

*Figure 41.6* A map depicting travel times by distance between fifty-two cities in 2012

*Source*: Re-drawn from the author's rough drafts.

cities in Asia the average family are having just one child. In China when given the choice to have two, many now choose to have one. Many more cities in Africa will be more populous. And will we still fly as much when we can soon meet as holograms projected across space to sit in meetings as if we were actually there?

Today many aeroplanes fly over the Arctic to connect some of the richest cities in the world but, if you were thinking of designing a London-underground-style cartogram of the globe, where each station is a mega-city, you might do well to start with the image above and then consider how to separate out the cities at the 'top of the world'; those which are now clustered so closely.

The map in Figure 41.6 is just a rough sketch, in effect drawn by hand. It will contain errors. Tehran is oddly placed, partly due to it being a little harder to fly to, partly due to the author being lazy towards the end of finishing this map. But try to imagine a map that does not yet exist. This is not a map of the future, but a map of 'now' that we cannot see. It is a world map showing travel times by ensuring that each distance shown on the map is very accurately proportional to travel time and not actual distance. It is a 2D map that bends and folds within a 3D space – it is what is called a manifold.

As more and more towns and cities are added to a map where distance is made proportional to travel time, the surface of the map has to become undulated. Hard-to-reach towns need to be drawn on mountaintops, these are of mountains that are only there in our minds and in the 'mountains of time' it takes us to travel between places. All this is a new cartography still waiting to be discovered. Figure 41.7 highlights the key parts of the travel time map idea.

In Figure 41.7 a surface is shown with two cities drawn upon it as peaks. Between those peaks is drawn a white 'rope'. This rope could be an intercity train line that only has stations within the centres of those cities and the time the train takes to travel is then proportional to the length of that rope. Beneath the two city centres is shown a grid of roads with eight roads meeting at several points. Travel time on the roads slows down as you head towards the city centres, and so the length of road segments grows and the city centres become higher and higher peaks. Two other cities are also shown, not connected to the specific city-train-network, and so their peaks are inverted. Two mathematicians a very long time ago showed that this surface could always be drawn. It has not been drawn yet. So much that is new is possible and that is just what we know of! The remainder of Figure 41.7 suggests new ways of mapping bidirectional flows upon this kind of map.

## Seeing the world anew

Maps can help us see the world in new and different ways, but until we see a new map it can be hard to imagine that it can tell us much that we did not already know. It is also only by looking back at maps made in the past that we can most easily understand how such maps were particular distortions designed to propagate a partial point of view (see Figure 41.1). Other maps use distortion more deliberately. Figure 41.8 shows how the coastline of Europe might look after sea level rises, flooding and storm surges. The projection used, however, is not a conventional topographic map (as used in many projections of sea level impacts) but is an equal-population cartogram that focuses attention on the numbers of people that might be affected.

Cartography is about making sense of the world. It can be conducted without conventional images, or simply by painting pictures using words or sounds (drawing a map in others' minds through what you say), but it is so much easier to describe how you believe different elements to be connected when you can use an image that fits your times and new explanations, and reach into other's minds through their eyes rather than their ears! Any description of the map in Figure 41.8 below, in words, can never substitute for the actual map.

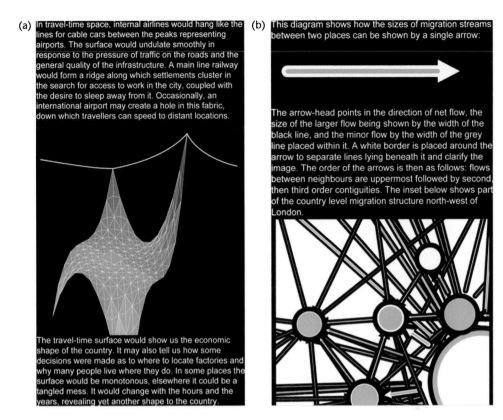

(a) In travel-time space, internal airlines would hang like the lines for cable cars between the peaks representing airports. The surface would undulate smoothly in response to the pressure of traffic on the roads and the general quality of the infrastructure. A main line railway would form a ridge along which settlements cluster in the search for access to work in the city, coupled with the desire to sleep away from it. Occasionally, an international airport may create a hole in this fabric, down which travellers can speed to distant locations.

The travel-time surface would show us the economic shape of the country. It may also tell us how some decisions were made as to where to locate factories and why many people live where they do. In some places the surface would be monotonous, elsewhere it could be a tangled mess. It would change with the hours and the years, revealing yet another shape to the country.

(b) This diagram shows how the sizes of migration streams between two places can be shown by a single arrow:

The arrow-head points in the direction of net flow, the size of the larger flow being shown by the width of the black line, and the minor flow by the width of the grey line placed within it. A white border is placed around the arrow to separate lines lying beneath it and clarify the image. The order of the arrows is then as follows: flows between neighbours are uppermost followed by second, then third order contiguities. The inset below shows part of the country level migration structure north-west of London.

*Figure 41.7*   **Travel time surfaces on which millions of flows could be shown**

*Source*: Graphical 'Textboxes' 6.3, 7.3 from 'The Visualization of Spatial Social Structure' (Dorling, 2012). Available at: *www.geog.ox.ac.uk/research/transformations/gis/books/visualisation/material.html#graphics* (Accessed: 10 July 2016)

Benjamin Hennig drew the map below (Figure 41.8), and all those that follow. The map in Figure 41.9 shows the now iconic image of the Earth at night, using December 2012 satellite data. It is a map of humanity and who is both best connected and most wasteful. The image is again drawn on an equal area population cartogram, where every very small grid square has its area enlarged or decreased so as to be shown in almost exact proportion to the number of people living in that area. The projection is made using a technique that ensures that all lines of latitude and longitude still meet at approximately 90 degrees. The map is conformal. It approximates the best unique solution.

Cartography should be about innovation (Hennig, 2013); it is when we innovate that we most change the world by changing how it is viewed. Begin to depict the world as being a place of inequalities in unnecessary pollution and you begin to suggest how we should change for the better. Figure 41.9 is a map of the amount of energy we waste by shining light up into the sky. Light generated at huge expense and requiring enormous amounts of carbon pollution in most cases to generate it.

On a standard map of the world the 'Earth at night' image appears to show where people are and where there are no people, or where there are people shedding little light. On the equal-population cartogram of the world there are people everywhere, in exactly equal density.

Where there is light it is most often where light is being wasted, shone up to the sky rather than down to a book.

Slight changes to cartography can promote great changes in how we conceive of our society or societies, whether we begin to see ourselves as one mass of 7 or 8 billion people, or whether we draw ourselves as supposedly more separated than that – we have choices to make.

*Figure 41.8* **Equal-population cartogram of Europe showing height above sea level**

*Source*: Dorling and Hennig (2015).

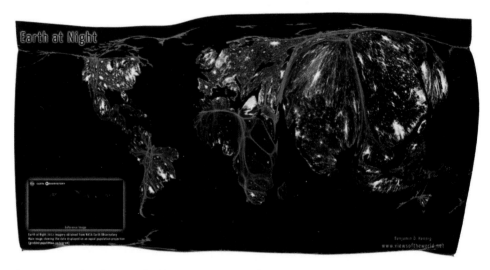

*Figure 41.9* **The sky at night drawn over an equal-population world cartogram**

*Source*: Hennig (2013).

Cartography is about choices. There is no single correct way to depict the world that we are a part of. How we choose to depict it will alter how we see it and treat others, the land, energy, air travel, how we see the seas of the Mediterranean, the lands of the Middle East, how we come to view Asia anew. It will affect whether we see a place as being in the middle, or as being east of somewhere we consider now more central (perhaps the Middle East should now be called the Middle West), and whether we worry more about volcanoes or the excesses of leopard print clothing, let alone what happened when you first built a city with no streets.

Cartography is both art and science, but also a part of the humanities and of popular culture, of new technology and of ancient history. 'Making maps or charts' is as old as humanity and changes as fast as we change. More new maps and charts will have been made in our lifetimes than over the course of the lives of every human who has ever lived. With new technology, it may also be possible for our children to be able to say the same thing again of us (that we say of our parents' and grandparents' cartography), and consign us to part of that strange and obscure history, no longer being at the forefront that we write so much about.

Always, what matters most is whether the sense that is being made of the world does itself make sense. If it does not, then we need new maps, more maps and better maps. The faster our world is changing the faster our maps will change. The east was deposed from the top of maps, including most Christian maps, when the compass suddenly became king and trade across oceans mattered most. The north had to be at the top when trade became more important than gods.

The Mercator projection became the new world 'imagined'; it dominated thinking in the seventeen to twentieth centuries, not just because it was useful at the start of those centuries to get ships to places they planned to visit, but because, as it became less useful as a chart it helped make the USSR (what is now mostly Russia) look like an enormous threat on the map. In between those centuries, the Mercator projection was used to pinpoint the embarkation points for hundreds of thousands of African slaves, and to plot the journey of those that survived the passage to the Americas, before the slave ships returned to Europe with sugar and cotton, and then travelled back down to Africa, often half empty.

Africa would be more populous today were it not for 400 years of slavery, war, colonialism and imperialism. China would be smaller today had not the Opium Wars destroyed traditional Chinese society and led to so much later population growth in China. There was also then much more population growth in India and its neighbours, from the places where the British harvested the opium they demanded the right to sell in China. Chinese students are taught this part of world history. British students are not. It is perhaps in our omissions, in the gaps left on our maps and the lacunae in our histories, that we create the greatest mis-education. It is not in what we say, but in what we choose not to say. It is not in what we map, but in what we choose not to map, that we most mislead.

Figure 41.10 below shows everyone on the planet, but now with the oceans removed. A small village in India is at the centre of this world. The giant new metropolis of Chongqing is clearly visible near the vastly expanded old Sichuan capital of Chengdu. In population space these areas now dominate in the east. Tokyo will soon be shrinking rapidly as its population ages. The threat of Russia boils down to just Moscow on this final new map. London is on the far north-west extreme. The USA is now on the edge of the map, as far away from the centre as are much of Africa or Indonesia.

Maps both change the world and show a changing world. Sometimes they can do both. We are where we tell ourselves collectively we live. We can aspire to become what we collectively tell ourselves we can be. Images are only images, but if they are repeated and reproduced enough times they become new versions of the truth. And the truth is a very powerful thing to create.

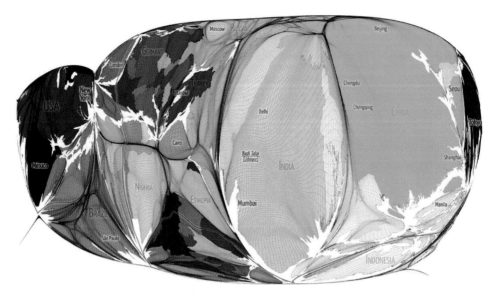

*Figure 41.10*   The world with the oceans taken out and all people given equal prominence

*Source*: Dorling and Lee (2016).

## References

Dorling, D. (2012) *The Visualization of Spatial Social Structure* Chichester, UK: Wiley.

Dorling, D. and Hennig, B.D. (2015) "Visualizing Urban and Regional Worlds: Power, Politics, and Practices" *Environment and Planning A* 47 pp.1346–1350.

Dorling, D. and Lee, C. (2016) *Geography: Ideas in Profile* London: Profile.

Hennig, B.D. (2013) *Rediscovering the World: Map Transformations of Human and Physical Space* Heidelberg, Germany: Springer. See also this posting made on 5 March 2012: 'Mapping the Anthropocene on the 500th birthday of the Cartographer Gerardus Mercator' Available at: *www.viewsoftheworld. net/?tag=anthropocene* (Accessed: 10 July 2016).

Truman, H. (1946) *U.S. Strategic Bombing Survey: The Effects of the Atomic Bombings of Hiroshima and Nagasaki, June 19, 1946* President's Secretary's File, Truman Papers: Decision to Drop the Atomic Bomb Documents. Available at: *www.trumanlibrary.org/whistlestop/study_collections/bomb/ large/documents/index.php?pagenumber=12* (Accessed: 10 July 2016).

Vujakovic, P. (2014) "Mapping the Octopus" *Maplines* Winter pp.10–11.

Note: An earlier (shorter) version of this chapter was published as Dorling, D. (2013) Cartography: Making Sense of our Worlds" *The Cartographic Journal* 50 (2) pp.152–154.

# 42

# Educating tomorrow's cartographers

*Beata Medynska-Gulij*

## The three threads in cartographic education: practice, theory, and art

In educating cartographers, the emphasis has always been placed on the practical dimension, which was directly linked to the technology of map design and map publishing. The beginning of professional cartography and of the profession of cartographer dates back to the sixteenth and seventeenth centuries, when over a span of a hundred years almost all of the world's map production was held in a single country by four publishing houses (Ortelius, De Jode, Mercator-Hondius-Janssonius, and Blaeu). A distinctive feature of their activity was specialization of the employees according to the successive stages of copperplate map design. The author (geographer, scientist) prepared the manuscript of cartographic content; then, the drawer produced the original sketch of the map on a sheet of paper using a quill; the sketch was subsequently transferred to a copper plate by the engraver; and finally, after printing, the colourist applied paint onto the sheet (Medynska-Gulij, 2013). One of the most famous cartographers of this period was Gerard Mercator (1512–1594), who was exceptional in that he combined all the aforementioned professions (Koeman *et al.*, 2007). The following centuries saw the introduction of new technologies of map creation, which further separated the successive scientific and conceptual activities from the technical jobs at the latter stages of production. It was, however, not until the advent of digital technology that a single person was enabled to create complete maps from scratch, i.e. from a raw concept to the final product. This newly appeared, exceptional position of the *map maker* works independently to create maps with the help of digital resources that have reduced the whole cartographic workshop to a single computer workstation.

In the course of the advance in cartographic research and the emergence of courses of cartography at universities in the second half of the twentieth century, the scientific aspect of educating cartographers has become equally important. Traditionally, universities teach cartography alongside geography, placing the emphasis on familiarizing students with natural and socioeconomic processes in the geographical space and, consequently, teaching them the skill of visualizing such processes. At technical universities, however, cartography is combined with geodesy and other technical disciplines. Yet the differences between geographical and technological approaches to teaching cartography pale into insignificance when one considers the new disciplines, such as geomatics, geoscience, geoinformatics, geoinformation, geomedia, and so on, whose emergence reflects the growing need for an interdisciplinary approach.

The most difficult to formalize and teach is the artistic aspect of map-making. In order to be perceived as an artistic image of space that is high in aesthetic value (see Kent, 2005), the map must be created with a particular sensitivity to the graphical means of expression. Teaching artistry in cartography is related to the analysis and comparison of graphical approaches adopted by publishing houses and old schools of cartography, especially in dealing with manuscript maps. Nowadays, such graphically attractive panoramic or perspective images abound mostly in materials presenting tourist areas and in virtual trips, especially in multimedia atlases and gaming maps (see Chapter 39).

A pragmatic approach to educating tomorrow's cartographers as placed in the international and multi-disciplinary context of the technology, science, and art of map-making, necessitates addressing the following issues:

- updating the profile of a candidate for studying cartography so that graduates pursuing careers in the profession are future-oriented;
- updating the range of knowledge and practical skills so that the discipline meets the needs of society;
- specifying the basic/primary level and the specialized level of education, defining the range of theoretical knowledge and its sources, as well as the practical skills to be obtained;
- capitalizing on the unique didactic possibilities that cartography offers, that is, teaching with the use of text as well as graphics, images, maps with the possibility of students monitoring their own progress on the basis of its tangible effects, e.g. maps and other visualizations that they have created on their own;
- helping the students to discover their aptitudes, be they scientific, technical, artistic, or of any other kind, and stimulating the students' creativity; and
- empowering the students to choose their individual education paths in accordance with their interests and with the prospective cartographic profession in mind.

## From the candidate for cartographer to a professional

At the beginning of the educational path stands a candidate for studying cartography: a person fond of maps, a user of school atlases, tourist maps, and open map portals, and often capable of nimbly navigating virtual maps in computer games. They are attracted to GIS-GPS technologies (see Chapters 17 and 18), multimedia presentations of geographical space, and mobile devices with augmented reality. The young person in question is willing to expand their knowledge about maps and wants to learn how to create them; their individual aptitudes are not yet revealed, and their expectations regarding the prospective profession are not yet identified.

At the end of the educational path, however, we see a cartography graduate ready to become a professional, having multiple job possibilities. According to the needs of society, the following cartographic professions that carry importance today and tomorrow can be distinguished:

- *Cartographer – Map Designer and Map Publisher;*
- *Cartographer – Geoviz-Creator and GIS Expert;*
- *Cartographer – Geomatics Expert; and*
- *Cartographer – Map Curator and Map Specialist.*

Each of these experts has academic knowledge about maps and map-making, but each prefers a different approach to work, according to different cartographic tasks.

The work of the *Cartographer – Map Designer and Map Publisher* most closely resembles the activities of a cartographer in the traditional sense, which involve taking care of the visual appeal of the created maps and cartographic visualizations, usually commercially. The result of such activity takes the form of cartographic products: maps, atlases, and multimedia visualizations that are high in aesthetic value and usability. Great importance is attached to original works created by individual designers or by teams of graphic designers using advanced graphics and multimedia editing tools.

The *Cartographer – Geoviz-Creator and GIS Expert*, however, is a profession very closely related to spatial data modelling on the basis of geoinformatic software with simplified editing functions. To this category also belong representatives of geographic information science who use geostatistical programs for industry and highly specialized software for data visualization used in many other fields that deal with spatial data.

The *Cartographer – Geomatics Expert* is a professional with greater knowledge of the whole spatial data management process: from gathering data in the field and from databases, through data modelling, saving, and sharing the data according to spatial data infrastructures, to visualizing the data in the form of official maps (e.g. topographic, geological, or cadastral maps). This specialization is focused on implementing national and official regulations on creating and maintaining the cartographic capital, as well as international standards of spatial data modelling.

*Cartographer – Map Curator and Map Specialist* is an indispensable profession in light of the significant need for conserving maps and sharing cartographic resources. There is an increasing demand for experts, not only in map departments, where they would look after the maps and cartographic databases currently in use by state institutions and offices, but also in huge collections of historical maps in libraries and state archives.

The aforementioned list of professions does not exhaust all the possibilities of professional or scientific development open to cartographers in the future. It is also possible to combine certain professions, that is, to remain open and flexible in undertaking various tasks and adapting to the needs of society.

## Two levels of education

Having described in general terms the profile of a candidate studying cartography and the profiles of professional cartographers, we might proceed to propose an adequate educational path which will lead a student of cartography to further their knowledge and develop their skills. This chapter will not discuss courses traditionally associated with academic teaching, such as Mathematics or Philosophy; rather, the focus will be on subjects directly related to the theory and practice of cartography.

Figure 42.1 shows a two-level process of cartographic education, at the end of which the graduates should be ready to tackle professional tasks in any of the four aforementioned professions. For each level, the range of theoretical knowledge to be gained and practical skills to be mastered has been presented. The knowledge is transferred by way of traditional academic lectures and the main sources are course textbooks and maps. The real potential in teaching cartography lies in the combined use of text and image (i.e. as graphics and maps). Academic practitioners of cartography have long used original graphical methods in their books, e.g. Erwin Raisz (1938) included his own handmade pen-and-ink drawings, while Eduard Imhof (1972) had a distinctive way of juxtaposing pairs of images to compare 'good technique' with 'poor technique'. Printed textbooks, due to their durability, still constitute an important method of recording cartographic knowledge, and, as can be seen from Figure 42.2, they can be used by students in a very active way.

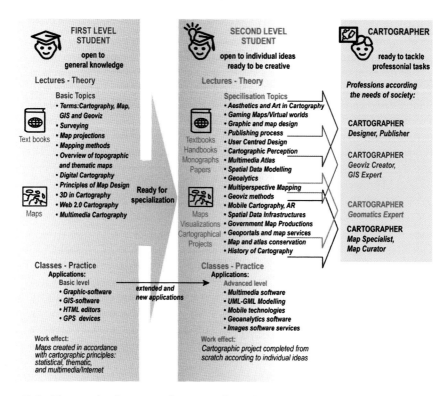

*Figure 42.1* The two-level process of cartographic education

*Figure 42.2* Active use of printed textbooks for cartography

A key element of the education process is practice, which in Figure 42.1 has been defined as the production of project-oriented work from the student's familiarity with the relevant applications. Students of cartography are able to apply the knowledge obtained to a practical end while making their own maps. In this case, practical skills are gained during computer classes supervised by a teacher and can then be individually honed with the help of tutorials. Tutorials are a preferable technique for teaching particular operations in cartographic applications because every student prefers to work at the computer at their own pace.

The successive maps created by students are a very good motivation for them to expand their knowledge, especially given the publishable quality of their final maps. Students willingly collect all their assignments and use the whole set to compile a portfolio, which can then be attached to job applications as documentation of the practical skills obtained.

## The first level of education: general knowledge and basic skills

The first level of education welcomes a student open to general knowledge. The goal at this level is to provide the student, in a comprehensive and complex way, with the theoretical bases of the discipline, and with the basic practical skills. Figure 42.1 outlines the subjects covered, starting from the explanation of terms and research trends, as well as the thematic scope of cartography. At the outset, the teacher has to discuss the definition of cartography, maps, and the links between cartography and other disciplines. He/she has to explain the relationship between cartography and visualization, geovisualization, geoinformatics, and geomatics, and so on. The following topics correspond to the teaching approach reflected in chapters of academic textbooks (e.g. Medynska-Gulij, 2015): Map Scale, Basic Geodesy, Coordinate Systems, Map Projections; A Review of Topographic and Thematic Maps, Mapping Methods and Principles of Map Design; Activities in Digital Cartography (Landscape and Cartographic Digital Model, Space Data Models; Space Data Acquisition; Cartographic Generalization). At this first level, the student is also familiarized with techniques of map design and with new trends in presenting the third dimension, as well as with Internet and multimedia cartography.

During classes, students familiarize themselves with computer applications to an adequate level so that each of them can create their first digital maps using various graphics and geoinformation software and an HTML (hypertext mark-up language) editor. Besides, the student learns how to collect data using a GPS (global positioning system) receiver and how to visualize it in an appropriate map application.

At this level, the teacher suggests the most appropriate presentation methods for various types of cartographic data, and may point out basic cartographic principles. A key task for the teacher is to provide a rationale for the application of cartographic principles to any given student's map project. The more maps and visualizations (even easy ones) that are concordant with cartographic principles but are based on different techniques that a given student creates during the first level, the better they will be equipped to sketch their individual path at level two. This method of teaching cartography by encouraging every student to create a portfolio of their own maps and visualizations is bound to bear fruit later, as the awakening of their creativity will definitely come in handy when they carry out their individual projects.

## The second level of education: a theoretical and practical specialization

A student embarking on the second level of education must be ready to acquire more of the pertinent theory and to further develop their practical skills. The choice is determined not only by interests and aptitude, but also by prospective career plans. The second column in Figure 42.1 lists the subjects covered, arranged in a top–down order so as to correspond to the subject matter of the four cartographic professions. The order and selection of subjects, however, are not fixed, because every cartographer-to-be is free to alter the scope and the priorities so that they fit their individual path. The current emphasis on 'User Centred Design' and 'Cartographic Perception', concepts associated with the key role of the map user in designing, and in testing the effectiveness of cartographic products, is likely to be a concern of every cartographer. On the other hand, the very specialized field 'Gaming Maps and Virtual Words' is bound to become solely the domain of *Designers*, although applications for mobile devices based on 'Augmented Reality' will be used more widely, also by *Geomatics Experts* and *Geoviz-Creators*.

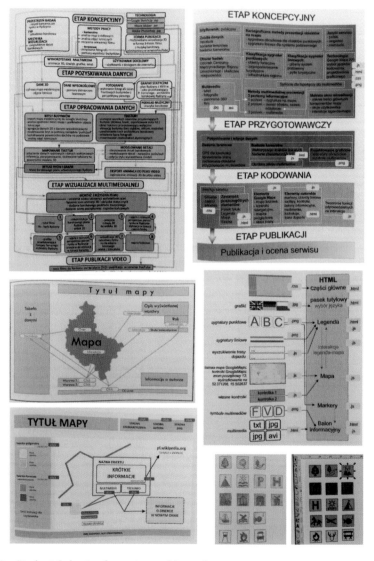

*Figure 42.3*   Students' charts of a cartographic project

Most of the subjects mentioned will correspond to the professions adjacent to one another in the chart, while others will be universally important to every cartographer. A given subject might also be incorporated into two rather different professions, e.g. the subject 'Aesthetics and Art in Cartography' might be worth fathoming by a *Map Curator*, but also possibly by a *Designer* wanting to draw inspiration from historical maps, who decides to include in their field of interest 'The History of Cartography'. Nevertheless, in the face of the limited choice, a student will usually construct their educational path around chosen subjects adapted for practical purposes, which in this case means the individual completion of a cartographic project supplemented with a written review of the relevant scientific literature. The subjects in Figure 42.1 do not constitute a complete, closed set; on the contrary, other topics might also be added, because an interdisciplinary approach has always been valued in cartography.

The list of theoretical sources will be widened at the second level to include not only basic academic textbooks, but also handbooks, monographs, and research papers. Furthermore, the accompanying maps from the first level will be replaced by larger sets of professional maps and advanced visualizations, geovisualizations, and other cartographic projects. The skill of using software enabling the creation of maps according to the guidelines and rules of cartographic design provided by the teacher, the rudiments of which were already introduced at the first level, will now be furthered to include the use of more advanced operations and tools. The students will have to learn to use new applications, depending on the aims of their individual projects.

The result of a student's work in this high-profile teaching stage may take the form of a Master's thesis, in which major significance will be given to the elaboration of an original, individual idea or concept, and shaping it into a form that is customized to fit the chosen technology. The individually developed idea can be presented in an activity chart, progressing from the raw concept stage to the public stage, and encompassing also the final evaluation of the product by a user, or by way of comparison with other maps/projects. The top of Figure 42.3 shows sample activity charts of a cartographic project, with arrows and lines showing connections between successive activities. The authors of the activity charts also provided the information regarding data formats and technologies that they would use to complete the project. Below, we can see other manifestations of the individual creativity of the prospective cartographer: layouts of a multimedia map as displayed on the screen, with events marked on them, a graph presenting the coding of map elements in HTML, and the manual design of pictorial signs to be developed later in a graphics program.

## The pragmatics of education

The description of the two levels of cartographic education lacks the information regarding the duration of each stage and the point at which a given student is ready to choose a specialization. Therefore, it is worth paying attention to the role of the teacher, their teaching methods, and their relationship with the student. First, the teacher passes on the knowledge and helps the student obtain practical skills so that the latter is ready to choose a specialization akin to their prospective profession or further scientific activity. Then, the role of the teacher changes; they become more oriented towards inspiring the student, boosting their creativity, and encouraging them to come up with their own ideas. At this point, the teacher is more of an adviser or a prompter to the student, who works on their cartographic project, following their own ideas. The teaching methods advocated by the author of this chapter as the most effective ones are the Socratic method, which involves asking students inspiring questions, and the method of scribbled ideas. The latter has the teacher schematically drawing the general idea of the project and its layout on a sheet of paper while discussing the cartographic project in question with its author (the student). The teacher scribbles frames with different elements, lines, arrows, captions, which they can later 'no longer decipher'. The very effort to decipher the notes is what unleashes the student's creativity at this key stage of map design (examples of the final concepts and layouts are shown in Figure 42.3).

The pragmatic approach to educating tomorrow's cartographers, which is advocated in this chapter, is an attempt to reconcile two objectives: educating qualified, employable professionals, and capturing the interest of the people studying cartography. This is made possible by the listing of potential cartographic professions and the division of the educational process into two levels. The main aims of the proposed approach are: at level one, for the student to obtain complex knowledge and practical, up-to-date, skills compatible with the latest technologies; and in stage two, for the student to receive specialized education oriented towards the choice of a prospective career, while at the same time accommodating the student's interests and aptitudes.

The constantly expanding capacity of the profession of cartographer, aimed at accommodating new technologies and accommodating the needs of society, also provides beneficial prospects for a young person. Most cartographers are likely to appreciate the satisfaction derived from marrying their work with their passion.

## References

Imhof, E. (1972) *Thematische Kartographie* Berlin: Walter de Gruyter.

Kent, A.J. (2005) "Aesthetics: A Lost Cause in Cartographic Theory?" *The Cartographic Journal* 42 (2) p.182–188.

Koeman, C., Schilder, G., van Egmond, M. and van der Krogt, P. (2007) "Commercial Cartography and Map Production in the Low Countries, 1500–ca. 1672" in Woodward, D. (Ed.) *The History of Cartography (Volume 3)* Chicago, IL: University of Chicago Press, pp.1296–1383.

Medynska-Gulij, B. (2013) "How the Black Line, Dash and Dot Created the Rules of Cartographic Design 400 Years Ago" *The Cartographic Journal* 50 (4) pp.356–368.

Medynska-Gulij, B. (2015) *Kartografia: Zasady i zastosowania geowizualizacji* (Cartography: Principles and Applications of Geovisualization, in Polish) Warsaw, Poland: Wydawnictwo PWN.

Raisz, E. (1938) *General Cartography* New York: McGraw-Hill.

# 43

# Drawing maps

## Human vs. machine

*William Cartwright*

---

## Preamble

We imagine.
We imagine how to represent geography through maps.

We transform.
We transform our imagined geography (a mental map) into a physical representation of geography – the represented world – by drawing a map.

The process of drawing enacts our transformation.

We draw.
We draw maps that represent geography.

This is what we do.

Through this process of drawing – holding pen or computer stylus – we make our marks that when read and interpreted, allow readers to share our imagined geography.

Our hands are the conduits through which we link the imagined to the representation of the imagined to the real world.

## This contribution

My approach to this contribution is to approach map drawing as something that I do. The act of drawing (and thus drawing maps) is a personal activity. I do this privately – to map out ideas and to place them in my 'personal' geography; and publicly – by constructing sketches and formal drawings to tell a geographical narrative – as rough drawings and as constructed digital media artefacts. I do this as individual or collaborative works.

The act of 'drawing' is a personal one. The drawing links one's cognitive views of the world and the ideas about how best to represent the world using various media – analogue or digital. I draw with pencil on paper, ink on plastic, plotting pen onto paper and pixel to screen. No matter the approach or the medium and method of choice, the act of drawing is a thing that we like to hold close, as we might whisper a thought or shout a revelation.

Therefore my approach here is to tell this 'story' of drawing in the first person, as I believe, as stated previously that the very process of drawing, and in fact thinking about what needs to be drawn, is the link between the cognitive map and the physical map. This process or drawing is what Woodward and Lewis (1998) refer to as 'performance cartography', the link between 'cognitive cartography' and material cartography / social construction. The outcome of our performance is the realization of the most apt representation of our view of a geography, real or imagined, set in a historic or a projected space and communicating a narrative in such a way that our narrative is understood.

## Map drawing and replication

Maps were initially drawn by hand, and then copied, again by hand. The cartographer had a direct link – from hand to paper, from hand to representation. However, as the process of map replication has always sought methods to speed up the copying process, printing technology became interposed between the cartographer's hand and the manuscript. Early printing techniques required cartographers to draw the map. First, woodcut printing, then copperplate printing and lithographic printing methods were employed. The cartographer still drew the map, but, rather than drawing the map directly onto a manuscript, the cartographer drew directly onto the printing plate. Impressions were then taken directly from the plate. The cartographer did draw the map, but as impressions were made from the plate they were essentially 'wrong-reading' images of the final printed map. The map was drawn as a mirror image.

These three different methods of manuscript preparation for subsequent map printing – and mirror-image cartographic drawing – are depicted in Figures 43.1a to 43.1c.

*Figure 43.1a* Woodcut map production. Available at: *http://mapmaker.rutgers.edu/356/ woodcutMapper.jpg* (Accessed: 9 July 2016)

*Figure 43.1b*  Printing plate produced by copper engraving. One of three plates produced for three-colour lithographic printing process used for printing topographic maps.

*Source*: USGS. Available at: *http://nationalmap.gov/ustopo/photos/j7-engraving-a-copper-plate.jpg* (Accessed: 9 July 2016). Image in the public domain

*Figure 43.1c*  Drawing on one of the three lithographic stones reprint of the United States Geological Survey (USGS) Donaldsonville, Louisiana, topographic map, 1915.

*Source*: USGS. Available at: *http://nationalmap.gov/ustopo/125history.html* (Accessed: 9 July 2016). Image in the public domain

Later, the drawing process was further removed from the final manuscript, whereby photographic processes were used to transform the hand-drawn map into printing plates for replication. The drawing process reverted to 'right-reading', whereby the cartographer saw the map how it would appear. But, for coloured maps, cartographers would draw each colour separation manuscript as a unique product. It was only after a proof or the final print was made that cartographers would see the outcomes of their endeavours. This process first generated positive images, as shown in Figure 43.2a and later negative drawings were produced by the scribing method, as per the example in Figure 43.2b.

With the application of the computer, everything changed. The mapping industry first used computers to facilitate computations related to surveying and map projections, then to guide drawing instruments as CAD (Computer-Aided Design) systems or as DeskTop Publishing (DTP), and finally as complete compositions to output systems, where maps are delivered via the Internet. With the application of computer power for map production the direct drawing of a map by a person stopped. The computer intervened. The link between the imagined map and the physical map ended. The direct engagement by the cartographer with the actual drawing of a representation of the Earth stopped. (This did appear again with the introduction of the computer digitizing tablet/pen as shown in Figure 43.2c, but the tactile link, that direct engagement with the drawing process did not.)

*Figure 43.2a*  Hand-drawing topographic map at USGS c.1942. Available at: *http://nationalmap.gov/ustopo/photos/j12-pen-and-ink-drafting.jpg* (Accessed: 9 July 2016). Image in the public domain

*Figure 43.2b* Scribing a USGS topographic map

*Source*: USGS. Available at: *http://nationalmap.gov/ustopo/photos/j13-scribing.jpg* (Accessed: 9 July 2016).
Image in the public domain.

*Figure 43.2c* Wacom digitizing tablet. Available at: *http://opendtect.org/rel/doc/User/base/
figures/3.Wacom_Tablet.png* (Accessed: 9 July 2016)

## Reflecting on my hand drawings of maps

I began drawing maps in the pen and ink era in the late 1960s. I began compiling and drawing field notes and survey plans associated with surveys linked to cadastral surveying. The field notes were initially drawn in the field, as a record of the measurements that had been taken. Subsequently, these field notes were copied and drawn on paper using black and red inks – red to depict the actual lines of the survey. Lines and lettering were drawn by hand and no mechanical aids, except a straight edge for drawing lines, were used. The process was completed when a survey plan was drawn, which had to accord with regulations associated with the depiction of cadastral elements. The map was drawn onto wax-impregnated linen, which was slightly opaque. This manuscript, holding ink-drawn lines and annotations (here drawn by hand and also with the assistance of mechanical lettering templates) was used for printing multiple copies of the maps using the diazo printing process.

Then, I moved into a geological exploration office, where I constructed and drew representations of geological and geophysical elements of the Earth. My mapping moved from depictions of elements above the ground to those below the ground. The underlying mapping procedures were still valid, but the drawing materials changed. The wax-impregnated linen of the survey office was replaced with drawing film (plastics) and lettering was undertaken using mechanical drawing templates – the *Leroy* mechanical guides and pantograph from the Keuffel and Esser Company. (See *http://en.wikipedia.org/wiki/Technical_lettering* for a description of these guides.) The drawings depended upon being skilled at completing the computations to develop the geographical grid that was the framework for recording the results of geological and geophysical surveys as well as mastering the various mechanical devices employed to ensure the positional rigour needed to place information in its correct gratular position. The drawings became more technical, the materials upon which the maps were drawn more stable, due to the advances in polymer film technology, and the outcome less 'personal' than the maps that had been drawn with fewer drawing aids and templates.

The last manual map drawing method was undertaken using the scribing method, described earlier. Here, maps were scribed into an emulsion that sat atop a stable-base plastic manuscript. Here, I drew negative manuscripts, rather than positive ones. Things like line widths and lettering were completed in an even more mechanical way, and the personal engagement with the mapping process became more removed than when drawing and writing on paper or linen directly. But, the maps were still hand-drawn.

My direct engagement with *drawing* maps really ended when I began to write computer programs to drive computer plotters. But, for a brief moment, with the purchase of the Hitachi *Peach* microcomputer, which allowed one to draw on the screen using a light pen stylus, the physical link between the representation (on screen or output via plotter or printer) – through the process of drawing – was sacrificed for efficiency and repeatability.

This link between cognitive cartography and material cartography was abandoned really when I began to experiment with other, non-print media with the arrival of multimedia and began to develop interactive applications.

## Traces

My hands in fact still bear the results of drawing manuscripts. The index finger of my right hand still has a hard piece of skin about 1 cm long, which is the result of many hours spent drawing maps. My hand has a permanent record of my engagement with the actual process of drawing maps, of my direct engagement with cartography.

## Postscript

We draw.
We draw maps that represent geography.
This is what we do.

## Reference

Woodward, D. and Lewis, G.M. (Eds) (1998) *The History of Cartography* (*Volume 2*) Chicago, IL: The University of Chicago Press.

# Index